The Spectroscopy of Heavy Nuclei 1989

The Spectroscopy of Heavy Nuclei 1989

Proceedings of the International Conference on the Spectroscopy of Heavy Nuclei held in Agia Pelagia, Crete, 25 June–1 July 1989

Edited by J F Sharpey-Schafer and L D Skouras

Institute of Physics Conference Series Number 105
Institute of Physics, Bristol, Philadelphia and New York

CODEN IPHSAC 105 1–486 (1990)

British Library Cataloguing in Publication Data

International Conference on the Spectroscopy of Heavy
Nuclei, (1989; Agia Pelagia, Crete)
1. Nuclear spectroscopy
I. Title II. Sharpey-Schafer, J. F. III. Skouras, L.
539.744

ISBN 0-85498-061-X

Library of Congress Cataloging-in-Publication Data are available

Published under The Institute of Physics imprint by IOP Publishing Ltd
Techno House, Redcliffe Way, Bristol BS1 6NX, England
335 East 45th Street, New York, NY 10017-3483, USA
US Editorial Office: 1411 Walnut Street, Philadelphia, PA 19102, USA

Printed in Great Britain by J W Arrowsmith Ltd, Bristol

Contents

Preface

This conference is one in a series of nuclear structure conferences that have been held in alternate years on the Greek island of Crete. The aim of this conference was to discuss 'New Aspects of Nuclear Spectroscopy'. That is, to address new theory and experimental results in topics as diverse as discrete gamma-ray spectroscopy and the search for the quark–gluon plasma. The aim of studying such a wide range of topics was to stress the unity and commonality of many of the problems facing nuclear physicists in different areas of the subject. Experimentally the problem is to devise new techniques to produce meaningful signals from the noise in a measurement or from very complex reactions. Ingenuity is required to devise new techniques if new phenomena are to be observed. That the international nuclear structure community is bursting with new ideas may be clearly seen from the contents of these proceedings.

This volume contains the texts of all invited and contributed talks given at the International Conference on the Spectroscopy of Heavy Nuclei held in Agia Pelagia on the Greek island of Crete from 25 June to 1 July 1989. From these proceedings the vigorous and healthy state of nuclear structure physics can be observed in the interplay between theory and experiment. New theoretical ideas inspire new experiments. New and unforeseen experimental results present new challenges to theory. The conference took place in what has become the expected stimulating scientific atmosphere enhanced by its location in such an idyllic part of the world.

Grateful thanks are due to the sponsors of the conference: the Hellenic Physical Society; the Greek Ministry of Culture and NRCPS Demokritos. Also, financial contributions from the Greek Tourist Organisation, Olympic Airways and Canberra Industries Inc. are gratefully acknowledged. The scientific strength of the conference was ensured by the wise guidance of the International Programme Advisory Committee whose members were:

W Andrjtschev, Sofia
P Blasi, Florence
F A Beck, Strasbourg
A Faessler, Tübingen
J D Garrett, Oak Ridge
W Gelletly, Daresbury
V Gillet, Saclay
C Gregoire, GANIL
B Herskind, NBI
K Heyde, Gent
B Jonson, Gothenberg
T L Khoo, Argonne
G Munzenberg, GSI
S Nagamiya, Columbia/BNM
E Nolte, TU München

R O Owens, Glasgow
T Paradellis, Demokritos
A Richter, TH Darmstadt
P Taras, Montreal
G Van Middlekoop, Amsterdam
J Verghados, Ioannina
D Ward, Chalk River.

The Organizing Committee consisted of:

J F Sharpey-Schafer, Liverpool (Chairman)
L D Skouras, Demokritos (Secretary)
C A Kalfas, Demokritos
S Kossionides, Demokritos
A C Xenoulis, Demokritos.

Inst. Phys. Conf. Ser. No 105
Paper presented at Int. Conf. on Spectroscopy of Heavy Nuclei, Crete, Greece, 1989

Developments in gamma-ray detection and channel selection devices

J. Simpson
SERC Daresbury Laboratory, Daresbury, Warrington WA4 4AD, UK
and
M.A. Riley
Oliver Lodge Laboratory, Department of Physics,
University of Liverpool, Oxford Street, Liverpool L69 3BX

Abstract

The development of large arrays of escape suppressed spectrometers is discussed. These arrays have revolutionised the field of γ-ray spectroscopy with their significant improvement in signal to background ratios enabling very weak photon lines to be identified. This enables many interesting nuclear properties to be studied for example at very high spin and recent results on ^{161}Er showing evidence for unpaired states are presented. Large arrays have been used with many channel selection devices to enable weak channels to be identified. Preliminary data on ^{49}Mn obtained using an array plus a recoil separator are discussed. Plans for the next generation of arrays which will have a sensitivity many orders of magnitudes greater than the current arrays are briefly mentioned.

SCIENCE AND ENGINEERING RESEARCH COUNCIL
DARESBURY LABORATORY

1. INTRODUCTION

In recent years the use of large arrays of escape suppressed spectrometers (ESS) have led to significant advances in the field of gamma-ray spectroscopy. These spectrometers have caused a renaissance in various aspects of nuclear spectroscopy in particular, the study of high spin states and the investigation of nuclei far from stability. Indeed, many results presented at this conference were obtained with these spectrometers. Several recent review articles contain data from ESS arrays, see for example[1-4]). This talk will review the development of arrays of escape suppressed spectrometers from their first use in the early 1980's and briefly mention the exciting plans and ideas for the next generation arrays in the 1990's. Preliminary results from a recent experiment using the TESSA3 spectrometer at Daresbury on very high spin states ($I \simeq 50\hbar$) in ^{161}Er will be presented to give an example of the quality of data obtainable at the present time. Large arrays have also been used in conjunction with channel selection devices of one form or another. Selection, in addition to the array, is often essential to study new phenomena and make advances in, for example, far from stability spectroscopy or actinide spectroscopy. Channel selection techniques are so numerous and diverse that it is impossible to mention them all and do them justice in this talk. Therefore the talk will concentrate on the techniques that have been employed with large arrays at Daresbury Laboratory. Comparable techniques and advances have been made at various laboratories around the world. Again as an example very preliminary data on the neutron deficient nucleus ^{49}Mn will be presented. These data were obtained using the Daresbury recoil separator and POLYTESSA array.

Before discussing the development of large arrays of ESS it is worth considering the impact these arrays have made. One measure of this is in high spin γ-ray spectroscopy with the intensity limit for the observation of discrete nuclear states. In the early 1960's yrast nuclear states were observed up to ~6^+ to 8^+ in rotational nuclei using a single sodium iodide crystal e.g.[5]). Decays from these states are of intensity 50-100% of the total intensity in the nucleus. By the early 1980's prior to the use of arrays of ESS the use of several germanium detectors in coincidence had reduced the intensity limit to 3-5% and discrete γ-rays from states of spin 30 were observed. Modern arrays of ESS have reduced the observable intensity limit to ~0.5% and enabled for example spin ~50[6]) to be observed in normal deformed rotors and spin ~60[7]) in the superdeformed phase of a few nuclei. Figure 1 shows some 'state of the art' spectra.

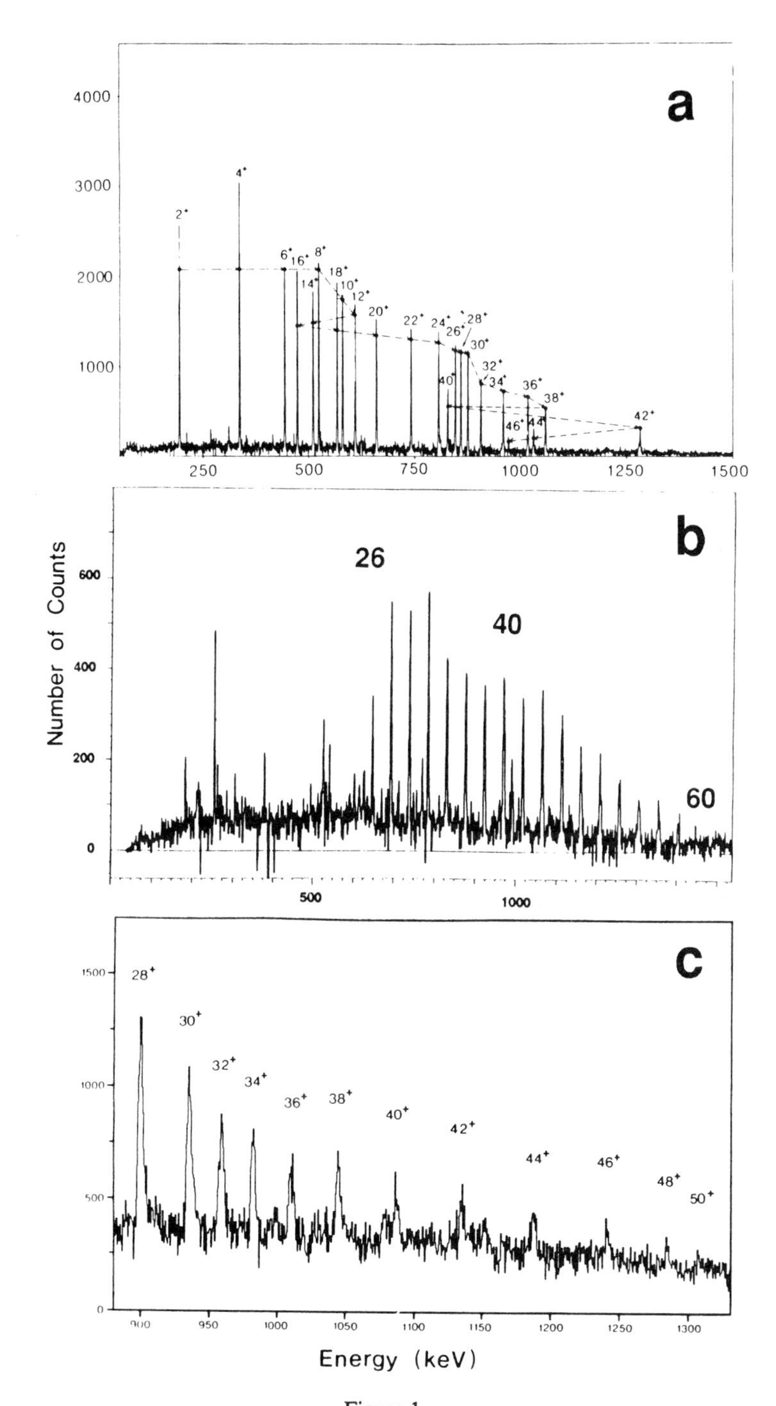

Figure 1.

Gamma-ray spectra showing (a) the yrast band of ^{158}Er (b) the superdeformed band in ^{152}Dy and (c) the lowest energy positive parity band in ^{160}Er.

2. DEVELOPMENT OF LARGE γ-RAY ARRAYS

2.1 Introduction

Early gamma-ray spectroscopy made use of single sodium iodide scintillation detectors to detect γ-rays from radioactive sources or following a nuclear reaction. A major step forward was taken with the development of reverse biased solid state germanium detectors in the mid-1960's. Germanium detectors have very good energy resolution which ranges from about 1 keV for low E_γ (~ 100 keV) to about 2 keV for E_γ of about 1 MeV. If large and complex decay schemes were to be constructed at least two detectors with their outputs taken in time coincidence were employed. The initial advance in spectroscopy of high spin was made by the Risø group of the Niels Bohr Institute in the late 1970's who realised that if the multiplicity M_γ was large, as it is for the decay of high spin states, then it was very advantageous to use as many germanium detectors as possible. If N detectors are used the γ-γ coincidence rate in proportional to N(N-1). Using 4 germanium detectors Riedinger et al[8]) established the detailed quasiparticle structure of ^{160,161}Yb and showed the statistical advantage of using arrays of detectors as opposed to just 2 (although just 2 detectors could produce classic results[9]). However the problem with these experiments was not lack of statistics but the poor signal to noise obtained. This problem was common to all experiments using bare germanium detectors. Also, prior to the use of ESS arrays, advances were made in high spin spectroscopy by the use of large multiplicity filters and sum spectrometers which were used to measure the number of γ-rays and their total energy following a heavy ion fusion evaporation reaction. The culmination of this technology was the construction of the spin spectrometer at Oak Ridge[10]) and the crystal ball at Heidelberg[11,12]).

The revolution of γ-ray spectroscopy came with the use of large numbers of germanium detectors in anticoincidence with escape suppression shields. The use of escape suppressed spectroscopy is not new[13-15]) but it is only in the last decade that arrays of ESS have been used extensively in heavy ion reactions. A typical germanium detector in isolation gives a spectrum for monoenergetic 1 MeV γ-rays in which 20% of the counts are in the photopeak and 80% are in a continuous background of lower energy caused mainly by γ-rays Compton scattered out of the volume of the germanium detector, see figure 2(a). The obvious solution to this problem of signal to background is to surround the germanium detector with a scintillator to detect the escaping scattered photons. A current ESS[16]) is shown in figure 2(c). When an escaping photon is detected the event may be suppressed from the spectrum electronically. The resulting spectrum now has ~65% of the spectrum in the photopeak, figure 2(b). This improvement in the peak to total ratio p is crucial in coincidence spectroscopy. In a double γ-γ coincidence experiment the signal to background

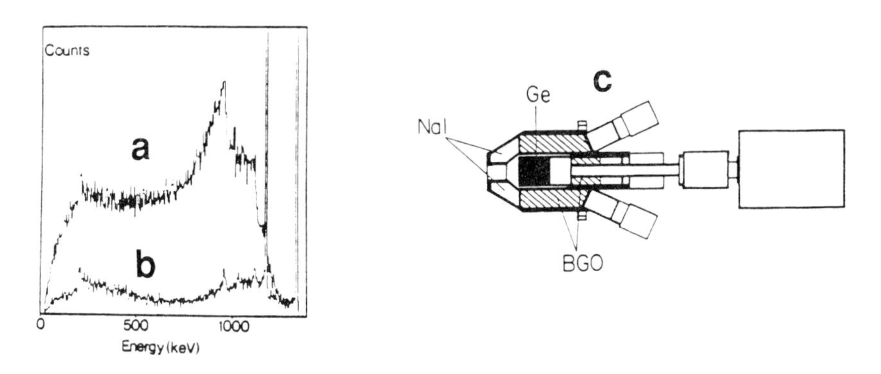

Figure 2.

Spectrum of γ-rays from a ^{60}Co source for (a) an unsuppressed germanium detector and (b) a suppressed germanium detector. (c) The current ESS used at Daresbury[16]).

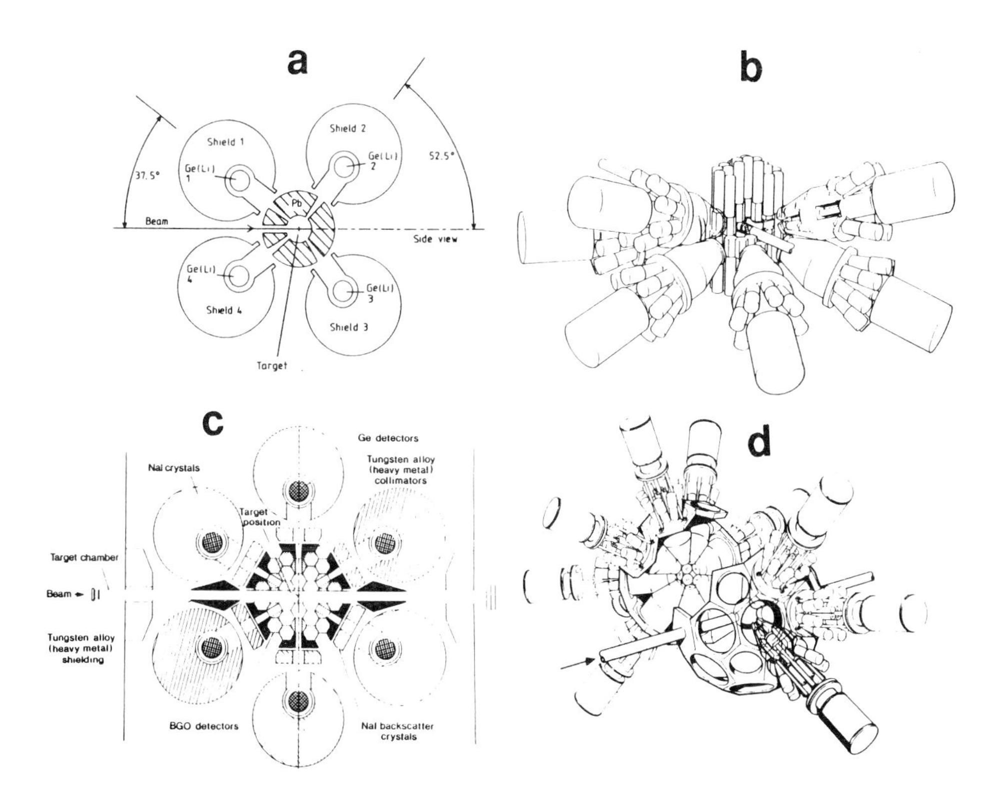

Figure 3.

The TESSA family of arrays. (a) TESSA0, (b) TESSA2, (c) TESSA3 and (d) POLYTESSA/ESSA30.

is proportional to p^2 so if at 1 MeV $p_u = 0.2$ for an unsuppressed spectrum and $p_s = 0.65$ for a suppressed spectrum, then $p_u^2 \sim 0.04$ and $p_s^2 = 0.42$. Therefore over 40% of suppressed coincidences contain useful information compared with 4% for unsuppressed coincidences. It is this factor of ten in improvement in signal to noise that has enabled ESS arrays to make such a dramatic contribution to nuclear spectroscopy.

2.2 NaI suppression arrays

The first array of escape suppressed spectrometers[17]), had the acronym TESSA (The Escape Suppressed Spectrometer Array) and consisted of four then five NaI(Tℓ) suppression shields and Ge(Li) detectors surrounding the target as illustrated in figure 3(a). This array was constructed as a joint Liverpool University/Niels Bohr Institute project and was installed at the Risø FN tandem in the autumn of 1980. A great deal of new and interesting physics was obtained with this array using the O, N and C from the tandem. In particular it allowed detailed spectroscopy of the non-yrast bands up to about spin 25 in many nuclei to be obtained (e.g. refs[17-28]). The level scheme of ^{163}Er (see ref.[29]) shows the power of this first array.

TESSA0, as this array is now referred to, did not have any channel selection and relied only on the reaction to enhance the nuclei of interest. The second stage of array development came with the addition of compact multiplicity filter/sum energy detectors which were used in coincidence with the ESS. The first of these consisted of 7 (soon after 14) hexagonal BGO detectors used in the TESSA array (TESSA1) which was moved in 1982 from the NBI to Daresbury Laboratory when heavy ion beams from the 20 MV tandem, the Nuclear Structure Facility, became available.

Soon after the array that firmly established the ascendancy of ESS arrays namely TESSA2[30]) began operation. This array figure 3(b), consisted of six NaI(Tℓ) suppression shields at 30°, 90° and 150° to the beam direction. The germanium detectors were Ge(Li) and hyper pure n-type detectors situated 270 mm from the target. The use of n-type germanium detectors as opposed to Ge(Li)'s in ESS was also a notable advance. These detectors have a relatively high tolerance to neutron damage and a very thin outer dead layer, ideal for ESS. TESSA2 also comprised an inner crystal ball of 62 hexagonal BGO detectors which were operated as a 50 element multiplicity/sum energy detector. This inner ball was used in coincidence with the ESS to provide the extremely useful channel selection.

TESSA2 was the first array to be used for high spin γ-ray spectroscopy in conjunction with an accelerator producing beams where the maximum amount of angular momentum that a nucleus could hold could be achieved with fusion-evaporation reactions. A typical data set using TESSA2 following a heavy ion fusion- evaporation reaction would contain 30-50 × 10^6

γ-γ-BGO ball coincidences. Although the statistics obtained are not remarkable by present day standards the extremely good signal to noise meant that a great deal of new physics was obtained (see review articles[1-4])).

2.3 BGO suppression arrays

The next major advance in ESS arrays came with the use of BGO rather than NaI as the suppression shield scintillator. For γ-ray absorption 1 cm of BGO is roughly equivalent to 1 inch of BGO which meant that BGO suppression shields could be made smaller and more compact allowing a greater number of ESS to be placed round a target increasing the detection efficiency. There are now several large BGO suppressed arrays in the world and these are briefly described.

The first BGO array was the High Energy Resolution Array (HERA) at Berkeley USA[31]). This array consists of 21 ESS and a recently added inner BGO ball. The design of this array was chosen to enhance multiple γ-ray coincidences and several $\times$ 10^8 triple (γ-γ-γ) coincidences are recorded in a typical experiment.

The OSIRIS spectrometer[32,33]) consists of six asymmetric and six symmetric ESS together with a central crystal ball. It was constructed as a collaboration between HMI Berlin, Bonn and Köln universities and KFA Jülich in the Federal Republic of Germany.

One of the most powerful and elegant arrays has been assembled at the Chalk River Laboratories in Canada by a collaboration between the Universities of Montreal and McMaster and Chalk River itself. This array, the 8π Spectrometer[34,35]) consists of 20 ESS surrounding an inner BGO crystal ball composed of 72 detectors.

An array of up to 12 ESS has also been constructed in the USA at the Argonne superconducting linac ATLAS by a collaboration between the University of Notre Dame and ANL. An array of up to eight ESS has been used by the Rochester group in Coulomb excitation experiments and a small array is in use at Stony Brook and Canberra.

The spin spectrometer at Oak Ridge USA has been adapted into a suppression shield array with elements of the spectrometer being replaced by 20 ESSs. Oak Ridge have also constructed a 'compact ball' consisting of 20 ESS placed close to the target designed to enhance triple and higher fold coincidence events.

An extensive Scandinavian collaboration has constructed an extremely sophisticated array NORDBALL[36]). Other non-nordic groups have joined in this project which has many options for combinations of different kinds of γ-ray, charged particle and neutron detectors. NORDBALL is located at the NBI Risø FN tandem and booster and will be described in detail at this conference by G. Sletten and the first results presented[37]).

At Daresbury the BGO successor to TESSA2 is TESSA3[38]) figure 2(c) with 16 ESS and a slightly modified inner BGO ball. The ESS used in all the arrays at Daresbury[16]) is shown in figure 2(c). Event rates of about 10^3 per second and data sets of 2 to 3 × 10^8 γ-γ-BGO ball coincidence events are routinely obtainable with this array. The 12 ESS version of TESSA3 was used during 1986 when the discrete line nature of the superdeformed phase of ^{152}Dy was discovered and spin 60 was observed for the first time[7]) figure 1(b). A review article by P.J. Nolan and P.J. Twin[39]) covers much of the experimental data on superdeformation from ESS arrays.

The array with the largest number of ESS to be assembled was the European Suppressed Spectrometer Array with 30 detectors (ESSA30). This array consisted of ESS from the UK, F.R. Germany, Scandanavia and Italy arranged in a soccer ball shaped framework and was sited at Daresbury in 1987. This array is shown in figure 3(d).

The current array that differs from all the others is the Chateaux de Crystal[40]) in France which consists of 12 germanium detectors and hexagonal barium fluoride crystals acting as both suppression shields elements and as a multiplicity filter/sum energy detector. This array was built as a collaboration between Bordeaux, Grenoble, Lyon, Orsay and Strasbourg. The recent inclusion of very large (~80% efficiency) germanium detectors and the fast timing properties of BaF_2 make this a powerful array.

2.4 Spin 50 in ^{161}Er and the new unpaired spectroscopy

The power of modern arrays is best demonstrated by presenting results from a recent experiment where the high resolving power of an array is needed to observe the weak γ-rays from very high spin states. Preliminary results are presented on ^{161}Er obtained using the TESSA3 array at Daresbury Laboratory. The 170 × 10^6 γ-γ-BGO ball coincidences were obtained following the bombardment of three stacked thin (400 μg cm^{-2}) foils of ^{130}Te, each with a thin (200 μg cm^{-2}) Au backing, by a 170 MeV ^{36}S beam. Figure 4 shows summed γ-γ coincidence spectra for each of the three main rotational bands observed to high spin. The spin labels assume the transitions are stretched E2 in nature. Thus the (+,1/2), (-,-1/2) and (-,1/2) sequences are observed tentatively up to spin ($101/2^+$), ($95/2^-$) and ($101/2^-$) respectively. Previous work[41]) using the TESSA2 spectrometer had observed these same structures up to spins $73/2^+$, $63/2^-$ and $65/2^-$ respectively. Decays from these very high spin states are $\lesssim$1% of the total intensity in ^{161}Er.

The observation of discrete states up to very high spins enables one of the most sought after phases of nuclei, the unpaired phase, to be investigated. It is now almost 30 years since it was predicted[42]) that pairing correlations in a deformed nucleus would be quenched if it was rotated at sufficiently high frequency. There has been a persistent search for definitive evidence marking the paired to unpaired phase transition in high spin nuclear studies. A suggested test[43]) for the existence of static pairing correlations at large angular momentum has been through the study of band crossings or "backbends". In regions of

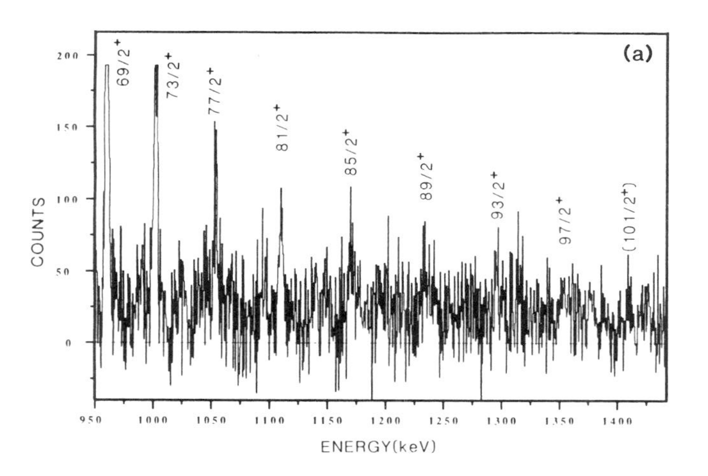

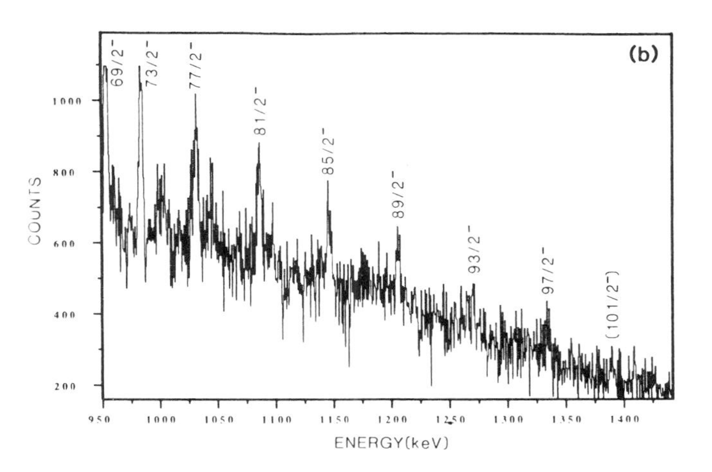

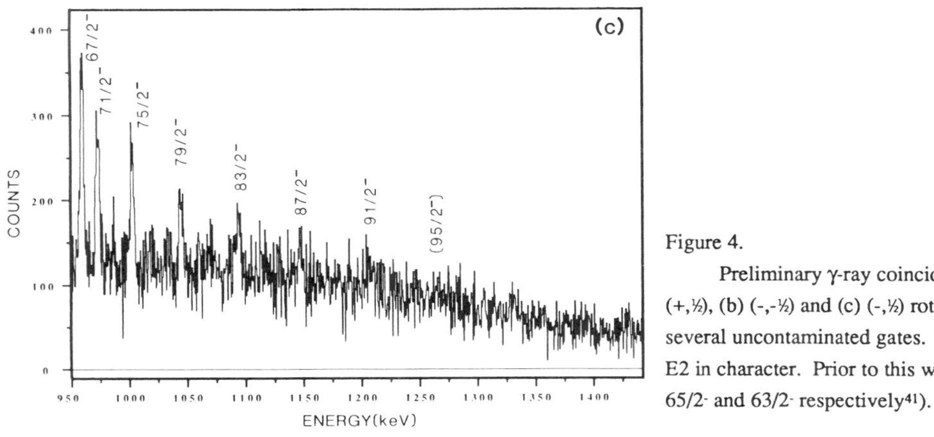

Figure 4.

Preliminary γ-ray coincidence spectra showing the high energy transitions in the (a) (+,½), (b) (-,-½) and (c) (-,½) rotational bands in ^{161}Er. Each spectrum is the sum of several uncontaminated gates. The spin assignments assume the transitions are stretched E2 in character. Prior to this work the three band had been observed up to spins $73/2^+$, $65/2^-$ and $63/2^-$ respectively[41]).

spin or rotational frequency where pairing is significant backbends or quasiparticle alignments occur at similar rotational frequency in all rotational bands where the alignment is not blocked. The crossing frequency is thus associated with the point where the Coriolis force on a particular pair of alignable quasiparticles compensates the effect of pairing correlations for this particular pair. Well over 500 backbending cases observed in deformed nuclei fit into this nice systematic pattern.

The absence of pairing correlations, however, does not exclude band crossings. Not only can band crossings occur based on the crossing of single-particle levels (not quasiparticle levels[44]) but a pair of particles with quantum numbers summing to $(\pi,\alpha) = (+,0)$ can replace another (+,0) pair of particles. Such a particle exchange would be quite specific and would not usually be expected to occur in several rotational bands at the same time. Thus a signature of the decline of static pairing correlations would be the observation of band crossings not correlated in rotational frequency. The first observation of this latter type of unpaired crossing was recently reported in high spin studies of $^{159,160}Er$[45]) (see also [46]). The object of this experiment on ^{161}Er was to test the proposed explanation in ref.[45]) and extend these systematics.

In figure 5 the experimental routhians[47]) for high spin decay sequences ($\hbar\omega > 0.4$ MeV) in ^{159}Er[48,49]), ^{160}Er[6]) and ^{161}Er are shown. These routhians are referred to a configuration with a constant moment of inertia ($\mathcal{J}_o$) of 72 $MeV^{-1}\hbar^2$. The decay sequences are labelled by the quantum number (π,α). The change in slope between $\hbar\omega = 0.4$-0.5 MeV in all sequences is attributed to the rotational alignment of the first pair of $h_{11/2}$ quasiprotons[49]) and is therefore evidence for the existence of significant proton pairing correlations up to these high rotational frequencies (see also [50])). It was the anomalous observation of the band crossing in the (-,1/2) sequence in ^{159}Er at $\hbar\omega = 0.56$ MeV, which could not be explained in terms of quasiparticle alignment but could be explained in terms of a suggested scheme of single neutron states in the absence of static neutron pair correlation that was the subject of ref.[45]). This simple scheme which has the general features expected for single neutron states in this region, shown in figure 6, compares remarkably well with unpaired neutron calculations[51,52]). Figure 7 shows the calculated[52]) single neutron energies for the light rare-earth nuclei as a function of rotational frequency. The rotational frequency which corresponds to the band crossing in the (-,1/2) sequence in ^{159}Er is indicated in figure 6 and is where it becomes energetically favourable for a pair of particles occupying the $(-,-1/2)_1$, and $(-,1/2)_2$ levels at lower frequency to move and occupy the $(+,\pm1/2)_1$ levels, at higher rotational frequency. Since both pairs couple to $(\pi,\alpha) = (+,0)$ the parity and signature of the decay sequence remains unchanged and a band crossing is observed.

The ordering and behaviour of the high frequency bands of $^{159,160}Er$ may be explained by the occupation of the levels in figure 6[45]). Can this picture be extended to ^{161}Er? In

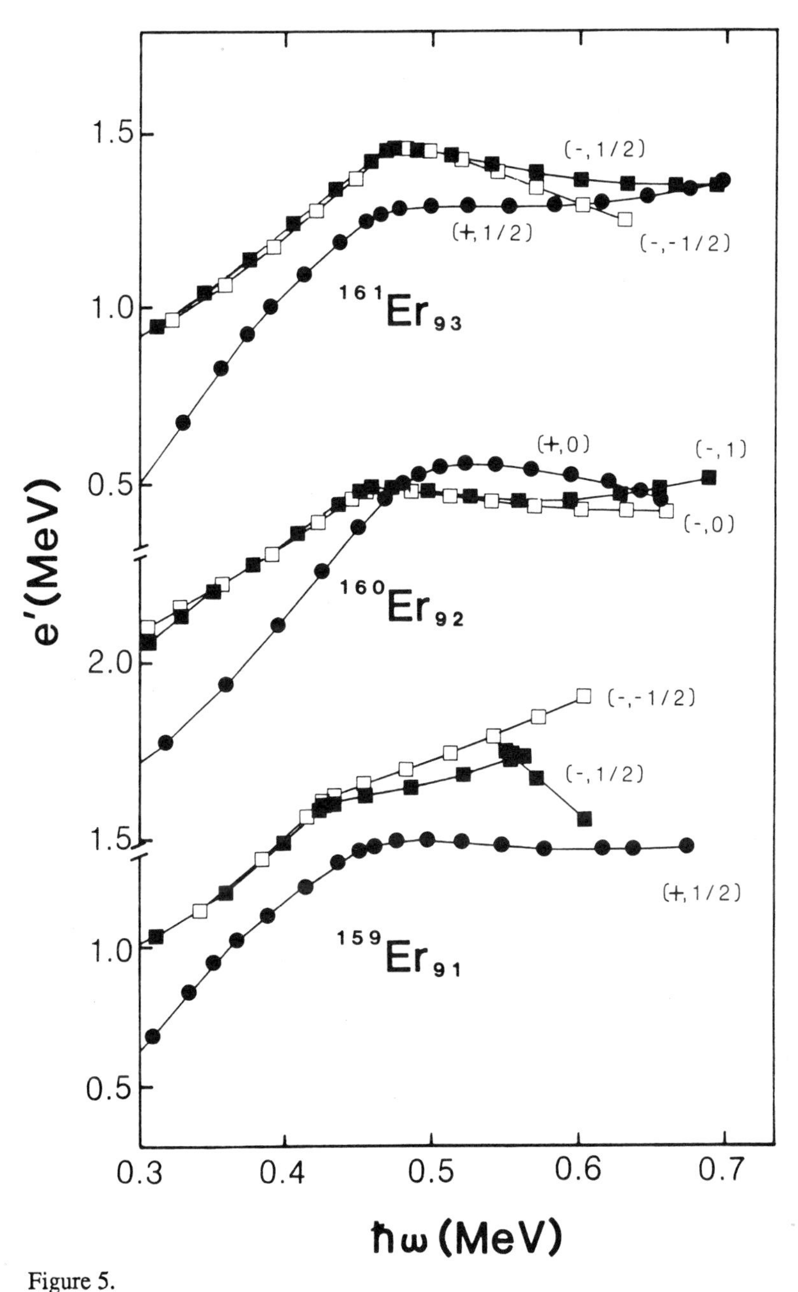

Figure 5.

Experimental routhians[47]) for high spin decay sequences ($\hbar\omega > 0.4$ MeV) in ^{159,160,161}Er. These routhians are refered to a configuration with constant moment of inertia ($\mathfrak{J}_0$) of 72 MeV$^{-1}\hbar^2$. The decay sequences are labelled by the appropriate quantum numbers (π,α).

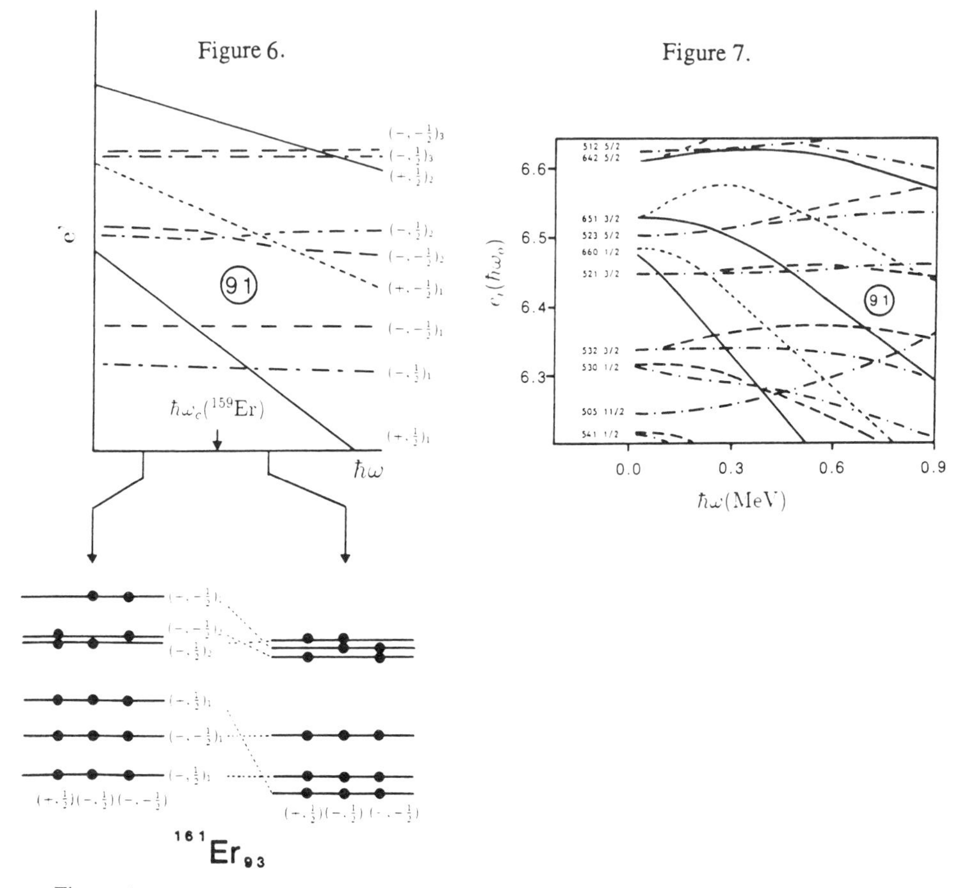

Figure 6.

The scheme of single-neutron levels that in the absence of static neutron pair correlations is able to explain the high rotational frequency behaviour of the band sequences in ^{159,160,161}Er. The rotational frequency which corresponds to the band crossing in the (-,½) sequence in ^{159}Er (see figure 4) is indicated. The explicit occupation is shown for the three low lying configurations in ^{161}Er at both lower and higher frequencies. This continues a similar analysis for ^{159,160}Er[45]).

Figure 7.

Calculated spectrum of single-neutron states in the light rare-earth region as a function of rotational frequency in the absence of a static neutron pair field[52]). The deformation parameters used are ε_2=0.19, ε_4=γ=0. At $\hbar\omega$=0 the orbitals are labelled by the asymptotic quantum numbers [$Nn_z\Lambda$]. The similarity to the simple schematic of figure 6 is striking.

order to do this and to explain the energy splitting between the (-,0) and (-,1) sequences in ^{160}Er, figure 5, it is necessary that the $(-,\pm1/2)_2$ levels invert at high frequency. This inversion at high rotational frequency is unusual but is necessary to fully explain the ^{159,160}Er data. Indeed such an inversion is predicted for these levels, see figure 7.

The explicit occupation of levels by the upper five valence neutrons at low and high rotational frequency is illustrated in the lower part of figure 6 for ^{161}Er. The behaviour of these configurations and level occupation can now be compared to the experimental data. Figure 6 predicts that the (+,1/2) band in ^{161}Er is yrast, owing to the energy difference between the $(+,-1/2)_1$ and $(-,\pm1/2)_2$ levels and that the two negative parity bands $(-,\pm1/2)$ lie close in energy. This is in good agreement with experiment. Above $\hbar\omega \approx 0.5$ MeV, i.e. above the crossing frequency of the band crossing in ^{159}Er, the ordering of the levels changes with the $(+,-1/2)_1$ level crossing the $(-,1/2)_2$ and then the $(-,-1/2)_2$ levels. Therefore at high frequency the negative parity sequences in ^{161}Er will lie lower in energy and in addition the inversion of the $(-,\pm1/2)_2$ levels will make the $(-,-1/2)_1$ sequence yrast. This is precisely what is observed experimentally. This same crossing of positive and negative parity levels occurs in ^{160}Er at a very similar rotational frequency, see figure 5.

Since no band crossings are observed in any one sequence in ^{161}Er the question must be asked are any predicted to occur within this simple model? The model predicts that the first crossing should occur in the (+,1/2) sequence when the occupation of the $(+,-1/2)_1$ and $(+,1/2)_2$ levels becomes more favourable than occupying the $(-,\pm1/2)_2$ levels. However, this exchange does not occur until very high rotational frequency, above the current experimental limit. Also the occupation of the $(+,-1/2)_2$ level in the two negative parity sequences in ^{161}Er blocks this crossing in these sequences. Thus these predictions are again in agreement with experiment where no crossings were observed in ^{161}Er between $\hbar\omega = 0.5$-0.7 MeV.

It is also interesting to point out that in ^{162}Er the even spin positive parity (+,0) yrast band is observed to spins in the mid-forties ($\hbar\omega = 0.6$ MeV) whereas the odd and even spin negative parity bands are only observed to near spin 30[41,53]. No high rotational ($\hbar\omega > 0.5$ MeV) band crossing is observed in the yrast band. Again using the schematic model depicted in figure 6 these observations can be explained. The 3 extra valence neutrons above the N=91 gap will occupy the $(+,-1/2)_1$, $(-,\pm1/2)_2$ levels at both high and low rotational frequency. The yrast band in ^{162}Er should have parity and signature (+,0) as is indeed observed. To produce negative parity the $(+,-1/2)_1$ valence neutron must be removed and placed in either of the $(-,\pm1/2)_3$ levels. It can be seen from figures 6 and 7 that since the $(+,-1/2)_1$, level is strongly downsloping as compared to the $(-,\pm1/2)_3$ levels, the positive parity yrast band in ^{162}Er will become increasingly favoured at high rotational frequency (and spin). Thus, in agreement with experiment, the positive parity yrast band in ^{162}Er will be observed to much higher spin than the negative parity sideband structures.

Also, no band crossings are expected to occur in the yrast band of ^{162}Er.

It appears therefore that the simple scheme of single neutron states in the absence of static neutron pairing correlations which was successful in explaining the high spin behaviour of ^{159,160}Er can be extended to ^{161,162}Er. The arrays at the present time are starting to examine the single neutron spectrum of states at the highest spins in these nuclei. These data provide a stringent test of current theoretical models and details of the theoretical developments in this and other regions are discussed by T. Bengtsson at this conference, see also [54,55]).

3. CHANNEL SELECTION DEVICES AND SPECTROSCOPY FAR FROM STABILITY

3.1 Introduction

It is very often the case that a large array of ESS in itself is insufficient to unravel the well hidden nuclear phenomena. In the study of proton rich nuclei far from stability, for example, heavy ion fusion evaporation reactions are generally used and these populate many final nuclei. The γ-ray decays from the nucleus of interest, usually produced with a very low cross section (σ << 1 mb), are hidden amongst γ-rays from other nuclei and some technique is needed to remove these from the spectrum. Also, in actinide spectroscopy, where fission is prevalent in heavy ion reaction, sophisticated techniques are needed particularly if high spin states are to be studied. The many and varied techniques that have been employed in just these two regions of spectroscopy are too numerous to mention them all in this talk. The study of prompt γ-ray from fission fragments will be discussed by J. Durell and some of the techniques involving gas avalanche detectors in conjunction with large arrays to study actinide nuclei will be discussed by P.A. Butler at this conference. In addition the channel selection devices that are available with NORDBALL will be discussed by G. Sletten and the Dwarf Ball used in the spin spectrometer by D. Sarantites. This talk will concentrate on the in-beam spectroscopy on the very neutron deficient side of the Segré chart and will be limited to developments made and results from Daresbury Laboratory.

3.2 Spectroscopy far from stability - The Daresbury Recoil Separator

The main means of production of proton rich nuclei is the heavy ion induced fusion-evaporation reaction and at Daresbury two techniques have been used to isolate the nucleus of interest. The first involved the measurement of prompt γ-rays in an array of ESS in coincidence with neutron and charged particle detectors[56]). This technique can be used to from the current arrays. In Europe there have been several observe γ-rays from nuclei produced at reaction cross sections of a few hundred μb. The second and cleaner technique is to measure the A and Z of the recoiling nuclei directly using a recoil separator in coincidence with the prompt γ-rays.

The Daresbury recoil separator [57]) has been described in detail by James et al 1988 and the design of recoil separators in general has been reviewed by Cormier[58]). Briefly the main elements of the Daresbury Separator are two crossed electric and magnetic devices Wien filters) and a 50° dipole bending magnet. The crossed field devices give velocity dispersion and deflect the primary beam while the bending magnet disperses the accepted reaction products in terms of their mass divided by charge state A/q. A position sensitive detector at the final focal plane determines A/q and a split anode ionisation chamber gives a measure of the atomic number Z. In addition, various quadrupole and multipole magnets are used to transport the recoils as efficiently as possible through the separator and produce a well focused image. The prompt γ-radiation is detected in an array of 20 ESS's POLYTESSA (15 of the type described by Nolan et al 1985[16]) and 5 of the type described by Simpson et al 1988[59])). The recoil separator plus POLYTESSA system has been used mainly to identify γ-rays in a given nucleus by measuring prompt γ-recoil (A/q and Z) coincidences. A highlight of this was the identification of γ-rays in the N=Z nuclei ^{68}Se[61]), ^{72}Kr[60]), ^{76}Se[61]) and ^{80}Zr[62]), see figure 8. The production cross-section of ^{80}Zr was only 10 ± 5 μb and represented a fifty fold increase in the sensitivity for measuring prompt γ-rays in a compound nucleus reaction. Identification of the isotope number of proton rich Nd nuclei containing highly deformed states was achieved[63]), octupole deformation was observed in the light Xe nuclei[64]), deformed nuclei in the mass 130 region were studied[65]) and sub-barrier transfer studies[66]) have taken place using this system.

The low production cross-section for the majority of these studies meant that only recoil-γ coincidences were statistically useful. Thus the indication from the data, for example, of the transition from extreme prolate deformation dominating around N=Z=40[62]) to shape coexistence in ^{72}Kr[60]) could not be confirmed or studied in any great detail. Such spectroscopy is in its infancy and a factor of at least 10 improvement in sensitivity is needed for the detailed spectroscopy of level scheme determination and lifetime measurements to take place. Improvements are in hand to increase the recoil separator efficiency, by ray tracing techniques and with the planned improvement in γ-ray detection efficiency, see section 4, detailed spectroscopy at the 10 μb level and simple spectroscopy below 1 μb should be possible in the near future.

3.3 Spectroscopy of ^{49}Mn

To give an example of data obtainable using the recoil separator and POLYTESSA array preliminary results on a recent experiment to study ^{49}Mn are given here. Prior to this work

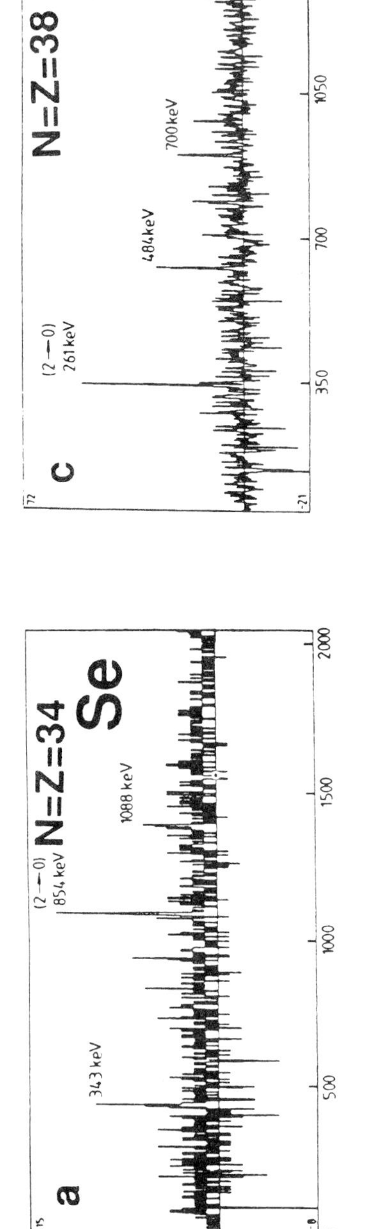

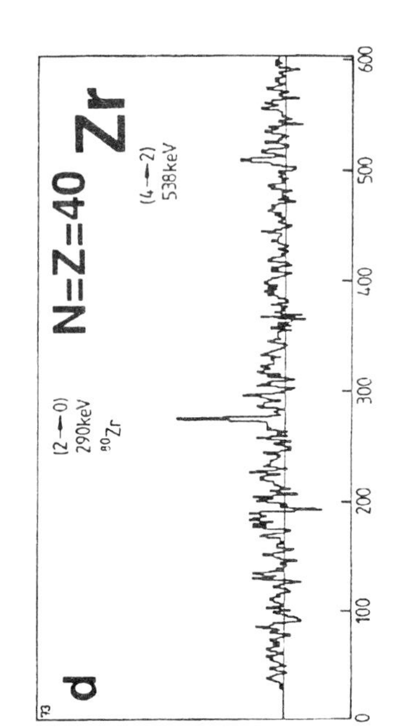

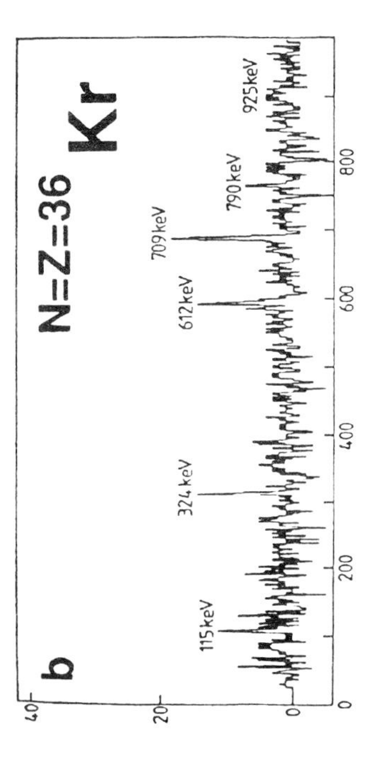

Figure 8.

Gamma-ray spectrum, obtained using the POLYTESSA and Recoil Separator system transitions in (a) $^{68}Se^{61}$), (b) $^{72}Kr^{60}$), (c) $^{76}Se^{61}$) and (d) $^{80}Zr^{62}$).

there were no γ-ray transitions known in ^{49}Mn. The object of this experiment was to determine the level scheme of ^{49}Mn and study the mirror and cross conjugate symmetry in the $f_{7/2}$ shell nuclei ^{47}V, ^{47}Cr, ^{49}Cr and ^{49}Mn.

The nucleus ^{49}Mn was produced by the p2n channel following the bombardment of a 0.5 mg cm^{-2} ^{12}C target by a 160 MeV ^{40}Ca beam. This reaction produces the nuclei ^{46}Ti, ^{48,49}V, and ^{49}Cr strongly and the full power of the recoil separator is needed to pull the ^{49}Mn γ-rays (produced with a cross section of ~ 1 mb) out of the spectra. The recoil separator focussed mass 49 nuclei at the focal plane and the split anode ionisation chamber was used to make a Z identification, possible in this reaction because of the high recoil velocity, v/c ~ 7%. A total of 20×10^6, 20×10^6 and 1×10^6 γ-γ, γ-recoil and γ-γ-recoil coincidences respectively were recorded.

The E vs ΔE spectrum obtained in the ionisation chamber, figure 9(a), shows separation of ^{49}V, ^{49}Cr, ^{49}Mn and ^{46}Ti recoils. The latter arises because of a A/q ambiguity. A pure spectrum of ^{49}Mn could easily be obtained from the γ-recoil coincidence data, figure 9(b). The level scheme of ^{49}Mn was then established up to about 6 MeV using both the γ-γ and γ-γ-recoil coincidence data. Although the γ-γ-recoil data was much weaker than the γ-γ data (the recoil separator efficiency being only ~5%) it proved invaluable for the determination of the level scheme since the γ-γ data were contaminated by the strongly populated nuclei ^{46}Ti, ^{48,49}V and ^{49}Cr. The value of γ-γ-recoil data was clearly evident in this experiment and such data in high statistics is a must for future progress in this type of spectroscopy. The results of this experiment are the subject of a forthcoming paper by J. Cameron et al[67]).

4. THE NEXT GENERATION OF GAMMA-RAY ARRAYS

There is a great deal of activity in the USA and Europe at the present time devoted to the next generation of large γ-ray arrays. Current γ-ray arrays have a total peak efficiency of less than 1% for the detection of 1 MeV γ-radiation and they have a detection intensity limit of ~0.5% in a nucleus using double γ-γ coincidence data. Although many nuclear physics phenomena have been observed for the first time and several questions answered, as many and more puzzles and riddles have been found. The physics programme that is envisaged in the near future requires a much more efficiency array and a significant lowering of the intensity limit. There are several approaches being considered to achieve this. The first steps towards the next generation were taken in the USA with the GAMMASPHERE proposal. This array which is now approved will consist of 110 large germanium detectors giving a total peak efficiency of approximately 7%. BGO crystals between the germanium will be used for Compton suppression with each crystal possibly suppressing events in two germanium detectors. This shared suppression mode enables the large increase in number of detectors

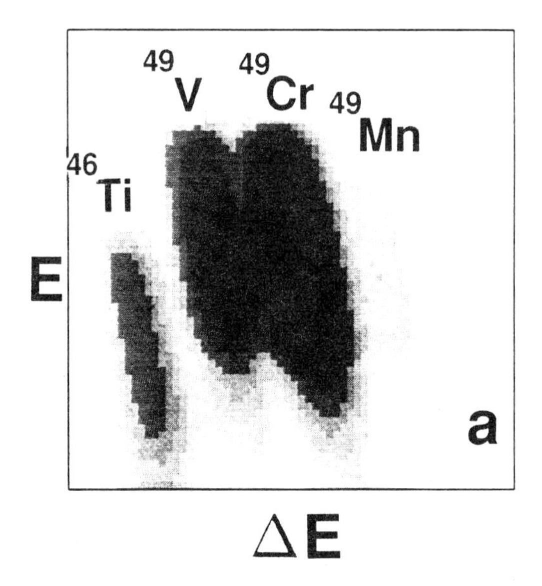

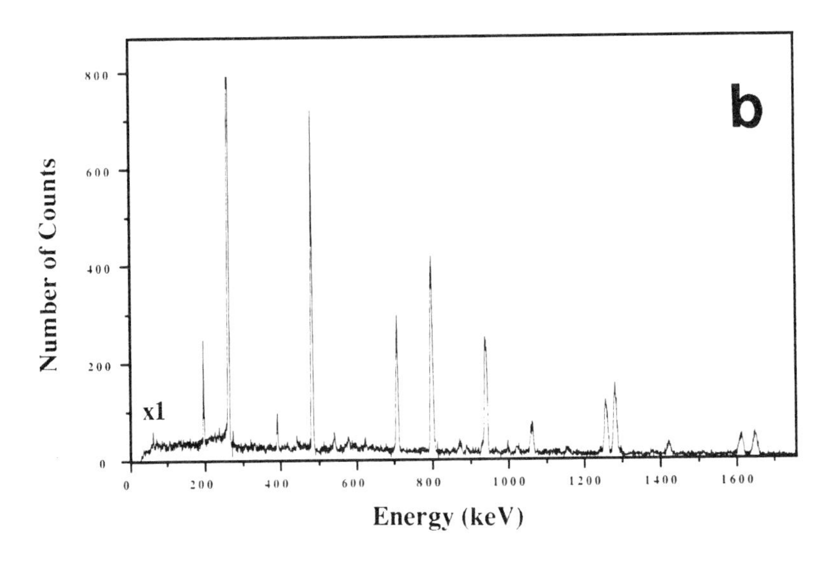

Figure 9.

(a) Spectrum of total energy E against ΔE obtained in the ionisation chamber of the recoil separator following the bombardment of a ^{12}C target by a 160 MeV ^{40}Ca beam. Events corresponding to ^{49}V, ^{49}Cr, ^{49}Mn and ^{46}Ti are indicated.

(b) Spectrum of γ-rays in ^{49}Mn obtained by selecting events in coincidence with the appropriate region of (a).

EUROBALL meetings to discuss the design and co-ordinate plans for a new array(s). The ultimate aim of this collaboration is to build an array with a total peak efficiency of ~20%. The first advance towards this may be to build an array of comparable efficiency to GAMMASPHERE and adapt/modify the array as germanium detector technology is improved. An array with a total efficiency of ~6% has been proposed in the UK. It is called GAMIC (GAmma-ray MICroscope). This array will be used initially in coincidence with the Daresbury recoil separator see section 3.2. In order to build an array with 20% peak efficiency, which would give at least two order of magnitude greater sensitivity than GAMMASPHERE or GAMIC, several approaches are being considered, some of which require germanium detector development. One approach is to cluster groups of 7 hexagonal germanium detectors together in the same cryostat with each cluster having a BGO suppression shield. Another is to mount several large diameter (100-120 mm) planar detectors in a single can producing a detector with a large full energy peak efficiency and build an array with these. Also closely packing over 500 germanium detectors in a germanium ball is being considered. Whatever approach is adopted an array with such a large efficiency is an exciting prospect and many physicists are currently working hard to achieve that objective.

5. CONCLUSION AND PERSPECTIVES

This talk has concentrated on reviewing the developments in large detector arrays which have caused a renaissance in γ-ray spectroscopy over the last decade. Although the talk is concentrated on techniques and results from Daresbury Laboratory developments have taken place in many laboratories worldwide. The talk has not had time to discuss much of the new and exciting physics that has been revealed by these arrays. Many of the papers presented to this conference will cover this. Highlights of the physics must be the discovery and mapping out of the superdeformed phase of nuclei near mass 150 and 130 (now 190), the spectroscopy of unpaired collective bands, the spectroscopy of shape changes at high spin with the observation of band terminations, the spectroscopy of reflection asymmetric-octupole deformed nuclei, the measurements on nuclear damping, spectroscopy of deformed nuclei far from stability and spectroscopy of N=Z nuclei up to mass 80 and many more. This talk has presented preliminary results on ^{161}Er using TESSA3 where evidence for unpaired states is observed and on ^{49}Mn using the recoil separator and POLYTESSA spectrometer.

In the present situation in spectroscopy it is necessary for experimentalists to have an ESS array available to them in one of its many forms. Currently proposals for even bigger arrays are being discussed in the USA and in Europe. Many interesting and vital problems in the physics of nuclear structure remain open which can be addressed using these magnificent spectrometers.

6. ACKNOWLEDGEMENTS

This work is supported by the UK Science and Engineering Research Council. A great many people have been involved in the developments and results described in this talk but special thanks must go to John Roberts and John Cameron who analysed and gave their permission for the presentation of the ^{161}Er and ^{49}Mn data prior to publication.

REFERENCES

1) J.F. Sharpey-Schafer and J. Simpson, Prog. Part. Nucl. Phys. 21 (1988) 293.
2) J.D. Garrett et al, Ann Rev. Nucl. Part. Sci. 36 (1986) 419.
3) I. Hammoto, High angular momentum phenomena, Ch.4 of Treatise on heavy ion science, ed. D.A. Bromley, Vol.3 (Plenum, New York 1985).
4) J.O. Newton to be published in contemporary physics.
5) H. Moringa and P.C. Gugelot, Nucl. Phys. 46 (1963) 210.
6) J. Simpson et al. J. Phys. G: Nucl. Phys. 13 (1987) L235.
7) P.J. Twin et al. Phys. Rev. Lett. 57 (1986) 811.
8) L.L. Riedinger et al Phys. Rev. Lett. 44 (1980) 568.
9) O.C. Kistner et al Phys. Rev. C 17 (1978) 17.
10) M. Jääskeläinen et al, Nucl. Instr. Meth. 204 (1983) 385.
11) R. Simon et al, J. de Phys. 41 (1980) C10.
12) V. Metag et al, Nucl. Phys. A409 (1983) 331C
13) A.J. Tavendale and G.T. Ewan, Nucl. Instr. Meth. 25 (1963) 185.
14) G.T. Ewan and A.J. Tavendale, Can. J. Phys. 42 (1964) 2286 and Nucl. Instr. Meth. 26 (1964) 183.
15) T.K. Alexander et al, Nucl. Instr. Meth. 65 (1968) 169.
16) P.J. Nolan et al, Nucl. Instr. Meth. A236 (1985) 95.
17) D.M. Todd et al, J. Phys. G10 (1984) 1407.
18) P.J. Nolan et al, Phys. Lett. 108B (1982) 269.
19) R. Aryaeinejad et al, J. Phys. G10 (1984) 955.
20) P.J. Smith et al, J. Phys. G11 (1985) 1271.
21) J. Simpson et al, J.Phys. G10 (1984) 383.
22) G.B. Hagemann et al, Nucl. Phys. A424 (11984) 365.
23) N. Roy et al, Nucl. Phys. A382 (1981) 125.
24) J. Kowacki et al, Nucl. Phys. A394 (1983) 269.
25) S. Jonsson et al, Nucl. Phys. A449 (1986) 537.

26) R.M. Lieder et al, Nucl. Phys. A375 (1982) 291.
27) A. Neskakis et al, Phys. Lett. 118B (1982) 49.
28) G. Sletten et al, Proc. XXII Winter Meeting on Nuclear Physics, Bormio, Italy 1984, p.668.
29) J.C. Bacelar et al, Phys. Lett. B152 (1985) 157.
30) P.J. Twin et al, Nucl. Phys. A409 (1983) 343c.
31) R.M. Diamond and F.S. Stephens, Proc. Int. Conf. on Instrumentation for Heavy Ion Research ed. O. Shapira, (New York: Harwood, 1984) p.259.
32) R.M. Lieder et al, Nucl. Instr. Meth. 220 (1984) 363.
33) R.M. Lieder, Proc. XXIII Int. Winter Meeting on Nuclear Physics, Bormio, Italy, 1985 p.276.
34) P. Taras, et al, The 8π Spectrometer, proposal 1983.
35) J.P. Martin et al, Nucl. Instr. Meth. A257 (1987) 301.
36) B. Herskind, Proc. 2nd Int. Conf. on Nucl. Nucl. Collisions, Visby, Sweden, 1988, Nucl. Phys. A447 395c.
37) G. Sletten, these proceedings.
38) P.J. Nolan, Proc. Int. Nucl. Phys. Conf. Harrogate, UK, eds, Durell, Irvine and Morrison, IOP Conf. Series 86 Vol.2 (1986) 155.
39) P.J. Nolan and P.J. Twin, Ann Rev. of Nucl. and Part. Sci. 38 (1988) 533.
40) F. Beck, Proc. Conf. on Instrumentation for Heavy Ion Research ed. D. Shapira, Vol.7 in Nucl. Sci. Research Conf. Series (New York: Harwood, 1984) p.129.
41) R. Chapman et al, submitted to Nucl. Phys.
42) B.R. Mottelson and J.G. Valatin, Phys. Rev. Lett. 5 (1960) 511.
43) S. Frauendorf, Nucl. Phys. A409 (1983) 243c.
44) C.X. Yang, et al, Phys. Lett. 133B (1983) 39.
45) M.A. Riley, et al, Phys. Rev. Lett. 60 (1988) 553; and J. Simpson et al, Proc. Int. Conf. on Nuclear Shapes, Crete, Greece, 1987, World Scientific, p.413.
46) J.D. Morrison et al, Europhys. Lett. 6(6) (1988) 493.
47) R. Bengtsson and S. Frauendorf, Nucl. Phys. A327 (1979) 139.
48) M.A. Deleplanque et al, Phys. Lett. 193B (1987) 422.
49) J. Simpson et al, J. Phys. G13 (1987) 847.
50) J. Simpson et al, Phys. Rev. Lett. 54 (1985) 1132.
51) J.C. Bacelar et al, Nucl. Phys. A442 (1985) 509.
52) R. Moricz, Lund report, 15 Nov.1984 and private communication.
53) J.A. Roberts private communication.
54) T. Bengtsson and I. Ragnarsson, Phys. Lett. 136B (1985) 31.

55) I. Ragnarsson and T. Bengtsson, Contrib. to Int. Conf. on Selected Topics in Nuclear Dubna, USSR, 1989.

56) L. Goettig et al, Nucl. Phys. A464 (1987) 159; and C.J. Lister et al Phys. Rev. Lett.

57) A.N. James et al, Nucl. Instr. Meth. A267 (1988) 144.

58) T.M. Cormier, Ann. Rev. of Nucl. and Part. Sci. 37 (1987) 537.

59) J. Simpson et al, Nucl. Instrum. Meth. A269 (1988) 209.

60) B.J. Varley et al, Phys. Lett. B194 (1987) 463.

61) C.J.Lister et al, Proc. 5th Int. Conf. on Nuclei far from Stability, Rosseau Lake, Ontario, Canada, (1987) p.354; and C.J. Lister et al, Proc. Int. Conf. on Contemporary Topics in Nuclear Structure Physics, Cocoyoc, Mexico, 1988, p.653.

62) C.J. Lister et al, Phys. Rev. Lett. 59 (1987) 1270.

63) R. Wadsworth et al, J. Phys. G13 (1987) L207.

64) S. Rugari et al, Bull. Amer. Phys. Soc. 32 (1987) 1096.

65) As an example see S.M. Mullins et al, J. Phys. G13 (1987) L201.

66) R.R. Betts et al, Phys. Rev. Lett. 59 (1987) 978.

67) J.A. Cameron et al, to be published, and private communication.

Inst. Phys. Conf. Ser. No 105
Paper presented at Int. Conf. on Spectroscopy of Heavy Nuclei, Crete, Greece, 1989

Diabatic orbitals in nuclear structure

tord bengtsson
Dept. of Math.Phys.
Lund Institute of Technology

Abstract: The importance of wavefunctions in detailed nuclear structure calculations is emphasized. A scheme to enable studies of the development of wavefunctions as the hamiltonian gradually changes is presented. By using this scheme, nuclear structure properties can be calculated with great detail.

1. INTRODUCTION

Theoretical nuclear physics is a challenging subject in many ways. A major difficulty is the fact that even if the bare nucleon-nucleon force was exactly known, present mathematical understanding does not provide any method to calculate nuclear properties with this force. Therefore the history of theoretical nuclear physics is a history of clever approximations — the liquid drop model, the mean field approach and their combination into Nilsson-Strutinsky type [St67, NT69] calculations, to mention one line of development.

At the present level of understanding, the major ingredients in a realistic mean-field calculation are rather well established: the bulk properties are taken as those described by a charged liquid drop [vW35, MS66]. To the bulk properties are then added the quantum-mechanical 'shell-correction', calculated as the deviation from the average properties of a nucleus in a mean field potential. Furthermore, to properly describe odd-even mass differences and some other observed properties, a short-range force must be present that gives an energy gain for two nucleons moving in the same spatial orbit. This 'pairing' effect is accounted for by a BCS calculation [BM58].

Using the general outline above, a wealth of nuclear properties have been calculated with astonishing precision. Some examples are ground state masses and moments [MN81], and the changes of nuclear structure with increasing angular momentum [AL76, NP76, DN85].

As the calculations have become more elaborate, however, some inherent limitations in the mathematical treatment have become more and more of overwhelming difficulty. We will study this in the examples below.

2. EXAMPLES OF DIFFICULTIES WITH PRESENT FORMALISM

2.1 Shape coexistence

A phenomenon that is rather common in calculations of nuclear ground-state properties is shape coexistence. For example, a nucleus with magic particle number is usually predicted to have a spherical ground state. But often exited 0^+ states can be found that are built by configurations that break the 'magic' and drive the nucleus deformed. A calculation that shows such properties is illustrated in fig. 1. The nucleus ^{186}Pb has a magic proton number, Z=82, and therefore the ground state is spherical. But at moderate deformations, configurations where protons are excited over the Z=82 shell gap can compete energetically with closed-shell configurations. This is most clearly visible at $\varepsilon > 0$ in fig. 1. But also at $\varepsilon < 0$, a change of slope is visible that might be an indication of an excited 0^+ state with excitation energy comparable to that of the prolate state.

Within the formalism used, it is very difficult to calculate the excitation energy and deformation in cases like the oblate state of ^{186}Pb . But a calculation that aims

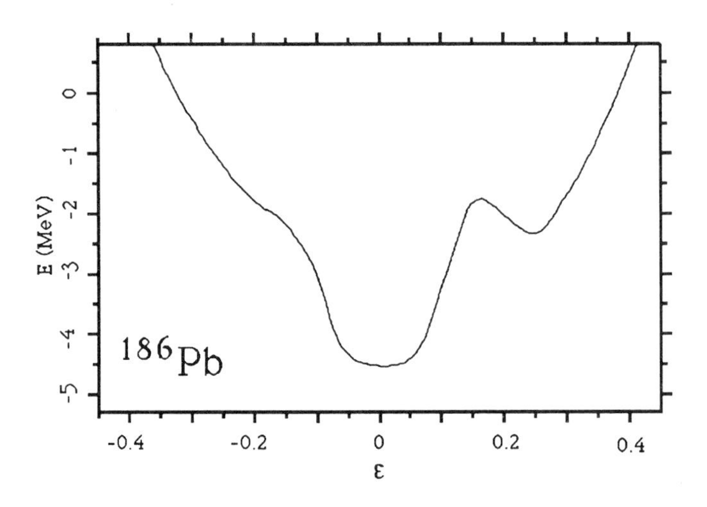

Fig. 1: Potential energy curve for the nucleus ^{186}Pb, resulting from a normal Nilsson-Strutinsky BCS calculation.

in predicting all possible 0^+ states ought to be able to find such a state. Even though the barrier for decay to the ground state seem to be absent in fig. 1, the oblate state is structurally different from the ground state to roughly the same degree as the prolate and therefore their decays should be hindered by roughly the same amount.

2.2 High-spin spectra

The investigation of nuclear high spin spectra is continuing to attract considerable effort and the use of large arrays of Compton-suppressed germanium detectors have lately given rise to a wealth of experimental data. On the theoretical side, the approach that has proven most powerful is very similar to the one used for ground state properties above. The sole difference is the introduction of a term in the hamiltonian to simulate the rotation of the nucleus, resulting in the cranking hamiltonian:

$$h^{\omega} = h - \omega \, j_x \, . \tag{1}$$

Up to medium-high spin states the pair force is still of importance as manifest by the occurrence of similar bandcrossings in neighboring nuclei. Usually, the

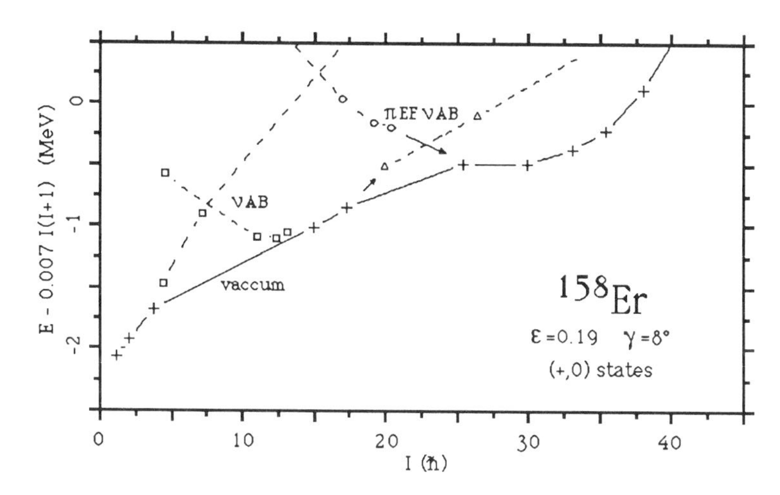

Fig. 2: Calculated total energy and spin for the yrast π=+, α=0 states in ^{158}Er.The energy is plotted relative to the energy of a rotating rigid body with $E_{rig}(I)$ = 0.007 I(I+1) MeV. This calculation is identical to the one for fig. 3. The symbols show the calculated values which are connected by straight lines. In addition to the vacuum states some excited quasiparticle configurations are plotted, which show the non-yrast continuation of the bands that form the yrast line.

cranking model with pairing included is only used to predict where these bandcrossings could occur and their nature [BF79]. This is generally only possible if the nuclear deformation is assumed not to change with angular momentum. For the experimental data now available this assumption is rarely justified. In order to predict the nuclear deformation, the potential energy of the nucleus must be calculated, similar to the example above but now for non-zero angular momentum.

As a high-spin example let us therefore calculate the yrast line of ^{158}Er. To simplify the understanding we will do this at a fixed deformation. The result is shown in fig. 2 and the appropriate quasi-particle diagrams in fig. 3. The most prominent feature of fig. 2 is probably that the bandcrossings in the proton and neutron systems are mainly manifest by large jumps in spin. In the spin region 10 to 25, where the neutron S-band is yrast, we find only two calculated states. If excited configurations are included (broken lines in fig. 2), we may follow the bands in a larger spin region. This is however only possible in simple situations like this one, where there are not too many interactions around the Fermi level (c.f. fig. 3). In the general case and especially when some selfconsistency is required, this is virtually impossible.

2.3 The source of the problem

Summarizing the above examples, we have seen that the lack of interest paid to the total nuclear wavefunction reduces the predictive power of the calculations. This is not primarily due to that oversimplified models have been used. On the contrary, the used 'standard' models have proven their predictive power in a wide range of similar applications, and there is no reason to believe that the ones above should be any exception.

The source of the problem is rather to be found in the mathematical treatment of the model. Only one state is calculated at each parameter value, the one of lowest

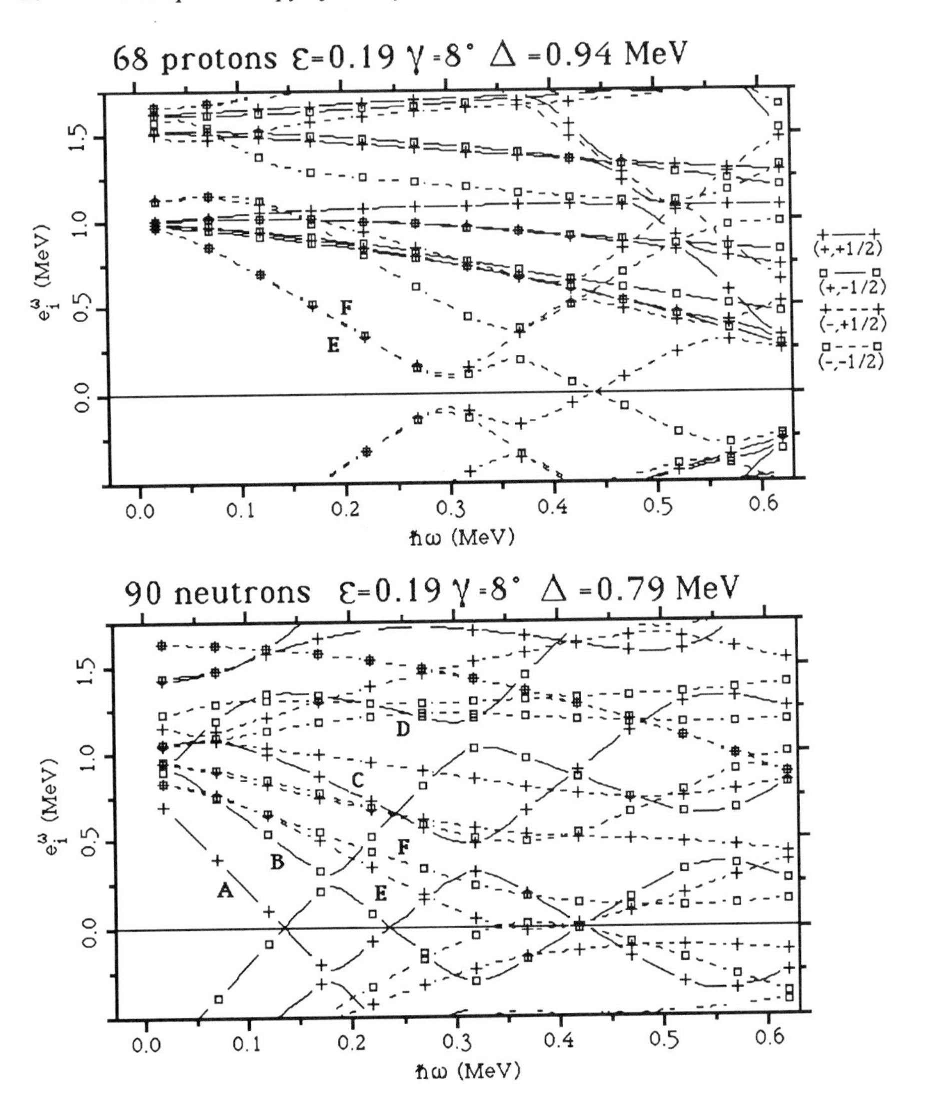

Fig. 3: Adiabatic quasi-particle energies as a function of cranking frequency ω, for protons and neutrons in the nucleus ^{158}Er. The values of Δ and λ are kept fixed as ω varies. The markers are placed at the frequencies where the actual calculations where performed, while the lines are drawn by a polynomial interpolation and does not reflect the actual size of interactions. The customary labels for the lowest quasiparticle orbitals are indicated.

energy, and states calculated at different parameter values does not necessarily bear any structural similarity. An improvement would be to calculate all excited states at each parameter value and then construct potential energy curves or rotational bands of those states that bear the greatest structural similarity.

Here, simplifications in the model gives rise to problems, though. In the simple BCS calculation used for ^{186}Pb above, the occupation of orbitals is allowed to change continuously, totally disregarding the fact that this occupation might introduce a major change in the nuclear structure. This is also true in the high-spin example, but here the fact that the hamiltonian is not rotationally invariant gives rise to additional problems. Without rotational invariance, states of different angular momentum are allowed to interact, giving rise to calculated states that are a mixture of rather different configurations. In both of these examples, simplifications in the model gives rise to interactions that are more or less unphysical or 'virtual'.

Therefore, the improved approach suggested above necessitates a rather complicated treatment in order to trace the nuclear structure when a parameter in the hamiltonian is changing. But this is one way to enable us to perform calculations of greater detail.

In the following, I will outline some ideas that enables the calculation of bands built on orbitals that show no drastic changes in their wavefunctions. Such 'smoothed' orbitals are usually referred to as *diabatic*, whereas the 'unsmoothed' eigenfunctions of the model hamiltonian are denoted *adiabatic*. In high-spin calculations, the importance of using diabatic orbitals were pointed out long ago [IH76], and quite a few attempts to enable their calculation have been tried (see [TB89] for a brief review). Although clearly showing the benefit of using diabatic orbitals, the hitherto used methods have all been rather specialized.

As a contrast, the method I will present here is applicable for a wide range of problems. A full description of the method would require much more than my allotted time and therefore I will restrict myself to a brief description of the main ideas, referring the interested reader to [TB89], containing a more complete description.

3. TRACING THE WAVEFUNCTION

In this description, let us write the hamiltonian as

$$h = h_0 + \delta V \ , \tag{2}$$

where h_0 is the unperturbed hamiltonian and δ is the parameter connected to the perturbing potential V. The problem is to construct a sequence of diabatic orbitals $\Psi_i(\delta)$ for a range of δ-values. These orbitals should as closely as possible resemble the eigenfunctions of h, but not show any abrupt changes in their wavefunction as δ changes.

In order to construct diabatic orbitals in calculations with pairing where furthermore some selfconsistency might be desired, there are some requirements to consider.

One is that the values of the hamiltonian parameters (pairing gap, Δ, and Fermi energy, λ) for which self-consistency is required, are not known a priori. Therefore the diabatic orbitals cannot be constructed from adiabatic ones that fulfill self-consistency as the self-consistency criteria might be affected by the removed interaction. Unfortunately this can happen precisely at those interactions that are physically important, namely those which connects unoccupied and occupied orbitals. This means that the construction of diabatic orbitals must be intimately connected to the actual calculation of orbitals.

Calculations with pairing use more quantities than just the energy and the expectation value of angular momentum of the orbitals. For example, the self-consistency condition on particle number requires knowledge of the particle amplitude, $<v^2>$, of the quasi-particle orbitals. Furthermore, in both paired and unpaired calculations, one might want to calculate other expectation values, for example quadrupole moments. Therefore a general method should treat the wavefunctions and not just expectation values.

Finally, an important requirement in connection with cranking calculations in general is that only interactions of rather small size should be removed. The main features of the adiabatic results should thus remain and the value of the smallest accepted interaction should preferably be controllable.

3.1 Brief description of a possible method

Two basic observations are helpful in understanding how interactions can be detected and removed:

— Realistic calculations are made using numerical methods that prohibits continuously changing parameters. Therefore a parameter must change in finite steps. This is to some extent a deficiency as results for intermediate parameter values are found by some interpolation. But an interaction strength does not only manifest itself by the minimal energy distance between interacting orbitals, it also affects the parameter range in which the orbitals are disturbed. Therefore this deficiency can be an advantage as the finite parameter step can be used as a measure of the maximal interaction strength that should be removed. One might also take the attitude that by using a specific parameter step, one is only interested in resolving details of the calculations that extends over a larger parameter range than this step. If one wants to resolve finer details, one has to use a smaller step.

— Another very basic observation is that the construction of non-interacting orbitals is straightforward when the interaction strength is small and happens to be in the middle of a parameter step. In this case the orbitals are unperturbed at the calculated parameter values and therefore a simple check of overlaps will define the diabatic continuation. A large number of virtual crossings can be treated in this way, especially if the parameter step is very much larger than the interactions that appear. By chance, though, calculations can be made at a parameter value where the interaction is non-negligible. The interaction might also be of such strength that it is not reasonable to use the large parameter step that is required. In such cases, the orbitals from the pure (adiabatic) calculation must be modified.

3.1.1 Detection and removal of interactions

Let us now suppose that we have a set of diabatic orbitals for a parameter value δ, $\Psi_i(\delta)$, and have solved eq. 2 with $\delta=\delta+\Delta\delta$ to obtain the adiabatic orbitals $\phi_j(\delta+\Delta\delta)$. From the observations above we may immediately decide that if the overlap

$$\Omega_{ij} = < \Psi_i(\delta) \mid \phi_j(\delta+\Delta\delta)> \qquad (3)$$

is large enough, the diabatic continuation of Ψ_i at $\delta+\Delta\delta$ is ϕ_j. The critical value of the overlap can be found from studies of a two level crossing model, which gives the result that a value of $\Omega_{cr} = 0.9$ effectively discriminates strong interactions from weaker.

If it is not possible to find an adiabatic orbital ϕ_j where $\Omega_{ij} > \Omega_{cr}$, the diabatic orbital Ψ_i is interacting at $\delta=\delta+\Delta\delta$. This interaction must be removed in order to find the continuation of Ψ_i. By making all the simple continuations, with $\Omega_{ij} > \Omega_{cr}$, we are thus left with two sets of orbitals, T_+ and T. The set T_+ is composed of the adiabatic orbitals at $\delta+\Delta\delta$ that does not have an overlap with any diabatic orbital $\Psi_i(\delta)$ larger than Ω_{cr} and the set T is the corresponding set of diabatic orbitals at δ.

Two different methods are employed to remove interactions — one quite simple and safe, however not as powerful as the other which requires knowledge of parts of the hamiltonian and might fail to find a unique continuation. Both methods are able to treat a large number of interacting orbitals at the same parameter value.

The simple method is based on the assumption that the sets T and T_+ cover approximately the same space. By the aid of the overlap matrix Ω_{ij} (eq. 3), we may then express an orbital from set T (the diabatic orbitals) in the basis of the set T_+ (the adiabatic ones):

$$\Psi_i(\delta+\Delta\delta) = \mathcal{N}^{-1} \sum_{j\in T+} \Omega_{ij} \, \phi_j(\delta+\Delta\delta) \, . \tag{4}$$

Where $\mathcal{N}$ is a normalization constant. This corresponds to a projection of the diabatic orbitals onto the space spanned by the adiabatic and this method might therefore be denoted the ‘projection method’. In this method, the interaction is found from the overlap matrix, Ω_{ij}, which has the advantage that no specific information about the hamiltonian is needed. A disadvantage is that the interaction is not as precisely determined as in the other method, this may cause unsatisfactorily results if there is a weak interaction in the supposedly diabatic orbitals, $\Psi_i(\delta)$.

In the other, more advanced, method it is assumed that if the hamiltonian can be written like in eq. 2, either h_0 or V, but not both, should be responsible for the interaction. Thus only one of these operators should have non-zero off-diagonal matrix elements. The interactions can then be removed simply by diagonalizing the other operator in the space spanned by T_+, to obtain candidates for diabatic wavefunctions at $\delta+\Delta\delta$: Ψ'_α. If, for each $\Psi_i(\delta) \in$ T, an α can be found so that

$$|\langle \Psi_i(\delta) \mid \Psi'_\alpha(\delta+\Delta\delta) \rangle| \geq 1/\sqrt{2} \tag{5}$$

the method was successful and $\Psi'_\alpha(\delta+\Delta\delta)$ is accepted as the diabatic continuation of $\Psi_i(\delta)$. If this is not the case, the ‘rediagonalization’ method have failed and the projection method must be used.

The decision which operator is causing the interaction is not crucial if both operators have substantial diagonal matrix elements. In this case the interaction can namely be viewed as caused by the cancellation of the diagonal matrix elements of h_0 and V.

3.1.2 Initiating the procedure

In the previous section we have seen how diabatic orbitals can be obtained from adiabatic ones by a comparison to known diabatic orbitals separated a distance $\Delta\delta$ in the parameter δ. To initiate the procedure, the adiabatic orbitals at some parameter value must be accepted as diabatic. This ‘setting’ parameter value must then be chosen with some care. For example, if V, the perturbing potential, is breaking a symmetry in h_0, $\delta=0$ cannot be used to set diabatic orbital, due to degeneracy.

As should be obvious from the description above, the parameter step, $\Delta\delta$, plays an important role in this scheme as it controls the size of accepted interactions. Therefore, to keep the major properties of the hamiltonian, the parameter step should not be too large, so that the treated spaces, T and T_+, are only a small fraction of the full calculated space.

4. DIFFICULT EXAMPLES REVISITED

Let us now try to use this method in the cases described in the introduction to see how the use of diabatic orbitals facilitates nuclear structure calculations.

4.1 Shape coexistence

We have noted that the excited minima for ^{186}Pb found in fig. 1, are due to configurations that are very different from the ground state. Primarily, they correspond to excitations of protons over the Z=82 shell closure. If pairing were excluded from the calculation, it would be relatively straightforward to find the potential energy dependence on deformation for the different configurations.

By the pairing interaction the excitation of a particle pair due to a changing deformation takes place by gradually changing the involved quasi-particle orbitals from being mainly of particle nature to hole nature and *vice versa*. The removal of the virtual interaction in this case will thus correspond to 'freeze' the particle-hole nature of orbitals that will pass through the pair gap until they have emerged on the other side. Thus the interaction is replaced by a quasiparticle-hole excitation.

In fig. 4, the result of an exploratory removal of this interaction is illustrated by thinner lines. The use of the rediagonalization method and diagonalizing the unpaired Hamiltonian, would in this case be equivalent to exclude the interacting orbitals from the pairing calculation. This is certainly a possible approach but here the projection method has been used. There are furthermore some special features that must be taken into consideration. It is for example important that the parame-

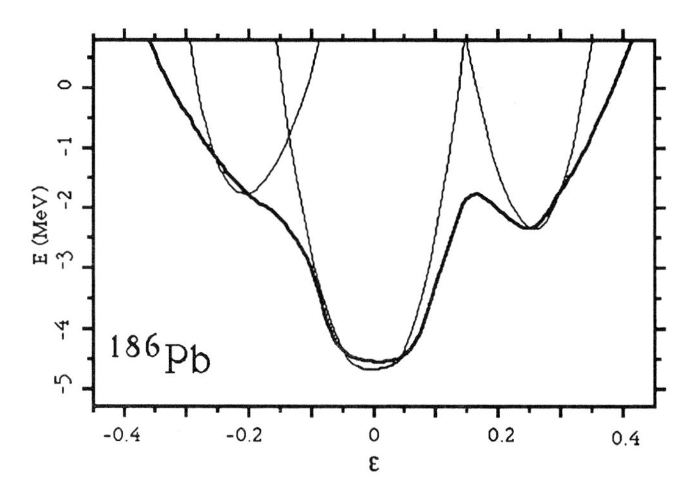

Fig. 4. Similar to fig. 1, but now with potential energy curves for dia-

ter step, $\Delta\delta=\Delta\varepsilon$, is large enough so that orbitals that will "pass through" the pair gap suffer enough change in their wave-functions to be detected for treatment. The corresponding unpaired orbitals should therefore roughly change their energy by Δ for one step in deformation. Self-consistency is required for λ but, as an approximation, the vacuum value for the pair gap parameter, Δ, is used even for excited configurations. The non-interacting potential energy has been calculated by accepting the eigenfunctions at some deformation as diabatic and then find the diabatic continuation of them at other deformations. For the spherical minimum, the diabatic orbitals are set at $\varepsilon = \pm 0.05$ as $\varepsilon = 0$ could not be used because of the high degeneracy. At the prolate deformation, $\varepsilon = 0.25$ and at the oblate $\varepsilon = -0.20$ has been used to define the diabatic orbitals. In ref. [TB89] the variation in setting point is somewhat explored. It is found that the self-consistent deformation, the curvature (zero-point energy) and the energy at the minimum is rather consistent for different starting points.

The result appears rather satisfactorily and allows us to predict the properties of the exited minima in greater detail. Removal of virtual interactions in deformation space is thus possible. At least two important cases where such a method is needed exist: situations like the oblate case considered here and situations where the potential energy is rather constant even for large variations of deformation parameters. In the latter case, it is possible to find out whether several configurations are involved. If this is the case, one can find their separate self-consistent deformations.

4.2 High-spin spectra

In high-spin calculations with the cranking hamiltonian (eq. 1), the j_x operator can be identified with the perturbing potential V of the schematic interaction hamiltonian (eq. 2). Thus we may use the rediagonalization method in this case (the projection method must be available as a last resort, however). A quasiparticle diagram, identical to that in fig. 3 but with weak interactions removed, is displayed in fig. 5. It is immediately obvious that the majority of interactions are removed and we are, for example, able to follow the ground configuration up to quite high frequencies. Some interactions remain, however, and rightly so as these have a strong interaction, stronger than $\Delta\omega = 50$ keV.

Proceeding further, the total energy and spin for the configurations that form the (π=+, α=0) yrast line at this deformation can be calculated, fig. 6. A comparison between fig. 2 and 6 reveals that most of the calculated values are the same but the yrast line structure are now much more clearly visible. Naturally, for a simple case like this, one may argue that the same information can be obtained from fig. 2 as from fig. 6. This is perfectly correct and is the way one usually performs paired high-spin calculations. When some selfconsistency is required, in pair field and/ or deformation, this becomes an extremely demanding and difficult task, however.

Let us also compare to the available experimental data from ref. [SR84, TD85], displayed in fig. 7. Up to roughly spin 30 to 35 we see a decent agreement between the calculated and observed (+,0) band. At higher spins the experimental data shows much more energy favored states, however. These form the 'terminating band' that was predicted [BR83], and later confirmed, to terminate at spin 46^+ in this nucleus.

The terminating bands are moderately collective bands that are built on configurations with a rather low angular momentum content (see [RX86] for a review). As the maximal angular momentum is approached, the nuclear shape gradually changes towards the oblate (γ=60°) axis, where the maximal angular momentum is energetically favored. The importance of using diabatic orbitals in high-spin calculations can be illustrated by the fact that the properties of terminating bands were calculated using the simple method for removal of virtual interactions described in [BR85].

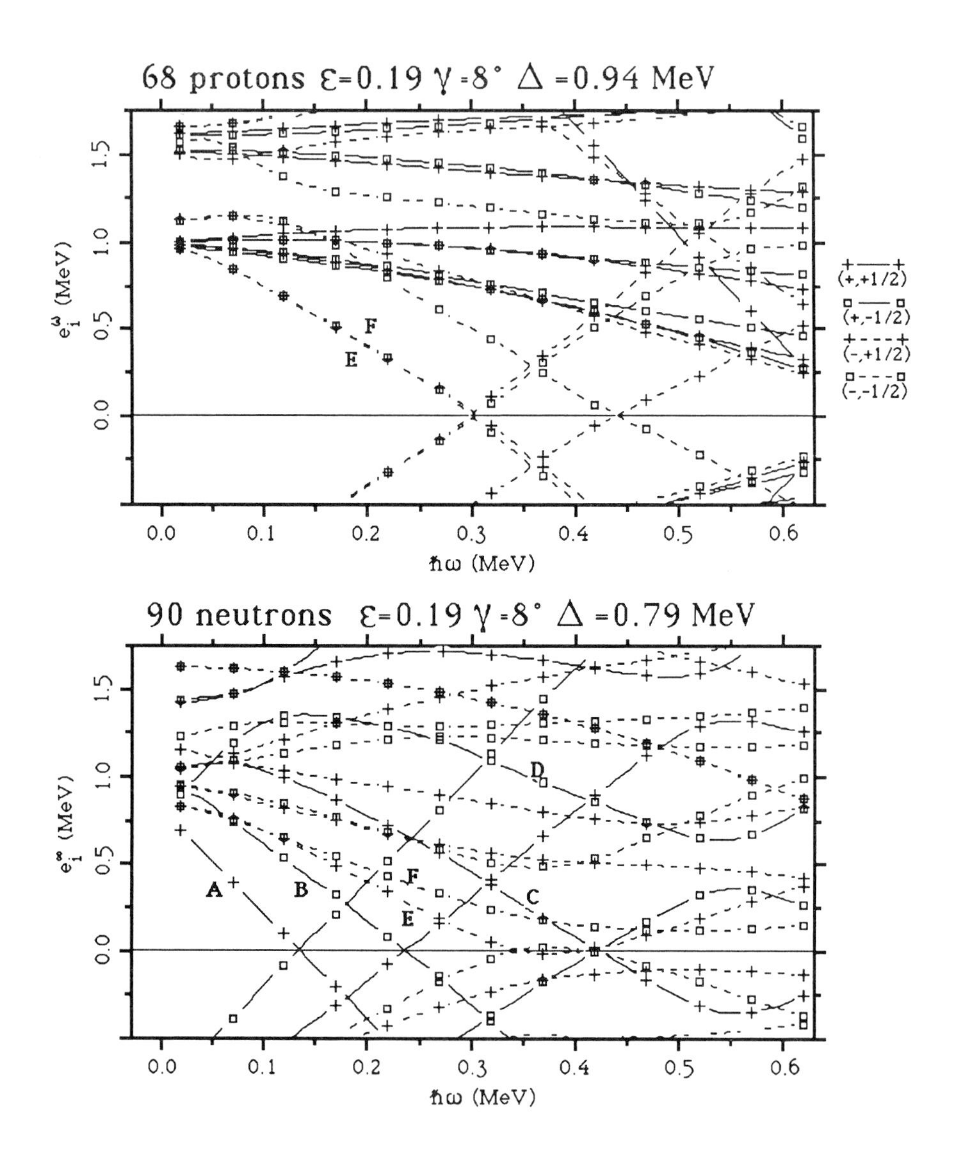

Fig. 5: Similar to fig. 3 but with weak interactions replaced by crossings.

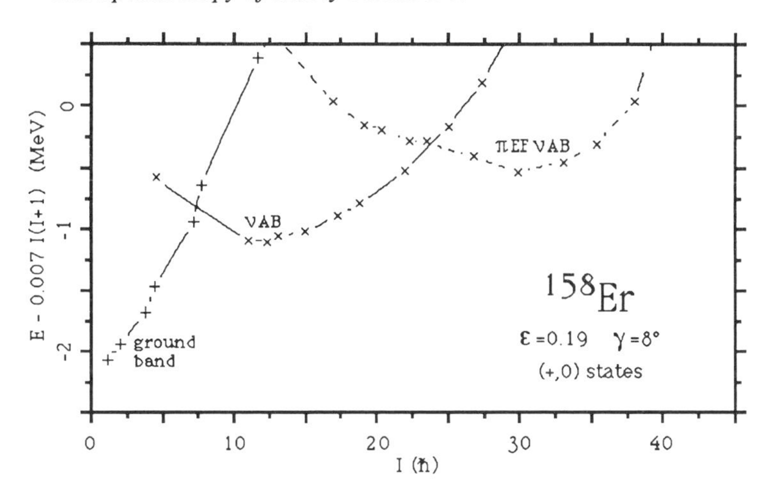

Fig. 6: Similar to fig. 2 but here the diabatic quasiparticle orbitals displayed in fig. 5 have been used.

4.3 A more elaborate high-spin calculation

From the considerations above, it becomes obvious that ^{158}Er was chosen as an example to illustrate a calculation of selfconsistent deformations with pairing. It should perhaps be mentioned, though, that the presented method is designed for pairing selfconsistency, an example of such a calculation is provided in [TB89].

Calculations that can predict the yrast line up to spin 50 require some extra considerations, especially when virtual interactions are removed. This is because the configurations that form the yrast line are no longer easily found, even for positive parity and even spins. They will depend on which virtual interactions that are replaced by crossings and this may change from deformation to deformation.

Therefore, some more steps must be introduced in the calculation:

— A procedure to find all possibly interesting configurations for protons and neutrons. This can be solved by finding the optimal (lowest energy) configuration at each cranking frequency, and then build some additional quasiparticle excitations upon it. A severe problem with pairing selfconsistency appears as the total quantities are desired for the entire ω range even for configurations that are found to be interesting at very high cranking frequencies. This means that total quantities can not be calculated selfconsistently with respect to pairing. The pair field parameters are the same for all configurations but with the aid of particle number projection the most severe deficiencies of this approach disappear.

— To obtain the yrast line, all proton and neutron configurations that have been found are combined and interpolated into the physical spin values as given by the total signature of the combined configuration. The yrast state for a specific parity and spin are then formed from the total configuration that have the lowest energy at this spin.

Only a few more words are needed to fully describe the deformation selfconsistent calculations: they are performed as described in [TB89] with the yrast lines calculated in a mesh in the (ε,γ) plane, ε_4=0. The pair field parameters Δ and λ are taken from a BCS calculation at ω=0, where the pair gap parameters are somewhat reduced, $\Delta_\pi = 0.85\Delta_{\pi BCS}$ and $\Delta_\nu = 0.75\Delta_{\nu BCS}$, to approximate the selfconsistent values for configurations with a few quasiparticles excited. As the 'standard' set of Nilsson model parameters [BR85] have been found to somewhat inappropriate close to the Z=64 gap, a modified parameter set [JR88] have been used. This latter parameter

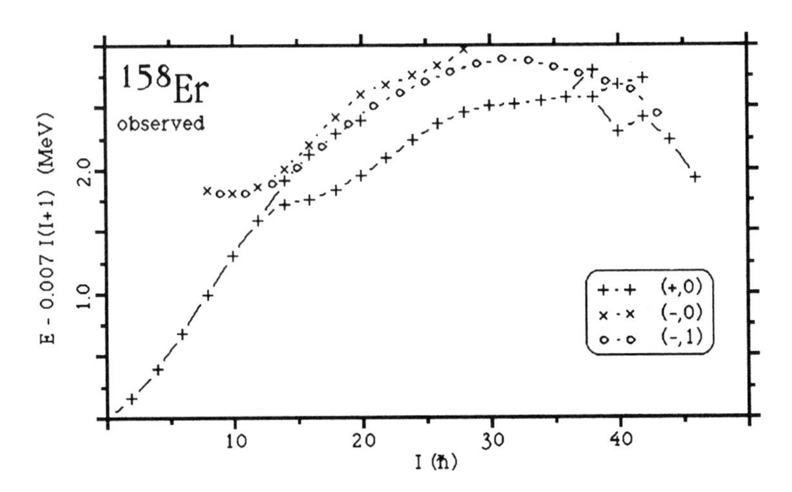

Fig. 7: Total energy versus spin for some observed, [SR84, TD85], high-spin bands in ^{158}Er, plotted as in figs. 2 and 6. The 0^{+} ground state energy is put to zero. Note change of spin scale.

A=120-140 region [JZ89].

The resulting total energies as functions of spin are illustrated in fig. 8. For the entire spin range, the deviation from experimentally found values does not deviate much more than 0.5 MeV. The ground — S-band crossing is predicted 4 spin units too low, probably because of the fixed pair gap parameters used in the calculation. Otherwise all observed features are reproduced fairly well: second band crossing due to protons in the (+,0) band observed around spin 26 (calculated at spin 24), the onset of termination at spin 38 with sub-termination at 40 and final termination at 46. In the negative parity bands, bandcrossings appear around spin 20 and spin 30, and the (−,1) band can be followed to the sub-termination at spin 43. The importance of the configuration finding procedure mentioned above can be exemplified by noting that the terminating states are calculated as configurations with ten quasiparticles excited, which furthermore only result in one spin state. Thus, to find all the non-collective yrast states requires that a large number of configurations are considered.

It may come as a due warning to experienced high-spin physicists that the calculated bandcrossings in the negative parity bands around spin 20 are not due to the neutron 'BC' or 'AD' crossings. Instead they are caused by the proton ground — S-band crossing, and the bandcrossing around spin 30 are due to the excitation of an additional pair of protons, where one proton have negative parity and the other positive. Alternatively, the negative parity bands in the spin range 20 to 30 could be built by a 2-quasiparticle negative parity proton configuration combined with the neutron S-band (AB). Both alternatives give an alignment gain of roughly 4 units in accordance with observations. The calculated energy difference between these alternatives is very small. The bands that correspond to neutron 'BC' or 'AD' excitations are in this calculation roughly 0.5 MeV higher in energy. Naturally, this cannot prove that the observed bandcrossings in the negative parity bands are solely due to protons but this emphasizes the importance of considering both particle systems when the nature of an observed band is to be determined.

Compared with the drastic deformation changes that occur due to band termina-

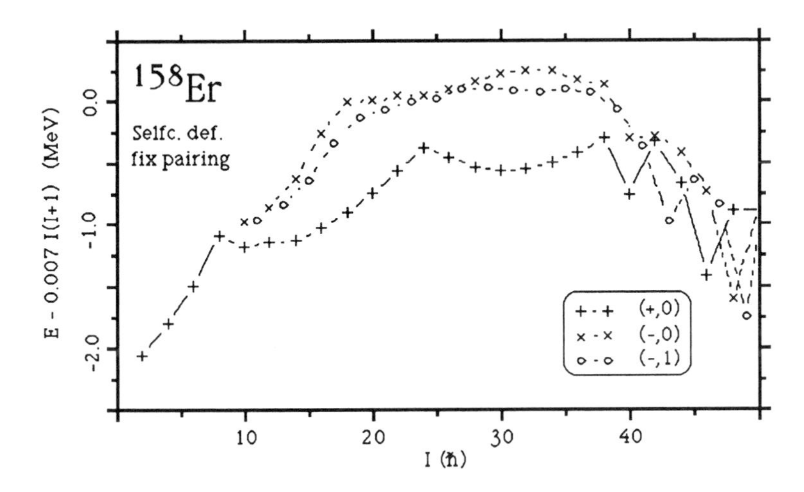

Fig. 8: Deformation selfconsistent high-spin states of ^{158}Er, calculated as described in text, plotted similarily to fig. 7. The (+,1) states have been omitted for clarity as they have no experimental counterpart.

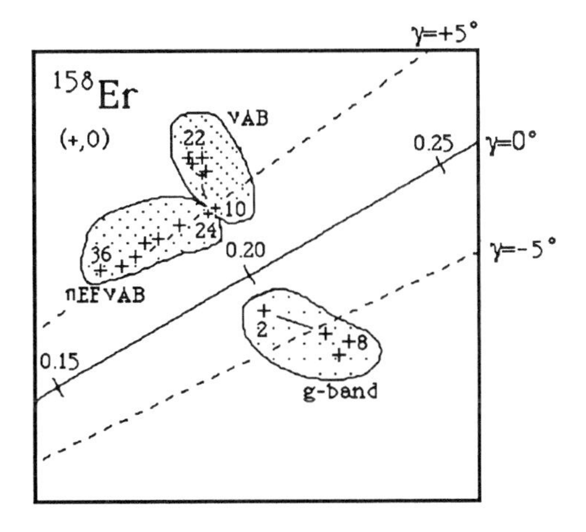

Fig. 9: Deformation development in the (+,0) yrast states. Numbers close to symbols indicate spin values. Note how well the three yrast configurations separate in the deformation plane as indicated by the shaded areas. Note also that this plot shows only a small fraction of the calculated (ε,γ) plane.

tions, the calculated shape changes in the collective bands are small but significant. I will here restrict the discussion to the (π=+,α=0) yrast band, for which furthermore the lifetimes have been measured [OJ86, BH88]. In fig. 9, the deformation development in this band is illustrated. The ground band prefers a deformation with ε = 0.21, γ = -6°, with some tendency to more negative gamma values for higher spins. The S-band (νAB) is also calculated to be fairly stable in deformation but now at positive γ-values: ε = 0.20, γ = 7°, whereas the πEFνAB band keeps roughly γ=5° but gradually decreases ε with increasing spin: from ε = 0.20 at spin 24 to 0.16 at spin 38.

From the experimentally observed lifetimes, transitional quadrupole moments can be obtained. These might be compared to values obtained from the calculated self-consistent deformations:

$$Q_t^{calc} = 8\, e\, Z\, r_0^2\, A^{2/3}\, \varepsilon(1 + \varepsilon/2)\, \cos(\gamma+30°)/(5\sqrt{3}) \qquad (6)$$

With r_0 = 1.2 fm, this quantity is in fig. 10 compared to experimental values, note that eq. 6 are not valid in bandcrossing regions and therefore these are marked in the figure. The calculations show the same major variations that may be read out from the observed values. In the ground band (note that the calculated ground band stops at spin 10 whereas the observed continues up to spin 14), the Q_t values are large and increasing, a reflection of the negative γ values obtained for this band. The calculated values do not increase as much with spin as the observed ones, though. In the spin range 14-24 (νAB band), the Q_t values are calculated to be rather constant which is in rough accordance with observations. Above spin 24, the πEFνAB configuration is yrast with a gradually decreasing ε and Q_t. This gradual decrease appears to be present also in the data, though with a smaller magnitude of Q_t.

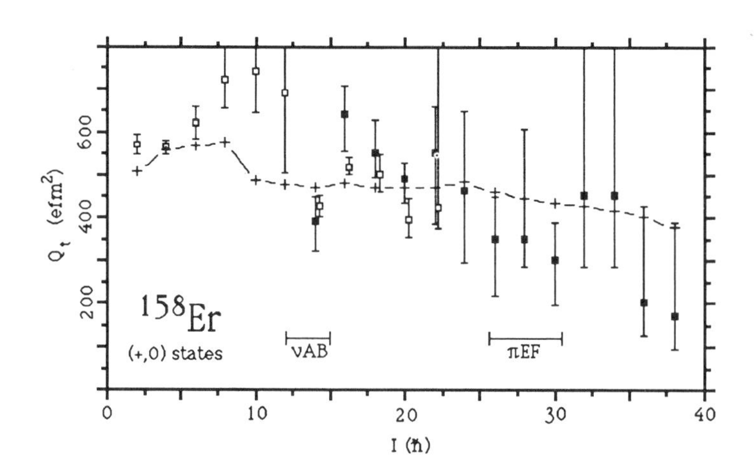

Fig. 10: Experimentally determined and calculated transitional quadrupole moments. Data from [OJ86] displayed with open squares and data from [BH88] with filled. The calculated values are obtained from selfconsistent deformations using eq. 6.

5. SUMMARY

In this contribution, I have tried to emphasize the importance of the configuration in nuclear structure calculations. If large changes in the total nuclear wavefunction is allowed, it is very difficult to obtain results with the detail that actually is possible within the used model.

One possibility to control the configuration is to use diabatic orbitals, which are approximate eigenfunctions of the model hamiltonian with weak interactions removed. By keeping a fixed configuration in the diabatic orbitals, it is ensured that the nuclear wavefunction do not show any drastic changes.

Thus the use of diabatic orbitals enables realistic and detailed calculations. Considering the wealth of experimental data that have been produced with the aid of Compton-suppressed arrays, the need for such calculations is great. One might say that the diabatic orbitals are the theoretical correspondence to an anti-Compton device.

This project is supported by the Swedish Natural Science Research Council.

References:

[St67] V.M. Strutinsky, Nucl. Phys. **A95** (1967) 420

[NT69] S.G. Nilsson, C.F. Tsang, A. Sobiczewski, Z. Szymanski, S. Wycech, C. Gustafson, I.-L. Lamm, P. Möller and B. Nilsson, Nucl. Phys. **A131** (1969) 1

[vW35] C.F. von Weizächer, Z. Physik **96** (1935) 431

[MS66] W.D. Myers and W.J. Swiatecki, Nucl.Phys. **81** (1966) 1

[BM58] A. Bohr, B.R. Mottelson and D. Pines, Phys.Rev. **110** (1958) 936

[MN81] P. Möller and J.R. Nix, Atomic data and Nuclear data tables, **26** (1981) 165

[AL76] G. Andersson, S.E. Larsson, G. Leander, P. Möller, S.G. Nilsson, I. Ragnarsson, S. Åberg, R. Bengtsson, J. Dudek, B. Nerlo-Pomorska, K. Pomorski and Z. Szymanski, Nucl. Phys. **A268** (1976) 205

[NP76] K. Neergård, V.V. Pashkevich and S. Frauendorf, Nucl. Phys. **A262** (1976) 61

[DN85] J. Dudek and W. Nazarewicz, Phys. Rev. **C31** (1985) 298

[BF79] R. Bengtsson and S. Frauendorf, Nucl. Phys. **A314** (1979) 27

[IH76] I. Hamamoto, Nucl. Phys. **A271** (1976) 15

[TB89] T. Bengtsson, Nucl.Phys. **A496** (1989) 56

[SR84] J. Simpson, M.A. Riley, J.R. Cresswell, P.D. Forsyth, D. Howe, B.M. Nyako, J.F. Sharpey-Shafer, J. Bacelar, J.D. Garret, G.B. Hagemann, B. Herskind and A. Holm, Phys. Rev. Lett. **53** (1984) 648

[TD85] P.O. Tjøm, R.M. Diamond, J.C. Bacelar, E.M. Beck, M.A. Deleplanque, J.E. Draper and F.S. Stephens, Phys. Rev. Lett. **55** (1985) 2405

[BR83] T. Bengtsson and I. Ragnarsson, Phys. Scr. **T5** (1983) 165

[RX86] I. Ragnarsson, Z. Xing, T. Bengtsson and M.A. Riley, Physica Scripta, **34** (1986) 651

[BR85] T. Bengtsson and I. Ragnarsson, Nucl. Phys. **A436** (1985) 14

[IR88] I. Ragnarsson, Proc. XXIII School on Physics, Zakopane, Poland, 1988

[JZ89] J.-y. Zhang, N. Xu, D.B. Fossan, Y. Liang, R. Ma an E.S. Paul, Phys. Rev. **C39** (1989) 714

[OJ86] M. Oshima, N.R. Johnson, F.K. McGowan, C. Baktash, I.Y. Lee, Y. Schutz, R.V. Ribas and

J.C. Wells, Phys. Rev. **C33** (1986) 1988

[BH88] E.M Beck, H. Hübel, R.M. Diamond, J.C. Bacelar, M.A. Deleplanque, K.H. Maier, R.J. McDonald, F.S. Stephens and P.O. Tjøm, Phys. Lett. **B215** (1988) 624

Inst. Phys. Conf. Ser. No 105
Paper presented at Int. Conf. on Spectroscopy of Heavy Nuclei, Crete, Greece, 1989

Alpha-particle and proton probes of nuclear shapes in the rare earth and mass 80 regions

D.G. Sarantites, N.G. Nicolis, V. Abenante, Z. Majka, T.M. Semkow
Washington University, St. Louis, MO 63130
and
C. Baktash, J.R. Beene, G. Garcia-Bermudez, M.L. Halbert, D.C. Hensley, N.R. Johnson, I.Y. Lee, F.K. McGowan, M.A. Riley, A. Virtanen
Oak Ridge National Laboratory, Oak Ridge, TN 37830
and
H.C. Griffin
University of Michigan, Ann Arbor, MI 48109

ABSTRACT: Low emission barriers and large subbarrier anisotropies in the alpha-particle decay with respect to the spin direction, of Sn and rare earth compound nuclei, are examined in the light of recent calculations incorporating deformation. To explore the possibility of a correlation between the proton emission barriers and nuclear deformation, we studied proton spectra from the $^{52}Cr(^{34}S,2p2n)^{82}Sr$ reaction. The proton spectra were observed with the Dwarf-Ball 4π CsI(Tl) array, in coincidence with 18 Compton suppressed Ge detectors operated in conjunction with the Spin Spectrometer, a 4π NaI(Tl) array. We found significant changes and shifts in the proton energy spectra as we selected gating transitions from bands of different moments of inertia or transitions from states of different spin in the same band. Substantial differences were also seen as a function of the γ-ray multiplicity. These results are discussed in terms of statistical model calculations incorporating deformation and structure effects of the emitting system.

1. INTRODUCTION.

The study of nuclear shapes at high angular momentum and excitation energy is a topic of current extensive theoretical and experimental interest in heavy-ion physics. It is well known that collective nuclei near the yrast line are deformed and their structure is well described by liquid-drop-Strutinsky cranked shell model calculations. A question of interest is the evolution of these shapes as the spin and the excitation energy (temperature) are increased. There is already considerable experimental evidence for the existence of superdeformed nuclei ($\beta = 0.6$) at high spins.[1] Theoretical calculations that explain these highly deformed shapes predict even higher deformations ($\beta \approx 0.9$) for nuclei close to the fission stability limit[2,3]. Temperature-induced noncollective rotation in nuclei, as well as shape changes have also been discussed in connection with predictions of mean field theories.[4] A number of experimental studies have tried to explore the effect of high excitation and/or angular momentum degrees of freedom on the nuclear shapes. This is made, on one hand, by exploiting the γ-decay properties (for example, study of giant resonances built on excited states) of the deexciting compound nuclei[5]. On the other hand, extensive searches are being made to find signatures of shape effects in the charged-particle decay properties of such systems[6–12].

The motivation for light charged particle studies lies in the well established fact that (fission-stable) compound nuclei, with the highest possible angular momentum, often decay by emitting alpha particles and protons. If the deexciting nucleus is deformed, it exhibits a lower evaporation barrier along the longer axis for charged-particle emission, compared to the spherical case. This results in strong enhancements of α and proton decay along the long axis, especially in the energy region below the evaporation Coulomb barrier[6,7,12]. Simulation studies along these lines have motivated a number of experiments consisting of the observation of α-particle spectra in heavy-ion fusion-evaporation reactions in a singles mode or in coincidence with evaporation residues[7,11,12]. The inability to reproduce the subbarrier part of the observed α-spectra, with statistical model calculations assuming spherical emission shapes, has been used as an indication of a deformation effect.

The desire for an unconstrained experimental study of these effects led us to the development of the spin alignment method with the Spin Spectrometer[13], a highly segmented 4π γ-ray detector system. In this method, the magnitude and orientation of the spin of the residual nuclei is deduced on an event-by-event basis. This makes possible detailed studies, such as the measurement of α-particle angular distributions with respect to the spin direction[14]. Furthermore, the γ-multiplicity selection with the spin spectrometer allows us to study these decay characteristics as a function of the evaporation residue spin, which is closely correlated to the compound nucleus spin. Therefore, the alpha-decay properties of different compound nuclear systems can be studied in detail[8–10].

In the first part of this paper, we report on the results of an earlier survey study[10] concering α-particle energy spectra and angular distributions with respect to the spin direction for a number of compound nuclear systems. The two typical cases of the closed shell $^{114}Sn^*$ and the rare earth $^{170}Yb^*$ deexciting compound nuclei are compared. The alpha emission properties of these systems are described by the anisotropy coefficients of the alpha particle emission with respect to the spin direction as a function of the α-particle energy and γ-ray multiplicity. Differences in the emission patterns, in the energy region below the evaporation Coulomb barrier, suggest nearly spherical and deformed emission shapes in the cases of $^{114}Sn^*$ and $^{170}Yb^*$, respectively.

Detailed statistical model calculations have been performed to clarify the deviations of the "deformed" versus the "spherical" behaviour. In the subbarrier region, the predictions of the statistical model for charged particle emission are sensitive to barrier penetration effects. These effects are expressed in terms of transmission coefficients resulting from an optical model calculation. Although our statistical model calculations describe closely the decay of $^{114}Sn^*$, we observe discrepancies in the other system. These descrepancies, appear as (a) an underestimate of the subbarrier 90° CM spectra, (b) underestimates of the multiplicity-gated 90° CM spectra, which increase with spin, and (c) deviations in the trend of the anisortopy coefficients from the one predicted in the $^{114}Sn^*$ case, which also increase with spin. We show that a simulation of deformation effects due to transmission accounts for these discrepancies. A detailed comparison with the $^{170}Yb^*$ data is made, which shows the angular momentum dependence of the deformation effect in the α spectra. However, more elaborate calculations are required to describe the corresponding effect in the anisotropy coefficients of these systems. The above findings are corroborated by both the known ground state properties and data of giant resonances built on excited states of similar compound nuclear systems.

In the second part of this paper, we report on experimental results of an attempt to explore the possibility of a correlation between the proton emission barriers and nuclear deformation and/or structure. Proton spectra from the $^{52}Cr(^{34}S,2p2n)^{82}Sr$ reaction at 130 MeV have been studied in coincidence with discrete γ transitions of selected exit channels. The proton spectra were observed with the Dwarf-Ball array[15], a 72 Cs(Tl) 4π detector, in coincidence with 18 Compton suppressed Ge detectors. This system operated in conjunction with the Spin Spectrometer which recorded the associated γ-ray multiplicity. We found significant changes and shifts in the proton energy spectra as we selected gating transitions from bands of different moments of inertia or transitions from states of different spin in the same band. Substantial differences were also seen as a function of the γ-ray multiplicity. The above results will be discussed in terms of statistical model calculations incorporating deformation and structure effects of the emitting system.

2. ALPHA EMISSION PROPERTIES AND DEFORMATION EFFECTS.

The experiments in this work were performed at the Oak Ridge Holifield Heavy-Ion Research Facility (HHIRF). The compound systems studied were $^{110}Sn^*(E^* =$ 93.9 MeV), $^{114}Sn^*$(79.5 MeV), $^{138}Nd^*$, $^{164}Yb^*$ (67.2 MeV) and $^{170}Yb^*$(134.8 MeV). The details of the experimental method and some of the general features of the data can be found in the current literature.[8,9,10,14] The complete study is in the process of publication. In the discussion below, we limit ourselfs to the presentation of distinct features shown in the deexcitation data of $^{114}Sn^*$ and $^{170}Yb^*$.

The compound nuclei $^{114}Sn^*$ and $^{170}Yb^*$ were produced in the reactions ^{64}Ni(250 MeV) + $^{50}Ti \rightarrow {}^{114}Sn^*$ and ^{20}Ne(176.6 MeV) + $^{150}Nd \rightarrow {}^{170}Yb^*$, respectively. In this study, self-supporting targets of high isotopic enrichment in each of the isotopes were used. The α-particles emitted in the deexcitation of the above compound nuclei were recorded by Si surface barrier telescopes positioned at the laboratory angles corresponding to $\sim 90°$ in the center-of-mass system.

The ΔE detectors had thickness of 65 μm and an acceptance cone of $\sim 6°$ half angle. The E detectors were 1500 μm thick and served as the triggers of the spin spectrometer. The spin spectrometer served as the γ-ray detector and measured simultaneously the γ-ray multiplicity, M_γ, the total γ-ray deexcitation energy and the γ-ray angular correlations. In these experiments the spin spectrometer provided a coverage of 95.8% of 4π sr.

For each compound nuclear system (A_{CN}, Z_{CN}), the alpha-particle events were transformed in the center-of-mass system for α + (A_{CN} - 4, Z_{CN} - 2), using two-body kinematics. The method used for determining the spin direction is based on the emission of γ radiation with a particular angular relationship to the spin direction.[8,14] The γ-cascades from rotational nuclei formed in heavy-ion fusion-evaporation reactions have a preponderence of stretched E2 transitions which exhibit a doughnut-like pattern about the spin axis. The spin direction is identified with the short symmetry axis of this pattern. This is close to the compound nucleus spin, i.e. perpendicular to the beam, provided that the misalignement caused by particle emission is small. The γ-pattern for each event was projected on a plane perpendicular to the beam direction and centroid-searching methods were used to determine the angle between the short symmetry axis and the direction of the emitted α-patricle.

In the following analysis, α-particle events corresponding to emission angles near $\theta_{CM} = 90°$ were sorted, imposing different γ-coincidence fold (k_γ gates, or angle with respect to spin (β) gates. Some useful remarks can be made from an inspection of multiplicity decomposed spectra. Fig. 1(a) shows, for $^{114}Sn^*$, the experimental $\theta_{C.M.} =$ 90° spectra corresponding to k_γ = 11-14 and k_γ = 27-33. The closed circles on the bottom show the total 90° center-of-mass spectrum integrated over k_γ (for $k_\gamma \geq 11$) and β. The corresponding α-particle spectra for $^{170}Yb^*$, under the same gating conditions, are shown in Fig. 1(b). In both cases, the selected k_γ bins correspond to α-particle emmision from nuclei with an average spin of $\approx 34\ \hbar$ and $\approx 64\ \hbar$, respectively, deduced from statistical model calculations[8,10].

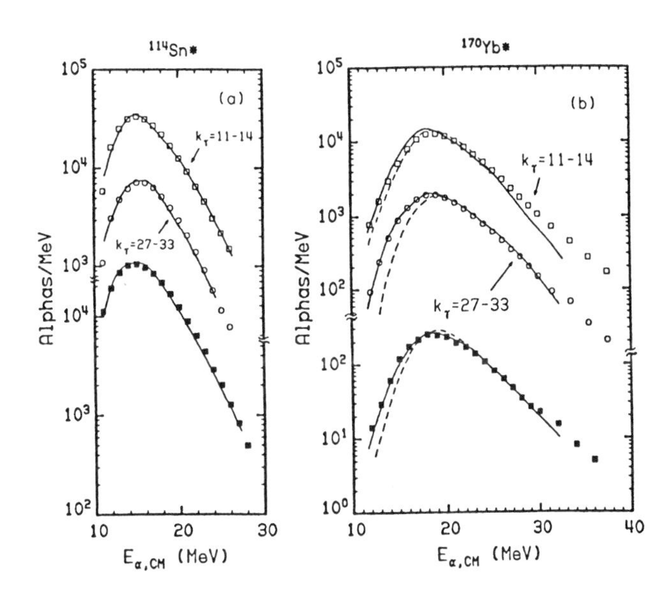

Figure 1. (a) Fold gated 90° center-of-mass alpha-particle spectra from the deexcitation of $^{114}Sn^*$. The open squares correspond to k_γ = 11–14 and the open circles to k_γ = 27–33. The closed squares show the 90° center-of-mass spectrum integrated over k_γ and β. (b) Multiplicity fold gated 90° center-of-mass alpha-particle spectra from $^{170}Yb^*$. The squares correspond to k_γ = 11–14 and the circles to k_γ = 27–33. The closed squares show the 90° center-of-mass spectrum integrated over k_γ and β. The solid and dashed lines are the results of calculations described in the text.

A common trend in both cases is that the k_γ = 11-14 spectra are slightly harder than the k_γ = 27-33 ones. This can be understood in terms of the higher excitation energy selection made by the low k_γ gate. However, the subbarrier trends of the spectra are quite different. For $^{114}Sn^*$, the low k_γ compared to the high k_γ spectrum has an excess of subbarrier alphas, whereas these two regions in the $^{170}Yb^*$ spectra are very similar. The solid lines in Fig. 1(a) are the result of a statistical model calculation with standard parameters.[8,16,17], We We see that there is a good agreement with the data in the whole energy range, for all of the gating conditions. The corresponding calculation for $^{170}Yb^*$ is shown, on Fig. 1(b), by the dashed lines. The k_γ = 11-14 spectrum is underpredicted in the subbarrier region as well as at high energies. In the k_γ = 27-33 bin, the discrepancy is only in the subbarrier region and it has increased. Similarly, the total alpha spectrum is underpredicted at subbarrier energies. In summary, although the behaviour of the alpha particle spectra from $^{114}Sn^*$ is well described by statistical model calculations, we observe a systematic underprediction of the subbarrier parts of the $^{170}Yb^*$ spectra, in a manner which increases with spin.

A striking differece is observed in the trend of the anisotropy coefficients of alpha emission with repect to the spin direction. The anisotropy coefficients A_2 (of a Legendre polynomial expansion) are plotted for the two systems as a function of $E_{\alpha,CM}$ for the indicated k_γ bins in Fig. 2. For $^{114}Sn^*$, we have monotonically decreasing A_2 coefficients (increasing anisotropies) with increasing E_α, in each of the k_γ bins. In contrast with these findings, the A_2 coefficients for $^{170}Yb^*$ have a maximum value (minimum anisotropy) at the evaporation Coulomb barrier ($\approx$ 20 MeV for $^{170}Yb^*$ assumed spherical) and become more negative (larger anisotropies, with stronger emission perpendicular to the spin direction) at lower and higher E_α values. Compared with the almost linear decrease with E_α observed for $^{114}Sn^*$, we see a deviation in the trend of the experimental correlations below the Coulomb barrier. The $^{170}Yb^*$ data suggest enhanced anisotropies which become larger with decreasing E_α. This enhancement increases with increasing spin.

Both of the data sets are compared with the results of a statistical model calculation with standard parameters in Fig. 2. The pairs of curves are FWHM boundaries of the calculated A_2 coefficients. For $^{114}Sn^*$, we see that the trend of the A_2 coefficients and their absolute magnitude are reproduced by the calculation.

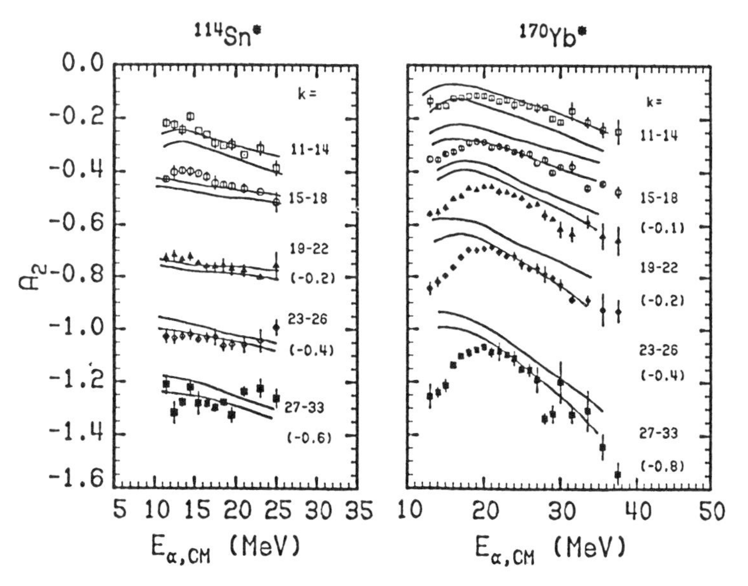

Figure 2. A_2 coefficients as a function of $E_{\alpha,C.M.}$ from the $^{114}Sn^*$ and $^{170}Yb^*$ systems. In both cases, the open squares, circles, closed triangles, diamonds and closed squares correspond to the k_γ bins of 11-14, 15-18, 19-22, 23-26 and 27-33, corresponding to $< I >_\alpha$ for α emission of 34, 43, 51, 59, and 64 $\hbar$, respectively. In some cases the data have been shifted along the A_2-axis by the indicated amount. The pairs of curves are FWHM boundaries of the A_2 coefficients from a statistical model calculation using transmission coefficients from a spherical optical model potential.

The calculated A_2 coefficients, for $^{170}Yb^*$, agree well with the monotonic decrease of the experimental A_2 values above the Coulomb barrier, but do not reproduce the decrease of A_2 at low E_α.

The fact that the observed deviation occurs at emission energies sensitive to barrier penetration effects, has suggested the nuclear deformation as a possible factor for the decrease of the measured A_2 coefficients at low E_α. If the emitting system is deformed with its longest axis perpendicular to the spin direction, the subbarrier α-particles will be emitted preferentially along this direction (because of the lower Coulomb barrier). This leads to decreasing A_2 coefficients with decreasing E_α. On the other hand, the α-particles above the barrier would not be affected much by the deformation since their emission is mainly determined by the level densities. Therefore, the observed deviation can be interpreted as a deformation effect which increases with spin. The same interpretation accounts also for the discrepancies observed in the multiplicity gated spectra of Fig. 1.

An interesting comparison of the above systems has been made with data from the decay of giant resonances built on excited states of similar compound nuclear systems. Giant resonance data from the decay of $^{166}Er^*$ (61.5 MeV) suggest a two-component resonance in contrast to the decay of $^{108}Sn^*$ (61.2 MeV) where a single resonance peak was observed.

A simulation of deformation effects in the statistical model code was performed in the case of $^{170}Yb^*$, in order to get an estimate of the effect. For this purpose we employed a variation of the method of equivalent spheres[18], which has been used successfully in the description of subbarrier fusion data with statically deformed targets.

The daughter nucleus was assumed to have a prolate axially symmetric shape described by the deformation parameter β, which was parametrized as $R(\theta) = R_0[1 + X(\beta) + \sqrt{5/4\pi}\beta P_2(cos\theta)]$, where θ is the angle with respect to the symmetry axis and $X(\beta) = -\beta^2/4\pi$ is the volume conservation term. Optical model transmission coefficients for protons and alphas were calculated, for all of the nuclei in the cascade, at 9 different angles from 5° to 85° in steps of 10°. At each angle, the optical model radii were scaled according to the above equation. The diffuseness of the of the Woods-Saxon nuclear potential of the spheroid was also modified, so that the normal derivative at each point on an equipotential surface is unaffected by the deformation. The alpha events from the evaporation calculations, emitted at each angle, were sorted and weighted according to the corresponding surface element of the spheroid: $2\pi R^2(\theta)sin\theta\Delta\theta/S$, where S is the nuclear surface: $S = 4\pi R_0^2(1 + a_2^2/5)$, $a_2 = \sqrt{5/4\pi}\beta$, including the first order correction term due to deformation.

The result of this calculation for the 90° center-of-mass spectra of $^{170}Yb^*$ is shown in Fig. 1, by the solid lines. A deformation of $\beta = 0.2$ was initially assumed. The result of this calculation for the total alpha spectrum is shown on the bottom of Fig. 1(b) by the solid line and provides a good description of the spin integrated spectrum.

On the same plot, the solid line for the k_γ=11-14 bin ($< I >_\alpha = 34\hbar$), shows the calculated spectrum with β=0.2. The subbarrier data points lie between the curves β=0 and β=0.2. In this case, a deformation somewhat smaller than β=0.2 is required to fit the spectrum. On the other hand, for the high spin case k_γ=27-33 ($< I >_\alpha = 64\hbar$), β=0.2 was insufficient to account for the excess of subbarrier alphas. The solid curve in the figure, corresponds to β=0.35 and fits closely the spectrum. These calculations demonstrate the extent and the angular momentum dependence of the effect. The originally deduced β=0.2 was based on the total 90° center-of-mass spectrum and represents an average over different deformed shapes.

The calculated A_2 coefficients, using the above logic, show a trend approaching the experimental data at subbarrier energies. However, no quantitative statement was made from such comparisons, because our present formalism for the calculation of the angular correlations is limited to spherically symmetric emitters. More refined calculations are required for this purpose.

Summarizing, the distinct differences in the alpha decay properties of the compound nuclei ^{114}Sn and ^{170}Yb have been interpreted in terms of deformation effects. For ^{170}Yb*, our calculations have demonstrated the angular momentum dependence of the effect in the 90° center-of-mass spectra, besides the one observed in the trend of the anisotropy coefficients of alpha emission with respect to the spin direction. One should keep in mind that the α-particle emission probe for studying nuclear shapes at high E* involves a broad range of initial excitation energies which contribute in low energy particle emission.[10]

3. STRUCTURE EFFECTS AND PROTON EVAPORATION SPECTRA

Compound nuclei that decay to residual nuclei with large deformations, such as superdeformed nuclei with discrete level structures, may be expected to have themselves significant deformations which persist to sufficiently high excitations of the order of at least one nucleon binding energy above the yrast line. In this case particle emission can be significantly influenced by nuclear shape effects. Furthermore, it is quite possible that structure effects can be observed on the shape of the charged particle evaporation spectra particularly near and below the emission barrier. The experimental observation of the existence of such effects is important both on its own merit and for providing important information about methods of populating the so called "hyperdeformed" structures in nuclei.

We report here on the results of an attempt to investigate the connection between proton emission from a decaying compound nucleus and underlying structural features in the final product. In this work we have selected for study ^{82}Sr as the final nucleus, because it has been predicted to be a good candidate for superdeformation.[19]

The experiment was performed at the Oak Ridge Heavy Ion Research Facility. The ^{82}Sr nuclei were produced by the ^{52}Cr(^{34}S, 2p2n)^{82}Sr reaction by bombarding a stack of self-supporting ^{52}Cr target foils with a 130 MeV ^{34}S beam. The experimental setup consisted of the ORNL Compton suppression spectrometer with 18 Ge detectors, which recorded the discrete γ-ray spectra from the reaction. The associated total γ-ray multiplicity and total energy were recorded with the Spin Spectrometer. The protons and α particles were detected with the 4π CsI(Tl) Dwarf Ball .[15] This system provided both high resolution γ-ray spectroscopic information and definite exit channel selection. The 72-element Dwarf Ball also provided light charged particle spectra and angular correlation information. The apparatus was triggered by two or more Ge detectors firindg in coincidence with any element of the Dwarf Ball. A total of 1.6×10^8 such events were collected and processed.

Figure 3 shows a scatter plot of the slow vs. tail map from a 42° CsI(Tl) detector. The γ-rays, protons and α particles are well separated. This was achieved by placing two time gates 400 ns (slow) and 1500 ns (tail) wide starting at times 0 and 1500 ns from the front of each pulse, respectively, and integrating the corresponding charge.

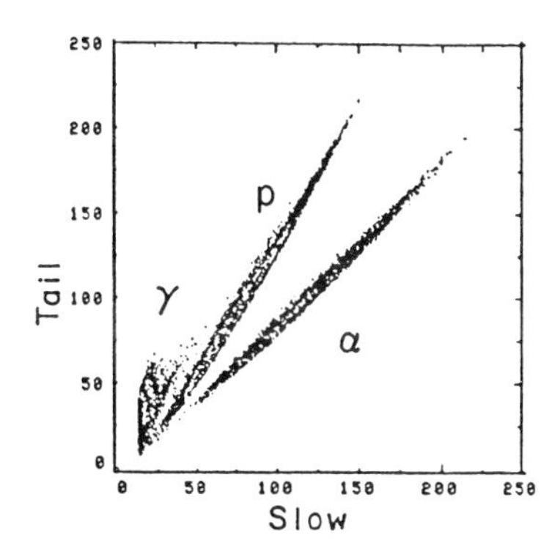

Figure 3. Scatter plot of the slow vs. tail light output of a CsI(Tl) detector in the Dwarf Ball showing the α, proton and γ-ray identification.

The particle identification in the Dwarf Ball utilizes the different time characteristics of the scintillation output of the CsI(Tl) (pulse shape discrimination). Excellent separation for all energies was achieved between proton and α pulses from each other and from γ-rays or neutrons for the detectors forward of 102° in the laboratory ($\sim$ 120° center-of-mass). The protons with subbarrier energies could be clearly identified for yields $\sim 1/20$ of that at the most probable value. The proton spectra sorted in this way contained $\sim$ 65% of the total proton yield. At larger angles due to kinematic forward focussing some of the subbarrier protons cannot be distinguished from α particles and therefore the detectors at these angles were used for channel selection, but not for particle spectra. The measured overall detection efficiency for proton detection was 85% of 4π (4 detectors out of 72 were removed for the beam, target rod and viewer port, and one detector did not operate). This caused $\sim$ 28% of the events with two protons to be identified as involving only one proton. The Ge energy spectra coincident with 2 protons involved primarily ^{82}Sr γ-rays with no contamination from the 1pxn or the αxn emitting channels. During the experiment, the most forward detectors were operated at a rate of $\sim$ 7000 c/s. A close examination of proton and α particle spectra recorded on different tapes revealed gain shifts up to 8% in the tail component, while the slow component was considerably more stable. Thus, gain shift corrections for every 2×10^6 events were applied for the tail component. This ensured good particle identification for the complete data set. The more stable slow component was calibrated for energy measurements using the $^{12}C(p,p')$ reaction at 9.0 and 20.0 MeV. The energy calibration uncertainties were estimated to be less than 3% for all but the 24° detectors for which uncertainty may be as large as 6%.

The particle energies were converted event by event to the center of mass system. The centroid angles for each detector in the laboratory system were used in deriving the center-of-mass energy and angle, assuming compound nucleus formation and decay to one proton. Proton energy spectra were sorted using the detectors forward of 102°, when 2 protons were identified in the complete Dwarf Ball and for three k_γ gates of 3-9, 10-14, and 15-25. Further selection was made by placing gates on discrete γ-rays associated with various rotational bands. The background to the proton spectra associated with the underlying Compton contribution in the Ge peaks were subtracted by placing equal width gates near each γ peak. Care was exercised to avoid peaks that are known to be doublets. Thus, for the four main bands, gates placed on single γ-ray peaks for transitions up to spin 10 gave spectra of good statistical quality. For higher spins, the proton spectra for two or three transitions were added to provide spectra of improved statistical quality, but in each case the spectra for each gate were examined for consistency.

Angular correlations of the coincident protons were recorded at 24°, 42°, 50°, 63°, 68°, and 78° in the laboratory, corresponding to angles ranging from $\sim$ 30° to 95° in the center of mass.

The level scheme for ^{82}Sr was constructed from a γ-γ matrix obtained by requiring that at least one proton was detected and that the γ-ray multiplicity exceeded 10. The matrix thus constructed was dominated by γ-rays from ^{82}Sr. Figure 4 shows a partial level scheme for ^{82}Sr constructed from these data. Two new bands were established and four previously known bands were extended from 20 to 27 $\hbar$. The even parity band 4 is yrast for spins between 10 and 22 $\hbar$. The odd spin band 2 becomes yrast at very high rotational frequencies ($I \geq 23$).

Proton spectra coincident with the $2^+ \rightarrow 0^+$ ground transition are shown in Fig. 5 for the three k_γ gates. For purposes of comparison, the spectra are shown normalized to the same total counts. The spectra at the most probable value have 1.1×10^4, 3.5×10^4, and 2.0×10^4 counts per 0.5 MeV interval. As one moves to the high k_γ gates, the spectra are seen to shift to lower energies, with a lower apparent emission barrier, but the slopes at high energies are similar. This is understood in terms of the decreasing available thermal excitation due to increased spin range in the entry

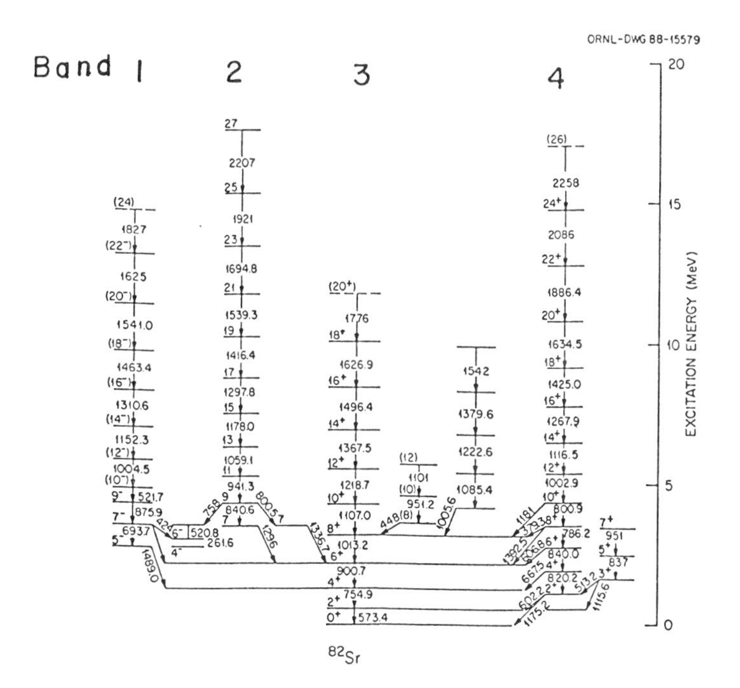

Figure 4. Partial level scheme for ^{82}Sr.

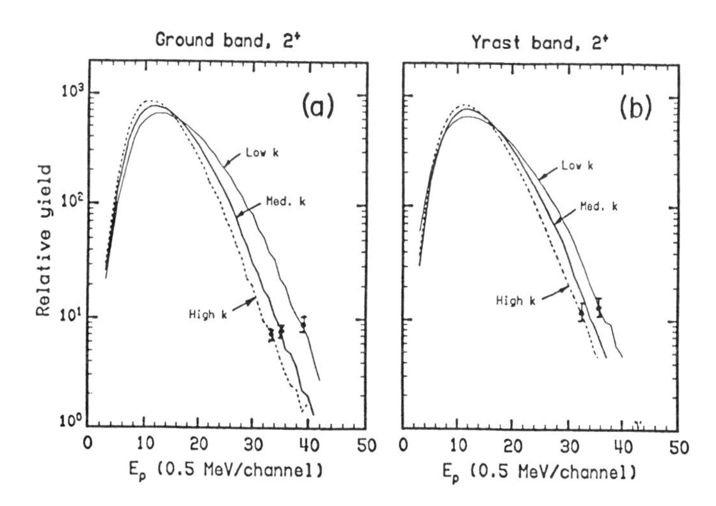

Figure 5. Panel (a) shows the proton spectra coincident with the $2^+ \rightarrow 0^+$ transition in the ground band (band 3) for the three k_γ gates. The spectra are normalized to the same total counts for comparison. It is seen that the proton emission barrier decreases as k_γ increases. Panel (b) shows the proton spectra coincident with the 2^+ level in band 4 (see Fig.4) for the three k_γ gates. Here the proton emission barrier is essentially independent of the k_γ gating, although the high energy yield decreases as k_γ increases.

region ($I \sim$ 4-19, 17-29, and 26-45 $\hbar$, at mean yrast energies of ~7, 14 and 23 MeV, respectively). In contract to this, the proton spectra associated with the 2^+ level of band 4 (non-yrast at this spin, but yrast at spins 12 to $22\hbar$) show similar emission barriers for the three k_γ gates (Fig. 5b). As the k_γ is decreased, the spectra shift somewhat to higher energies, but the slope parameters at high E are similar. The differences between the spectra coincident with the different 2^+ states suggest that some structural effects and/or feeding patterns may be responsible.

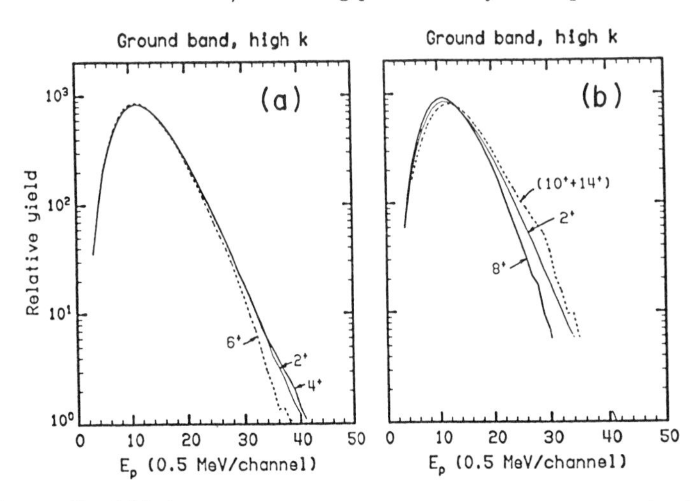

Figure 6. Panel (a) shows proton spectra coincident with γ-rays from the $2^+, 4^+$, and 6^+ levels in the ground band 3 for the high k_γ gate. The spectra are normalized to the same total counts for ease of comparison. The spectra are remarkably similar. Panel (b) shows proton spectra coincident with the 8^+ and the sum of $(10^+ + 14^+)$ levels in the same ground band 3 and for the high k_γ gate. The 2^+ spectrum is also shown for comparison. Significant differences are seen (see text).

In order to explore this possibility, we compare in Fig. 6(a) the proton spectra coincident with the γ-rays from the 2^+, 4^+, and 6^+ states in the ground band 3 and for the high k_γ gate. Clearly, the spectra are very similar, with a hint of softening of the spectra for the 6^+ state. We note that the 2^+ and 4^+ and for the most part the 6^+ levels receive practically all the feeding from all the bands in ^{82}Sr, whereas the 8^+ level is only fed from higher spin members of the ground band and to a significant fraction by the yrast band 4. Levels above the 8^+, however, are only populated by the decay of the levels of the continuation of the ground band. The proton spectra coincident with the 8^+ and the sum of 10^+ and 14^+ levels are compared in Fig. 6(b) with the 2^+ spectrum. It is seen that the spectra for the 8^+ and $10^+ + 14^+$ levels shift toward lower and higher emission barriers, respectively. This unexpected behaviour is explained below by considering spectra associated with high spin states in the four major bands in ^{82}Sr.

Next we note that the proton spectra from individual levels within each band having spins higher than $10\hbar$ were found to be similar to each other, but differ considerably for different bands. This is shown in Figs. 7(a) and (b) for protons associated with the levels indicated. There are three striking features in these spectra: (1) for the same k_γ gate the emission barrier shifts by as much as 1 MeV toward lower values in going from the ground band to the yrast band; (2) the higher energy region also shows differences with the ground band having excess high energy protons; and (3) for the high and medium k_γ gates only the higher energy part of the spectrum changes, while the spectral shapes at and below the barrier are the same for each band. The latter is to be contrasted with the 2^+ spectra of Fig. 5(a) where both the low and high energy portions of the spectra show differences as k_γ is increased. The above features are at first surprising, since in the standard statistical evaporation picture one should expect no differentiation between bands, because the entry state population is at spins greater than $\sim 17\hbar$ and $26\hbar$ for the medium and high k_γ gates, respectively, which are well above the gating discrete transitions.

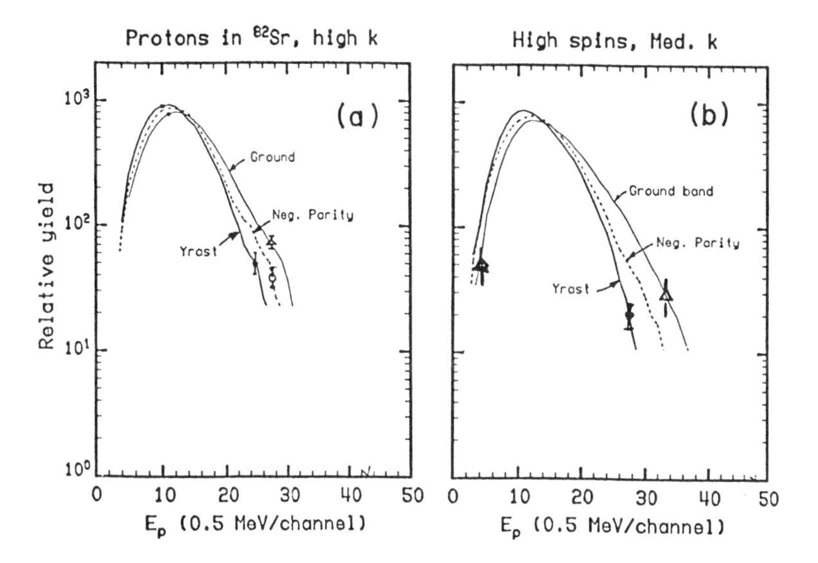

Figure 7. Panel (a) shows proton spectra from the high k_γ gate coincident with transitions from the $(10^+ + 14^+)$ levels of the ground band 3 (thin line), from the $(14^- + 16^-)$ levels of the negative parity band 1 (dashed line), and from the $(14^+ + 16^+)$ levels in the yrast band 4 (thick line). Shifts as large as 1 MeV are seen in going from the ground to the yrast band gates. Panel (b) shows proton spectra from the medium k_γ gate. The curve labels are for the same as in (a).

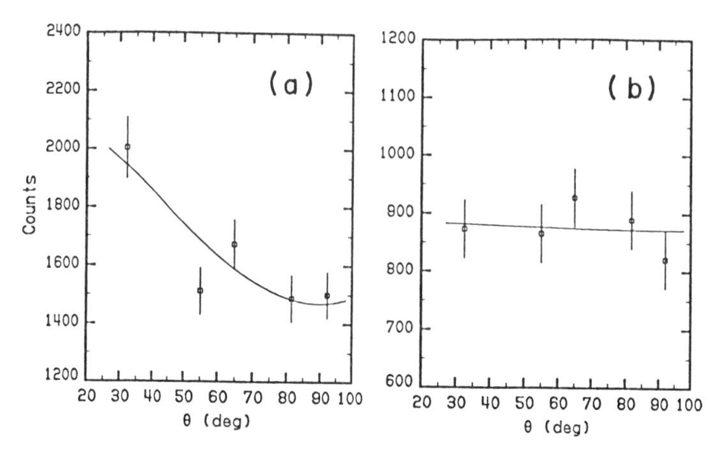

Figure 8. Angular correlations in the center-of-mass of the protons in the energy range 4.0-5.0 MeV for the high k_γ gate coincident with the $(14^+ + 16^+)$ tranistions of the yrast band 4 (left panel), and the $(10^+ + 14^+)$ transitions of the ground band 3 (right panel).

A careful examination of the measured k_γ distribution coupled with energy balance considerations require that the entry line for this channel be only a few MeV above the yrast line. This suggests that a sizeable fraction of the population of the yrast band may occur by direct proton emission at or close to the yrast line. This mechanism would explain items (1) and (3) mentioned above. The observation in point (2) can be understood in terms of differences in thermal energy across the k_γ gate, provided one can argue that the ground band at high spin is preferentially populated by the connecting "statistical" γ-rays only from the higher I and E* region in the k_γ gate. This mechanism cannot, however, account for point (3) above, namely the independence of the subbarrier shape or k_γ gate.

If this picture with significant direct population of the yrast line by stretched proton emission is correct, then we would expect the subbarrier region to show a substantial enhancement of the proton angular correlations with respect to the beam direction.[20] This is indeed supported by the angular correlations. The angular correlations were constructed by selecting 5 proton energy gates, which give comparable counts. As an example, the correlation for high k_γ and the 4.0-5.5 MeV proton bin are shown in Fig. 8 for the yrast and ground band $(14^+ + 16^+)$ and $(10^+ + 14^+)$ gates, respectively. The correlations were fitted by least square procedure to the expression $A_0\,[1 + A_2 P_2(cos\,\theta)]$, where $P_2(cos\,\theta)$ is the Legendre polynomial. Then the $W(0°)/W(90°)$ anisotropies were calculated from the fit and these were then divided by the proton anisotropy associated with the $2^+ \rightarrow 0^+$ ground transition. This procedure reduces considerably any systematic errors due to uncertainties in the proton energy calibrations. The resulting anisotropy ratios for the yrast band $(14^+ + 16^+)$ and the ground band $(10^+ + 14^+)$ are shown in Fig. 9(b). The horizontal bars give the range of the proton energy gates.

In Fig. 9(a) the A_0 coefficient expressed as counts/MeV are shown for comparison. Thus, a significant enhancement of the yrast band anisotropy at and below the barrier is clearly seen relative to that of the ground band. This substantiates the above interpretation.

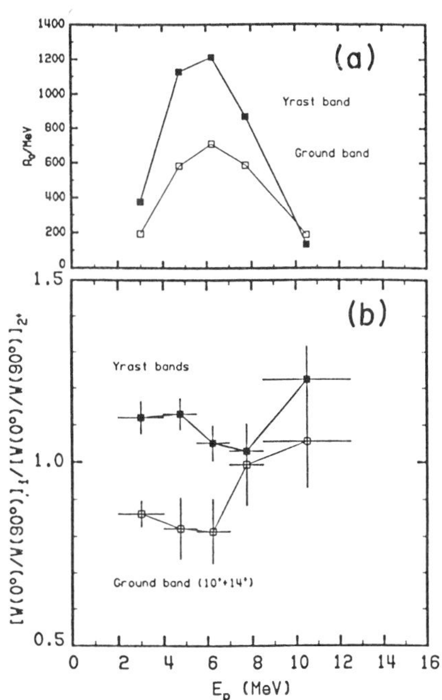

Figure 9. The lower panel shows the anisotropy ratios for the protons from high k_γ gate coincident with the high spin transitions (Fig. 8) in the yrast band 4 (full squares) and the ground band 3 (open squares), relative to that for the $2^+ \rightarrow 0^+$ transition to the ground state. The horizontal error bar give the widths of the proton energy gates used. Thes were chosen to give comparable statistics in the correlations. The upper panel shows the corresponding portions of the proton energy spectra used in the energy gates for the angular correlations. The quantity plotted are the A_0 coefficients per MeV obtained from the angular correlation fits.

The angular correlations for the middle k_γ gate show a similar, but somewhat reduced enhancement, but the anisotropy for the ground band is not reduced significantly over that for the $2^+ \rightarrow 0^+$ ground transition.

This can come about from the instability of nuclei near the yrast region with large spins toward emission of nucleons. Such instabilities could be connected to the occupation of $h_{11/2}$ resonance states in the nucleus.[20] These effects may in turn be related to the competing decay modes by nuclear emission from superdeformed nuclei.

REFERENCES

1. P.J. Twin, B.M. Nyako, A.H.Nelson, J. Simpson, M.A. Bentley, H.W. Cranmer-Gordon, P.D. Forsyth, D. Howe, A.R. Mokhtar, J.D. Morrison, J.F. Sharpey-Schafer and G. Sletten, Phys. Rev. Lett. **57** (1986) 811.
2. J. Dudek, T. Werner and L.L. Riedinger, Phys. Lett. B**211**, 252 (1988).
3. S. Cohen, F. Plasil, and W. Swiatecki, Ann. of Physics, N.Y., **82**, 557 (1974).
4. L.A. Goodman, Phys. Rev. C**37**, 2162 (1988).
5. K.A. Snover, Ann. Rev. Nucl. Part. Sci., **36**, 545 (1986).

6. M.Blann and T.T.Komoto, Phys. Scr. **24**, 93 (1981).
7. J.M. Alexander, D.Guerreau and L.C. Vaz, Z. Phys. A **305**, 313 (1982).
8. F.A. Dilmanian, D.G. Sarantites, M. Jääskeläinen, H. Puchta, R. Woodward, J.R. Beene, D.C. Hensley, M.L. Halbert, R. Novotny, L. Adler, R.K. Choudhury, M.N. Namboodiri, R.P. Schmitt, and J.B. Natowitz, Phys. Rev. Lett. **49**, 1909 (1982).
9. Z. Majka, D.G. Sarantites, L.G. Sobotka, K.J. Honkanen, E.L. Dines, L.A. Adler, Ze Li, M.L. Halbert, J.R. Beene, D.C. Hensley, R.P. Schmitt and G. Nebbia, Phys. Rev. Lett. **59**, 322 (1987).
10. N.G. Nicolis, D.G. Sarantites, L.A. Adler, F.A. Dilmanian, K.J. Honkanen, Z. Majka, L.G. Sobotka, Z. Li, T.M. Semkow, J.R. Beene, M.L. Halbert, D.C. Hensley, J.B. Natowitz, R.P. Schmitt, D. Fabris, G. Nebbia and G. Mouchaty, '*The Variety of Nuclear Shapes*', edited by J.D. Garrett, C.A. Kalfas, G. Anagnostatos, R. Vlastou, (World Scientific, 1987) pp. 526–536. (the complete study has been submitted for publication in Phys. Rev. C).
11. Z. Majka, M.E. Brandan, D. Fabris, K. Hagel, A. Menchaca-Rocha, J.B. Natowitz, G. Nebbia, G. Prete, B. Sterling and G. Viesti, Phys. Rev. C **35**, 2125 (1987).
12. L.C. Vaz and J.M. Alexander, Z. Phys. A **318**, 231 (1984).
13. M. Jääskeläinen, D.G. Sarantites, R. Woodward, F. A. Dilmanian, J.T. Hood, R. Jääskeläinen, D.C. Hensley, M.L. Halbert, and J.H. Barker, Nucl. Instr. Meth., **204**, 385 (1983).
14. K.J. Honkanen, F.A. Dilmanian, D.G.Sarantites, and S.P. Sorensen, Nucl. Instr. Meth. **257**, 233 (1987).
15. A CsI(Tl) version of the Dwarf Ball system described in: D.G. Sarantites, L.G. Sobotka, T.M. Semkow, V. Abenante, J. Elson, J.T. Hood, Z. Li, N.G. Nicolis, D.W. Stracener, J. Valdes and D.C. Hensley. Nucl. Instr. Meth. **A264**, 319 (1987).
16. A. Gavron, Phys. Rev. C **21**, 230 (1980); modification PACE2S by J.R. Beene.
17. N.G. Nicolis and D.G. Sarantites, submitted in Phys. Rev. C.
18. R.G. Stokstad and E.E. Gross, Phys. Rev. C **23**, 281 (1981).
19. W. Nazarewicz *et al.*, Nucl. Phys. **A435**, 397 (1987).
20. T. Dossing, S. Frauendorf, and H. Schulz, Nucl. Phys. **A287**, 137 (1977).

The population and decay of superdeformed bands

J. Gascon[1] K. Schiffer[2] and B. Herskind

Tandem Accelerator Laboratory, Niels Bohr Institute, Risø, 4000 Roskilde, Denmark

Abstract The population of superdeformed bands is studied with a simple model, where two type of states (normal-deformed and superdeformed) are separated by an energy barrier, and decay via collective E2 or statistical E1 transitions. The strong E2 rates in the superdeformed bands and the different level density of the two type of states can explain the general behavior of the population of these bands as a function of spin.

1 Introduction

Superdeformed bands, corresponding to the collective rotation of nuclei having a shape with a 1:2 axis ratio, have been observed in a number of light Gd (Deleplanque *et al* 1988; Haas *et al* 1988 and Fallon 1989), Tb (Deleplanque 1989 and Fallon 1988) and Dy (Rathkee *et al* 1989; Twin *et al* 1986 and Waddington 1989). All share common features: the bands are fed at very high spin, with an intensity exceeding that of any other "normal-deformed" structure. For example, the intensity of the superdeformed (SD) band in ^{149}Gd reaches up to 2 % of the total yield of the channel already at a spin of $\sim$ 50 $\hbar$. The population proceeds down to spin 20 - 30 $\hbar$ with a remarquably constant intensity. At these spin values, the intensity suddenly drops over an interval of a few spin units. To understand the basic mechanisms explaining these features, a simple statistical model has been proposed by Herskind *et al* (1987). The superdeformed structure is assumed to compete with a coexisting "normal"-deformed structure, separated by a energy barrier. At each step of the γ-ray cascade, the decay can occur either via a collective E2 or a statistical E1 transition within each well, or by tunnelling through the barrier to the other structure. Using the predicted barrier shapes and other relevant parameters, the model successfully reproduce the observed general population pattern and lifetimes (Schiffer *et al* 1989). The model has been generalized by Schiffer *et al* (1988) to study the effect of direct population of the band via neutron emission or via high-energy E1 transitions. Also, the simulations are useful tools for the study of the superdeformed E2 continuum. On all those subjects, the model helps the interpretation of the observed features, and suggests interesting experimental observables.

2 The Model

For a more detailed description of the model and a more complete discussion of the choice of the parameters, we will refer the reader to Herskind *et al* (1987) and Schiffer *et al* (1988 and 1989). The different parameters used in the simulation of ^{152}Dy can be found in the table 1 of Schiffer *et al* (1989). In the simulations of ^{149}Gd , the parameters adopted for ^{152}Dy have been subsituted by those measured by Taras *et al* (1989a), when available. Ragnarsson *et al* (1980,1986) and Åberg (1982) predicts that, from ^{152}Dy to ^{149}Gd, there will be a small change in the energy barrier that separate the superdeformed states from the other states. In addition, Åberg (1988) and Dudek *et al* (1988) predict a sizeable change in the level density of

[1]Present address: Université de Montréal, Montréal, Québec, Canada
[2]Present address: Canberra University, Australia

superdeformed states between the two nuclei (see below). These two changes have also been included in the simulations of ^{149}Gd.

The essential ingredients of the model are the following. The model assumes the coexistence of two type of rotational states: "normal" (label N) and "superdeformed" (label S), corresponding to the collective rotation of nuclei having a quadrupole deformation (β) of the order of 0.1 — 0.3 or 0.5 — 0.6, respectively. Each type of structure has its own "yrast" configuration; they cross each other at spin 54 $\hbar$. The density of N and S states depends on the excitation energy above the corresponding yrast band (U_N and U_S, respectively). It is given by the expression:

$$\rho_{N/S} = \frac{1}{24} \frac{2I+1}{(U_{N/S} + U'_{N/S})^2} (\frac{\hbar^2}{2J^{(2)}})^{3/2} \sqrt{a_{N/S}} \exp{(2\sqrt{a_{N/S} U_{N/S}})} \qquad (1)$$

The parameter a_N is taken to be A/7 (A being the mass), in accordance to the measurements of Holzmann *et al* (1987). The parameter a_S is taken from calculations: according to Åberg (1987), it is A/10 in the case of ^{152}Dy . Åberg (1988) predicts that the level density of S states in ^{149}Gd is a factor 3 to 5 smaller than in ^{152}Dy , for excitation energies up to 10 MeV above the yrast line. This corresponds to a value of a_S of about A/12 for ^{149}Gd.

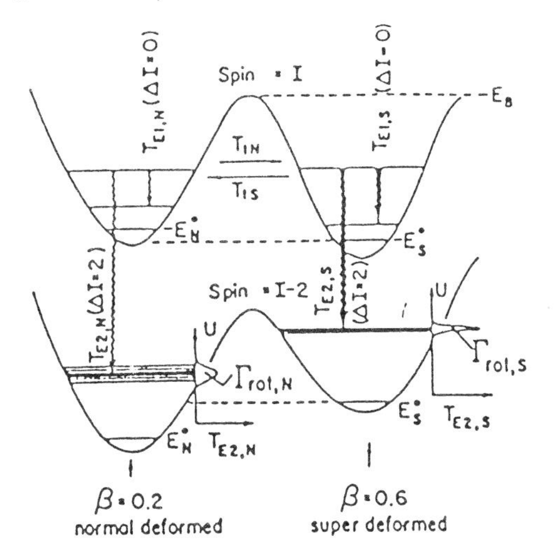

Figure 1. *Possible decay modes for normal (N) and superdeformed (S) states in the model, from Schiffer* et al *(1988). The potential energy barrier is plotted at two spin values (I and (I-2). The E1 and E2 transitions keep the population in the same well (S or N). The E2 transitions remove two units of angular momentum, and the E1 transitions, none. Tunneling occurs between two states of the same spin and energy, but in different wells; hence, for large tunneling rate, the two type of states will be fully mixed.*

The two structures are separated by an energy barrier, whose shape as a function of spin, is taken from calculation of potential energy as a function of β and γ deformation, performed by Åberg (1982), Ragnarsson (1980 and 1986) and Dudek (1988). In the original version of the code, only the gamma-ray decay is considered. Three decay modes are taken into account: stretched E2 transitions, statistical E1 decay, and tunnelling through the barrier (figure 1). When the excitation energy is comparable or superior to the height of the barrier, the tunneling rate becomes very large, and the S and N states are fully mixed.

The simulation starts with an entry point distribution of gaussian shape with respect to both energy above the yrast line and spin. The decay is then followed step by step, for 10,000 to 25,000 cascades. At each step the E1, E2 and tunneling rates are calculated, and the actual decay is chosen by the Monte Carlo method.

3 Feeding by high-energy E1

Herskind *et al* (1987) suggested that the increased yield of the discrete SD band relative to the "normal" structures could be explained by the different level densities of the N and S states in the continuum. This effect originates from the dependence of the E1 rate ($T(E1)$) on the level density and on the energy above the yrast configuration (U_i):

$$T(E1, U_i) = \int_0^{U_i} \frac{\rho(U)}{\rho(U_i)} \cdot f_{GDR(U_i - U)} \cdot (U - U_i)^3 \cdot dU \qquad (2)$$

Here, f_{GDR} is the strength function of the giant dipole resonance: its U dependence (see Herskind *et al* 1987, or Schiffer *et al* 1988) is smoother than that of the factors $\rho(U)/\rho(U_i)$ or $(U - U_i)^3$. The lower level density gradient of the S states and the lowering of the SD yrast band relative to the normal states at high spins result in a larger E1 rate for the excited S states. In the simulations, this larger E1 rate explains the larger population of the superdeformed discrete band relative to any other normal-deformed band at equal spins. This preferential feeding of the SD yrast band stops at lower spins, when the superdeformed structure is no longer energetically favored and simultaneously ρ_N becomes larger than ρ_S at any excitation energy. Thus, this mechanism explains also why the discrete SD bands are only fed at high spins.

An additional consequence of the lower level density gradient of the S states and the lowering of the yrast SD band at high spins is that, for a given U_i, the E1 γ rays emitted by a S state have on average a larger energy than those emitted by N states. According to Herskind *et al* (1987) and Schiffer (1988), in a simulation of ^{152}Dy , with an average entry energy of 7 MeV above the yrast, a gate on γ-rays above 5 MeV produces an enhancement of the yield of the discrete S band relative to that of low-spin N transitions by a factor 2 to 3. Such an enhancement has been observed by Twin (1987) with the reaction ^{108}Pd(^{48}Ca,4n)^{152}Dy at 205 MeV. In contrast, in the case of the reaction ^{124}Sn(^{30}Si,5n)^{149}Gd at beam energies of 150 MeV (Taras *et al* 1989a), 155 and 160 MeV (Haas 1989), a gate on high energy gammas does not produce any enhancement.

This may be explained by the low excitation energy in the residual nuclei for the later reactions: not enough energy is available for the high-energy E1 transition. Furthermore, Åberg (1988) predicts that the level density of the continuum of ^{149}Gd is 3 to 5 times smaller than in the case of ^{152}Dy . Thus, the probability of populating a highly excited S state is lower in ^{149}Gd than in ^{152}Dy, and this may explain why Taras *et al* (1989a) could not feed the S band in ^{149}Gd using reactions yielding excitation energies in the final nucleus higher than 5—10 MeV.

The non-observation of the E1 enhancement in several cases is an indication that the earlier version of the model was incomplete. However, the large population of the SD band relative to any other N structure still requires the existence of a special mechanism feeding preferentially the SD band. It will be shown in the following section that this mechanism could be the effect of the level density on the neutron evaporation spectrum.

The observation of an enhancement of the SD band in the reaction ^{108}Pd(^{48}Ca,4n)^{152}Dy at 205 MeV by Twin (1987) is so far the most direct proof that superdeformed states lie at large excitation energies in the continuum. The enhancement factor itself is an indirect measure of the density of S states, and more experiments on the feeding of the SD band in ^{152}Dy in a regime of high excitation energy would be important.

4 Neutron feeding

The evidence that SD bands can be strongly fed at very high spin without the high-energy E1 feeding mechanism was firmly established by Taras *et al* (1989a) and Haas (1989). Taras *et al* (1989a) have proposed that these band are fed directly by the neutron evaporation process. That possibility was investigated by Schiffer (1988) using an analytical model, where the neutron evaporation process is governed by the same type of competition between the S and N states that was assumed for the γ-ray decay (figure 1). Consequently, the neutron decay will also be affected by the same factors that lead to the preferential feeding of the SD band in the case of E1 decay, namely the lowering of the SD band relative to the normal states and the different level density gradients. In the same way that, for a same excitation energy, the energy of a E1 transitions emitted by a S state is larger than that emitted by a N state, the average kinetic energy of a neutron is larger if it is emitted by a S state rather than a N state. As a consequence, the distribution of entry points in the final nucleus will depend on the type of state (N or S) which is populated.

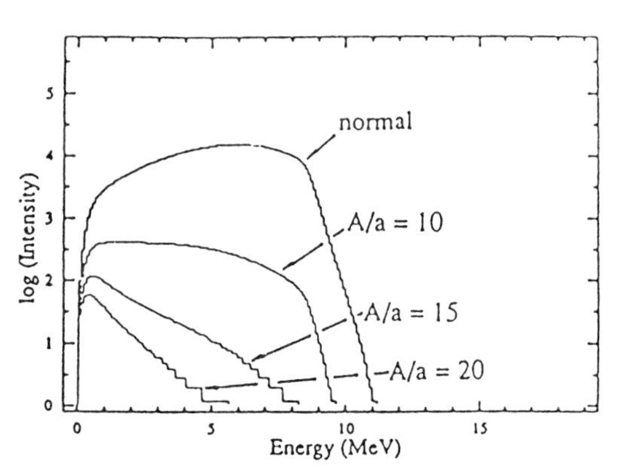

Figure 2. *Entry point population (after the evaporation of the last neutron) calculated by Schiffer* et al *(1988), for different values of level density gradient A/a. The curve labeled "normal" correspond to A/a = 7.*

The distributions of excitation energies in the final nucleus, calculated for different level density parameters, are shown in figure 2. The calculation was done for $I = 54\hbar$, at which spin the normal and SD yrast configuration have the same energy, and thus $U_N = U_S$. The centroid of the curve for normal states (A/a = 7) is higher than the corresponding centroid for SD states (A/a $\geq$ 10). The initial population of the S states is concentrated at the lower excitation energies, leading to an enhancement of the yield of the SD yrast band. Simultaneously, the population of S states at high excitation energies is seriously depleted, explaining the disappearance of the high-energy E1 enhancement effect. The neutron feeding effect is more pronounced when the level density is lower, and may explain why Taras *et al* (1989a) and Haas (1989) could not populate the SD band in ^{149}Gd via reaction leading to high excitation energies in the final nuclei.

In the same way that it was proposed to use a gate on high-energy E1 to enhance the relative yield of the SD band, a gate on high-energy neutrons could enhance the population of the low-lying SD bands. According to Schiffer *et al* (1989), the shift in average kinetic energy could be as large as 0.9 MeV. The study of this effect would yield information on the relative level densities of N and S states at very large excitation energies, and on their mixing.

5 Saturation of the total yield of the band at high beam energy

In the preceding sections, it was shown that the accelerated E1 or neutron evaporation rates provide a satisfactory explanation of the larger population of the SD yrast band, relative to the normal structure. Schiffer *et al* (1989) have also shown that, using relevant parameters, the calculated yields can reproduce the observed yields. In this section, we will discuss the variation of the total yield on the SD discrete band as a function of the initial spin and excitation energy in the final nucleus.

Several experiments have been performed, where the total intensity of a SD discrete band was measured as a function of beam energy. The intensity of the band in ^{149}Gd relative to the total yield of that channel has been measured in the reaction ^{124}Sn(^{30}Si,5n)^{149}Gd at beam energies varying from 140 to 160 MeV by Taras *et al* (1989a) and Haas (1989). Figure 3a shows the results. It is found that the yield increases as the beam energy increases up to 150 MeV. Then, it reaches a plateau, or slightly decreases. Twin (1987) observes an identical trend in ^{152}Dy (figure 3b). It is thus interesting to see whether the model can reproduce this trend.

This comparison of simulations with experiments requires that the relation between the beam energy and the average initial spin and excitation energy in the final nuclei (I_0 and U_0, respectively) be known. In the experiment performed by Taras *et al* (1989a), the multiplicity and the total energy of the γ-ray cascades in coincidence with transitions in the normal and superdeformed yrast bands were measured with a 71-element BGO array covering a solid angle close to 4π. The measured values are shown in figure 4. From these data, it is possible to establish that for every increase of beam energy of 5 MeV, 1.5 transitions are added to the cascades recorded in coincidence with SD transitions, adding 2.5 MeV to the excitation energy. A similar increase is observed for cascades in coincidence with transition in the normal yrast band. Assuming that the added transitions are stretched E2, and taking into account the slope of the yrast band, the excitation energy above the yrast and the initial spin in the final nuclei increase by about 1 MeV and 3 $\hbar$, respectively, for each step of 5 MeV in beam energy. This fixes the relative behavior of U_0 and I_0. There is a larger uncertainty on the absolute values of U_0 and I_0, but they are constrained by the available energy left after the evaporation of 5 neutrons, and the spin of entry in the SD band. This larger uncertainty will not affect our main conclusions, as it will be shown that the essential features depend much more on the slope (U_0 versus I_0) of the set of entry points composing the excitation function, rather than the precise values of U_0 and I_0.

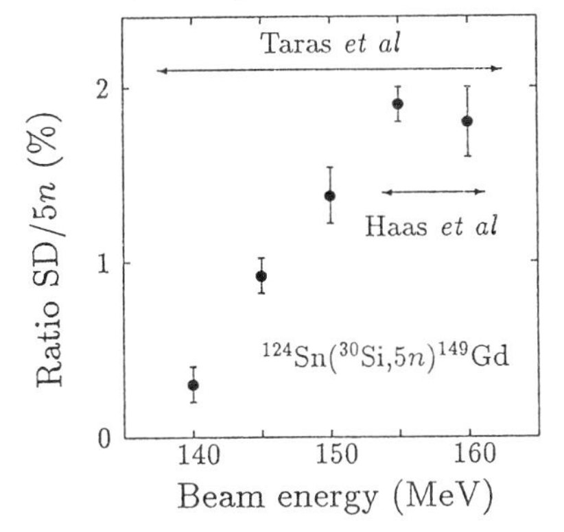

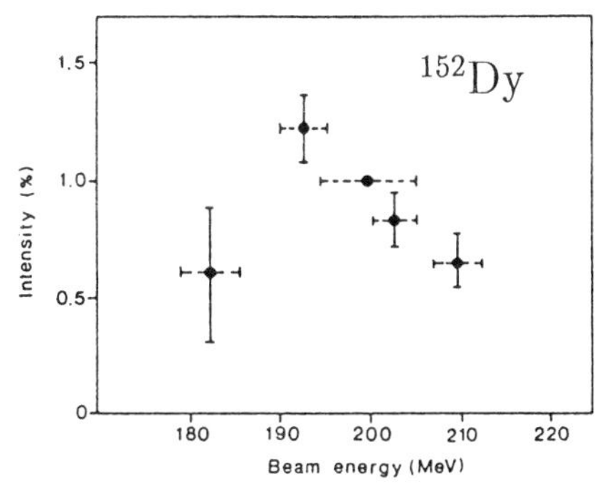

Figure 3. *Population of the superdeformed discrete band relative to the total yield of the nucleus. Left (a): Results of Taras* et al *(1989a) and Haas (1989) for* ^{149}Gd. *Right (b): Results of Twin (1987) for* ^{152}Dy, *using the reaction* $^{108}Pd(^{48}Ca,4n)^{152}Dy$.

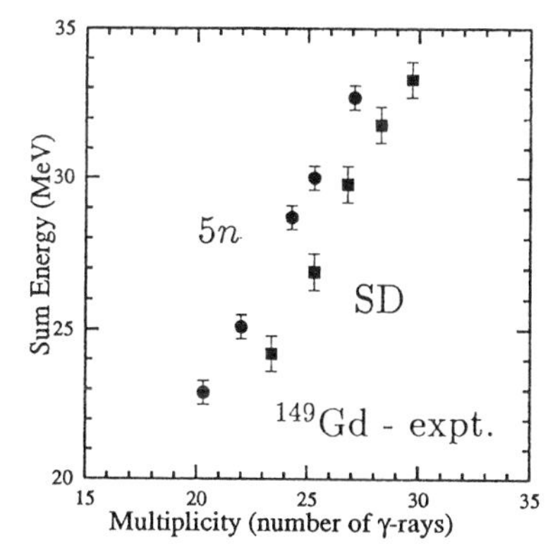

Figure 4. *Sum energy and multiplicity of the* γ*-ray cascades in coincidence with transitions with normal (circles) and superdeformed (squares) transitions in* ^{149}Gd *, as measured by Taras* et al *(1989a), following the reaction* $^{124}Sn(^{30}Si,5n)^{149}Gd$ *at beam energies from 140 to 160 MeV in step of 5 MeV. The lowest (highest) energies and multiplicities are obtained at the lowest (highest) beam energy for both type of cascades.*

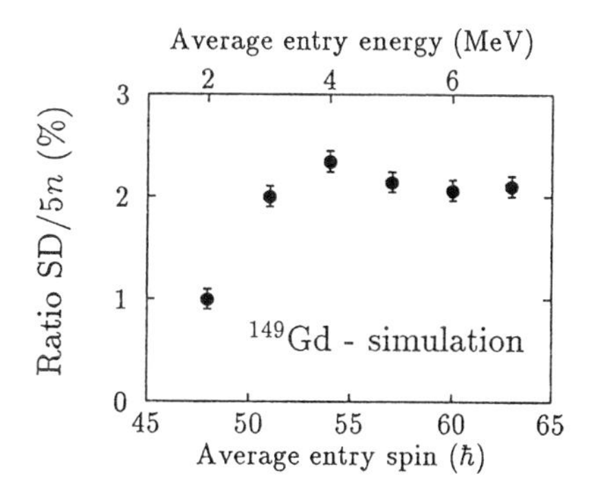

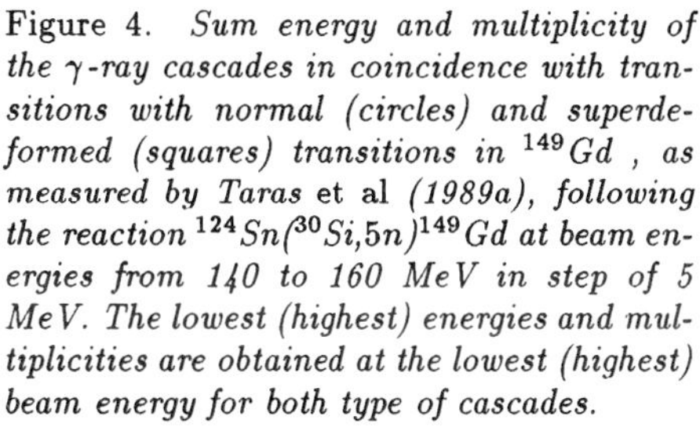

Figure 5. *Simulated total yield in the superdeformed band of* ^{149}Gd *relative to the total yield of* ^{149}Gd *, as a function of the centroid of the entry cloud; in spin (lower scale) or excitation energy (upper scale). The lowest point correspond to the reaction* $^{124}Sn(^{30}Si,5n)^{149}Gd$ *at a beam energy of 140 MeV, and the highest point, 165 MeV.*

The population of ^{149}Gd was simulated for values of I_0 and U_0 that correspond to ^{30}Si beam energies varying from 140 to 160 MeV, in step of 5 MeV. Figure 5 shows the resulting total intensities in the SD band, relative to the total yield of the nucleus. As we are more interested in the general trend than the actual values, the width of the initial distribution of I_0 and U_0 were set to narrow values (FWHM of 10 $\hbar$ and 5 MeV, respectively). The real distributions may be wider, and thus the observable effect may be less pronounced than the results of the calculation. The experimental trends are basically reproduced: *i)* first, the intensity increases with beam energy, and then *ii)* decreases above 150 MeV, and *iii)* the rate of the increase is larger than the rate of the decrease.

In the model, the rapid increase at low spin is a consequence of the lower level density of the S states relative to the N states, with a corresponding steeper gradient $d\rho_S/dU$. Above an excitation energy of 2 to 4 MeV, the energy barrier between the S and N structures dissolves, and the states are fully mixed. The S component in states in the continuum is given by the relative importance of ρ_S and ρ_N. If ρ_N is much larger than ρ_S, it is unlikely that a state will decay to the SD discrete band. The regions where $\rho_N < \rho_S$ and where $\rho_N > \rho_S$ are shown in figure 6. As the entry cloud moves up in spin and approaches the border between the two regions, there will be an increase of the population of the SD yrast band. The steepness of the variation is due to the exponential dependence of ρ on U (equation 1).

To understand the saturation, and subsequent decrease, at higher beam energies, we must first look at the relative strength of the E1 and E2 rates in the high-spin region of figure 6. In the figure is shown the limit at which the E1 rate is 10 % of the E2 rate. Below that line (region III), the E1 rate is smaller, and the decay follows a line parallel to the SD yrast band. The population trapped in these excited SD bands have a larger chance of reaching the region where $\rho_N > \rho_S$, and subsequently decay down to the normal yrast, than of reaching the discrete SD band. At each step of the γ-ray cascades in these excited bands, only a small fraction decays down directly to the SD yrast. One would expect that populating these bands at higher spins would increase the number of steps in that cascade, and thus improve in proportion the chances of population the SD yrast band. However, this effect is largely compensated by the E_γ^5 dependence of the E2 rate, and the supplementary steps added at high spins do not contribute significantly to the probability of reaching the SD yrast band. This leads to an attenuation of the rapid increase of the population, observed at lower beam energy.

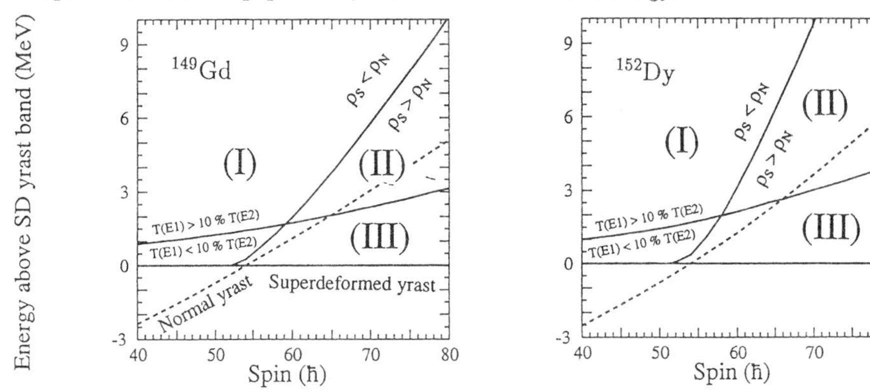

Figure 6. *(a) The three regions of the continuum above the superdeformed band of ^{149}Gd discussed in the text. In region I, the density of N state is large than the density of S states, and the population has a high probability of decaying into N states. In region II and III, ρ_S is larger than ρ_N, so the probability of decaying into a S state is high. In region III, the E1 rate of S states is less than 10 % of the E2 rate, and the population is trapped in the excited SD bands and has a small chance of feeding the SD yrast band. Above, in region II, the E1 rate of S states is larger than 10 % of the E2 rate, and an important fraction of the population will "cool down", either to the SD yrast band or to excited SD states. (b) Same as (a), but for ^{152}Dy.*

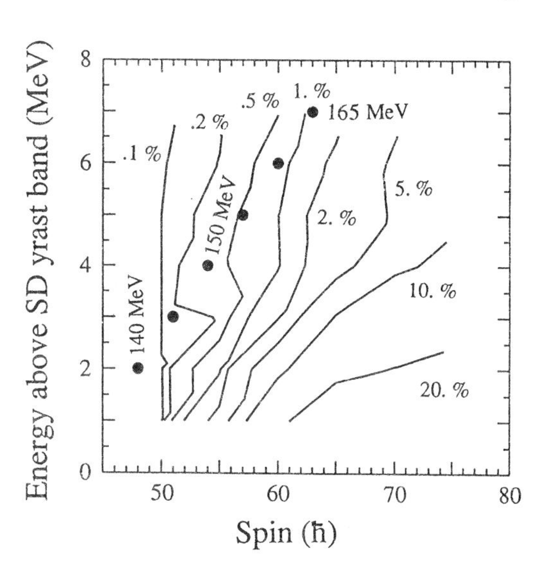

Figure 7. *Calculated (see text) probability of populating the discrete SD band in ^{149}Gd, as a function of the initial spin and the excitation energy above the band. The lines join points with equal probability. The solid dots represent the centroids of the entry cloud of the simulated excitation function discussed in the text.*

The key factor that explains the actual decrease of the yield of the SD band above beam energies of 150 MeV is that the gain in spin at each step of the excitation function is accompanied by an increase in excitation energy. The (I_0,U_0) values corresponding to the centroids of the entry clouds for each step of the simulated excitation function are shown as solid dots in figure 7. The figure also illustrate the probability of populating the superdeformed discrete band from any point above this band, from spin 50 to 75 $\hbar$ and U_0 from 1 to 7 MeV. It has been obtained from simulations where the width of the entry clouds were set to a fraction of $\hbar$ and a few 100 keV, and the central values of I_0 and U_0 were varied in steps of 5 $\hbar$ and 1 MeV, respectively. The plotted contours represent lines of equal probability. This value include the probability of *i)* occupying a S state at the particular value of U_0 and I_0, and of *ii)* eventually decaying into the SD discrete band, at any spin. It can be seen that the probability decreases with U_0. Part of this is due to the increasing level density of N states, but the reduced probability of decaying down to the SD yrast in one step is also in cause. As more γ rays are added to the cascade that "cools down" the nucleus, the probability of tunnelling to a N state at one of the supplementary step is increased. But figure 7 shows that the decrease is more pronounced at higher spins, and this is an indication that the larger E2 rates add more E2 transitions to the "cooling down" cascade, and thus more steps in the cascade and more chances of tunnelling to the N states, or simply to get trapped in an excited S band.

Figure 7 shows that the increase in U_0 at each step of the excitation function is large enough to produce a decrease in the total yield of the SD discrete band. It also shows that the same trend will be observed for any excitation function where the increase of I_0 is accompanied by an increase of U_0. Furthermore, the rate of the decrease of the SD discrete yield in the region II and III of figure 6 is smaller than the rate of the increase in region I, so that the behavior observed in the simulation of ^{149}Gd should be a general feature for SD bands in neighboring nuclei (see figure 6a and b for a comparison of the relative E1 and E2 rates in ^{149}Gd and ^{152}Dy).

In the present model, the fission barrier is not treated explicitly. The fission cuts off the distribution in spin, but not in excitation energy, and in an excitation function, this would increase the slope dU_0/dI_0. If the fission cut-off manifests itself in the experimental excitation function of ^{149}Gd, then this effect is included in our simulations, because the slope dU_0/dI_0 is taken from the measured energy sums and multiplicities. Figure 4 shows a quite regular increase of the energy sum and multiplicity with increasing beam energy, and it is more likely that the present experiment is not very sensitive to the fission barrier. In particular, larger average multiplicities and sum energies are observed in the $4n$ channel (^{150}Gd). In any case, the model offers a simple interpretation of the eventual effect of the fission barrier: the more pronounced dU_0/dI_0 slope would accentuate the decrease of intensity in the SD yrast band at high beam energies.

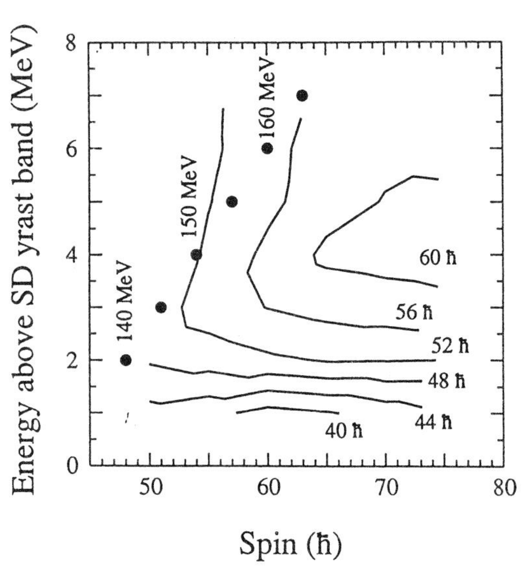

Figure 8. *Average spin at which a level feeds the superdeformed discrete band of* ^{149}Gd *(simulation), for levels in the continuum above that band. This value* ($< I_{in} >$ *), corresponds to the spin at which the intensity in the SD discrete band is equal to half of its maximum value, in simulations where the entry cloud is concentrated at one point of excitation energy and spin. The lines join points with an equal value of* $< I_{in} >$ *. The solid dots represent the centroids of the entry cloud of the simulated excitation function discussed in the text.*

6 Saturation of the average entry spin in the SD yrast band

Another interesting experimental observable is the average spin at which the SD yrast band is populated. This quantity, $< I_{in} >$, is simply the spin at which the intensity of the gamma rays in the SD yrast band is equal to half the maximum value. Haas (1989) pointed out that this quantity is remarquably stable for beam energies from 150 to 160 MeV for the reaction ^{124}Sn(^{30}Si,5n)^{149}Gd (about 56 $\hbar$), and is similar to the value for ^{152}Dy , populated in a completely different reaction. Indeed, the values of $< I_{in} >$ obtained in the simulated excitation function present this saturation effect: with an average I_0 varying from 47 to 63 $\hbar$, the $< I_{in} >$ values ranges from 54.9 to 57.2 $\hbar$. Figure 8 shows the values of $< I_{in} >$ obtained by populating the nucleus ^{149}Gd at spins varying from 50 to 75 $\hbar$, and excitation energies from 1 to 7 MeV. Throughout the entire range, $< I_{in} >$ never exceeds 61 $\hbar$; for the region above 2 MeV, it is never lower than 52 $\hbar$.

The causes of this effect are similar to those of the saturation of the total intensity, i.e. the low level density of S states in the region I of figure 5, and the strong E2 rates in the SD bands parallel to the yrast in regions II and III. At low spins, the population directly decays into the normal continuum. Only the high-spin tail of the entry cloud can populate the yrast SD states. At higher spins (region II and III of figure 5), the strong E2 rates in the excited SD bands increase the probability that several E2 transitions will be emitted before the nucleus cools down to the SD yrast band. The net result is that the value of $< I_{in} >$ lies close to the boundary between region I and III. From figure 5, it is also clear that very similar condition will apply to ^{152}Dy , and the similar values of $< I_{in} >$ in both nuclei are no surprise.

In other words, $< I_{in} >$ is roughly constant because the transitions that are added when the initial spin is increased are added in the continuum, rather than in the SD yrast cascade itself. This agrees with the sum energies and multiplicities measured by Taras *et al* (1989a) (figure 4), which indicates that the increase of beam energy above 150 MeV causes an increase of the multiplicity of the cascades in coincidence with transitions in the SD yrast band, without revealing any new discrete transition in that band. It also agrees with the fact that more SD transitions were observed with ^{40}Ar induced reactions leadind to ^{152}Dy (de Voigt *et al* 1987) than with the original ^{48}Ca induced experiment by Twin *et al* (1986).

7 Intensity in the ridges

Although the intensity of the discrete SD band is never more than a few percent of the total population of the nucleus, the simulations predict that the majority of the cascades starting at high spin and excitation energy will go through at least one S state. Of course, in order to identify a "superdeformed" γ-ray transition, two S states must be populated consecutively. This is less likely, because the S and N states are fully mixed a few MeV above the yrast. The probability that three S levels will be populated consecutively, and thus the coincidence between the two γ rays will appear along the so-called "first ridges", is even lower. Another unknown is the effect of the rotational damping, which should remove the correlation between the energies of the two successive γ-ray at high excitation energies. Twin (1987) and Taras *et al* (1989b) have compared the intensity of the ridges and of the discrete SD band for ^{152}Dy and ^{149}Gd, respectively. The results for ^{152}Dy are shown in figure 9a. The results for ^{149}Gd are quite similar: the maximum intensity in the ridge is lower, but of the same order, than the intensity in the discrete bands. Also, the ridge intensity disappears at a spin higher than the one at which the discrete band decays out.

Schiffer *et al* (1988) have studied the ridge intensities in simulations of the nucleus ^{152}Dy . The results are shown in figure 9b. The fact that the ridges are depopulated at higher spin than the discrete band is easily understood, as the energy barrier only offers a reduced protection to the excited S bands, which can more readily mix with the continuum of N states. Also, the excited bands encounter the line at which $\rho_N = \rho_S$ at a higher spin than the discrete band (figure 6). In the present simulations, the ridge intensities are reproduced, assuming that rotational damping sets in at excitation energies of 1 to 2 MeV above the yrast. As the ridges intensities depend also on ρ_N, ρ_S, the initial population and possibly on the detailed structure of low-lying excitation above the yrast, more experimental data, taken in different conditions of initial spin and energy, are needed.

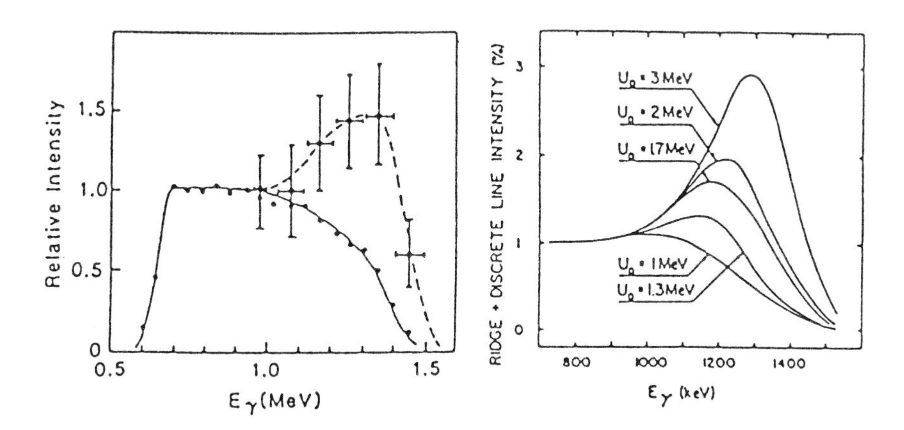

Figure 9. *Experimental (a) and calculated (b) density of the superdeformed ridge structure, normalized on the intensity of the discrete band at lower spin, for the nucleus ^{152}Dy populated via the reaction ^{108}Pd(^{48}Ca,4n)^{152}Dy at 205 MeV. The experimental data is from Twin (1987), the calculation from Schiffer* et al *(1988). The calculations are performed with different values of U_0, the excitation energy above which the correlation of the energy of the successive γ rays is washed out because of the rotational damping. A value of U_0 from 1 to 2 seems to reproduce the data quite well.*

8 Decay rates and decay mechanism

In this section, we will discuss the process leading to the decay out of the SD discrete bands at low spin. In addition to the energy of the in-band E2 transitions, two other pieces of experimental information are available: the lifetimes of the states, and the branching ratio of the in-band E2 transitions. We can combine these data to extract the absolute rate of the process that governs the depopulation of the SD bands, in an almost model-independent way,

The branching ratios can be obtained from the intensities of the E2 transitions in γ spectra gated by transitions lying above in the cascade. As the intensity pattern above the region of the out-of-band decay is flat, the E2 branching ratios for ^{149}Gd and ^{152}Dy can be directly extracted from the intensity pattern published by Taras *et al* (1989a) and Twin *et al* (1986), respectively. The lifetime data must be interpreted more carefully. For these two respective nuclei, Taras *et al* (1989a) and Bentley *et al* (1987) have only attempted to fit an average transition quadrupole moment Q_t to the entire band. No measures of each state lifetimes are available. Fortunately, the largest attenuation of the Doppler shifts are observed in the region of the out-of-band decay, and thus the fitted Q_t are heavily weighted by these states. From figure 10, where the observed shifts for ^{149}Gd are compared with calculations assuming a constant or a variable value of Q_t, we can deduce an upper limit on the variation of the transition quadrupole moment of the lowest spin members of the SD discrete band. A systematic variation of more than 10 % would be unlikely, and variation of more than 25 % for a single state would not be compatible with the present data. Thus, the transition quadrupole moments seem to be rather constant, down to spins where already 35 % of the initial population has decayed out of the band.

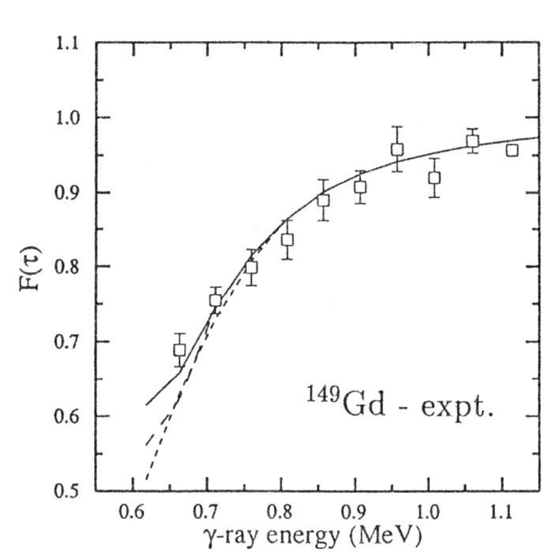

Figure 10. *Attenuation of the Doppler shift observed for transitions of the superdeformed band of ^{149}Gd populated via the reaction ^{124}Sn(^{30}Si,5n)^{149}Gd at 150 MeV, measured by Taras* et al *(1989a). The lines represent the best fit to the data assuming* i) *a constant value of the transition moment Q_t (solid line),* ii) *that the three last values of Q_t are reduced by 7 % (short-dashed line) and* iii) *the last value of Q_t is reduced by 25 % (long-dashed line).*

With those reasonable assumptions, we can obtain the absolute out-of-band decay rate (T_{out}), without any assumption concerning the nature of this process:

$$T_{out(I\to I-2)} = \frac{Y(I+2\to I) - Y(I\to I-2)}{Y(I\to I-2)} \cdot \frac{E_\gamma^5 \cdot Q_t^2}{2.19} \tag{3}$$

where Q_t is in eb, E_γ in MeV, $Y(I\to I-2)$ is the γ-ray yield of the transition between the states I and $I-2$, and the rate T_{out} is expressed in ps^{-1}. The experimental values are plotted in figure 11, for ^{149}Gd and ^{152}Dy. The most important feature is the sizeable decrease at a rotational frequency of ~ 0.325 MeV$\cdot\hbar^{-1}$ ($E_\gamma \sim 0.65$ MeV), observed in both data sets. On the same figure, this behavior can be compared with the much smoother spin dependence of the rate of the E2 transitions. Unfortunately, an upper and a lower bound on the error on the experimental T_{out} can be obtained only in the cases where the in-band intensity is comparable to the out-of-band intensity, and it is hard to decide whether the steep slope of T_{out} extends from the lowest to the highest spins, or is just limited to a single jump of a factor $\sim$ 10 at a rotational frequency of

~ 0.325 MeV$\cdot\hbar^{-1}$, the out-of-band decay rate remaining constant above and below this point. The ^{152}Dy data seems to be more compatible with the latter hypothesis. The overall similarity of the absolute values of T_{out} is also remarquable.

We can conclude that the rapid decay out of the discrete superdeformed band is caused by an increase by a factor ~ 10 of the rate of the out-of-band decay process, while the B(E2) values remains much more constant. This is compatible with the interpretation of Herskind *et al* (1987), according to which the out-of-band decay result from a barrier penetration effect, accelerated at low spin by the onset of pairing. In this model, T_{out} is a smooth function of the spin, except at the onset of pairing. In contrast, other explanations, in particular those involving a resonant effect, would yield T_{out} rates uncorrelated with the rotational frequency. To discriminate between the two hypothesis, the present type of analysis should be applied to other nuclei. This will require precise lifetime and branching ratios measurements, in superdeformed bands in neighboring nuclei.

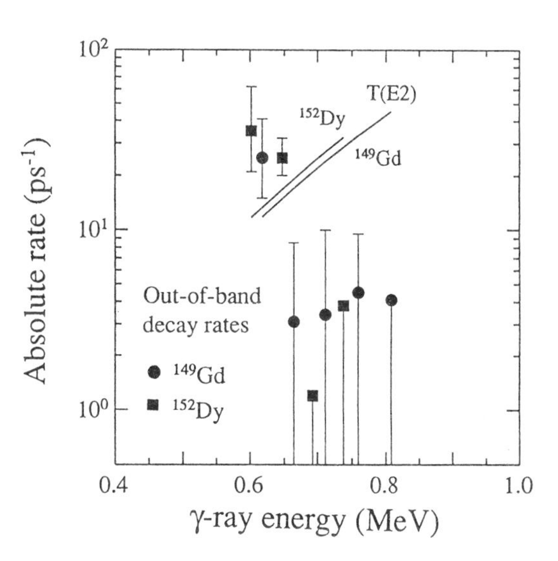

Figure 11. *Absolute rate of the decay out of the superdeformed band of* ^{149}*Gd (circles) and* ^{152}*Dy (squares), calculated from the lifetime and intensity measurements of Taras* et al *(1989a), Twin* et al *(1986) and Bentley* et al *(1987). The values are plotted as a function of the* γ*-ray energy of the competing E2 transition, which is roughly twice the rotational frequency. The rapid decrease around* $E_\gamma = 0.65$ *MeV should be compared with the smooth behavior of the E2 rate (solid lines).*

9 Conclusion

Many aspects of the population of the superdeformed bands can be understood in term of the competition of two structures, separated by an energy barrier, and differing by their quadrupole deformation and their level density. The model presented here successfully reproduces *i)* the larger population of the SD band relative to normal structures at high spins, *ii)* the intensity pattern of the band, *iii)* the variation of this pattern as a function of the average spin and excitation energy in the residual nuclei, and *iv)* the intensity in the ridges. The relative simplicity of the model facilitate its interpretation, and this constitute a useful basis for establishing links between the physicals quantities (level densities, decay rates, energy barrier, etc...) and the experimental observables. Precise measurements of the intensity of the band relative to the total yield of the channel, the average spin at which the discrete band is populated, and the intensity of the ridges in different regimes of initial spin and excitation energy, and of the branching ratios of the in-band transitions and the lifetimes of the states in the region of the decay out of the band would greatly improve our understanding of these intriguing rotational bands.

This work was partially financed by a NATO postgraduate fellowship (JG) and by the Feodor-Lynen program of the Alexander-von-Humbolt Stiftung (KS). Supports from the Danish Natural Science Research Council are acknowledged.

References

Åberg S 1982 *Physica Scripta* **25** 23
Åberg S 1987 *Proc. XXV Int. Winter Meet. on Nucl. Phys.* Bormio, Italy
Åberg S 1988 Private communication.
Bentley M A, Ball G C, Cranmer-Gordon H W, Forsyth P D, Howe D, Mokhtar A R, Morrison J D, Sharpey-Schaffer J F, Twin P J, Fant B, Kalfas C A, Nelson A H, Simpson J and Sletten G 1987 *Phys. Rev. Lett.* **59** 2141
Deleplanque M A, Beausang C, Burde J, Diamond R M, Draper J E, Duyar C, Macchiavelli A O, McDonald R J and Stephens F S 1988 *Phys. Rev. Lett.* **60** 1626.
Deleplanque M A 1989 *Workshop on Nuclear Structure at High Spins, Bad Honnef* (unpublished)
Dudek J, Herskind B, Nazarewicz W, Szymanski Z and Werner T J 1988 *Phys. Rev. C* **38** 940
Fallon P 1988 *Workshop on Nuclear Structure, Copenhagen* (unpublished)
Fallon P 1989 *Workshop on Nuclear Structure at High Spins, Bad Honnef* (unpublished)
Haas B, Taras P, Fliobotte S, Banvile F, Gascon J, Cournoyer S, Monaro S, Nadon N, Prevost D, Thibault D, Johansson J K, Tucker D M, Waddington J C, Andrews H R, Ball G C, Horn D, Radford D C, Ward D, St. Pierre C and Dudek J 1988 *Phys. Rev. Lett.* **60** 506
Haas B 1989 *Workshop on Nuclear Structure at High Spins, Bad Honnef* (unpublished)
Herskind B, Lauritzen B, Schiffer K, Broglia R A, Barranco F, Garlardo M, Dudek J and Vigezzi E 1987 *Phys. Rev. Lett.* **59** 2416
Holzmann R, Ahmad I, Dichter B K, Emling H, Janssens R V, Khoo T L, Ma W C, Drigert M W, Garg U, Radford D C, Daky P J, Grabowski Z, Helppi H, Quader M and Trzaska W 1987 *Phys. Lett.* **B195** 321
Ragnarsson I, Bengtsson T, Leander G and Åberg S 1980 *Nucl. Phys.* **A347** 287
Ragnarsson I and Åberg S 1986 *Phys. Lett.* **B180** 191
Rathkee G E, Janssens R V F, Drifert M W, Ahmad I, Beard K, Chasman R R, Garg U, Hass M, Khoo T L, Körner H J, Ma W C, Pilotte P and Taras P 1988 *Phys. Lett.* **B209** 177
Schiffer K, Gascon J and Herskind B 1988 *Symposium on High Spin Nuclear Spectroscopy* Tokyo, Japan
Schiffer K, Herskind B and Gascon J 1989 *Z. Phys. A* **332** 17
Taras P, Flibotte S, Gascon J, Haas B, Pilotte S, Radford D C, Ward D, Andrews H R, Vall G C, Banville F, Cournoyer S, Horn D, Johannson J K, Monaro S, Nadon N, Prevost D, Pruneau C, Thibeault D, Tucker D M and Waddington J C 1989a *Phys. Rev. Lett.* **61** 1348
Taras P, Flibotte S, Gascon J, Haas B, Pilotte S, Radford D C, Ward D, Andrews H R, Banville F, Johansson J K and Waddington J C 1989b *Phys. Lett.* bf B222 357
Twin P, Nyako B M, Nelson A H, Simpson J, Bentley M A, Cranmer-Gordon H W, Forsyth P D, Howe D, Mohktar A R, Morrison J D, Sharpey-Schafer J F and Sletten G 1986 *Phys. Rev. Lett.* **57** 811
Twin P 1987 *Proc. Int. Conf. on Nucl. Phys., Crete* World scientific publishers
de Voigt M J A, Bacelar J C, Beck E M, Deleplanque M A, Diamond R M, Draper J E, Riezebos H J and Stephens F S 1987 *Phys. Rev. Lett.* **59** 270.
Waddington J C 1989 *Workshop on Nuclear Structure at High Spins, Bad Honnef* (unpublished)

Inst. Phys. Conf. Ser. No 105
Paper presented at Int. Conf. on Spectroscopy of Heavy Nuclei, Crete, Greece, 1989

Neutron–proton pairing in a cranked, deformed single-j shell

J A Sheikh, N Rowley, M A Nagarajan and H G Price

SERC Daresbury Laboratory, Warrington WA4 4AD, UK

ABSTRACT: Neutron-proton pairing should not be important if neutrons and protons occupy different single-particle levels. In the mass-80 region, however, increases in spin are seen which are due to the almost simultaneous alignment of $g_{9/2}$ protons and $g_{9/2}$ neutrons (e.g. Price et al 1983). We have performed some model calculations to see if there should be any experimental signature of such a force. Crossing frequencies are more number dependent than with $H_{np} = 0$ and we predict a band which has a superposition of an aligned neutron pair and an aligned proton pair. This should affect the experimental g-factors.

1. INTRODUCTION

Cranked shell-model (CSM) calculations of rotational bands in deformed nuclei do not generally take into account neutron-proton correlations except in that the same or similar deformations are taken for both types of nucleon. However, in nuclei where N and Z are roughly equal, pairing correlations between neutrons and protons should be important. Surprisingly there seems to be little evidence for the need to take a non-zero H_{np} and even in the mass-80 region, a reasonable description of many properties may be obtained with $H_{np} = 0$ (e.g. Dudek, Nazarewicz and Rowley, 1987).

Some recent data (Gross et al) on the Kr isotopes, however, show a very strong dependence of the band-crossing frequencies on neutron number - an effect which it was not possible to explain in terms of the conventional CSM. In the case of ^{76}Kr a pair of neutrons and a pair of protons appear to align at precisely the same frequency ($\hbar\omega \approx 0.565$ MeV) whereas in ^{78}Kr two distinct alignments occur separated in $\hbar\omega$ by 0.35 MeV. The object of this work is to investigate the possible number dependence of these frequencies which may be introduced by including a neutron-proton force, and to see if any other experimental properties may give a signature of n-p correlations.

2. MODEL

We have basically extended some earlier single-j shell calculations (Pal, Rowley and Nagarajan 1987) to include neutrons and protons both with and without an n-p interaction. Our single-particle hamiltonian is

$$h = h_{def} - \omega j_x \qquad (1)$$

to which we add an attractive delta-function $H_{12} = - G\,\delta(\underline{r}_1 - \underline{r}_2)$ for nucleons in the $g_{9/2}$ shell. In order to have isospin as a good quantum number (and thus reduce the dimensionality of our problem), we ignore Coulomb forces and take $H_{nn} = H_{pp} = H_{np}$.

Figure 1(a) shows the eigenvalues of H_{12} for two $g_{9/2}$ nucleons coupled to spin J and

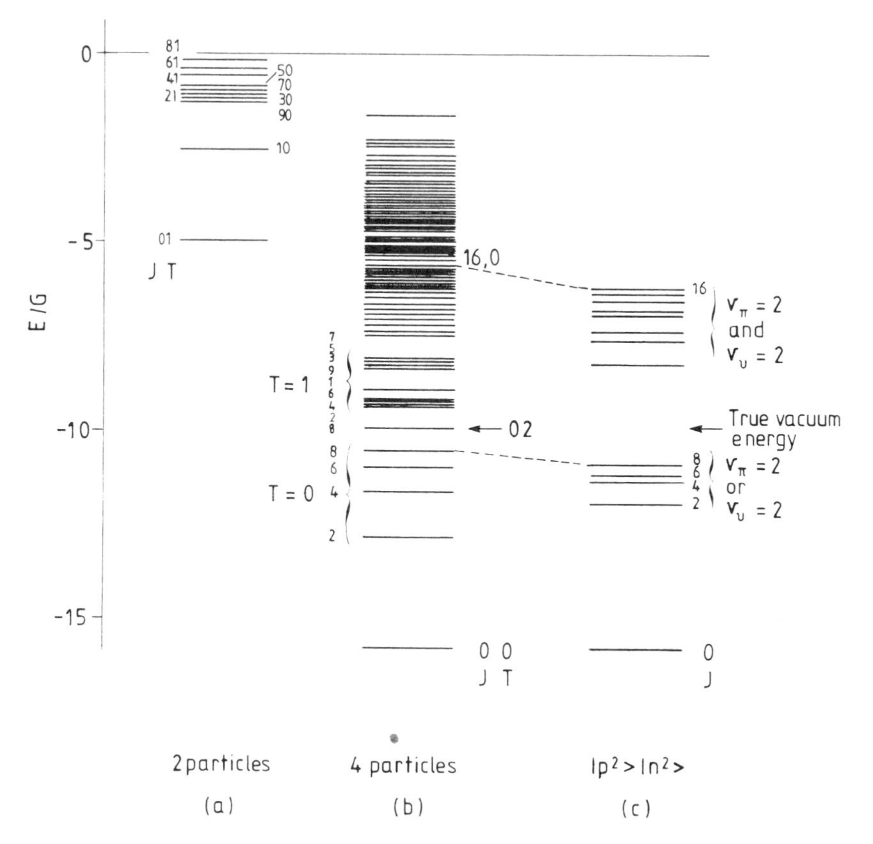

Fig. 1. (a) The spectrum of two $g_{9/2}$ nucleons interacting through a charge-independent delta-function force is shown in units of the strength G of that force. The spin J and isospin T of the states are indicated. The same force leads to the four-nucleon spectrum shown in (b). The lowest state allowed for four identical nucleons (T = 2) is marked by the arrow. The states below this all have T = 0 and are allowed only in the 2n-2p system. Part (c) of the figure shows the much simpler 2n-2p spectrum obtained if H_{np} is set to zero. (The vacuum energy has been put equal to that in (b).)

isospin T. For two identical nucleons only the T = 1 states are allowed and one sees that the J = 0 state is then depressed considerably below the others. This is the justification for the simplifying monopole pairing force generally used in CSM calculations. The n-p system, however, has the J = 1 and J = 9 (T = 0) as its lowest states and so we have chosen to do our calculations with no truncation of the two-body matrix elements. Since we shall also be discussing the relatively small effects of differences of crossing frequencies between calculations with and without H_{np}, we do not wish to introduce any errors through the use of the Hartree-Fock-Bogoljubov approximation (Goodman, 1976) and we shall, therefore, as far as possible, perform exact diagonalisations of our hamiltonian. This is the reason we constrict ourselves to a single-j shell.

Even so the dimensionality of the problem increases rapidly with particle number if both neutrons and protons are admitted. We shall, therefore, initially discuss the cranked spherical problem in which J and M_X are good quantum numbers.

3. SPHERICAL CASE

Figures 1(b) and (c) show the non-deformed 2n-2p spectra calculated both with (b) and without (c) a neutron-proton force. (The vacuum in (c) has been dropped by about 60% to equal that in (b)). On cranking these systems the states with high M_x (thus high J) but low excitation energy E* will fall and ultimately go below the $\omega = 0$ vacuum since the routhian in this case is simply

$$R = E^* - \omega M_x \qquad (2)$$

The states with one broken pair which will align most strongly are the J = 8 states. In (c) these are doubly degenerate since they may have either neutron or proton seniority v = 2. For $H_{np} \neq 0$, the T = 0, J = 8 state in (b) lies lower than the J = 8 state with T = 1. The most aligning state with 2 broken pairs is the J = 16 state which is again marked in figures 1(a) and 1(b).

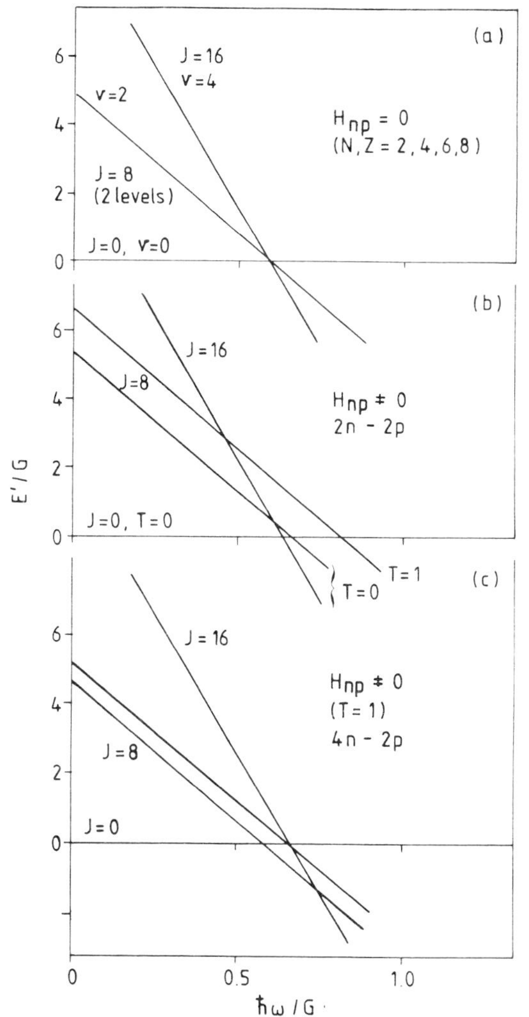

Fig. 2. The routhians of the cranked spherical problem are shown as a function of the cranking frequency for the most aligning states with one broken pair (J = 8) and two broken pairs (J = 16). In (a) we show the results for $H_{np} = 0$. These results are the same for any even value of N and Z. If $H_{np} \neq 0$ the results become number dependent and (b) and (c) show results for the 2n-2p and 4n-2p systems respectively.

The behaviour of these states at finite ω is shown in figures 2(a) and 2(b). In figure 2(a) we see that both the v = 2, J = 8 states and the J = 16 state cross the vacuum at exactly the same frequency. This is easily understood since for $H_{np} = 0$, the J = 16 state has an energy which is just twice that for J = 8. In figure 2(b) we see that the non-zero H_{np} splits the two J = 8 states (T = 0, 1) and causes J = 16 to have an energy less than twice that of the lowest 8^+ state. It thus crosses the vacuum first and causes the double alignment effect shown by the solid step-function in figure 3(c).

In figure 2(c) we show the routhians analogous to those in figure 2(b) but now for the 4n-2p system. Now the lowest 8^+ state crosses the vacuum before the 16^+ giving two distinct alignments in the yrast configuration as shown by the broken step functions in figure 3(c). The 4n-2p results with $H_{np} = 0$ are identical to the 2n-2p results of figure 2(a) since proton and neutron seniority is conserved in that case.

We thus see that a strong number dependence of the type observed in the ^{76}Kr and ^{78}Kr data may arise when $H_{np} \neq 0$ and seniority is violated.

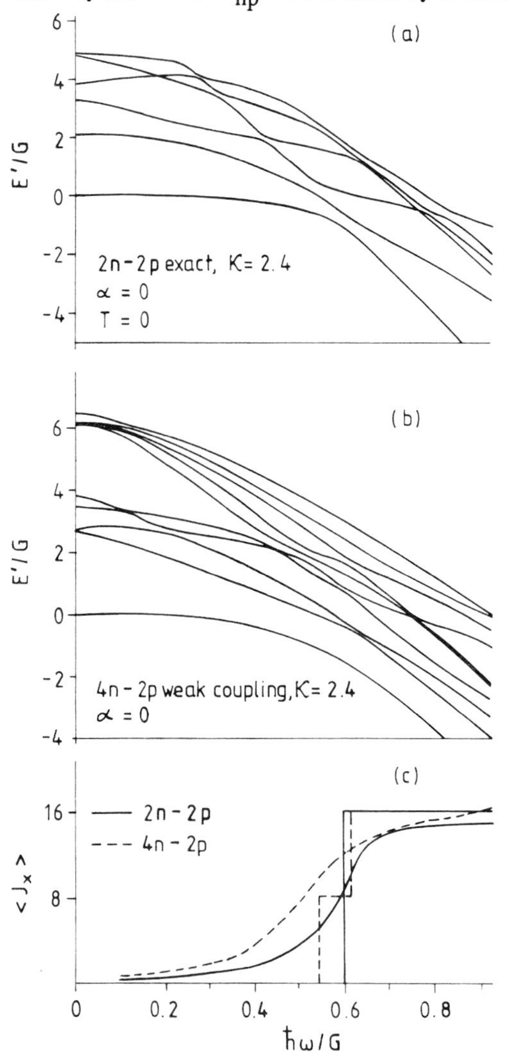

Fig. 3. For $H_{np} \neq 0$ and a finite deformation κ we show the routhians of (a) the 2n-2p system and (b) the 4n-2p system for signature $\alpha = 0$. The calculations in (a) are exact whereas those in (b) are performed in the weak-coupling approach. The alignments corresponding to the yrast states of (a) and (b) are shown by the smooth curves in (c). The step-functions show the same quantities in the zero-deformation limit.

4. DEFORMED RESULTS

Figures 3(a) and 3(b) again show the 2n-2p and 4n-2p results with a neutron-proton force but for a finite deformation. The results of figure 3(a) are exact whereas those in figure 3(b) are weak-coupling calculations which will be described elsewhere (Sheikh et al 1989). It can be seen from the alignment plots in figure 3(c) that the deformation has smeared out the two distinct crossings of the 4n-2p problem (dashed line) into an apparently single alignment. This, however, is an artefact of our small model space as we shall discuss elsewhere (Sheikh et al 1989).

5. CONCLUSIONS

Our model can give interesting effects in the crossing frequencies of bands containing two or four aligned particles. In particlar the order of the crossings may be reversed for different particle numbers. This effect should persist in a larger model space and may well account for the anomalies seen in the Kr isotopes. In any case a strong neutron-proton interaction causes admixtures in the band containing two aligned nucleons which should approximately be 50% aligned neutrons and 50% aligned protons. This should have an important effect on the g-factors of the yrast states in the corresponding frequency range.

REFERENCES

Dudek J, Nazarewicz W and Rowley N 1987 Phys. Rev. C35 1489

Goodman A L 1976 Nucl. Phys. A265 113

Gross C J et al Band crossings and near-rigid rotation in ^{76}Kr and ^{78}Kr, to be published

Pal K F, Rowley N and Nagarajan M A 1987 Nucl. Phys. A470 285

Price H G, Lister C J, Varley B J, Gelletly W and Olness J W 1983 Phys. Rev. Lett. 51 1842

Sheikh J A et al 1989, work in progress

Inst. Phys. Conf. Ser. No 105
Paper presented at Int. Conf. on Spectroscopy of Heavy Nuclei, Crete, Greece, 1989

A study of the superdeformed band in ^{152}Dy

A.Alderson[†], P.J.Twin[†], M.A.Bentley[††], A.M.Bruce[††], P.Fallon[†], P.D.Forsyth[†], D.Howe[†], J.W.Roberts[†] and J.F.Sharpey-Schafer[†]

† Oliver Lodge Laboratory, Dept. of Physics, University of Liverpool, Liverpool, L69 3BX.

†† Daresbury Laboratory, Daresbury, Warrington, WA4 4AD.

Abstract

The superdeformed continuum radiation has been measured for the nucleus ^{152}Dy formed at two different 'temperatures'. The results are presented here along with a discussion into the striking difference observed between the two cases. Comparison is also made with the superdeformed continuum measured in ^{149}Gd.

1. INTRODUCTION.

The discovery[1] in recent years of a discrete line rotational band of nineteen evenly spaced γ-rays extending up to spin $\simeq 60\hbar$ in the nucleus ^{152}Dy, corresponding to a superdeformed prolate rotor with a 2:1 major-to-minor axis ratio, aroused considerable interest and generated much experimental activity into the mass 150 region. To date a further eight such bands have been observed (^{146,148,149,150}Gd[2,3,4,5], ^{150,151}Tb[6,5] and ^{151,153}Dy[7,8]) in this region, all being fed over only a few transitions. Clearly it is vital to understand the mechanism involved in the population of these bands. Unresolved non-yrast superdeformed bands are predicted[9] to exist giving rise to a superdeformed continuum radiation which may shed some light onto the matter. In this talk I shall present data obtained using the heavy ion fusion evaporation reaction ^{108}Pd(^{48}Ca,4n)^{152}Dy* at a bombarding energy of 197MeV and make comparisons with previous data from the same nucleus at a higher excitation energy (or 'temperature') as well as data[10] from ^{149}Gd.

2. THE SUPERDEFORMED CONTINUUM AND THE E_γ-E_γ COINCIDENCE MATRIX.

Figure 1a is a schematic representation of the scenario in the excitation energy versus angular momentum plane of the final nucleus after the last neutron has 'boiled' off in a heavy ion fusion evaporation reaction. De-excitation of the final nucleus may occur by γ-ray emission via normal states or through the superdeformed continuum states (either in-band or from one band to another) to the superdeformed yrast band, or some combination of the two.

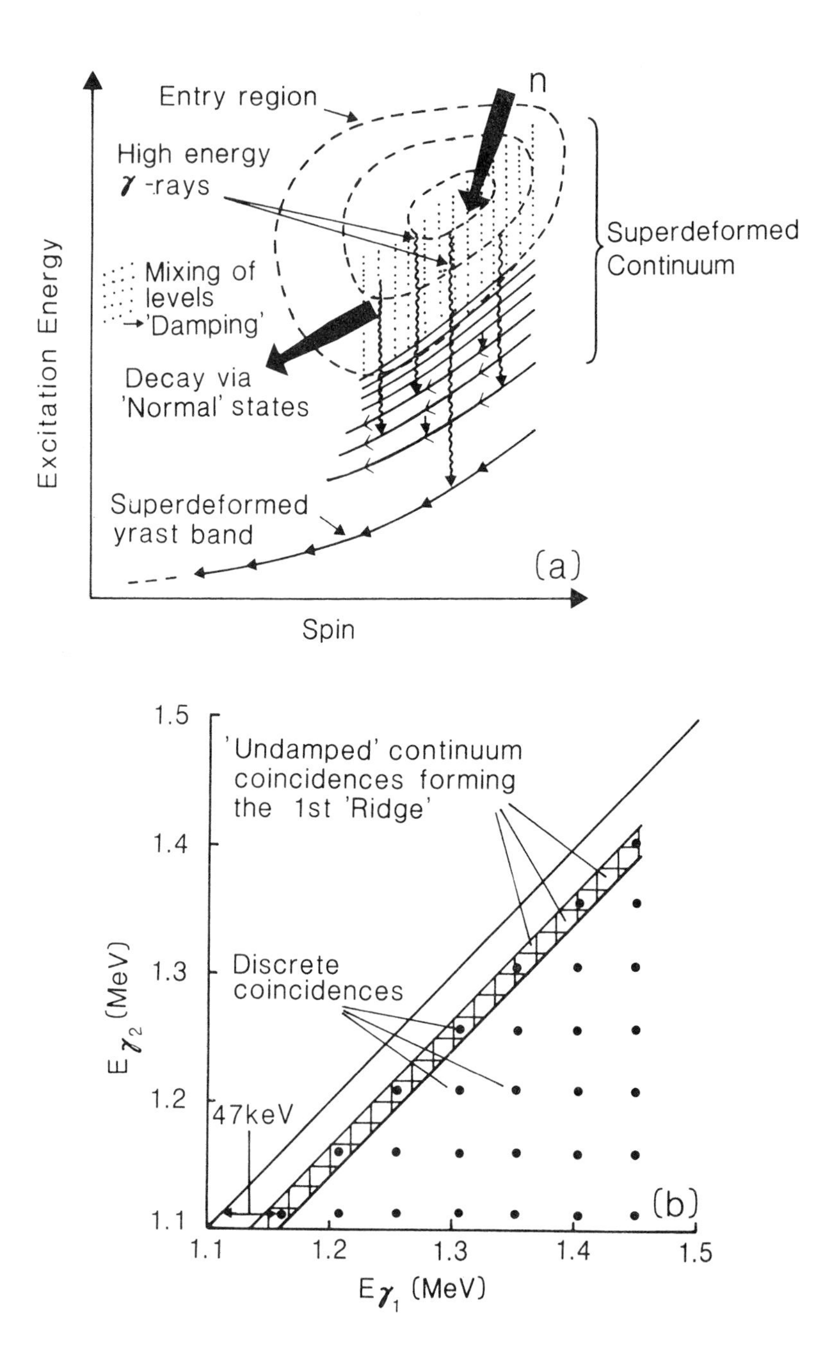

Figure 1. (a) A schematic showing the possible methods of decay of the final nucleus in a heavy ion fusion evaporation reaction. (b) Discrete coincidences, characteristic of the decay down a rotational cascade (and with $\Delta E_\gamma \sim$ 47keV for ^{152}Dy), are shown in an E_γ-E_γ coincidence matrix. Decays amongst the undamped continuum states in 1(a) will contribute to ridge structures in the matrix (see text).

Consider the situation where the decay goes through the superdeformed states. In the case of ^{152}Dy the energy difference between successive superdeformed γ-rays in the yrast band is $\Delta E_\gamma \sim 47$keV, so coincidences between any two consecutive transitions in this band will lead to a series of evenly spaced points along a line parallel to the diagonal of an E_γ-E_γ coincidence matrix and at a distance of 47keV from it (see figure 1b). Similarly a coincidence between a γ-ray and the next but one γ-ray will lie on a line at 94keV from the diagonal. Now if there are similar coincidences in the continuum states then these too will contribute to the lines parallel to the E_γ-E_γ matrix diagonal to form 'ridges', with consecutive coincidences contributing to the first ridge, alternate ones to the second ridge and so on. At some stage the superdeformed continuum states are expected to become mixed leading to uncorrelated γ-rays ('damping') which will not therefore contribute to any of the ridge structures.

3. THE EXPERIMENT.

A series of experiments using the tandem at the Nuclear Structure Facility (NSF) at Daresbury to investigate the conditions for the selection of the band revealed that the reaction ^{108}Pd(^{48}Ca,4n)^{152}Dy*, using a 197MeV beam, gave the largest intensity of the band relative to the total population intensity in ^{152}Dy. These conditions were chosen for a high statistics γ-γ coincidence experiment which was also carried out at the NSF at Daresbury. The TESSA3[(11,12)] spectrometer was used to detect the 270 million γ-γ suppressed coincidence events recorded onto magnetic tape, with a target consisting of $2 \times 500\mu$gcm^{-2} self-supporting foils of ^{108}Pd.

The TESSA3 array consists of 16 high resolution germanium detectors, each surrounded by a sodium iodide and/or bismuth germanate (BGO) anti-compton suppression shield, and an inner calorimeter of 50 high efficiency, low resolution hexagonal BGO detectors arranged in a 'ball' configuration such that a solid angle close to 4π is covered. A further eight BGO detectors placed downstream of the target (and out of focus of the 16 compton suppressed germanium detectors) measured decays from the 60ns isomeric state in ^{152}Dy (17^+ level). An isomer timing circuit was started by a γ-γ-BGO(isomer) coincidence and stopped by a delayed 'prompt' signal generated by at least 2 BGOs firing at the same time.

4. ANALYSIS AND RESULTS.

A fold condition was set on the BGO 'ball' of $8 < \text{fold} < 16$, with each element of the 'ball' having a threshold of 500keV. At the bombarding energy of 197MeV the isomer and fold filters enabled a large improvement (a factor of 3 compared to previous experiments which used a bombarding energy of 205MeV) in the selection of the superdeformed states relative to the ^{152}Dy channel (see figure 2). The resulting E_γ-E_γ coincidence matrix, after the selection of ^{152}Dy, had a total of 34 million counts. A ridge structure associated with a superdeformed shape was observed in the energy region from 600 to 1500keV, at approximately 47keV from the diagonal. No further ridges were observed.

A series of 2keV wide diagonal slices were taken across the first ridge, whereby the lowest of the two coincident γ-rays was projected out. Integrations were performed over the limits of each superdeformed peak, and over the regions inbetween the peaks,

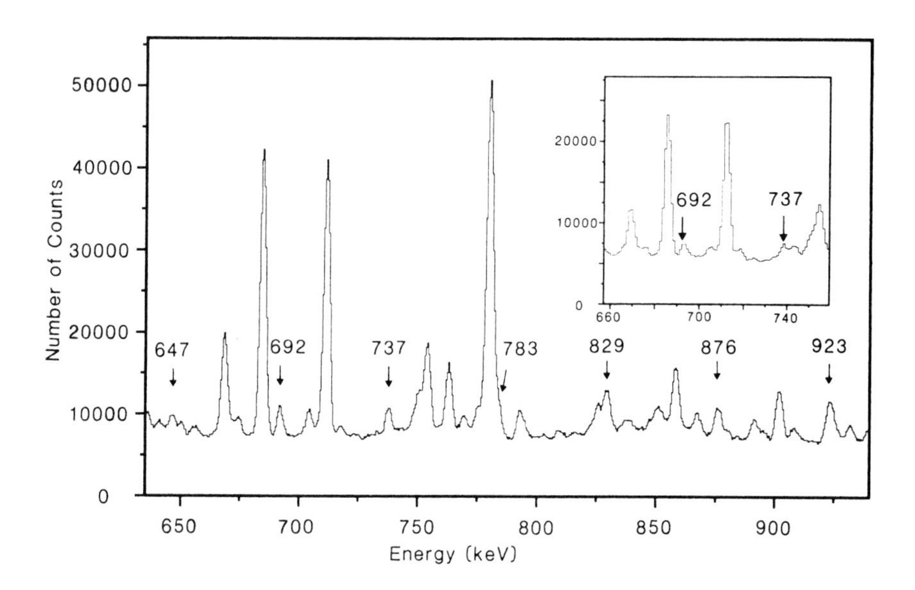

Figure 2. A section of the total projection of the E_γ-E_γ matrix gated by isomer and high-fold conditions from the reaction $^{108}Pd(^{48}Ca,4n)^{152}Dy^*$ at a beam energy of 197MeV. Some members of the superdeformed band are labelled. An improvement by a factor of 3 on previous data at a bombarding energy of 205MeV (inset) in the selection of the superdeformed states relative to the ^{152}Dy channel was achieved.

for each slice. The peak integrations gave the normalisation to fit the data to the intensity curve for the band and indeed a good fit was obtained to that generated from a 'sum of superdeformed gates' coincidence spectrum. The data clearly show very little continuum intensity in the first ridge (see figure 3). This was rechecked by taking a series of cross-diagonal slices (thin slices over the peaks, thick slices inbetween peaks) where the energy difference between the two coincident γ-rays was projected out. Peaks were fitted to the ridge and integrated over for each cross-diagonal slice, yielding the same results as the diagonal slicing technique.

Data from a previous experiment (same reaction but with a 205MeV beam) performed at Daresbury were analysed in exactly the same way and are also presented in figure 3. Here a very much larger continuum contribution to the first ridge intensity can be seen, indicating that the average entry region of the final nucleus after the last neutron has 'boiled' off is critical to the amount of superdeformed continuum feeding. It also suggests that the point at which damping occurs in the superdeformed continuum of ^{152}Dy is at a fairly high excitation energy above the yrast superdeformed band (at least $\sim$ 2MeV). This compares well with a computer simulation[13] where an E_γ-E_γ coincidence matrix was generated similar to experiment with a value of U_0, the energy above yrast marking the onset of damping effects, of 2MeV, see figure 3.

Data on ^{149}Gd yield a large superdeformed continuum which is in contrast to the ^{152}Dy

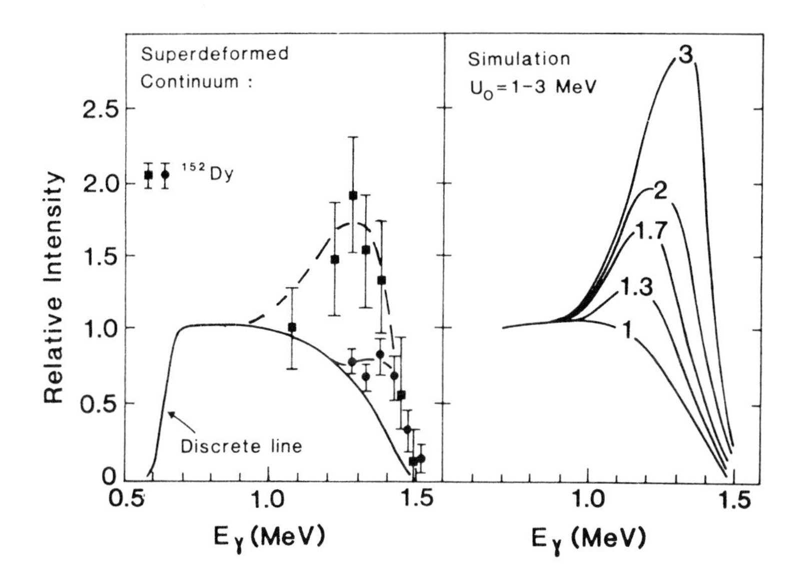

Figure 3. The measured relative intensity of the superdeformed continuum in the first ridge for ^{152}Dy at 197MeV (circles) and 205MeV (squares) beam energies in the reaction ^{108}Pd(^{48}Ca,4n)^{152}Dy* in comparison with the results of a computer simulation by Schiffer and Herskind[13] varying the value of U_0, a parameter defining the onset of damping (see text).

data. This indicates a much smaller energy gap between the yrast and non-yrast superdeformed states in ^{149}Gd compared to ^{152}Dy, and/or a larger density of states in the minimum of ^{149}Gd below the damping cut-off relative to that in ^{152}Dy, which alters the ratio of yrast to continuum states being fed.

The fact that no further ridges were observed in the 197MeV data means that any non-yrast superdeformed bands are much shorter (2 or 3 transitions, say) than the yrast band.

5. CONCLUSIONS.

The average entry region of the final nucleus appears to be critical to the number of superdeformed non-yrast states that are fed. This may be expected since the 'cooler' the final nucleus is formed the fewer the non-yrast superdeformed states there may be available to excite. In the case of good conditions for formation of the yrast superdeformed band a possible scenario is that the band is fed almost entirely directly by high energy γ-rays from the entry region, hence by-passing the few non-yrast superdeformed states available. Another possible explanation is that the band could be fed directly by the last neutron boiled off from the fusion evaporation reaction.

Damping effects in the superdeformed continuum of ^{152}Dy do not appear to arise until at least $\sim$ 2MeV above the yrast band.

No second or higher ridges were observed, hence the non-yrast superdeformed bands in ^{152}Dy must be short.

Comparison with ^{149}Gd data raises the question of differences in the energy gap between yrast and non-yrast bands in different nuclei. If, as may be the case in ^{149}Gd, the gap is relatively small, then it would be difficult to form the yrast superdeformed band without seeing a significant superdeformed continuum. This may not necessarily be true if the gap is somewhat larger, as possibly indicated by the ^{152}Dy data. Another possibility is that the density of states in ^{149}Gd is somewhat larger in the region of continuum excited than that in ^{152}Dy.

This work was funded by grants from the United Kingdom Science and Engineering Research Council from whom four of us were recipients of postgraduate studentships (A.A.,M.A.B.,P.F. and J.W.R.), with two of us (A.M.B. and D.H.) supported by post-doctoral fellowships.

REFERENCES.

1. P.J.Twin *et al.*, Phys. Rev. Lett. 57 (1986) 811.
2. G.Hebbinghaus *et al.*, Phys. Rev. Lett. 59 (1987) 2024.
3. M.A.Deleplanque *et al.*, Phys. Rev. Lett. 60 (1988) 1626.
4. B.Haas *et al.*, Phys. Rev. Lett. 60 (1988) 503.
5. P.Fallon *et al.*, Phys. Lett. B 218 (1989) 137.
6. M.A.Deleplanque *et al.*, Phys. Rev. C 39 1651 (1989).
7. G.E.Rathke *et al.*, Phys. Lett. B 209 (1988) 177.
8. J.C.Waddington, talk given to Nuclear Structure workshop, Bad Honnef, 1989.
9. B.Herskind *et al.*, Phys. Rev. Lett. 59 (1987) 2416.
10. P.Taras *et al.*, private communication.
11. P.J.Twin *et al.*, Nucl. Phys. A409, 343c (1983).
12. P.J.Nolan *et al.*, Nucl. Instrum. Methods Res., Sect. A236, 95 (1985).
13. K.Schiffer and B.Herskind, NBI preprint, 1987.

Inst. Phys. Conf. Ser. No 105
Paper presented at Int. Conf. on Spectroscopy of Heavy Nuclei, Crete, Greece, 1989

Non-yrast structure in tellerium nuclei

J A Cizewski, D Barker,* R G Henry, and C S Lee

Department of Physics and Astronomy, Rutgers University
New Brunswick, New Jersey 08903 USA

ABSTRACT: We have measured the transitional tellerium nuclei via Sn(α,xnγ) reactions. The yrast structure of ^{122,124}Te show coexistence between weakly collective and two quasiproton excitations. The present work does not support a more deformed structure for the higher spin yrast states.

1. INTRODUCTION

The study of heavy nuclei with relatively few valence particles, and in particular the non-yrast excitations in these transitional nuclei, can provide sensitive tests of our understanding of nuclear structure, and the interplay between collective and single-particle degrees of freedom. The tellerium nuclei with only two valence protons outside of the Z=50 closed shell of tin provide a special challenge to nuclear models. While an intermediate coupling scheme[1], the pairing-plus-quadrupole model[2], the interacting boson model (IBA) (both invoking limiting symmetries[3,4] and complicated mixed configuration calculations[5]), and the dynamic deformation model[6] predictions have been applied to understand the excitations in N>66 Te nuclei, none of these models has been fully successful in understanding the complicated structure exhibited in nature.

The unique structure of the Te nuclei is further illustrated in Fig.1. The ratio of yrast state energies can be a signature of the appropriate structure. For example, in a harmonic quadrupole vibrator the $R(4/2)=E(4^+)/E(2^+)$ energy ratio is 2.0, and R(6/4)=1.5; or for a rotor R(4/2)=3.33, R(6/4)=2.10. Defining the onset of collective motion when the energy ratios are those expected for a vibrator, divides a plot of R(6/4) vs R(4/2) into four regions. When R(4/2)<2.0, R(6/4)<1.5, the excitations are typical of single-particle configurations; when R(4/2)≥2.0, R(6/4)≥1.5, the excitations are collective. However, R(4/2)≥2.0, but R(6/4)<1.5, or R(4/2)<2.0, but R(6/4)≥1.5, would not be expected in a naive description because such energy ratios signal a competition between collective and single-particle structure in the low-lying yrast excitations. In fact, in a recent survey of energy ratios in nuclei[7] it was found that the energy ratios for most collective nuclei with A>60 lie along the solid line in Fig.1. This highlights the anamolous structure of the Te nuclei with N>68: while the R(4/2) values are representative of collective structure, the R(6/4) ratios indicate single-particle structure for the 6^+ states. In addition to the non-collective nature of the 6^+ states in Te, there is also evidence for intruder 4p-2h excitations across the Z=50 shell gap, yielding a very complicated low-lying structure of this transitional element.

2. EXPERIMENTS

To understand better the excitations in Te nuclei, we have recently completed measurements of the Sn(α,xnγ) reactions for A=116-124 targets, and x=1,2. These were some of the final experiments at the Rutgers FN-1 Tandem Van de Graaff Accelerator. The measurements of standard gamma-ray spectroscopy were done as a function of beam energy to populate both yrast and non-yrast excitations in even- and odd-A Te nuclei. An example of our excitation function measurements is presented in Fig.2 for the ^{120}Sn target. The 1-n channel peaks at about 17 MeV; the 2-n channel yrast excitations require the highest beam energies available. For lighter Sn targets the excitation function moves to higher beam energies. However, for the heavier Sn targets, the excitation function moved to lower energies and, therefore, it was not possible to study the 1-n channels for the ^{122,124}Sn targets. For our ^{120}Sn measurements, three detector γ-γ coincidence measurements were made at 18 and 24 MeV to populate higher-spin excitations in ^{123}Te and yrast and non-yrast excitations in ^{122}Te. For example, we have been able to populate excited 0^+ states, providing the first in-beam studies of these excitations, which may be candidates for intruder configurations. We have also been able to identify higher angular momentum excitations in both the even- and odd-A Te isotopes.

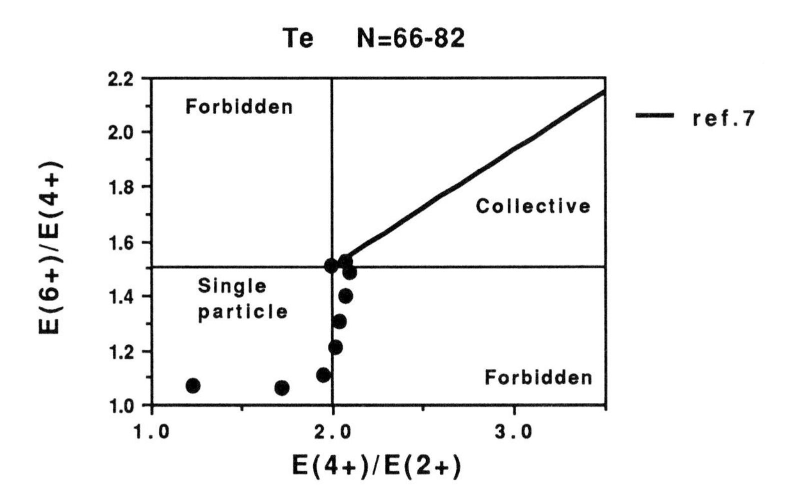

Fig.1 Energy ratios in Te nuclei. The solid line is representative of ratios in collective nuclei, see ref.7.

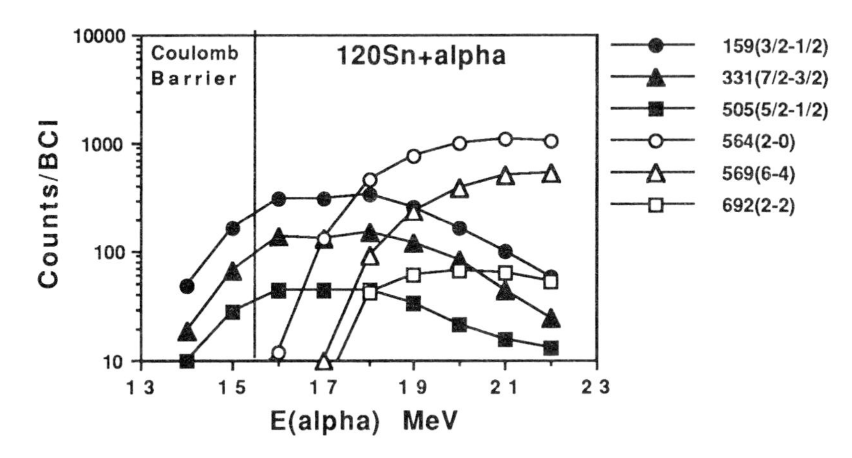

Fig.2 Excitation function of the ^{120}Sn + α reaction. The solid symbols are transitions in ^{123}Te; the open symbols are transitions in ^{122}Te.

3. DISCUSSION

To illustrate the typical structure of the heavier even-A Te cores, a partial level spectrum of ^{122}Te with N=70 is presented in Fig.3. Here R(4/2)=2.09, R(6/4)=1.48. (In addition to the levels presented here, we have populated additional low-spin positive-parity, and natural and unnatural negative-parity states.) In a simple scheme the valence protons occupy the $g_{7/2}$ and $d_{5/2}$ orbitals; for the neutrons the lower orbitals are filled, the $h_{11/2}$ orbital about half-filled. Given the valence orbitals, it is easy to construct 2^+ states, yielding a low-lying, collective 2^+ excitation, while the 6^+ state is predominantly two-quasiprotons. At the 8^+ state, a new, possibly more collective structure becomes yrast. This structure has been proposed as a 4p-2h proton excitation across the Z=50 closed shell: the six valence protons allow a more collective, possibly deformed structure.

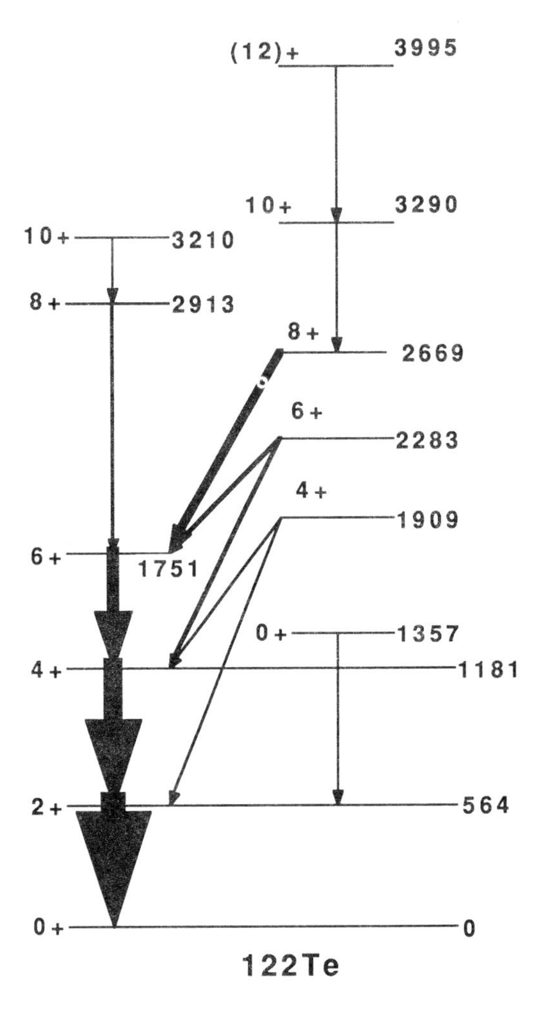

Fig. 3 Partial deexcitation spectrum of 122Te obtained in our present 120Sn+alpha measurements.

To study further the structure of this "intruder" band, we searched for the intraband transitions; only upper limits on the intensities could be extracted. With these values conservative estimates of the relative intraband/interband B(E2) transition probabilities give ratios less than 4:1. In other words, the proposed "intruder" band is no more collective than the ground-state band, with "interband" and "intraband" transition rates of the same order of magnitude. Considerable mixing between the configurations could give rise to the observed relative E2 strengths. However, for two-state mixing, the mixing matrix element is at most one-half the final energy spacing. For the 10^+ states in ^{122}Te the mixing matrix element is, therefore, at most 40 keV, rather than the 100 keV value used for mixing calculations in Cd[8]. (In ^{124}Te similar arguments limit the mixing matrix element between the 10^+ states to less than 9 keV). Unless the mixing matrix element is spin-dependent, there is no evidence that the "intruder" band is of a different structure than the ground-state band.

The positive-parity excitations in the Te nuclei can, therefore, be treated as a single structure, with a predominantly collective nature for the lower spin states. In fact, both anharmonic vibrational (SU(5) limiting symmetry of the IBA) or γ-unstable rotational (O(6) IBA symmetry) structure can be assigned to these nuclei (although the deexcitation of excited 0^+ states favors a vibrational picture). To probe further the structure of the cores one can study the particle-core coupling in the adjacent odd-A nuclei. For our initial study we have concentrated on the negative-parity excitations, built on the relatively pure $h_{11/2}$ isomer; the systematics of these excitations in $^{117-123}$Te are presented in Fig.4. To compare the particle-core coupling for a variety of core shapes, we have chosen the framework of the interacting boson-fermion (IBFA) model, where the boson-fermion interaction is modeled as a monopole+quadrupole+exchange interaction. The strength of the quadrupole interaction will depend upon the occupation probabilites, which can be taken from experiment[9]. In our initial calculations, we found no need for a monopole term, and have used constant values for the overall strength of the quadrupole and exchange interactions. The challenge is to reproduce the rather constant negative-parity excitations and, in particular, the location of the non-yrast states as a function of neutron number. Further calculations are in progress.

4. CONCLUSIONS

In summary, no single model appears to be able to understand the complicated structure of the N>68 Te nuclei. There is coexistence between single-particle and collective excitations, where the collective structure may be vibrational or γ-unstable. The hope is that by studying the odd-A Te nuclei, the particle-core coupling may shed further light on the structure.

We would like to thank H Dorsett, L P Farris, J W Gan, G Kumbartzki, H Li, J Szczepanski, and R Tanczyn for their assistance in the data acquisition and analysis. This work was partly supported by the U.S. National Science Foundation.

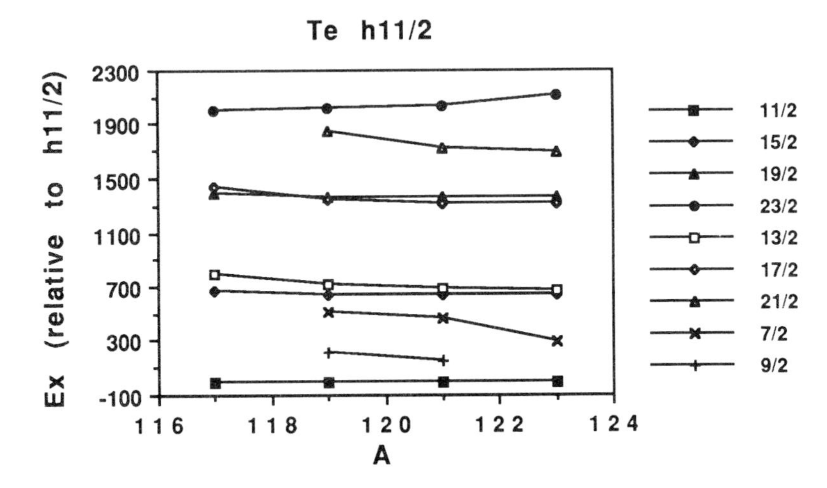

Fig.4 Negative-parity systematics for $^{117-123}$Te.

* Present address: Logica, Cobham, Surrey, U.K.

REFERENCES

[1] Fernandes M A G and Rao M N 1977 J.Phys. **G3** 1397

[2] Walters W B and Meyer R A 1976 Phys. Rev. **C14** 1925

[3] Robinson S J, Hamilton W D, and Snelling D M 1983 J.Phys. **G9** L71; **G9** 961

[4] Jolie J, in Nuclear Structure, Reactions, and Symmetries, R A Meyer and V Paar (ed.), World Scientific (1986) p.205; Jolie J, Heyde K, Van Isacker P, and Frank A 1987 Nucl.Phys. **A466** 1

[5] Rikovska J, Stone N J, and Walters W B 1987 Phys. Rev. **C36** 2162

[6] Park P et al. 1985 J. Phys. **G11** L251 ; Subber A R H, Hamilton W D, Park P, and Kumar K 1987 J. Phys. **G13** 161

[7] Cizewski J A 1989 Phys. Lett. **219B** 189

[8] Aprahamian A, Brenner D S, Casten R F, Gill R L, and Piotrowski A 1987 Phys. Rev. Lett. **59** 535

[9] Rodland T, Lien J R, Løvhoiden G, Thorsteinsen T F, and Vaagen J S 1987 Nucl.Phys. **A469** 407, and references therein

Inst. Phys. Conf. Ser. No 105
Paper presented at Int. Conf. on Spectroscopy of Heavy Nuclei, Crete, Greece, 1989

Neutron and proton $i_{13/2}$ alignments in ^{174}Os

L. Hildingsson[1], J. Gizon[2], C.A. Kalfas[3], W. Klamra[1], S. Kossionides[3], Th. Lindblad[1], C. Papadopoulos[3] and R. Vlastou[3].

1) The Manne Sieghbahn Institute of Physics, Stockolm, Sweden.
2) N.R.C.P.S. Democritos Athens, Greece.
3) Institut des Sciences Nucléaires, Grenoble, France.

ABSTRACT: Results from a spectroscopic study of high spin states in ^{174}Os is presented. The ground state band has an $i_{13/2}$ neutron alignment at $\hbar\omega_c = 0.30$ MeV with a strong interaction. Possible assignments for the negative parity sidebands are discussed in terms of both neutron and proton $i_{13/2}$ alignments.

Nuclei in the rare earth region are known to possess stable nuclear shapes with prolate deformation and their structure has successfully been described within the crank shell model approach. However, when going away from this region toward Z = 82 the nuclear potential becomes soft and and the structure of the yrast and near yrast states become less clear. For this reason a systematic study was undertaken with the ESSA 30 detector system at the Daresbury tandem laboratory, a study which covered nuclei with N $\leq$ 100. The aim was to evaluate the structure of the yrast region in these nuclei and to establish the nature of the band crossings. We here present the preliminary results from a part of this study - the high spin structure of ^{174}Os

The reaction ^{32}S(^{146}Nd,4n) at 166 MeV was used to populate high spin states in ^{174}Os and levels up to spin $I = 35\hbar$ were observed. In the same reaction ^{173}Re and ^{173}Os were produced as bi-products and their level schemes could also be determined. This offers the possibility to compare the structure of ^{174}Os with its odd-proton and odd-neutron neighbours. The level scheme of ^{174}Os obtained from the present investigation is presented in Figure 1. Compared to previous studies (Durell *et al* 1982, Gascon *et al* 1987) the data extends to higher spin and in addition a new sideband has been established.

Experimental spin alignmentments i_x have been extracted and they are shown in Figure 2. Since different quasiparticle configurations are expected to have different shapes there exist no unique reference. The absolute values of i_x are thus uncertain but the figure gives a good idea of the general behaviour of the alignment as a function

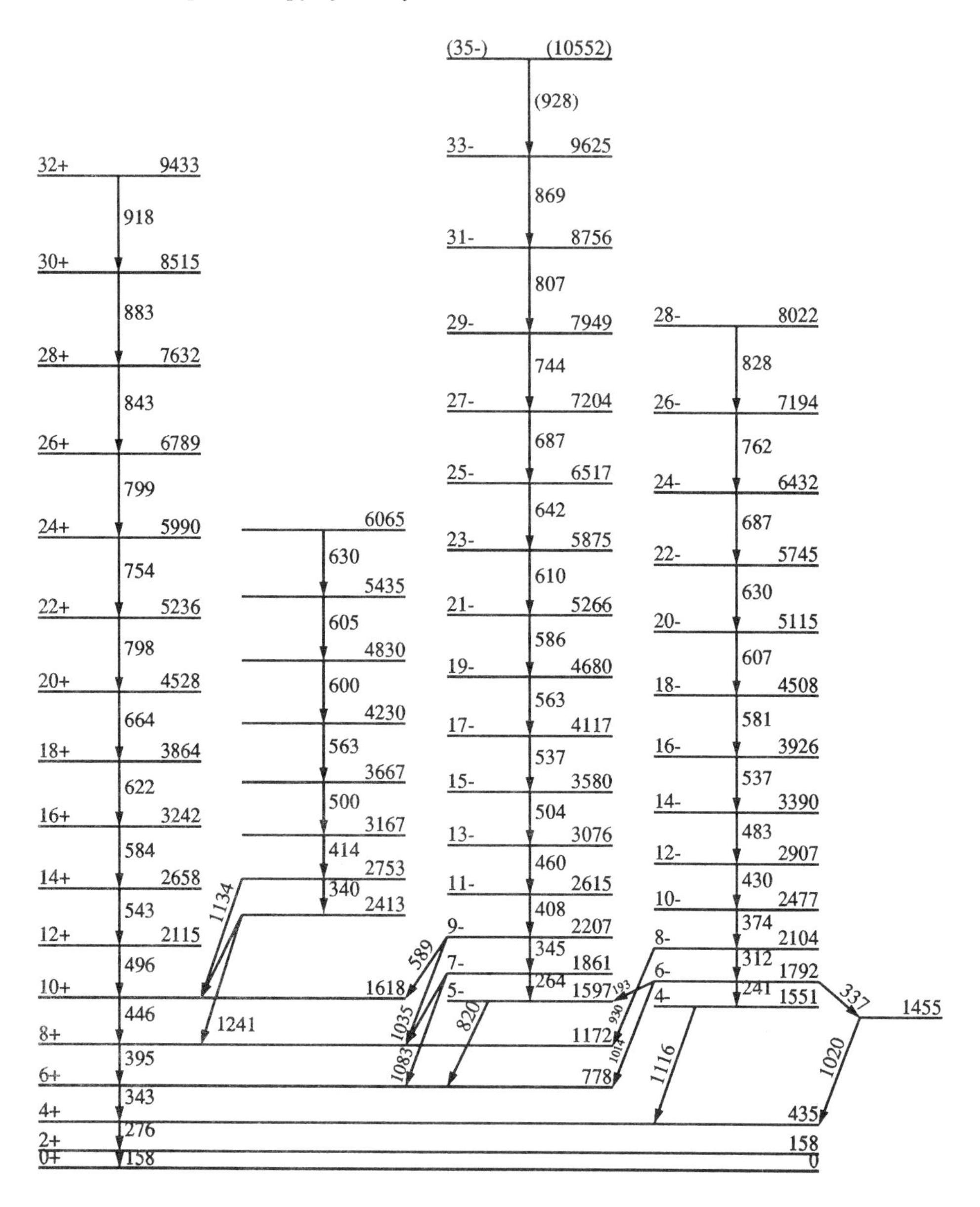

Figure 1. Level scheme of ^{174}Os obtained from this work.

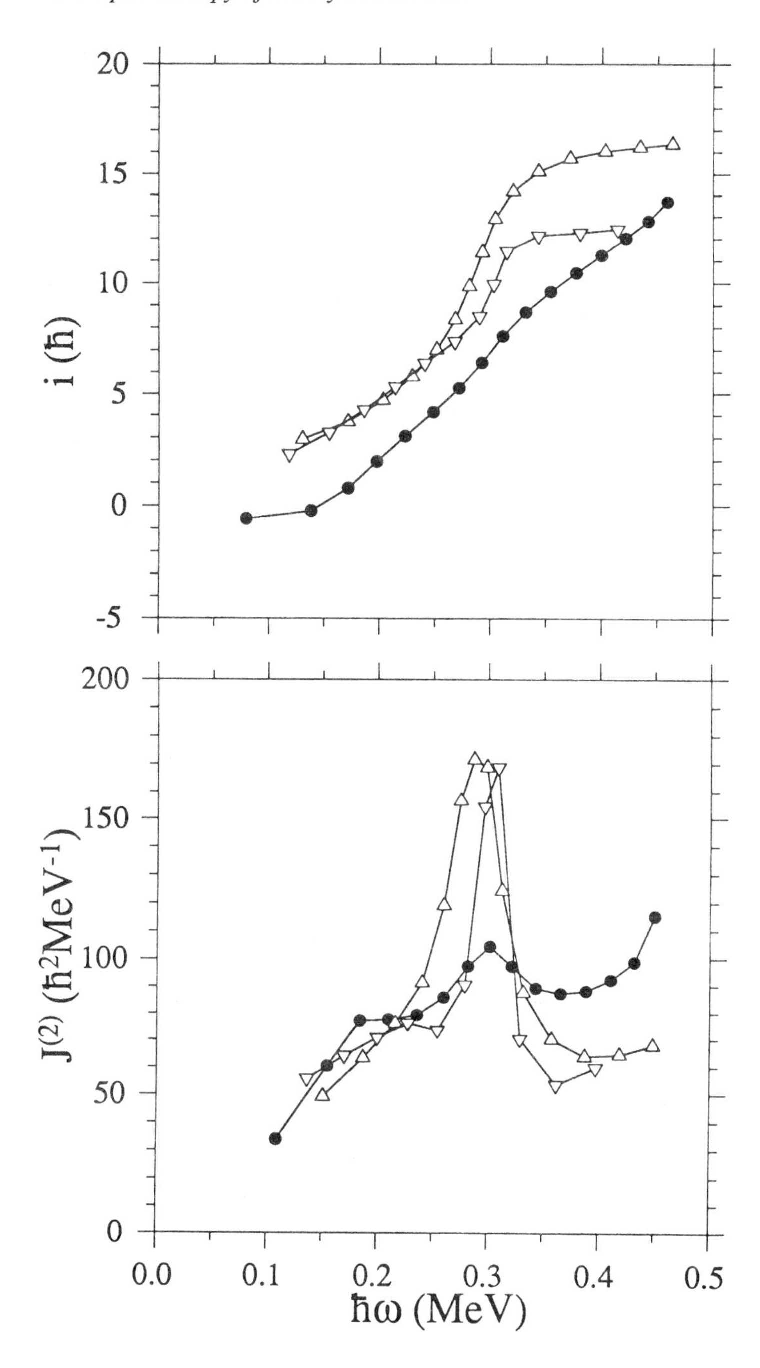

Figure 2. Experimental alignments, i_x (top) and second moment of inertia, $\mathcal{J}^{(2)}$ (bottom) for ^{174}Os. A reference has been subtracted subtracted based on a moment of inertia given by the Harris parameters $\mathcal{J}_0 = 25.8\hbar^2 MeV^{-1}$ and $\mathcal{J}_1 = 61.8\hbar^4 MeV^{-3}$.

of spin. The second moment of inertia $\mathcal{J}^{(2)}$, which is independent of any reference, has also been extracted and it is shown in the lower part of Figure 2.

To be able to compare the observed values with theoretical predictions Total Routhian Surface (TRS) calculations have been performed to find the equilbrium deformation of the nucleus as well as the character of the yrast states. Routhians have been calculated using a Woods-Saxon potential and the results from such a calculation is shown in Figure 3.

The ground state band (GSB) show a gradual increase in the alignment with $\hbar\omega$. It is only in the $\mathcal{J}^{(2)}$ vs $\hbar\omega$ plot that any irregularities show up. The peak in $\mathcal{J}^{(2)}$ at 0.30 MeV can be interpreted as due to an $i_{13/2}$ neutron alignment (the AB alignment). This neutron alignment has been observed in several nuclei in this mass regon. The interacion strength between the ground state band and the S-band show a maximum for the N=98 isotones which explains the smooth increase in the alignment in ^{174}Os.

The calculations support the $i_{13/2}$ neutron interpretation of the alignment although the observed $\hbar\omega_c$ is higher than the predicted AB crossing frequency and the interaction strenght is underestimated in the model. This can be due to a larger deformation than the predicted $\beta \approx 0.21$. The first predicted proton alignment occures at ≈ 0.5 MeV were the first negative parity proton orbitals align. The upbend in $\mathcal{J}^{(2)}$ at 0.46 MeV can be this proton alignment but the neutron CD crossing is also possible.

Two side bands starting at spin 4 and 5, respectively, with presumably negative parity were identified. They both have peaks in the $\mathcal{J}^{(2)}$ curve at 0.30 MeV similar to the GSB, indicative of the same $i_{13/2}$ neutron alignment. However, there is a difference in the high spin behaviour of the two bands.The odd spin band (5^- band) starts the alignment at somewhat lower frequency, $\hbar\omega = 0.28$ MeV and has after the crossing a larger alingment than the 4^- band

Several interpretations are possible for these two bands. One possibility is to use blocking arguments and the fact that the crossing frequency is the same as that of the neutron $i_{13/2}$ crossing in the GSB. This excludes a $i_{13/2}$ neutron in the low spin part of the side bands. This would imply a two quasiproton band, possibly based on the $\pi h_{9/2} \otimes \pi 5/2^+[402]$ configuration. The extra alignment observed for the odd spin band members can then be caused by the $i_{13/2}$ proton orbital that comes down in energy with increasing spin. It is the $\alpha = +1/2$ component of the $\pi i_{13/2}$ orbital that come lowest and crosses the $\alpha = +1/2$ signature component of the $5/2^+[402]$ orbital. The odd spin band has thus after the crossing a $\pi h_{9/2} \otimes \pi i_{13/2} \otimes (\nu i_{13/2})^2$ configuration. The corresponding $\pi 5/2^+[402] - \pi i_{13/2}$ crossing is observed in ^{173}Re at $\hbar\omega_c = 0.19$ MeV.

However, blocking arguments should be used with care - changes in the shape and in the neutron strength will influence the character of the neutron $i_{13/2}$ crossing.

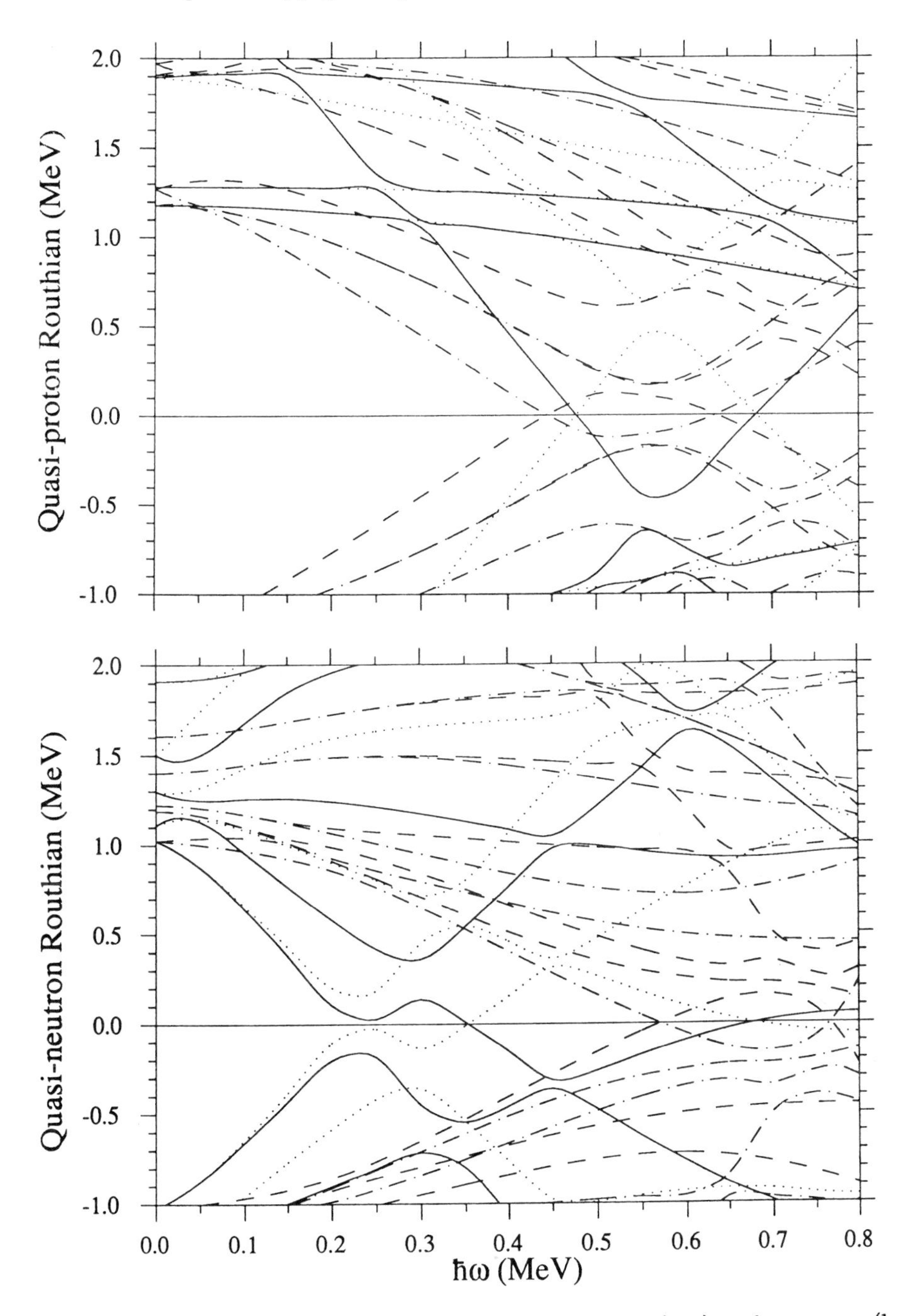

Figure 3. Calculated quasiparticle Routhians for protons (top) and neutrons (bottom) using a Woods-Saxon potential with $\beta_2 = 0.24$, $\beta_4 =$ -0.01 and $\gamma = 1°$. Legend: solid line $(\pi, \alpha) = (+, +1/2)$; dotted $(+, -1/2)$; dash-dotted $(-, +1/2)$; dashed $(-, -1/2)$.

Assuming a $\nu i_{13/2} \otimes \nu 5/2^-[523]$ configuration for for the sideband the neutron pairing will be reduced which will reduce the crossing frequency for the second neutron alignment (the BC alignment). This can then happen at the same frequency as the AB crossing in the ground state band. An inspection of Fig. 3 also supports the neutron assignment. A small reduction in Δ_n might reduce the neutron BC crossing frequency to a value close to that of the unblocked AB crossing. It is also the $i_{13/2}$ neutron orbitals that first come low in energy and are expected to form the near yrast states. The difference in the two signatures at high spin can be an effect of different shapes.

References:

Durell J L, Dracoulis G D, Fahlander C and Byrne A P 1982 Phys. Lett. **115B** 367

Gascon J, Banville F, Taras P, Ward D, Alexander T K, Andrews H R Ball G C, Horn D, Radford D C, Waddington J C and Christy A 1987 Nucl Phys **A470** 230

Inst. Phys. Conf. Ser. No 105
Paper presented at Int. Conf. on Spectroscopy of Heavy Nuclei, Crete, Greece, 1989

Evidence for $h_{11/2}$ proton alignment in ^{127}Ce

B.M. Nyakó[x§], J. Gizon[x], V. Barci[x], A. Gizon[x], S. André[x], D. Barnéoud[x], D. Curien[+], J. Genevey[x], J.C. Merdinger[+]

[x]Institut des Sciences Nucléaires, 38026 Grenoble Cedex, France
[+]Centre de Recherches Nucléaires, 67037 Strasbourg Cedex, France

ABSTRACT : Excited states have been identified for the first time in ^{127}Ce using γ-ray spectroscopic techniques. A band developed up to spin 51/2 is proposed to be based on an $h_{11/2}$ neutron-hole configuration. The band crossing frequencies and aligned angular momenta indicate that the observed backbend is produced by the alignment of an $h_{11/2}$ proton pair. Another band structure is found whose configuration is very likely $g_{7/2}$ or $d_{5/2}$ neutron-hole.

1. INTRODUCTION

For transitional nuclei situated between the Z = 50 shell closure and the rigidly deformed rare-earths, the main collective excitations at low and intermediate rotational frequencies are generated by $h_{11/2}$ quasi-particles. Odd-A, odd-Z nuclei are characterized by $\Delta I = 2$ decoupled bands produced by quasi-protons situated at the bottom of the $h_{11/2}$ proton shell [1] while odd-A, even-Z nuclei exhibit $\Delta I = 1$ rotation-like bands induced by the coupling of $h_{11/2}$ quasi-neutron holes to triaxial cores [2]. This $\nu h_{11/2}$ configuration is well known in odd-A Ce isotopes (Z =58) : it has been found first in ^{133,135}Ce [2] and then in ^{129,131}Ce together with a $g_{7/2}$ neutron-hole band [3]. Later on a backbend caused by $(h_{11/2})^2$ proton alignment was identified in both configurations [4]. Similar positive and negative parity bands have been recently observed in ^{125}Ce [5] but there was a complete lack of data concerning ^{127}Ce when we started our experiments to search for collective excitations in this nucleus and to complete the systematics of odd-A Ce isotopes.

2. IDENTIFICATION OF ^{127}Ce

The two bands shown in Figure 1 are assigned to ^{127}Ce from the following arguments : 1) By means of the Daresbury recoil separator, James et al [6] assigned three γ-rays to ^{127}Ce with energies of 126, 163 and 228 keV. 2) An

---

$ Talk given by J. Gizon

§ Permanent address : Institute of Nuclear Research of the Hungarian Academy of Sciences, H-4001 Debrecen, Hungary.

experiment has been performed at Grenoble with the He jet coupled to the on-line isotope separator. In the ^{92,94}Mo + ^{40}Ca reactions at 272 MeV, a 126 keV γ-ray shows up in the spectra selected for mass A = 127. This γ-line which coincides with the Ce X-rays is associated with the decay of ^{127}Pr. Its measured internal conversion coefficient $\alpha_K = 0.50 \pm 0.05$ indicates a magnetic dipole character [7].
3) γ-rays belonging to the observed bands coincide with Ce X-rays.

3. EXPERIMENTS

The first indications for collective bands in ^{127}Ce were obtained by bombarding 4 stacked self-supporting ^{93}Nb targets (1.26 mg.cm^{-2} total thickness) with a 155 MeV ^{37}Cl beam. Another experiment was performed by using the ^{94}Mo(^{37}Cl,p3n) reaction at 160 MeV with 2 stacked self-supporting enriched ^{94}Mo targets of a total thickness of 1.25 mg.cm^{-2}. The γ-rays were detected by means of the French facility "Château de cristal" installed at the M.P. Tandem accelerator at C.R.N. Strasbourg.

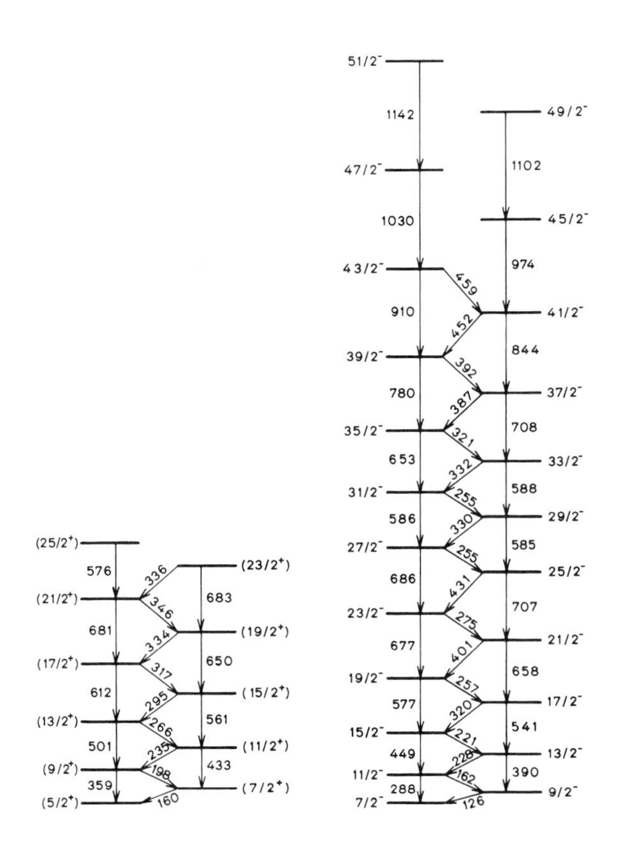

Fig. 1. Collective bands observed in ^{127}Ce

Eleven Compton-suppressed coaxial germanium detectors placed at 30°, 90° and 150° to the beam direction (six of them having efficiencies ranging from 70% to 80%) and one planar Ge detector placed at 30° have been used. 340 million γ-γ-fold-sum energy events have been stored onto Exabyte video cassettes.

4. OBSERVED BANDS AND $h_{11/2}$ PROTON ALIGNMENT

Two level sequences have been assigned to ^{127}Ce. The first one which is shown on the right hand side of Figure 1 is strongly populated. It is made of 23 levels connected by M1 and E2 transitions. From comparisons with heavier odd-A Ce isotopes (see Figure 4) we assign it as the $\nu h_{11/2}$ neutron-hole band based on a $I^\pi = 7/2^-$ state. The $h_{11/2}$ neutron-hole character of its low spin part agrees with the signature energy splitting appearing in the experimental Routhians (Figure 2). These Routhians as well as the aligned angular momenta (Figure 3) are plotted as a function of the rotational frequency by choosing a reference with Harris parameters $J_0 = 17.0\ \hbar^2\ MeV^{-1}$ and $J_1 = 25.8\ \hbar^4\ MeV^{-3}$.

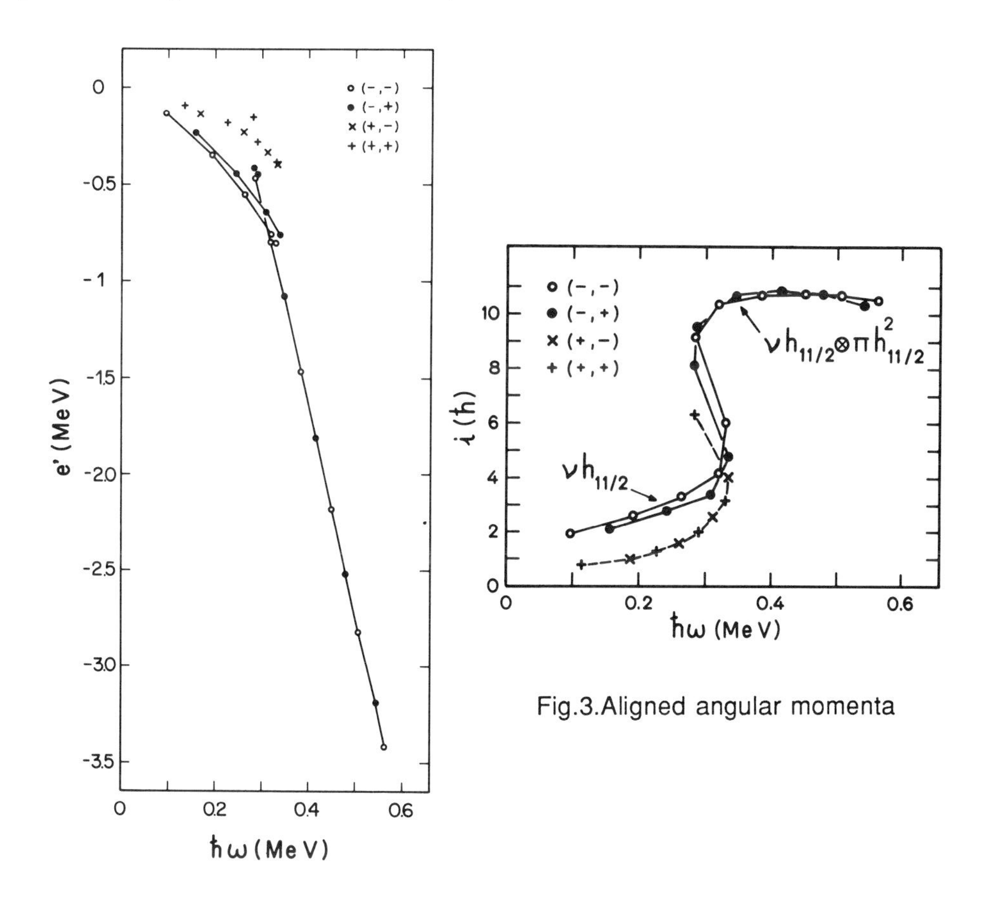

Fig.2. Experimental Routhians

Fig.3. Aligned angular momenta

Both signature branches of the negative parity band show a backbend at spins 25/2⁻ and 27/2⁻,respectively. The large aligned angular momenta (i ~ 9ħ)indicate that the aligned quasi-particles are on low Ω orbital i.e. that the backbend is due to $(h_{11/2})^2$ proton alignment. The band crossing frequencies at 0.312 and 0.305 MeV for the α = - 1/2 and α = +1/2 signature branches, respectively, also agree with the alignment of an $h_{11/2}$ proton pair. Indeed, they are very close to the values measured for ^{131}Ce[4] and ^{129}Ce [8]. The $\nu h_{11/2} \otimes \pi h_{11/2}{}^2$ nature of the three quasi-particle band is also confirmed by blocking arguments and by theoretical model calculations which predict that $h_{11/2}$ protons align before $h_{11/2}$ neutrons.

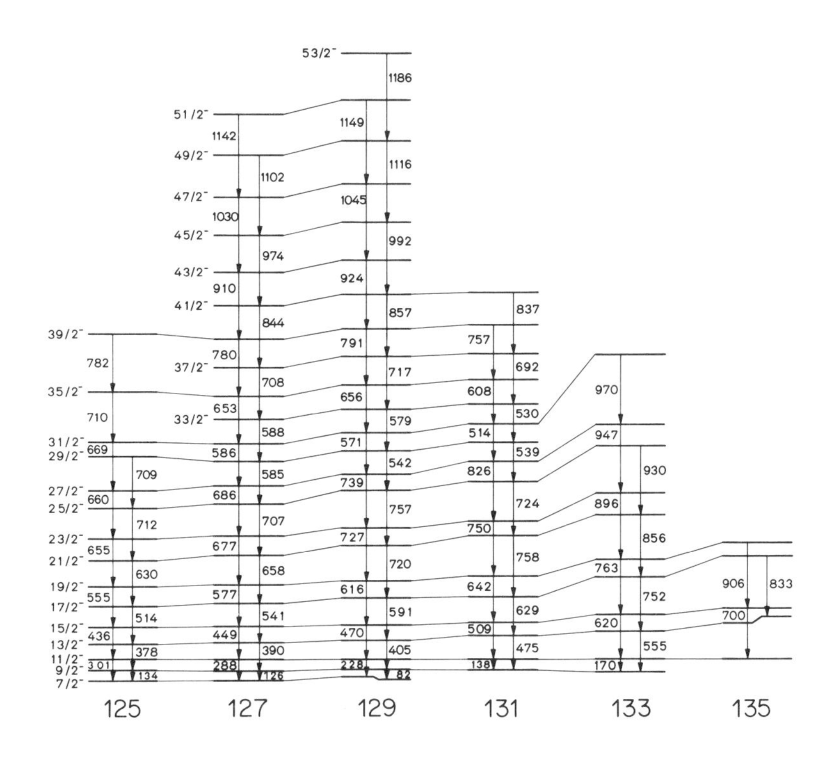

Fig. 4. Systematics of the $h_{11/2}$ neutron band in odd-A Ce nuclei

The neutron-deficient Ce nuclei are known to be triaxial with negative γ values in their $\nu h_{11/2}$ configuration (γ ~ - 20° near A = 130) [9] with a tendency towards prolate shape (γ ~ 0°) when going to lighter masses. The alignment of an $h_{11/2}$

proton pair drives the shape of A ~ 130 nuclei towards oblate shapes with $\gamma = +60°$ while the γ deformation of lighter ones is driven to slightly positive values. This shape polarization effect of aligned quasi-particles is well known and has been studied in detail for even-even Ce, Ba, Xe isotopes[10]. The effect of aligning two $h_{11/2}$ protons is visible in the experimental Routhians where the signature splitting which is of the order of 65 keV at $\hbar\omega = 0.300$ MeV disappears almost completely after the band crossing. This is confirmed by Total Routhian Surface calculations [11] based on a non-axially deformed Woods-Saxon potential and a monopole residual interaction [12] which show that the γ deformation changes from slightly negative ($\gamma = -0.9°$) before the band crossing to slightly positive ($\gamma = +1.4°$) after. The shape polarization effect of the $h_{11/2}$ proton alignment in ^{127}Ce is weak because of the rigidity of the core.

The $\Delta I = 1$ band represented on the left of Figure 1 is less populated and not well developed. From comparisons with the positive parity band identified in heavier odd-A neighbouring isotopes [3] [8], we assign $5/2^+$ to its base state. Its configuration is $g_{7/2}$ or $d_{5/2}$ neutron-hole as proposed for the first time in ^{129}Ba [9]. This configuration is associated with the absence of signature splitting. We expect a band crossing around $\hbar\omega = 0.30$ MeV as shown by the rapid increase of the aligned angular momentum in the $\alpha = +1/2$ signature branch. This approximate value for band crossing frequency is an indication of $h_{11/2}$ proton alignment but more levels are needed for a definite conclusion.

This set of data on ^{127}Ce added to other experimental information on odd-A Ce isotopes, namely A = 135 [2], 133 [2] [13], 131 [3] [4] [14], 129 [3] [8], 125 [5], gives the possibility of a systematic study of this negative parity band (Figure 4). The evolution of the β and γ deformations can be followed continuously with the neutron number varying from 77 to 67 i.e. when the neutron Fermi energy penetrates deeply into the $h_{11/2}$ shell. Collective excitations, in particular a $\nu h_{11/2} \otimes \pi h_{11/2}$ configuration, have been found recently in doubly-odd ^{126}La [15]. By using our data for ^{127}Ce with those already available for ^{125}Ba, $^{125,127}La$ it is now possible to calculate the residual neutron-proton interaction in the odd-odd nucleus ^{126}La.

5. CONCLUSION

Collective excitations have been identified in ^{127}Ce for the first time. A band observed up to spin 51/2 is proposed to be based on an $h_{11/2}$ neutron-hole configuration. The backbend which shows up in both signature branches is due to $(h_{11/2})^2$ proton alignment. Another level sequence is also found which exhibits no signature splitting. It could be generated from a $g_{7/2}$ or $d_{5/2}$ neutron-hole. In addition the data obtained for ^{127}Ce can be used together with those known for other odd-A isotopes in a systematic study of the variations of β and γ deformations as a function of neutron number. The knowledge of the $\nu h_{11/2}$ configuration in ^{127}Ce will be useful also to calculate the residual neutron-proton interaction in doubly-odd ^{126}La.

REFERENCES

[1] Stephens F S, Diamond R M, Leigh J R, Kammuri T and Nakai K 1972 Phys. Rev. Lett. 29 438
[2] Gizon J, Gizon A, Maier M R, Diamond R M and Stephens F S 1974 Nucl. Phys. A222 557
[3] Gizon J, Gizon A, Diamond R M and Stephens F S 1977 Nucl. Phys. A290 272
[4] Nolan P J, Todd D M , Smith P J, Love D J G, Twin P J, Andersen O, Garrett JD, Hagemann G B and Herskind B 1982 Phys. Lett. 108B 269
[5] Ying K L, Bishop P J, James A N, Kirwan A J, Morrisson T P, Nolan P J, Watson D C B, Connell K A, Love D J G, Nelson A H and Simpson J 1986/87 Nuclear Structure Appendix to Daresbury Annual Report p27
[6] James A N, Morrisson T P, Nolan P J, Watson D, Ying K L, Connel K A and Simpson J 1985/86 Nuclear Structure Appendix to Daresbury Annual Report p103
[7] Genevey J, Gizon A, Barci V, Barnéoud D, Blachot J, Gizon J, Liang C F , Paris P, Weiss B, Béraud R, Duffait R, Emsallem A, Meyer M and Redon N to be published
[8] Aryaeinejad R, Love D J G, Nelson A.H, Nolan P J, Smith P J, Todd D M and Twin P J 1984 J. Phys. G : Nucl Phys. 10 955
[9] Gizon J, Gizon A and Meyer-ter-Vehn J 1977 Nucl. Phys. A277 464
[10] Wyss R, Granderath A, Bengtsson R, von Brentano P, Dewald A, Gelberg A, Gizon A, Gizon J, Harissopoulos S, Johnson A, Lieberz W, Nazarewicz W, Nyberg J and Schiffer K to be published
[11] Wyss R, 1989 private communication
[12] Nazarewicz W, Dudek J, Bengtsson R, Bengtsson T and Ragnarsson I 1985 Nucl. Phys. A435 397
[13] Ma R, Paul E S, Beausang C W, Shi S, Xu N and Fossan D B 1987 Phys. Rev. C36 2322
[14] Luo Y X et al, private communication
[15] Nyakó B.M, Gizon J, Barnéoud D, Gizon A, Józsa M, Klamra W, Beck F A and Merdinger J C 1989 Zeit. Phys. A332 235

Inst. Phys. Conf. Ser. No 105
Paper presented at Int. Conf. on Spectroscopy of Heavy Nuclei, Crete, Greece, 1989

Lifetimes of high spin states in ^{168}Yb

J C Lisle, D Clarke, R Chapman, F Khazaie, and J N Mo
University of Manchester, England

H Hübel, W Schmitz and K Theine
University of Bonn, West Germany

J D Garrett, G B Hagemann, B Herskind and K Schiffer
Niels Bohr Institute, Copenhagen, Denmark

ABSTRACT: Lifetimes of high spin states in ^{168}Yb have been extracted using a Doppler Shift Attenuation Method and a loss of collectivity has been found at the highest spins. A possible explanation, involving a reduction of deformation due to changes of intrinsic structure, is discussed. In addition a Doppler Shift Attenuation Method has been applied to the ridge structure resulting from the E2 quasi-continuum in an $E_{\gamma,x} - E_{\gamma,y}$ matrix. Strong collective enhancement is found to persist for transitions a few MeV above the yrast line, but the transition rates of γ-rays with energies of approximately 1 MeV are modified by a gain in alignment, which is probably associated with the creation of two quasi-protons.

1. INTRODUCTION

Several investigations of E2 transition rates at high spins in rotational bands of rare earth nuclei have been carried out in recent years, for example[1–3].

In most cases a loss of collectivity has been observed at the highest spins. For nuclei with neutron numbers $N \approx 90$, a change of shape from prolate at low spins to oblate ($\gamma = 60°$) at high spins is predicted to occur, and the loss of collectivity is consistent with this change of shape[1]. However, for higher neutron numbers, the nuclear shape is predicted to remain rather stable even at the highest spins[4]. In spite of this, several nuclei in this class have been found to lose collectivity at the highest spins[2,3]. Possible explanations offered include an unpredicted change of shape and complex band crossing phenomena. In this paper we present new data on E2 transition rates in ^{168}Yb, a nucleus with neutron number N = 98 which is predicted to be exceptionally stable against deformation changes.

The results are discussed within a framework of changing intrinsic structure as a result of coriolis effects due to rotation.

For many years it has been recognised that $\gamma - \gamma$ coincidence data for a rotational nucleus, if presented in the form of a 2-dimensional spectrum, will exhibit a characteristic ridge-valley structure[5]. A valley is expected along the $E_{\gamma x} = E_{\gamma y}$ diagonal, which is bounded by parallel ridges resulting from the energy correlations of the γ-ray transitions associated with an ideal rotational sequence. At high spins rotational nuclei are known to decay through an E2 quasi-continuum associated with many parallel bands. For γ-ray energies in excess of 800 keV the major part of the intensity in the ridge structure is associated with this quasi-continuum.

In this paper we describe how the Doppler shift attenuation method may be applied to the 2-dimensional ridge structure. The technique provides a method of exploring transition rates in the quasi-continuum. In particular it is shown that the technique provides a probe for monitoring changes in transition rate as the decay proceeds from a high spin, high excitation entry point to states near the yrast line. It will be seen that it is possible to infer changes of intrinsic structure from the measurements.

2. EXPERIMENTAL METHOD AND DATA ANALYSIS

High spin states in ^{168}Yb have been excited using the reaction ^{124}Sn(^{48}Ca,4n)^{168}Yb. The 201 MeV ^{48}Ca beam was obtained from the Nuclear Structure Facility at the Daresbury Laboratory. Data were taken with a 1 mg cm^{-2} ^{124}Sn target on a 15 mg cm^{-2} gold backing and with a thin unbacked target. The γ-rays emitted were detected with the ESSA30 European multi-detector array, which consisted of thirty escape-suppressed germanium detectors. The detectors were positional at six different angles with respect to the beam diretion, namely 37°, 63°, 79°, 101°, 117° and 143°. Approximately 5×10^8 $\gamma - \gamma$ coincidence events were recorded with the backed targets and about 1×10^8 events of fully Doppler shifted data with the thin targets.

In the analysis of the discrete line lifetimes the backed target data was sorted into four 2-dimensional matrices corresponding to γ-rays detected at 37°, 67°, 117° and 143° respectively in coincidence with γ-rays at all other angles. The gains of the 30 detectors were matched for stopped transitions. Coincidence spectra were obtained for the yrast (+,0) band and the most intense sideband (−,1) of ^{168}Yb at each of the four angles by placing gates on the $22^+ \rightarrow 20^+$ and $20^+ \rightarrow 18^+$ transitions and the $23^- \rightarrow 21^-$ and $21^- \rightarrow 19^-$ transitions respectively.

The Doppler shifted lineshapes of high spin transitions in both bands were fitted using a set of codes developed by Bacelar et al[3]. Initially a Monte Carlo method was used to determine the magnitude and direction of the velocities of the recoiling ions as a function of time as they slowed down in the target and backing materials; 10000 histories were followed. The electronic stopping power data were taken from the tabulation of Northcliffe and Schilling[6] with the corrections, proposed by Chu and Ziegler[7], applied. The method of Lindhart, Scharff and Schiott[8] was used to determine the nuclear stopping. The next stage was to construct a 2-dimensional array of the distribution of velocities with respect to the detector as a function of slowing down time for each of the four detector angles. Finally the experimental lineshapes were fitted using these velocity distributions and a model to simulate the γ-ray decay of the nucleus from its formation by the fusion–evaporation reaction into the discrete line bands. The decay model extended the known rotational states in the band under consideration to higher spin by extrapolation using the known moment of inertia. The purpose of the artificially introduced transitions is to simulate a feeding time and their transition rates were adjusted to give the best fit to the highest observed lineshape.

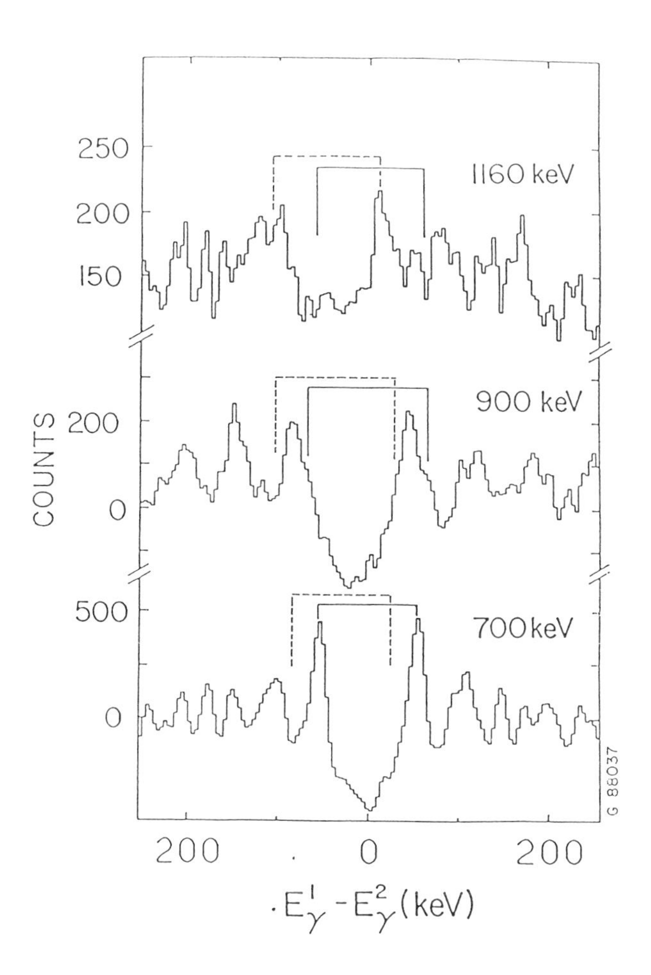

Figure 1: Cuts perpendicular to the diagonal in the forward–backward (37°, 143°) spectrum at cut energies of 700keV and 1160 keV. The position of the fully shifted and stopped ridges are marked by dashed and full lines, respectively.

The lineshapes of the lower transitions were then fitted in order starting, from the top, by introducing individual state lifetimes, which were varied. Side–feeding at each step in the cascade was simulated by considering a further band with the same moment of inertia as the band being investigated. The decay rates of each of these sidebands were determined by assigning individual transition quadrupole moments to them, which were treated as parameters in the fits. Side–feeding intensities were derived from the observed discrete line intensities up the band, taken from the unbacked data. In essence the calculation involved solving coupled Bateman differential equations for the cascade of transitions from the initial starting point. The velocity distribution of the recoiling ions, as a function of the time of the decay of transition being fitted, was folded in.

For the continuum investigation the backed data were sorted into three 2–dimensional matrices corresponding to the angle combinations (36°, 143°), (63°, 117°) and (79°, 101°). In addition a fully Doppler shifted matrix was prepared from the unbacked data. In order to enhance the correlated structures in these matrices, relative to the smooth background, a background subtraction procedure ("COR" treatment [5]) was employed. Spectra were then projected from the 2–dimensional matrices perpendicular to the diagonals such that $0.5(E_{\gamma x} + E_{\gamma y})$ is constant for a range of values of $0.5(E_{\gamma x} + E_{\gamma y})$. Typical projections taken from the (37°, 143°) backed matrix are shown in fig. 1. It is seen that at the higher energies (1160 keV) the ridge structure lies close to the fully shifted positions, as determined from the unbacked data. For lower values of $0.5(E_{\gamma x} + E_{\gamma y})$ (700 keV) the ridges are observed near to the fully stopped positions.

3. DISCUSSION OF DISCRETE LINE DATA

As discussed in the previous section fits of the individual lineshapes involved varying both the state lifetime τ_i and the side–feeding transition probability lifetime parameter Q_i. In general it was not possible to fully decouple these two variables, and it was necessary to constrain the side–feeding parameter to vary in a smooth way with transition number. Examples of the fits obtained for high spin states in the (–,1) sideband of ^{168}Yb are shown in fig. 2.

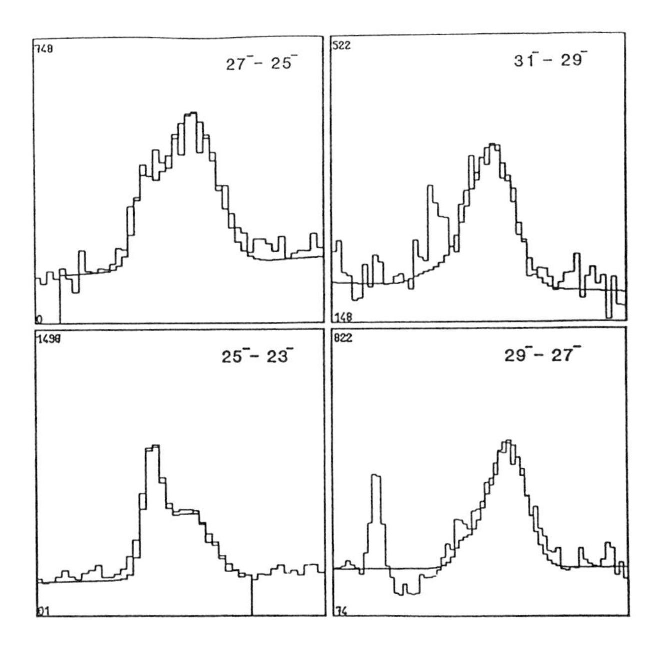

Figure 2: Lineshapes for transitions in the negative parity, signature 1 band of ^{168}Yb.

Light lines indicate experimental data and heavy lines indicatge the best calculated fits.

From the measured lifetimes τ_i, values of B(E2) expressed in single particle units, can be extracted using the relation :-

$$B(E2) = [0.33 < I_i \mid I_f >^2 \ \tau_i \ E_\gamma^5]^{-1}$$

E_γ is measured in MeV and τ_i in picoseconds. Values of the transition quadrupole moment Q_t are then determined from the measured B(E2) values using the standard rotational model prescription. In fig. 3 Q_t is plotted as a function of spin I for states between 26^+ and 34^+ in the yrast band and for states between 25^- and 37^- in the (–,1) sideband. The values of Q_t are presented as a ratio of the rotational model predictions, based on the B(E2) for the $2^+ \rightarrow 0^+$ transition determined by Coulomb excitation. It is apparent that, while Q_t has essentially the full rotational value at lower spins, there is a marked loss of collectivity at the highest observed spins in both bands. A similar loss of collectivity has been observed by Bacelar et al. [3] for the neighbouring nucleus ^{166}Yb.

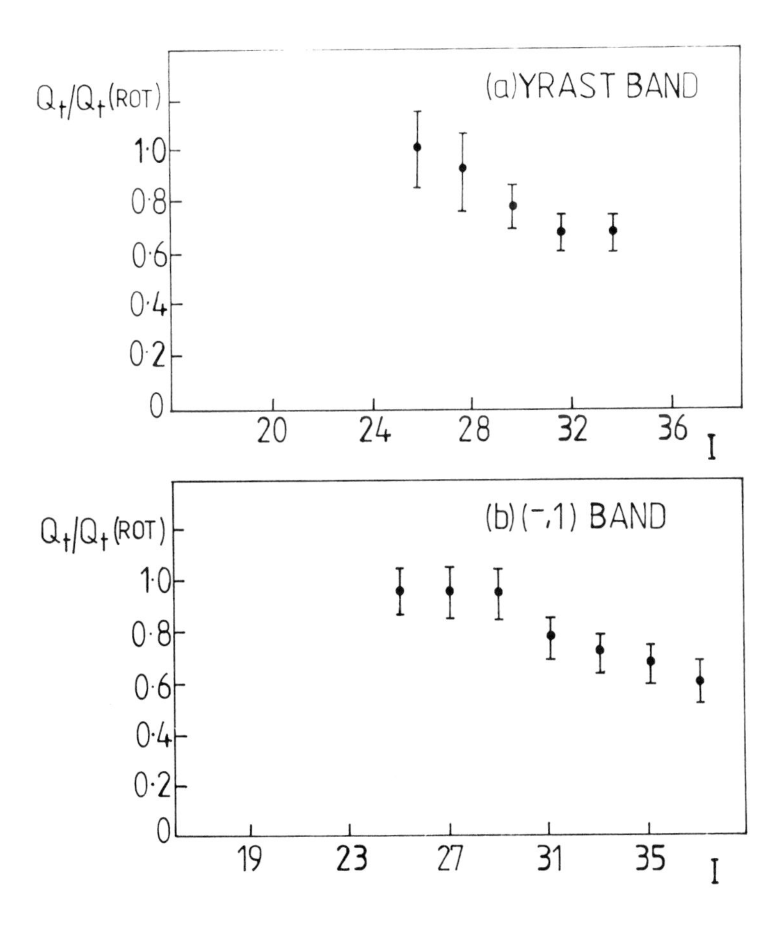

Figure 3: Plots of $Qt/Qt(rot)$ as a function of spin I for (a) the yrast band and (b) the (–,1) band of ^{168}Yb.

The simplest explanations for the loss of collectivity at high spin involve changes of nuclear shape. The observed fall in Q_t implies a reduction of approximately 30% in the deformation ε_2 or a change in the triaxiality parameter γ from 0° (prolate) at low spin to approximately +20° at the highest observed spins. Cranked Nilsson–Strutinski calculations without pairing by Bengtson and Ragnarsson[4] predict that ε_2 reduces from 0.275 at lower spins to 0.245 at $I \sim 40\hbar$. However these calculations suggest that the nucleus remains nearly prolate ($\gamma = 0°$) at the highest spins. Recently calculations, which include pairing have been performed by Garrett[9]. Both calculations are unable to reproduce the loss of collectivity in terms of shape changes, although the calculations which include pairing predict a slightly bigger reduction.

Garrett[9] has proposed an explanation for the reduction in ε_2 at the higher spins. Essentially it can be understood as an effect of the deoccupation of high j, low Ω orbitals in which the neutrons or protons are moving in the opposite direction to the rotation. The energies of these unfavoured states with a large signature splitting rise with increasing rotational frequency and pass above the Fermi level. This explanation involves changes of intrinsic structure within a rotational band. It is, therefore, no longer appropriate to use the rotational model arguments for calculating B(E2)'s, which depend purely on shape parameters following the model's assumption of unchanging intrinsic structure. A proper theoretical treatment must involve the evaluation of the E2 matrix element with realistic wavefunctions of well defined spin. Unfortunately such wavefunctions are not readily available from the current generation of cranked shell model calculations, nor have other shell model approaches been successfully applied to this region of the periodic table.

4. DISCUSSION OF CONTINUUM DATA

It is well known that the ridge structure observed in 2–dimensional γ–ray spectra is associated with coincidences within a rotational band of well defined moment of inertia[5]. In particular the first ridge observed on either side of the diagonal from the bottom left to the top right of a 2–dimensional spectrum is produced by coincidences between successive transitions in a band. Generally for higher transition energies ($\geq$ 800 keV), corresponding to higher spins, the intensity of the ridge structure is weak. Typically it is found that only 10–20% of the total intensity contributes to the ridge structure[10]. It has been established that for rotational band lying more than 1 MeV or so above the yrast line, the rotational energy correlations are partially damped and do not contribute to the ridge structure[11]. It is, therefore, to be expected that only near yrast bands will form a ridge structure. For ^{168}Yb the intensities of known discrete lines above 800 keV are found to account for 30–50% of the total intensity of the ridge structure. The remaining 50–70% is presumably associated with the quasi continuum produced by parallel bands at relatively small excitations above the yrast line. Thus it is to be expected that Doppler shift measurements will throw light on the decay of these low lying quasi–continuum bands and their mode of feeding.

An effective dynamical moment of inertia for the quasi–continuum can be determined from the separation of the ridges on either side of the diagonal. In fig. 4a $J^{(2)}$ is plotted as a function of gamma ray energy or spin. Corresponding values of $J^{(2)}$ taken from discrete line data for selected bands in ^{167,168}Yb are shown for comparison. The

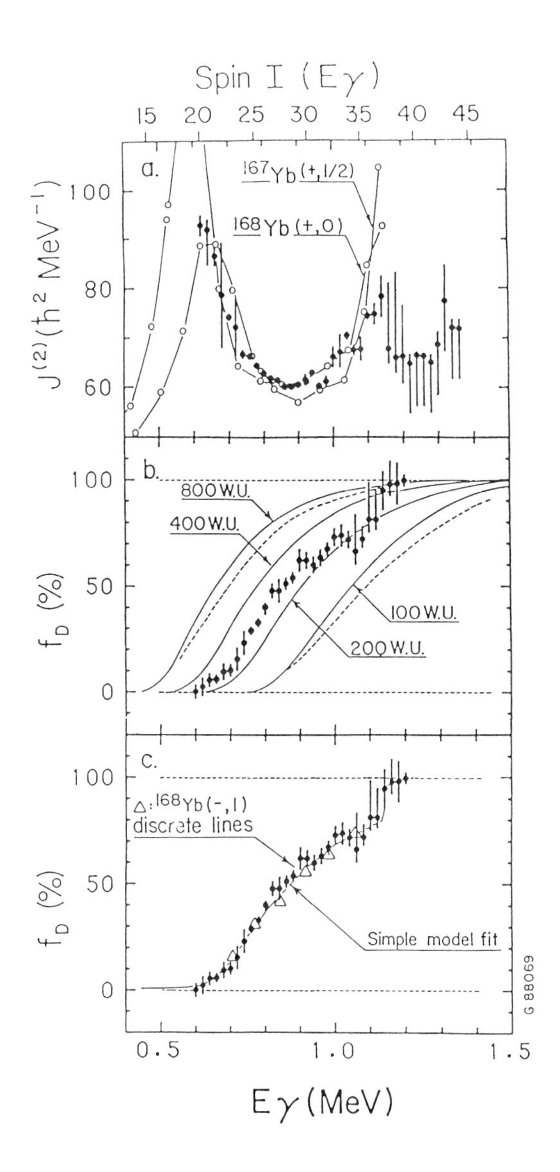

Figure 4: Part (a) is a plot of inertia $\mathcal{J}^{(2)}$ versus gamma ray energy as extracted from the mean separation of the ridges. Shown with dots and full lines are the $\mathcal{J}^{(2)}$ moments of inertia as observed from some discrete bands in ^{167}Yb and ^{168}Yb. In parts (b) and (c) of the figure the fraction f_D of the full Doppler shift of the first ridges, as observed from the main ridge components in the cuts perpendicular to the diagonal, is shown versus gamma ray energy, $0.5(E_{\gamma1} + E_{\gamma2})$. In (b) the lines indicate the shift predicted by the simple model without alignment for different B(E2) values of 100, 200, 500 and 800 WU. The fit obtained by including $10\hbar$ of alignment with B(E2) $=$ 300 WU is shown in (c).

rise in $J^{(20}$ at $E_\gamma = 1.1$ MeV, seen in both the discrete line and quasi-continuum data, is of particular interest. This corresponds to the rotational frequency (0.5 - 0.6 MeV) at which $h_{9/2}$ and $i_{13/3}$ quasi-proton alignment is predicted to occur by Bengtson and Ragnarsson[4].

From cross-diagonal projections of 2-dimensional spectra, as described in section 2, the experimental Doppler shifts of the centroids of the ridges as a fraction of the maximum Doppler shift (F_D) were determined as a function of γ-ray energy. The results are presented in fig. 4b. A simple cascade model was developed to describe these results. It assumed that the rotational cascades started from an entry spin of $50\hbar$ and that decay proceeded down an ideal rotational band with dynamic moment of inertia $J^{(2)} = 65\hbar^2$ MeV^{-1}. In calculating the fractional Doppler shift f_D the code of Bacelar et al.[3] described in section 2 was used to determine the velocity of the recoiling ^{168}Yb ions as a function of time in the stopping material. The full lines in fig. 4b show examples of the calculated Doppler shift for a number of B(E2) values between 100 WU and 800 WU. The experimental values of f_D below $E_\gamma = 1$ MeV are quite well fitted by a B(E2) value of 300 WU. This is to be compared with 400 WU for the Coulomb excited $2^+ \rightarrow 0^+$ transition. However, above approximately 1 MeV the calculation seriously fails to reproduce the experimental data.

The fact that the experimental values of f_D show a marked discontinuity at $E_\gamma \approx 1.1$ MeV suggests that the gain in alignment associated with the decoupling of protons may play an important role. The major result of an increase in alignment is a reduction in γ-ray energy which, because of E_γ^5 weighting of the E2 transition probability, is dominant irrespective of the exact value of B(E2). The relation between γ-ray energy and alignment i is given by

$$E_\gamma(I \rightarrow I-2) = \frac{2(I-i)}{J^{(2)}}$$

A further calculation has been performed in which an alignment gain of $10\hbar$ is introduced over 2–3 transitions at spin $40\hbar$; in all other respects the calculation is identical to the one described in the previous paragraph. The results of this calculation with B(E2) = 300 WU are shown in fig. 4c; it is seen to give a remarkably good description of the experimental data.

Two further pieces of experimental information support the introduction of an alignment gain at $E_\gamma \sim 1$ MeV. Firstly the diagonal cut of the 2-dimensional spectrum taken at 900 keV (see fig. 1) exhibits two components in the ridge structure, a strong component with a fractional Doppler shift of 50% and a weak slow component. Secondly the high energy cut-off of the E2 continuum occurs at $E_\gamma \sim 1300$ keV. This is approximately what is expected for an entry spin of $50\hbar$ if $10\hbar$ of additional alignment is introduced, but is approximately 200keV too low without the alignment.

5. CONCLUSIONS

A reduction in the collective E2 transition rate is found in the yrast and odd parity, signature 1 bands of ^{168}Yb, which is quite analogous to the situation previously observed in ^{166}Yb. This implies a reduction in the deformation parameter ε_2. It is suggested that this may be due to changes in the intrinsic structure of a band resulting from Coriolis effects associated with rotation. Large signature splittings are expected for high j, low Ω Nilsson orbitals, such that the unfavoured signature states become deoccupied at high rotational frequencies. The reduction in deformation is a result of particles (both neutrons and protons) moving from these strongly deformation driving orbitals into other orbitals which favour smaller deformations. Although the reduction in deformation is only expected to be about 10%, larger reductions in B(E2) are expected because progressive changes in intrinsic structure lead to a poorer overlap between initial and final states in a transition. A quantitative analysis of the experimental data can only be made when realistic wave functions describing the intrinsic structure become available.

Studies of the fractional Doppler shifts of the ridge structure in 2-dimensional spectra lead to a better understanding of the collectivity at very high spins and high excitations. The present results are consistent with an average B(E2), measured in single particle units, of 300 WU. Structure observed in the plot of fractional Dopper shift against γ-ray energy is apparently associated with quasi-proton alignment. The results of these ridge studies and their interpretation are discussed in greater detail in a recent paper by Schiffer et al.[12].

References

1. P.O. Tjøm, R.M. Diamond, J.C. Bacelar, E.M. Beck, M.A. Delaplanque, J.E. Draper and F.S. Stephens,
   Phys. Rev. Lett. **55** (1985) 2405

2. M.N. Rao, N.R. Johnson, F.K. McGowan, I.Y. Lee, C. Baktash, M. Oshima, J.W. McConnel, J.C. Wells, A. Larabee, L.L. Riedinger, R. Bengtsson, Z. Xing, Y.S. Chen, P.B. Semmes and G.A. Leander,
   Phys. Rev. Lett. **57** (1986) 667

3. J.C. Bacelar, A. Holm, R.M. Diamond, E.M. Beck, M.A. Delaplanque, J. Draper, B. Herskind and F. Stephens,
   Phys. Rev. Lett. **57** (1986) 3019

4. T. Bengtson and I. Ragnarsson,
   Nucl. Phys. **A436** (1985) 14

5. O. Anderson, J.D. Garrett, G.B. Hagemann, B. Herskind, D.L. Hullis and L.L. Riedinger,
   Phys. Rev. Lett. **43** (1979) 687

6. L.C. Northcliffe and R.F. Schilling,
   Nucl. Data. Tables **7** (1970) 233

7. F.J. Ziegler and W.K. Chu,
   Nucl. Data Tables **13** (1974) 463

8. J. Lindhart, M. Scharff and H.E. Schiott,
K. Dav. Vid. Selsk. Mat.-Fys. Medd. **33** (1963) 14

9. J.D. Garrett, Private communication (1989)

10. D.J.G. Love, A.H. Nelson, P.J. Nolan and P.J. Twin,
Phys. Rev. Lett. **54** (1985) 1361

11. J.C. Bacelar, G.B. Hagemann, B. Herskind, B. Lauritzen, A. Holm, J.C. Lisle and P.O. Tjøm,
Phys. Rev. Lett. **55** (1985) 1858

12. K. Schiffer, G.B. Hagemann, B. Herskind, K. Theine, W. Schmitz, H. Hübel,
R. Chapman, D. Clarke, F. Khazaie, J.C. Lisle and J.N. Mo,
Phys. Lett. B **219** (1989) 52

Inst. Phys. Conf. Ser. No 105
Paper presented at Int. Conf. on Spectroscopy of Heavy Nuclei, Crete, Greece, 1989

Observation of superdeformation in ^{191}Hg

E.F. Moore[a], R.V.F. Janssens[a], D. Ye[b], R.R. Chasman[a], I. Ahmad[a], K.B. Beard[b], Ph. Benet[c], J.A. Cizewski[d], M.W. Drigert[e], Z.W. Grabowski[c], T.L. Khoo[a], F.L.H. Wolfs[a]

[a]: Argonne National Laboratory, Argonne IL 60439, USA.
[b]: University of Notre Dame, Notre Dame IN 46556, USA.
[c]: Purdue University, West Lafayette, IN 47907, USA.
[d]: Rutgers University, New Brunswick, NJ 08903, USA.
[e]: Idaho National Engineering Laboratory, EG&G Idaho Inc., Idaho Falls, ID 83415, USA.

Abstract: The first observation of superdeformation in the A≃190 mass region is reported. A rotational band of 12 transitions with an average energy spacing of 37 keV, an average moment of inertia of 110 $\hbar^2$ MeV^{-1}, and an average quadrupole moment of 18 ± 3 eb has been observed in ^{191}Hg. These results are in excellent agreement with a calculation that predicts an ellipsoidal axis ratio of 1.65:1 for the superdeformed shape in this nucleus. Evidence for another discrete superdeformed band and superdeformed structures in the quasi-continuum was also found in the data.

Superdeformation was first proposed (Strutinsky 1967) some twenty years ago to explain the fission isomers observed in some actinide nuclei (Polikanov et al. 1962). Fission isomers are found in nuclei trapped in a metastable minimum associated with very elongated ellipsiodal shapes (axis ratio of roughly 2:1). The interest in the mechanisms responsible for these exotic shapes (i.e., mainly shell effects) has increased enormously with the discovery of a superdeformed band of nineteen discrete lines in ^{152}Dy (Twin et al. 1986) and in several neighboring nuclei (Hass et al. 1988, Deleplanque et al. 1988, Rathke et al. 1988, Fallon et al. 1988). Nuclei with large deformations have been reported (Kirwan et al. 1987) at high spin in the A≃135 mass region and fragmentary evidence exists around A≃180 and 105 (Burde et al. 1988).

In this contribution we report the discovery of a rotational band of twelve transitions in the nucleus ^{191}Hg having properties consistent with superdeformation. This study was motivated by the results of cranked Strutinsky calculations by Chasman (Chasman 1989), who found that deep secondary minima in the total energy surface exist for many nuclei in the region A≃186-205. The axis ratios in these nuclei are calculated to be ~1.65:1. These minima were found to become yrast at spins above 30 $\hbar$ in some cases and were shown to persist even to the lowest spins. Superdeformed shapes with axis ratios >1.45:1 have also been obtained at I=0 for some nuclei in this region in the calculations of (Girod et al. 1988).

The experiment was carried out at the Argonne superconducting linear accelerator ATLAS using the Argonne-Notre Dame BGO γ-ray facility which consists of 50 hexagonal BGO elements surrounded by 12 Compton suppressed Ge spectrometers (CSG's). The states in ^{191}Hg were populated by the reaction ^{160}Gd(^{36}S,5n) at a beam energy of 172 MeV. The target consisted of two isotopically enriched 500 μg/cm^2 self-supporting foils stacked together. Under the experimental conditions the compound nucleus is formed with an excitation energy of 71 MeV and a maximum angular momentum in excess of 50 $\hbar$. In the analysis, only events where at least 14 of the array detectors fired in coincidence with the CSG's were considered. The final γ-γ coincidence matrix contained 95 x 10^6 events, of which 60% were in the 5n channel. The remaining events belong mainly to the 6n channel; no measurable yield was found for 4n evaporation.

A band of twelve coincident transitions, extending from 350 to 754 keV with an average energy spacing of 37 keV, corresponding to an average dynamic moment of inertia of 110 $\hbar^2$ MeV^{-1}, was observed and is shown in Fig. 1. The stretched E2 character of the γ-rays was established

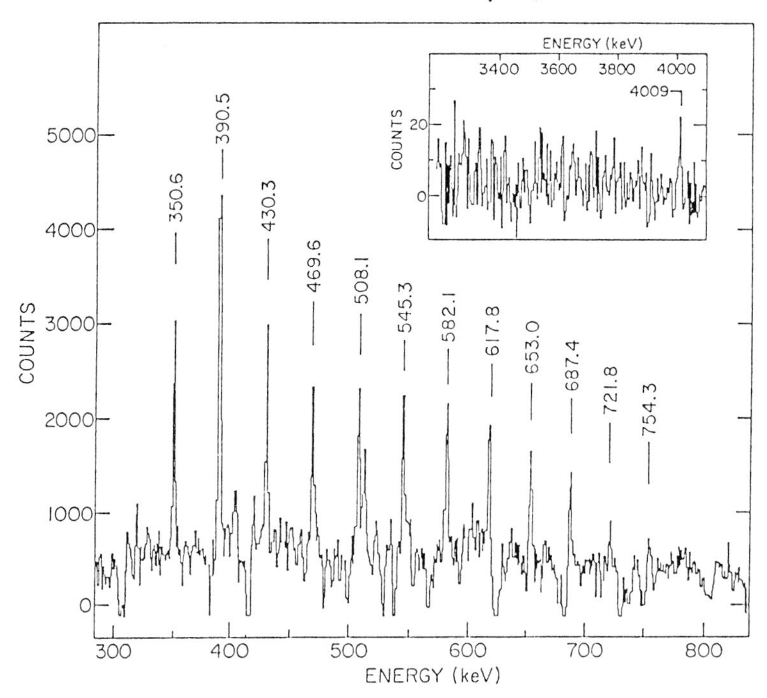

Fig. 1. γ-ray spectrum in ^{191}Hg obtained by summing coincidence gates on selected transitions (351, 471, 508, 545, 582, and 653 keV). The γ-ray at 514 keV is an identified contaminant (seen only in the 508-keV gate). Inset: The high-energy end portion of this spectrum, with the 4009-keV line discussed in the text.

from the angular correlations. Under the multiplicity condition described above, the flow through the SD band represents 2% of the ^{191}Hg intensity. One of the transitions in the band has the same energy (390.5 keV) as the previously assigned (Hübel et al. 1986) $17/2^+$ - $13/2^+$ transition in ^{191}Hg. The intensity of this transition is about 25% greater than any other transition in the band and it is proposed that the excess intensity arises from the decay of the band, at least partly, through the $17/2^+$ state. This result supports the assignment of the band in ^{191}Hg. The fold and sum-energy distributions measured in coincidence with transitions in the band peak at values only slightly larger than those of known γ-rays in ^{191}Hg: this result is also consistent with an assignment in ^{191}Hg.

Figure 2 presents the intensity pattern for the transitions in the new band derived from the analysis of the coincidence gates. The relative

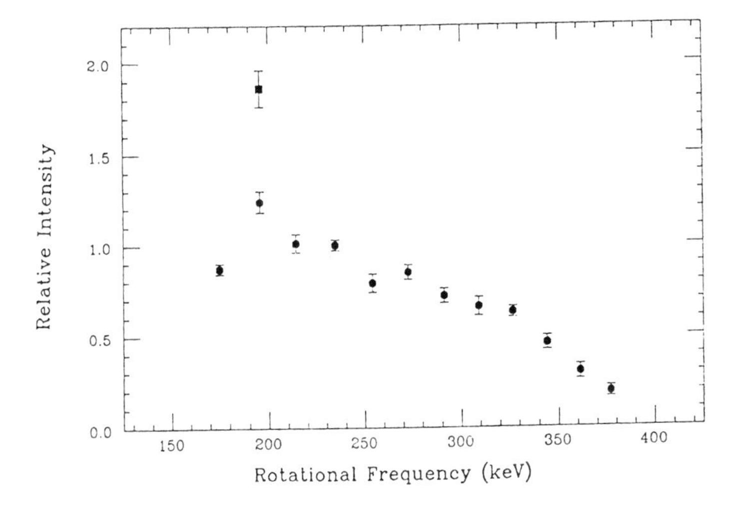

Fig. 2. Relative intensity of the γ-rays in the new band of ^{191}Hg as measured in the thin-target experiment. The intensity of the 390-keV line in the thick-target measurement is also given (■).

intensity is seen to decrease with increasing γ-ray energy. At the bottom of the cascade, the intensity remains essentially constant over the last 3-4 transitions, with the exception of the excess strength in the 390 keV γ-ray as discussed above. The decay out of the band is abrupt; no coincident γ-ray having an energy less than 350 keV and an intensity of 5% of the 350 keV line was observed. Thus, the main decay out of the band towards the yrast line occurs from the lowest transition observed in the cascade.

As discussed above, the average dynamic moment of inertia $\mathfrak{J}^{(2)}$ for the new band is consistent with a superdeformed shape. However, since single particle alignment coupled with a strong interaction can result in large apparent values of the moment of inertia, a more direct determination of the nuclear shape is required. A second experiment was performed to measure the transition quadrupole moment in the band using the Doppler shift attenuation method (DSAM). In this experiment, the target consisted of a 1.0 mg/cm^2 ^{160}Gd foil on which 14 mg/cm^2 of Au was evaporated; all other experimental conditions were identical to those outlined above.

The same procedure and approximations made in lifetime studies of the superdeformed bands in ^{152}Dy, ^{149}Gd, and ^{132}Ce (Twin et al. 1986, Hass et al. 1988, Kirwan et al. 1987) were used here. Figure 3 presents the measured fraction of the full Doppler shift F for the highest

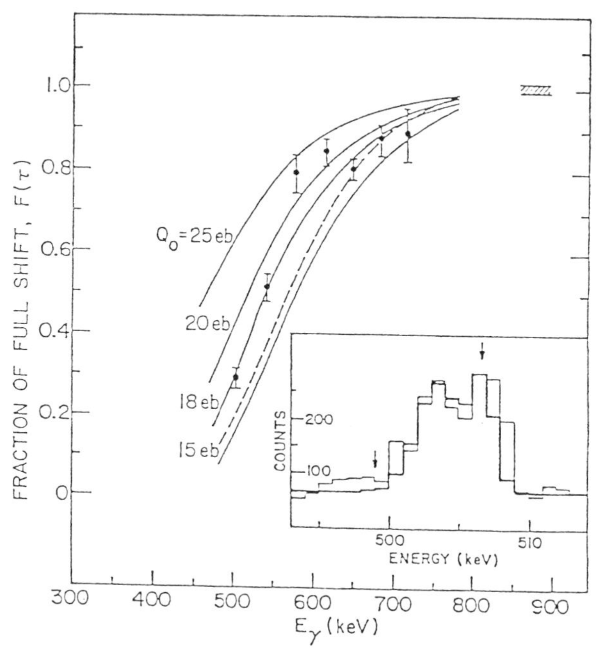

Fig. 3. Measured fraction of the full Doppler shift for all transitions in the band with $508 \lesssim E_\gamma \lesssim 721$ keV. Calculated shifts are given for various quadrupole moments; solid curves include side feeding (see text), dashed curve assumes infinitely fast side feeding with Q_0=15 eb. The shaded area shows the spread in full shift due to the slowing down of the beam across the target. Inset: The line shape of the 508-keV transition measured at 146^0; experiment (thin line) and calculation with Q_0=18 eb (thick line); arrows indicate energies corresponding to full and zero Doppler shift.

transitions in the band. The lowest transitions were fully stopped or contaminated by other γ-rays in the spectrum. The calculated curves in Fig. 3 represent values of F for various quadrupole moment Q_0 values under the assumption of a constant deformation. The data are consistent with Q_0=18±3 eb, where the errors include uncertainties in the slowing down process and in the side-feeding intensities. The full curves in Fig. 3 take into account side feeding; the side-feeding lifetime into a state of spin I was assumed to be equal to the lifetime of the I+2 state in the band. Similar results were obtained when a constant side-feeding time of 30 fs was assumed - the difference between the highest data points and the calculated full shift for recoils formed in the center of the target suggests a delay in the feeding of these states of this order. In any case, the measured shifts clearly indicate very fast transitions and impose a lower bound on the quadrupole moment of $Q_0 \sim 15$ eb [obtained by comparison to calculations with infinitely fast side feeding-see dashed curve in Fig. 3]. The inset in Fig. 3 shows the broadened line shape for the 508 keV transition which compares nicely with the calculated line shape obtained when a Q_0 value of 18 eb is used. Using the relation between Q_0 and β given in (Löbner et al. 1970), the measured value of Q_0 implies a deformation of β=0.55, in excellent agreement with the calculated value of (Chasman 1989). We conclude that this band is based on a superdeformed configuration in ^{191}Hg.

One question of considerable importance is the location in energy of the SD band with respect to the yrast line and its decay towards the ground state. The following information was obtained in this study: (i) An excess intensity was observed in the 390 keV γ-ray [see Fig. 2]. Since this excess is only 25% and not 100% in the thin target experiment, one must conclude that either part of the decay proceeds directly to the ground state via unobserved transitions, or the lowest state in the new band has a lifetime long enough to allow the ^{191}Hg nuclei to recoil out of focus of the CSG's before the full intensity of the decay is observed. (ii) The latter explanation is favored by the thick-target measurement in which the excess intensity in the 390 keV γ-ray is 86% [square in Fig. 2]. (iii) The spectra from individual coincidence gates in the thin target measurement contain evidence for a 4009 keV γ-ray (inset in Fig. 1) which may be a link between the SD band and the $17/2^+$ yrast state. Its intensity relative to that of the SD band (16±9)% is consistent with this assignment, but the uncertainties are large due to the reduced efficiency of the CSG's at such high energies. The thick target data yielded no new information on the 4009 keV γ-ray primarily due to lower statistics. Hence the placement of this line is still tentative. Clearly, more experimental work is necessary to study the decay out of the SD band in detail and it is not possible to assign definite spins to the members of this band at this time.

The coincidence data contain evidence for a second band of transitions with energy spacings similar to those in ^{191}Hg, but with an intensity smaller by a factor of ~2.5. No firm placement could be made, although there is tentative evidence for an assignment in ^{190}Hg (Ye et al.). Figure 4 presents the dynamic moment of inertia $\mathfrak{J}^{(2)}$ as a function of rotational frequency $\hbar\omega$ for the band in ^{191}Hg and for this weaker band. Both bands display very similar behavior, with a gradual increase in the $\mathfrak{J}^{(2)}$ value with increasing $\hbar\omega$. The solid line in Fig. 4 shows the calculated[10] value of $\mathfrak{J}^{(2)}$. Although the calculations do not reproduce the gradual increase in $\mathfrak{J}^{(2)}$, the average value of 110 $\hbar^2$ MeV^{-1} is in excellent agreement with the data.

The possibility of populating excited superdeformed bands which appear in the quasi-continuum (non-discrete superdeformation) was also investigated. An E_γ-E_γ correlation matrix was constructed from the thin

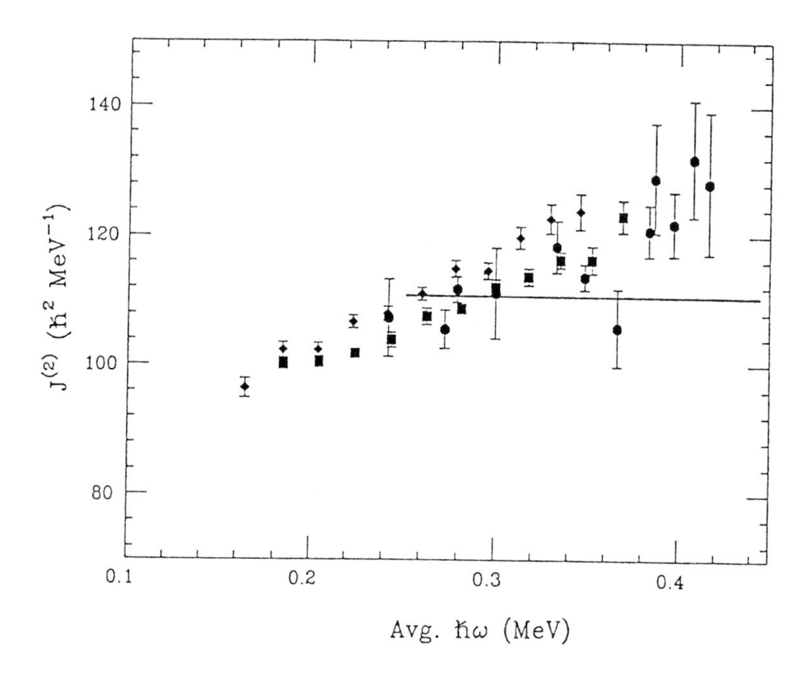

Fig. 4. Dynamic moment of inertia $\mathcal{J}^{(2)}=4\hbar^2/\Delta E_\gamma$ for the superdeformed band in ^{191}Hg (■), the weaker discrete band discussed in text (◆), and the continuum SD γ-rays (•). The line is the result of the calculations of (Chasman 1989) and corresponds to a β value of 0.55.

target data. Uncorrelated events were subtracted out using the method proposed in (Anderson et al. 1979), modified according to (De Voight et al. 1981) to account for the large photoefficiency of the CSG's. The symmetrized matrix was sliced along the x-axis and the resulting spectra were offset in such a way that channel zero corresponds to the $E_{\gamma 1}=E_{\gamma 2}$ main diagonal. The location of the first ridge in the spectra obtained in this manner will be at the position corresponding to the separation of the γ-ray energies in the cascade. The advantage of slicing in this manner is that it is a simple matter to set gates at energies corresponding to the known lines in the discrete SD band(s) and in between them, as well as to avoid gating on strong contaminant lines. Gates were set on the energies in between the known discrete SD lines in 2-3 keV steps over a wide energy range. The slices revealed intensity in the first ridge for energies of $480 \lesssim E_\gamma \lesssim 840$ keV (discrete SD lines are found for $30 \lesssim E_\gamma \lesssim 750$ keV). No statistically significant intensity in the second ridge was observed in the slices gated on the continuum regions. The absence of a second ridge in the continuum gates implies that the non-discrete SD cascades are quite short, consisting on average of two transitions. From the position of the first ridge in the gated slices, the dynamic moment of inertia $\mathcal{J}^{(2)}$ was determined. Figure 4 presents the $\mathcal{J}^{(2)}$ values for the non-discrete superdeformed bands as a function of $\hbar\omega$. Although the error bars are large, the overall trend is remarkably similar to that seen in the discrete SD bands. In particular, the smooth increase of $\mathfrak{J}^{(2)}$ as a function of $\hbar\omega$ seems to continue beyond the frequency corresponding to the last discrete SD line.

To investigate the mechanisms responsible for the population of superdeformed bands in this mass region, an excitation function is currently being performed at Argonne. The beam energy for the second step in the excitation function was 167 MeV, the other experimental conditions were identical to those described above for the thin target run at 172 MeV. Figure 5 shows the relative population of states in the

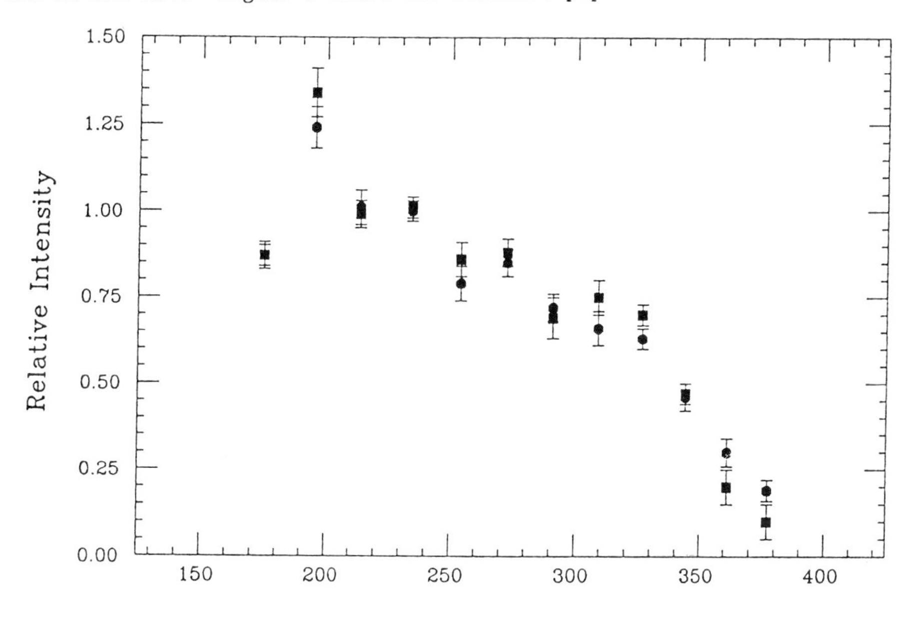

Fig. 5. Relative intensity of the γ-rays in the SD band in ^{191}Hg as measured at 172 MeV (•), and at 167 Mev (■). The relative intensities at each energy were obtained from the analysis of the same coincident gates.

SD band in ^{191}Hg at both 172 and 167 MeV. Within the error bars, the two patterns are identical with the possible exception of the two highest γ-rays. The overall population of the SD band in ^{191}Hg, however, decreases from 2.0% to about 1.2% at the lower beam energy. In contrast to the behavior in the SD band in ^{191}Hg, the high spin members of the weaker band described above are populated much less strongly than at 172 MeV. This result is also consistent with an assignment in ^{190}Hg but at this stage in the analysis it is not possible to draw any definite conclusions. Although the analysis of the excitation function data is still in progress, the results already obtained for the SD band in ^{191}Hg are consistent with those reported (Nolan and Twin 1988) for the SD band in ^{152}Dy.

As a byproduct of the search for superdeformation in the Hg isotopes, considerably more information has been obtained on the 'normal' spectroscopy of these nuclei. One of the more interesting results is the observation of level structure consistent with single-particle excitations in ^{191}Hg (Ye et al.). The previously published (Hübel et al. 1986) level scheme for ^{191}Hg has been interpreted in terms of the collective rotation of a weakly deformed oblate system. Figure 6

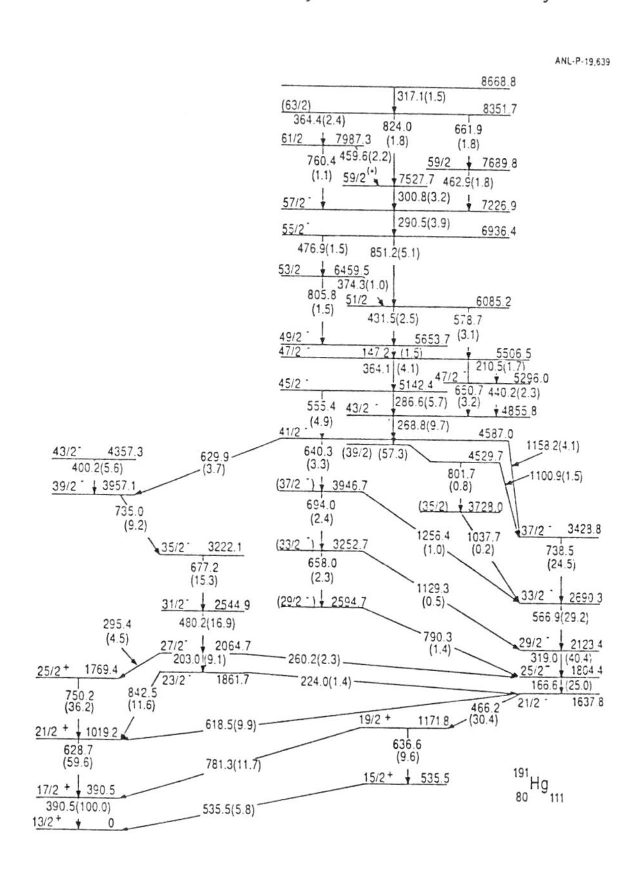

Fig. 6. Partial level scheme of ^{191}Hg showing the new level structure discussed in the text. The uncertainty on the γ-ray energy varies from 0.2 keV for the strong transitions to 0.5 keV for the weakest lines. Gamma-ray intensities are given in parentheses for all levels associated with the new sequence.

presents the partial level scheme relevant to the discussion below. For spins lower than 41/2, the band structure shows the regular level patterns and connecting E2 transitions characteristic of collective rotation. Above spin 41/2, however, the level scheme displays very irregular level spacing in marked contrast to the other band structures in the nucleus. In addition, a large number of dipole and quadrupole transitions compete in the decay. Furthermore, the $41/2^-$ state was found to be isomeric with a mean lifetime of 5±1 ns and is seen to decay to several rotational structures. These features point to a very different character for the levels in this band when compared to the rotational yrast states at lower excitation energy. These properties are very similar to those observed in nuclei in the beginning of the rare earth region. For example, the yrast lines of ^{152}Dy (Khoo et al. 1978) or ^{148}Gd (Piiparinen et al. 1987) are irregular and several isomers with decay rates typical of single-particle transitions are present. In these cases, the angular momentum is generated by the alignment of the spins of individual nucleons along a symmetry axis. We propose that the new level structure in ^{191}Hg is also of single-particle character. The present study represents the first experimental observation of this mode of excitation in this mass region.

In summary, a band of twelve coincident transitions has been seen in ^{191}Hg with an average moment of inertia $\mathcal{J}^{(2)}$ of 110 $\hbar^2$ MeV^{-1} and an intrinsic quadrupole moment of 18±3 eb. The data are in excellent agreement with the results of cranked Strutinsky calculations and provide strong support for the existence of the large new region of superdeformation as discussed in (Chasman 1989). There is evidence in our data that superdeformation in this mass region is not limited to the case of ^{191}Hg. This is provided by the existence of another, much weaker, band with similar properties to the SD band in ^{191}Hg which has very tentatively been assigned to ^{190}Hg. Our data also contains evidence for superdeformed continuum γ-rays. These non-yrast superdeformed structures have moment of inertia values that show very similar trends to those seen in the discrete SD bands. Thus, it appears that excited SD bands are being populated in the present reaction. Finally, a great deal of new information has been obtained regarding the "normal" spectroscopy of ^{191}Hg. Specifically, a new band in ^{191}Hg has been found with properties consistent with single-particle excitations. These results, when taken together, show that the nucleus ^{191}Hg exhibits the competition between very different shapes and modes of excitation. Therefore, this nucleus and presumably others in this mass region should provide a very fruitful testing ground for the predictions of various nuclear models.

This work was supported by the U.S. Department of Energy, Nuclear Physics Division, under Contracts No. W-31-109-ENG-38, No. DE-AC07-76ID01570, No. DE-FG02-87ER40346 and the National Science Foundation under Grants No. PHY88-02279 and No. PHY87-19661.

References

Anderson O et al., Phys. Rev. Lett. **43**, 687 (1979)
Burde J et al., Phys. Rev. C **38**, 2470 (1988); A.O. Machiavelli et al., ibid. **38**, 1088 (1988)
Chasman R R, Phys. Lett. B **219**, 227 (1989)
Deleplanque M A et al., Phys. Rev. Lett. **60**, 1626 (1988)
Fallon P et al., Phys. Lett. B **218**, 137 (1988)
Girod M et al., Phys. Rev. C **38**, 1519 (1988); M. Weiss and S. Krieger (private communication).
Hass B et al., Phys. Rev. Lett. **60**, 503 (1988)
Hübel H et al., Nucl. Phys. **A453**, 316 (1986).
Khoo T L et al., Phys. Rev. Lett. **41**, 1027 (1978).
Kirwan A J et al., Phys. Rev. Lett. **58**, 467 (1987)
Löbner K E G et al., Nucl. Data, Sect. A 7, 495 (1970).
Nolan P J and Twin P J, Ann. Rev. Nucl. Part. Sci. **38**, 533 (1988).
Piiparinen M et al., Phys. Lett. B **194**, 468 (1987).
Polikanov S M et al., Zh. Eksp. Teor. Fiz. **42**, 1464 (1962) (Sov. Phys. JETP **15**, 1016 (1962))
Rathke G.-E. et al., Phys. Lett. B **209**, 177 (1988)
Strutinsky V M, Nucl. Phys. **A95**, 420 (1967)
Twin P J et al., Phys. Rev. Lett. **57**, 811 (1986); M.A. Bentley et al., ibid. **59**, 2141 (1987)
De Voight M J A et al., Phys. Lett. B **106**, 480 (1981).
Ye D et al. (to be published).

Inst. Phys. Conf. Ser. No 105
Paper presented at Int. Conf. on Spectroscopy of Heavy Nuclei, Crete, Greece, 1989

Hyperdeformed pear-shaped nuclei

J. Blons and D. Paya
Service de Physique Nucléaire - Basse Energie
CEN Saclay, 91191 Gif-sur-Yvette Cedex, France

ABSTRACT: Several wells are found in the potential energy surface of the actinides. The first well corresponds to the ground state deformation ($\varepsilon_2 \simeq 0.2$). The second well accomodates fission isomers with a larger deformation ($\varepsilon_2 \simeq 0.6$). A third well is calculated to occur at the $\varepsilon_2 \simeq 0.9$, $\varepsilon_3 \simeq \pm 0.2$ quadrupole and octopole deformation, respectively. In this paper, we present experimental evidence for the existence of hyperdeformed pear-shaped thorium nuclei in this third well.

1. INTRODUCTION

Fission isomers were discovered at Dubna in 1962 (Polikanov et al 1962). They were explained, in the mid 60's, by the existence of a second well in the potential energy surface (PES) of heavy nuclei. According to the Strutinsky's calculations (Strutinsky 1967), this second well is due to the shell effects and arises at large quadrupole deformation ($\varepsilon_2 \simeq 0.6$). For comparison, the ground state deformation for the actinides in the first well is only $\varepsilon_2 \simeq 0.2$. Ten years later (Specht et al 1972), the observation of a rotational band for the ^{240}Pu isomer brought a beautiful confirmation of this interpretation : indeed, the moment of inertia was found twice as large as the one measured in the ground state band. A model-independent proof of the existence of such a superdeformation was given by a measurement of the electric quadrupole moment, $Q = 36 \pm 4$ b, of the 8 μs isomer of ^{239}Pu (Habs et al 1977). Also, the theoretical values were found very close to the experimental ones for both the moment of inertia and the quadrupole moment.
When theoretical calculations became more sophisticated, it was established that nuclear shapes could be more complicated as the fission process evolves (Möller et al 1974). In fact, it turned out that the PES presents more favorable paths when certain symmetries of the nucleus are lost. For instance, at the second maximum, the potential energy decreases with an associated finite loss of the right-left symmetry. For the thorium isotopes, the loss of the right-left symmetry even causes the creation of a rather shallow (~ 1 MeV) third well at very large deformation ($\varepsilon_2 \simeq 0.9$). Following the method used for the identification of the superdeformed second well, the experimental challenge was to detect rotational bands in this hyperdeformed third well and to extract the moment of inertia.

Such an experimental research is presented in this paper for ^{231}Th and ^{233}Th. In section 2, we describe the experimental technique. Some experimental data and analyses are given in section 3. Discussions and conclusions are presented in section 4 and 5, respectively.

2. EXPERIMENTAL TECHNIQUE

The existence of rotational hyperdeformed states trapped in a third well could be experimentally shown by the specific properties of resonances in the fission probability (P_f). It is obvious that the measurements must have an energy resolution better than the rotational energy spacings. A second condition is to assign the J, K and π quantum numbers to the individual levels. These quantum numbers can be deduced from the fission probability and fission fragment angular distributions (FFAD).

Until now, the best energy resolution for the fission probability near the fission threshold, was obtained with the neutron time-of-flight technique. First experiments were performed using a linear electron accelerator (GELINA) and a 51 or 101 meter long flight path. In these conditions, the energy resolution was 1.7 keV at $E_n \simeq$ 720 keV for the ^{230}Th(n,f) reaction, and 2.3 keV at $E_n \simeq$ 1.6 MeV for the ^{232}Th(n,f) reaction (Blons et al 1984).

However, if the neutron-induced reactions lend themselves to high energy resolution measurements, they are, nevertheless, of limited use because : i) They are limited to excitation energies just above the neutron binding energy. ii) They only bring a small angular momentum into the fissioning nucleus, since only neutron energies below $\approx$ 2 MeV are of practical use.

Appropriate (d,p) reactions, leading to the same fissioning nucleus, do not suffer from the above drawbacks. P_f and FFAD of several isotopes of thorium and uranium were investigated with the deuteron beam of the Saclay Tandem Van de Graaff accelerator (Blons et al 1988). The overall energy resolution was about 7 keV. The experimental lay-out is schematically represented in Figure 1. The experimental procedure consisted in measuring the energy of the proton emitted at 130° in the laboratory in coincidence with a fission fragment emitted at an angle, θ, with respect to the recoil direction of the fissioning nucleus.

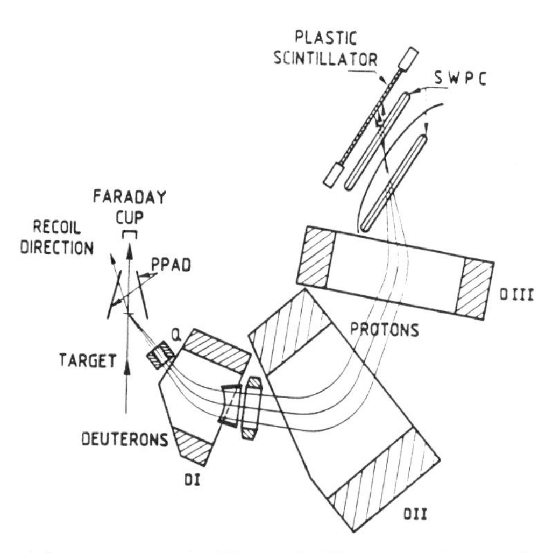

Fig. 1. Schematic representation of the experimental set-up used for (d,pf) reactions.

The fission fragments were detected by two parallel plate avalanche detectors (PPAD). The anodes of the PPADs were divided into seven strips as shown in Figure 2, each one corresponding to a 10° (or 5°) span in polar angle, θ. Each strip was connected to its neighbours by a 2 ns delay line and the angle identification was made by measuring the time delay between the PPAD signals at the two ends of the total delay line.

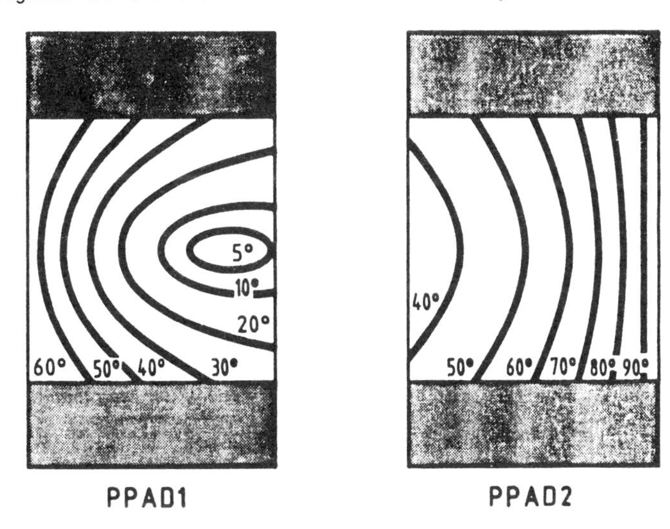

Fig. 2. Lay-out of the PPAD anodes. The lines, labelled by the polar angle, constitute the separations between adjacent strips.

The proton energy was measured in the Q3D magnetic spectrometer. The focal surface of the Q3D was equipped with two single wire proportional counters (SWPC) and a plastic scintillator. The SWPCs were used as position sensitive detectors by means of a charge division applied to both ends of the resistive wire. The use of these two detectors allowed, firstly, the curvature of the focal surface to be corrected and, secondly, the proton trajectory in the Q3D to be identified. The plastic scintillator, set behind the SWPCs, stopped protons and gave a fast signal. This signal was used as the "start" input for a time to amplitude converter (TAC), and the "stop" input subsequently came from the PPAD's cathodes.

3. EXPERIMENTAL RESULTS AND ANALYSES

The excellent energy resolution available at GELINA has shown, for the first time, a "fine structure" in the $E_n \simeq 720$ keV resonance of $^{230}Th(n,f)$ cross section (Blons et al 1978). The same "fine structure" was also found in the Los Alamos $\sigma_{nf}(E_n;\theta)$ results, obtained at two angles, θ=100° and θ=125°, with respect to the neutron beam (Veeser et al 1981). Furthermore, the ensuing experimental ratio, $R(E_n) = \sigma_{nf}(E_n;125°)/\sigma_{nf}(E_n;100°)$, gives accurate information on FFAD. In addition, several FFAD were measured for this resonance, using monoenergetic neutron beams. The last results obtained at Lucas Heights (Boldeman et al 1985) seem the most accurate ones.

All the experimental data have been analysed according to the Hauser-Feshbach's theory (Hauser et al 1952). In the compound nucleus model, it is generally assumed that a nuclear reaction proceeds in two independent steps : the first is the formation of a compound nucleus in the first well and the second is its decay in which fission competes with all other exit channels like γ, or possibly, neutron emission. The compound nucleus cross section is calculated in the DWB approximation while the fission transmission coefficient is calculated as the transmission coefficient of a one-dimensional barrier. Due to the shallowness of the third well, only low-lying levels can be trapped. These levels are expected to be rotational and must have well defined J, K and π quantum numbers. The energy sequence of a rotational band can be expressed as

$$E_{JK}^{\pi} = E_{KK}^{\pi} + \frac{\hbar^2}{2\mathfrak{J}} \left[J(J+1) - K(K+1) + a^{\pi}\left[(-1)^{J+1/2}(J+1/2) + 1\right]\right]$$

where E_{KK}^{π} is the rotational band-head energy for parity π, J is the angular momentum of the rotational level, K is the J-projection on the nuclear symmetry axis, $\mathfrak{J}$ is the moment of inertia of the deformed nucleus perpendiculary to the symmetry axis, and a^{π} is the Coriolis force decoupling parameter which vanishes for all $K \neq 1/2$.

All analyses of the neutron data undertaken by different groups agree on the necessity of using two rotational bands with opposite parities. Figure 3 shows a simultaneous best fit of the 720 keV resonance of the ^{230}Th(n,f) cross section when the FFAD are included in the fitting procedure. The K-quantum number is K = 1/2. The deduced rotational band parameters are

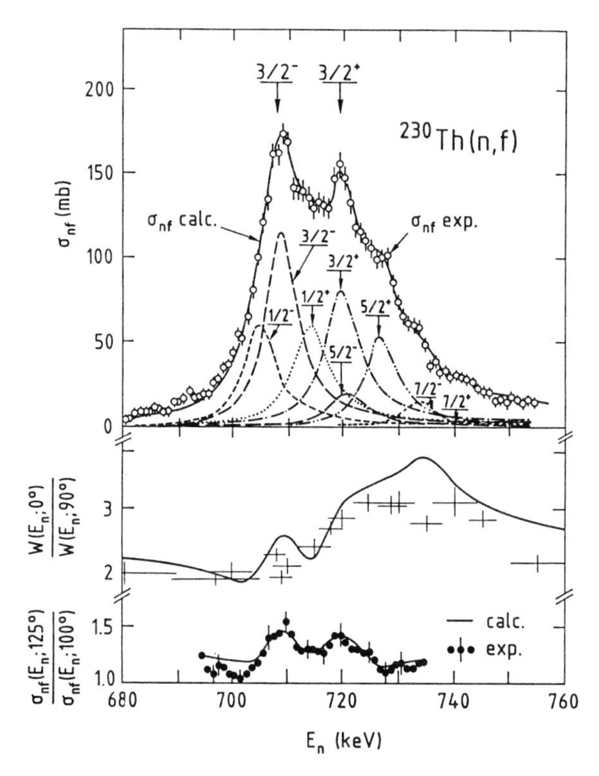

Fig. 3. "Simultaneous best fit" to the experimental cross-section, FFAD ratio and $R(E_n) = \sigma_{nf}(E_n;125°)/\sigma_{nf}(E_n;100°)$ for the $E_n \simeq 720$ keV resonance in ^{230}Th(n,f).

$(\hbar^2/2\mathfrak{J})^+ = 1.9 \pm 0.3$ keV $\qquad a^+ = 0.2 \pm 0.2$
$(\hbar^2/2\mathfrak{J})^- = 2.1 \pm 0.1$ keV $\qquad a^- = -0.3 \pm 0.2$

$$\Delta E_{+,-} = E_{KK}{}^+ - E_{KK}{}^- = 8.5 \text{ keV}$$

As seen in Figure 3, the $E_n \simeq 720$ keV neutrons do not bring enough angular momentum to feed nuclear states with spin, J, larger than 7/2. A (d,pf) reaction provides more angular momentum in the compound nucleus. This is illustrated in Figure 4 where the predicted ^{230}Th(d,pf) probability is calculated on the basis of the two-parity band analysis of the ^{230}Th(n,f) data. This comparison shows an expected large reinforcement of the (d,pf) probability on the high-energy side of the (n,f) structure, in particular the $9/2^+$ and $11/2^-$ components are now clearly separated.

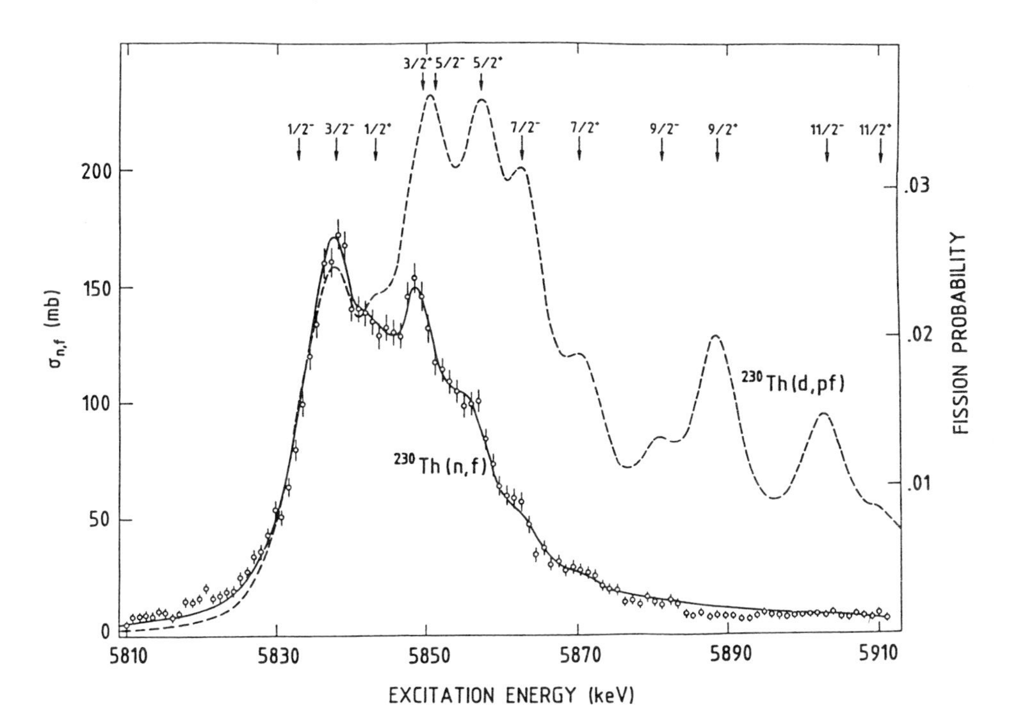

Fig. 4. The experimental, ϕ, and calculated, —, ^{230}Th(n,f) cross section , compared with the ^{230}Th(d,pf) calculated probability, ---.

So, in order to extend the observation of the rotational levels beyond J = 7/2, an additional experiment on the (d,pf) reaction was undertaken at Saclay (Blons et al 1988). Figure 5 shows a simultaneous best fit to the ^{230}Th(n,f) cross-section and the ^{230}Th(d,pf) probability at forward angles ($0° < \theta < 30°$). The deduced rotational band parameters are the same (within the error bars) as those obtained in fitting the neutron data only. Indeed, this feature strongly confirms the conclusions drawn from the analysis of the neutron data. It is particularly rewarding to see all the expected members of the two K = 1/2 rotational bands from J = 1/2 up to J = 13/2.

Nevertheless, the calculated P_f in the sideward direction ($60° < \theta < 90°$), is appreciably smaller than the experimental ^{230}Th(d,f) data (Blons et al

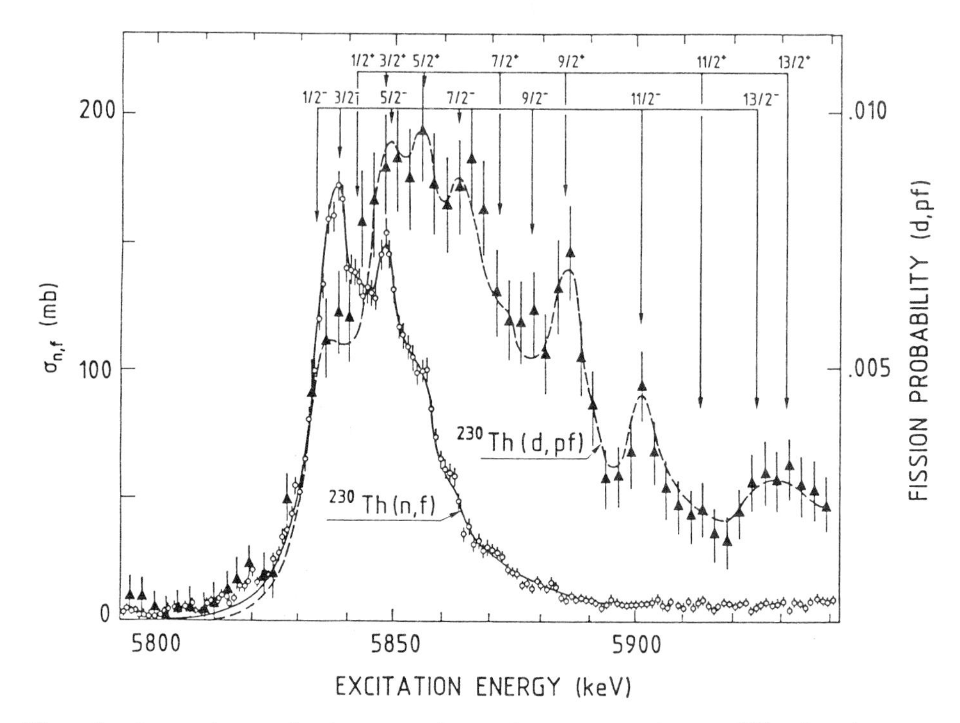

Fig. 5. Comparison of the experimental and calculated ^{230}Th(n,f) cross-section with the experimental and calculated ^{230}Th(d,pf) probability.

1988). Such a difference was attributed to a higher-value K quantum number contribution. Additional K=5/2 or K=7/2 rotational bands restore equally well the experimental probability. A K = 5/2 band is predicted by theoretical calculations (Bengtsson et al 1987). However, taking the FFAD also into account, a better fit was obtained with K = 7/2. The result of the fit is indicated by the full lines in Figure 6. A K = 7/2 rotational band has no effect on the σ_{nf}, since 700 keV neutrons do not bring enough angular momentum into the compound nucleus to feed nuclear states with spins as large as J = 7/2 and more. Thus, the adjunction of such high K rotational bands does not affect the analysis of the (n,f) data.

Following the success achieved in ascribing the observed fine structures to class III levels of the fissioning nucleus ^{231}Th, similar studies were also undertaken for ^{233}Th. The neutron-induced and the (d,p)-induced fission cross sections of ^{232}Th, as well as the associated fission fragment anisotropies, were measured. The fission subthreshold resonances appear also fine-structured in both (n,f) and (d,pf) probabilities. A simultaneous best fit similar to the one applied to the ^{230}Th data indicates also the necessity of involving two rotational bands with K = 3/2 and opposite parities with an energy separation, $\Delta E_{+,-}$ = - 22 keV.

Again, the inertia parameter was found equal to 2 keV for the two bands. The fits of the calculated fission probabilities to the experimental data around 6.4 MeV excitation energy are shown in Figure 7. As expected, the levels with spins higher than 5/2 are more populated in the (d,pf) than in the (n,f) data. The good agreement obtained on the fission probabilities,

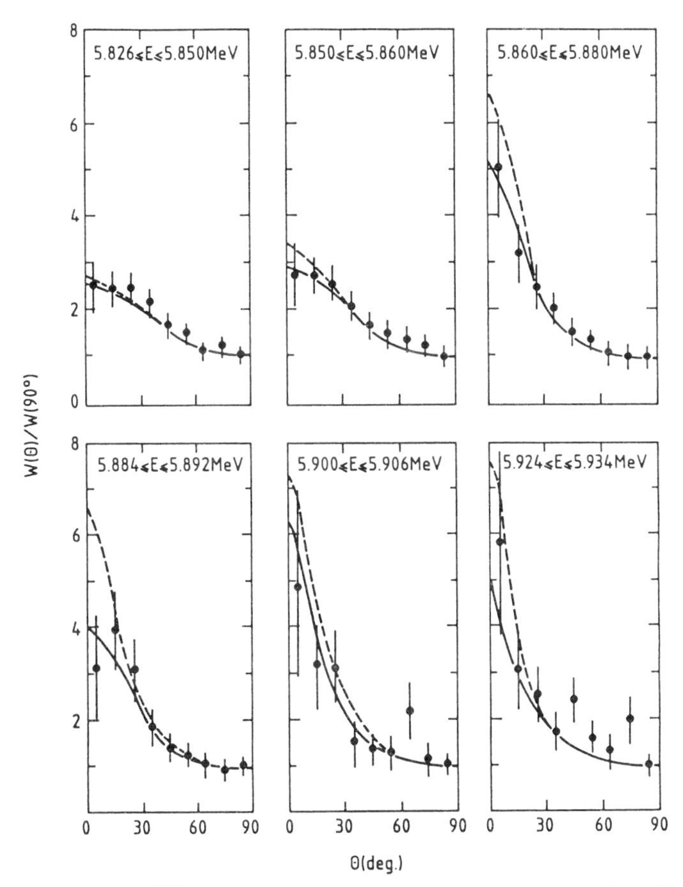

Fig. 6. Experimental ^{231}Th fission fragment angular distributions compared with the calculated values. The dashed lines represent the $K^{\pi} = 1/2^{\pm}$ contributions alone and the full lines represent the sum of the $K^{\pi} = 1/2^{\pm}$ and $7/2^{\pm}$ contributions.

as well as on the angular distributions, is also a clear indication of the presence of a third well in the potential energy surface of ^{233}Th (Blons et al 1984 and 1988).

Some indications of two rotational bands with opposite parities were also found in the ^{229}Th(d,pf) (Blons et al 1988) and ^{231}Pa (n,f) cross sections (Plattard et al 1981). As one comes to the uraniums, the first hump of the triple-humped fission barrier becomes as high as, or even higher than, the other two. In these conditions, the intrusion of the class II levels produces a well known intermediate structure effect which renders the observation of rotational bands in the third well pratically impossible.

4. DISCUSSION

For ^{231}Th and ^{233}Th, both analyses of (n,f) and (d,pf) data give a value of the moment of inertia, $2\mathfrak{J}/\hbar^2$ = 500 MeV^{-1}. Moments of inertia have been calculated with the Nilsson and Woods-Saxon potentials as a function of

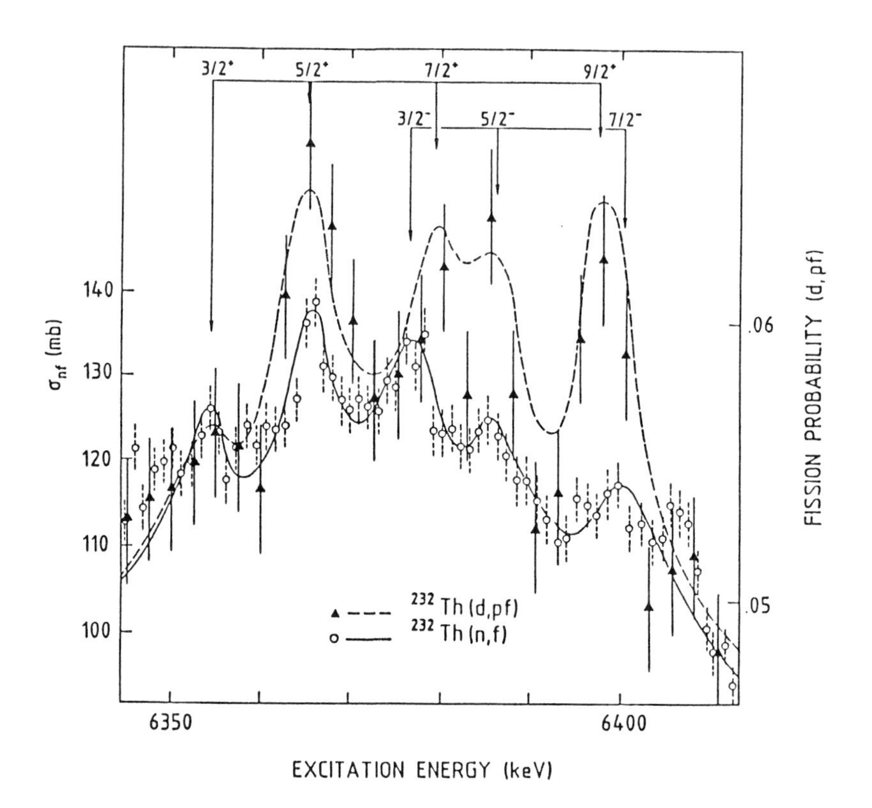

Fig. 7. Experimental and calculated cross sections in the ^{232}Th(n,f) and ^{232}Th(d,pf) reactions.

deformation (Bengtsson et al 1987). For large deformation, $\varepsilon_2 \sim 0.9$, the results are very close to rigid-body values. Also the influence of the octupole deformation or of an odd particle seems rather small. In these conditions, as illustrated in Figure 8, the experimental values are consistent with the moment of inertia at the third minimum. This deformation corresponds to a 3 : 1 axis ratio (hyperdeformation). By comparison, in the second well where the moment of inertia is 300 MeV^{-1} (Specht et al 1972), the axis ratio is 2 : 1 (superdeformation).

Moreover, a typical property of nuclei in the third well is their reflection asymmetric shape. For such a shape, the parity is no longer a good quantum number. Nevertheless, definite-parity wave functions can be constructed from each wave functions $\phi(\varepsilon_3)$ and $\phi(-\varepsilon_3)$ by forming the linear combinations

$$\psi^+ = \sqrt{1/2}\,[\psi(\varepsilon_3) + \psi(-\varepsilon_3)] \qquad \text{and} \qquad \psi^- = \sqrt{1/2}\,[\psi(\varepsilon_3) - \psi(-\varepsilon_3)]$$

An examination of the potential energy surface within the framework of the triple-humped barrier shows that there are, in fact, two third wells at positive and negative values, ε_{III}, of the deformation coordinate ε_3 describing the octupole deformation. Without coupling between the two octupole third wells, the theory predicts a parity doubling of all rotational states (Bohr et al. 1975). A separation between the parity doublets arises

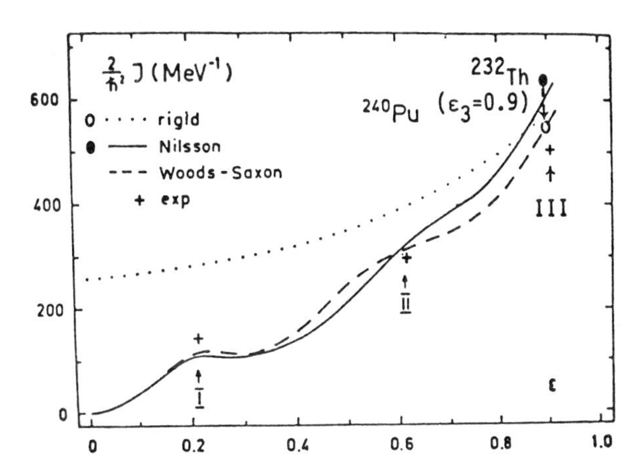

Fig. 8. Moment of inertia, J, calculated as a function of deformation for ^{240}Pu (reflexion symmetric shapes) and for ^{232}Th at the third well . The experimental values of J, at the first and the second minimum for ^{240}Pu and at the third minimum for ^{231}Th are denoted by crosses.

when the coupling between the two wells is switched on. The energy splitting, $\Delta E_{+,-}$ between the heads of the two rotational bands depends on the transmission coefficient of the barrier between the two third wells. Theoretically, $\Delta E_{+,-}$ can be expressed in a quasi-classical procedure (Landau et al 1963).

$$\Delta E_{+,-} = \hbar\omega/\pi \; \exp\left[-\frac{1}{\hbar}\int_{-a}^{a} \sqrt{2\mu(V(\varepsilon_3)-E_{III})}\; d\varepsilon_3\right]$$

where $\pm$ a are the values of ε_3 at the turning point, ω and E_{III} are the frequency of vibration and the energy of the state in one potential well, respectively. μ is the mass parameter and $V(\varepsilon_3)$ the height of the barrier between the two octupole third wells.

As we have shown in section 3, experimental results on $^{231,233}Th$ can only be fitted with pairs of rotational bands. The energy splitting, $\Delta E_{+,-}$, is found equal to 8.5 keV for ^{231}Th and -22 keV for ^{233}Th. It is difficult to compare this value to the theory in a static calculation since the mass parameter, μ, is unknown. If the mass parameter is assumed to be independent of deformation, μ = 0.0540 $A^{5/3}\hbar^2$ MeV^{-1} the calculated value when assuming a tunneling at constant value of the quadrupole deformation is largely smaller than the experimental one (Paya 1980). Nevertheless, it is possible to find, in the two dimensional ε_2, ε_3 space, a path with a lower action integral and giving $\Delta E_{+,-}$ value closer to the experimental one. For instance, in Figure 9, $\Delta E_{+,-}$ is 0.01 keV for the path labelled c, while $\Delta E_{+,-}$ is 5 keV for the path labelled b'.

5. CONCLUSION

Today, a third minimum is found in both theoretical macroscopic-microscopic and purely microscopic calculations. Experimentally, when high energy resolution is achieved, a third minimum is also observed in the PES. The hyperdeformation of the class III states is consistent with theoretical predictions (Bengtsson et al 1987). Also the observation of parity doublets indicates the occurence of pear-shaped deformation. If the ^{231}Th nucleus is, by far, the most striking test ground for the third well hypothesis, a similar work was undertaken for the two other thorium isotopes 230 and 233 and lead to the same conclusions (Blons et al 1988). Indeed in both cases, the analyses indicate the necessity of involving two rotational bands of opposite parities and inertia parameter equal to 2 keV.

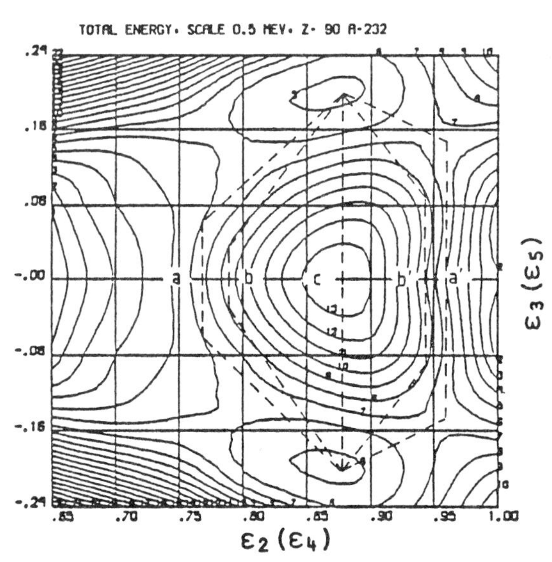

Fig. 9. Potential energy surface as function of coordinates ε_2, ε_4, ε_3 and ε_5 (Möller et al 1972). The different paths used in the calculation of $\Delta E_{+,-}$ are also shown.

REFERENCES

Bengtsson R, Ragnarsson I, Åberg S, Gyurkovich A, Sobiczewski A and Pomorski K 1987 Nucl. Phys. **A473** 77

Blons J, Mazur C, Paya D, Ribrag M and Weigmann H 1978 Phys. Rev. Lett. **41** 1282

Blons J, Mazur C, Paya D, Ribrag M and Weigmann H 1984 Nucl. Phys. **A414** 1

Blons J, Fabbro B, Mazur C, Patin Y, Paya D and Ribrag M 1988 Nucl. Phys. **A477** 231

Bohr A and Mottelson B R 1975 Nuclear structure, vol. 2 (Benjamin, Reading, Mass)

Boldeman J W and Walsh R L 1985 Proc. Int. Conf. Santa Fe, New Mexico, vol. 1, p. 317

Habs D, Metag V, Specht H J and Ulfert G 1977 **38** 387

Hauser W and Feshbach H 1952 **87** 366

Landau L D and Lifshits E M 1963 Kvantovaya mekhanika (Moskva) chap. 7

Möller P 1972 Nucl. Phys. **A192** 529

Möller P and Nix J R 1974 Physics and chemistry of fission 1973 (IAEA, Vienna) vol. 1, p. 103

Paya D 1980 Proc. Symp. on physics and chemistry of fission, Julich 1979 (IAEA, Vienna) p. 207

Plattard S, Auchampaugh G F, Hill N W, de Saussure G, Harvey J A and Perez R 1981 Phys. Rev. Lett. **46** 633

Polikanov S M, Druin V A, Karnauchov V A, Mikheev V L, Pleve A A, Skobelev N K, Subotin V G, Ter-Akapian G M and Fomichev V A 1962 Zh. Eksp. Theor. Fis. **42** 1016

Specht H J, Weber J, Konecny F and Heunemann D 1972 Phys. Lett. **41B** 43

Strutinsky V M 1967 Nucl. Phys. **A95** 420

Veeser L R and Muir D W 1981 Phys. Rev. **C24** 1540

Inst. Phys. Conf. Ser. No 105
Paper presented at Int. Conf. on Spectroscopy of Heavy Nuclei, Crete, Greece, 1989

First physics with NORDBALL

G.Sletten, J.Gascon and J.Nyberg.
The Niels Bohr Institute, University of Copenhagen, Denmark.

ABSTRACT The NORDBALL is an array of 20 Compton suppressed Germanium spectrometers in a close packed geometry forming a truncated icosahedron; 12 pentagonal and 20 hexagonal surfaces inscribing a sphere. A variety of selective devices are introduced to work in conjunction with the Ge-spectrometers in order to provide unique sensitivity in nuclear structure studies. Charged particle detection as well as neutron multiplicity measurements in coincidence with Ge-spectrometers will be discussed. Two specific experiments both involving lifetime measurements by the DSAM method will be discussed. A first report of a strongly deformed rotational band in ^{131}Ce populated by oxygen beams and a report on deformation dependent transition rates in ^{157}Ho will be presented.

1. Introduction.

The NORDBALL is an array of Compton-suppressed Germanium spectrometers dedicated to nuclear structure research. The nordic countries Denmark, Finland, Norway and Sweden decided to pool their resources within this field in order to create an instrument with the highest possible sensitivity to photon spectroscopy in its broadest sence; the name of the instrument became obvious.

Later the project was joined by several other groups each one contributing with further instrumentation and ideas. Today the participating groups represent Holland, Italy, Japan and West-Germany in addition to the nordic countries. The total investment from all groups is at present more than 4 million $ not counting the host VAX 8650 computer and its peripherals.

The geometry of the array is a truncated icosahedron with 12 pentagonal surfaces and 20 hexagonal surfaces inscribing a sphere with a diameter of 660 mm. Figure 1. Each hexagon supports a Ge-spectrometer and each pentagon a quintet of BaF_2 scintillators or other selective devices. The present paper will describe the overall layout of the array with a discussion of the efficiency and performance of the Ge-spectrometers and show examples of the selective devices which can be implemented in a modular way.

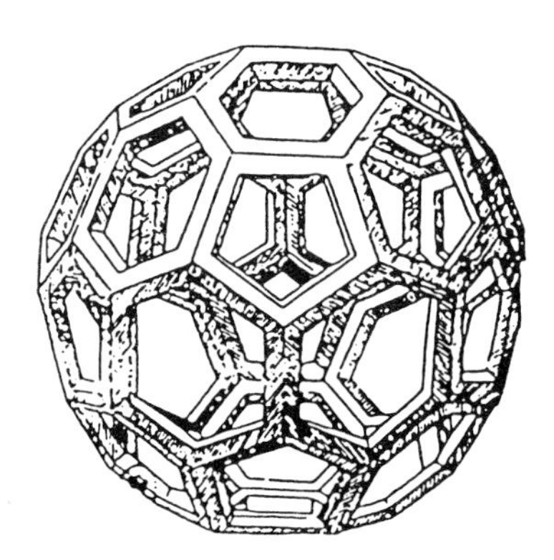

Figure 1. The NORDBALL frame structure designed by Leonardo da Vinci in the beginning of the 16th century. Later also used in football design.

After the first year of experimental activity, with all the trivial childrens deseases that a complex instrument will have, the first datasets have been analyzed, discussed and are presently being submitted to scientific journals. Two of these investigations, both dealing with rotational structure of deformed nuclei will be presented in this paper in a condensed form.

2. The Germanium Spectrometers.

The Germanium spectrometers are Compton suppressed by a BGO scintillator in coaxial symmetry with the Ge-crystal. Figure 2. A detailed report on the design and performance of this anti-Compton shield is given by Moszynski (1989). The Germanium detectors are all of n-type with efficiencies ranging from 20% to 48% relative to 3" x 3" NaI.

A main goal in the design of such an array is to obtain high counting rates of higher fold coincidenses and at the same time maintain the spectral resolution of the instrument. Here the factors that have to be considered are the geometrical solid angle, the relative efficiency of the detector, the photopeak to total ratio, the Doppler effects and the granularity of the array.

The interesting numbers to the first order are the product of the relative detector efficiency and the geometrical solid angle and the peak to total ratios. The experiments performed so far has been with a total efficiency of 0.93%, but a recent replacement of some smaller detectors with deep, large diameter detectors increases this number to 1.25%. An upgrading of the total efficiency will start during the coming year through the collaboration with groups in Milano and Legnaro. This will imply 10 detectors with relative efficiencies of about 80% and result in a total total efficiency for Nordball of 4%. The numbers given in this discussion all refer to reactions with oxygen beam at 86 MeV and it is clear that Doppler effects will enforce a somewhat smaller geometrical solid angle and correspondingly smaller total efficiency.

The peak to total ratio of the spectrometers have been measured at 1.33 MeV and give an average value of 0.55 for all Nordball spectrometers (Moszynski 1989). Regarding the granularity the spin and multiplicity range encountered in the oxygen and fluorine induced reactions gives a single hit probability practically equal to 1. In future experiments with 225 MeV ^{48}Ti beam this number will be smaller.

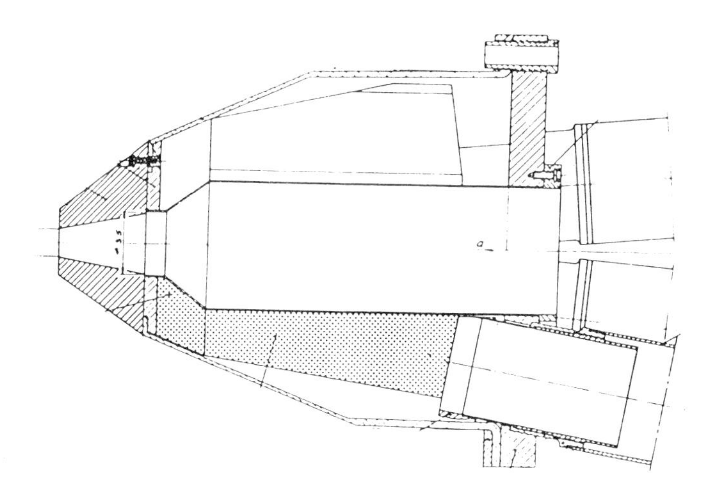

Figure 2. Cross section of the Nordball anti-Compton shield. The 2" photomultiplier tubes are not shown. The scintillator is BGO and divided in 6 units optically separated from each other. Each unit is coupled to a separate photomultiplier.

The Nordball Ge-array can also be used as a high energy photon detector as described by Bruce (1988). By this technique the compton scattered photon detected in the surrounding suppression shield is added to the energy seen by the Ge-counter. A high resolution spectrum with high efficiency up to 20 MeV and with the stability determined by the Ge-counter is obtained. A more powerful configuration is to replace Ge-spectrometers in the hexagons with 15 cm x 17.5 cm BaF_2 detectors. There are 8 such detectors available. The remaining Ge-spectrometers can then be used as gating devices for determining the final nucleus in a cascade involving a high energy photon.

The geometry of the array is oriented in space with the horizontal beamline entering and exiting through a pentagon. This orients the hexagons into 4 rings perpendicular to the beam axis with angles 37^0, 79^0, 101^0 and 143^0. Coincidences between detectors in a given ring with those of other rings determine angular correlations of the detected gamma-rays. In the evaluation of data separate gamma-gamma coincidence matrices were sorted for combinations of these angles.

3. Electronics and Data Acquisition.

The electronics associated with the Nordball is with a few exceptions standard commercially available components. A highest possible degree of integration has been the guiding line in building the analogue and the logic parts. Both NIM and ECL components are used and access to individual channels and components by computer has been used as much as possible. A further description of the electronic schemes is outside the scope of the present report. The data acquisition, reading of NIM and CAMAC components is facilitated by a VME multiprocessor system described by (Jerrestam 1989).

4. Selective Devices.

The terms Selectivity and Sensitivity have been proverbs in the development of the Nordball instrumentation. It has been demonstrated that the higher folds of gamma-gamma coincidences increases the sensitivity to minuscule details in nuclear structure (Herskind 1988), (Nolan 1989). Increased sensitivity may also be achieved by combining the gamma-gamma coincidence with detection of charged particles, neutrons, residual nuclei, isomers, total gamma energy, gamma-ray multiplicity and combinations of these. The selection of gamma-gamma coincidences by the additional coincidence with a light charged particle or the neutron multiplicity will specify the final nucleus where the photon cascade occurs and in many cases correspond to the sensitivity of the next higher fold of gamma-gamma coincidences. An optional or additional claim on the total gamma energy and multiplicity will have a similar effect.

4.1 Measurement of Neutron Multiplicity

The neutron multiplicity filter consists of 16 liquid scintillator detectors occupying ten hexagon and six pentagon positions in the Nordball foreward hemisphere Figure 3. These 16 detectors containing altogether 48.7 litres of BC-501 liquid covers 42 % of 4π geometrically. The instrument is intended primarily for studies of neutron deficient nuclei far from stability or studies of nuclei in the N = Z region where a large number of exit channels (25-30) are open. (Arnell 1989).

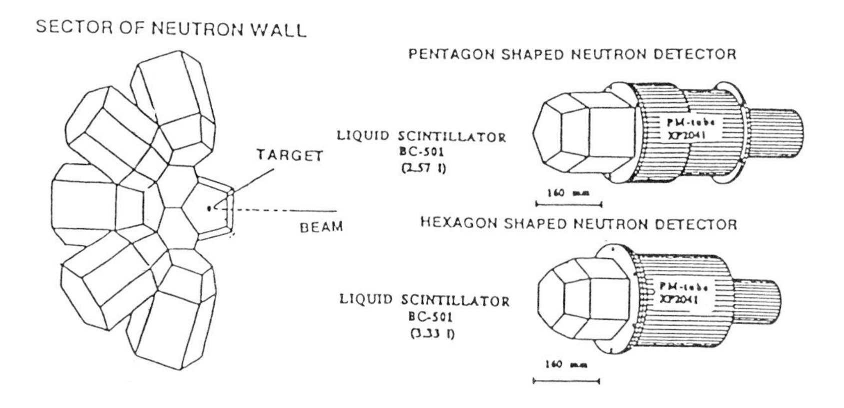

Figure 3. Sector of the 2 π neutron multiplicity filter and the two types of detectors used.

Experimental configurations involving this selective device are manyfold. In combination with the Ge-spectrometers in the backwards hemisphere of Nordball, 10 suppressed Ge-spectrometers can be gated by the neutron and gamma multiplicity from the neutron wall since these are uniquely separated. The number of Ge-spectrometers can be increased to 20 by filling pentagon positions in the backwards hemisphere with Ge-spectrometers, removing the 5 neutron detectors at 78^0 and introducing additional Ge-spectrometers there. Because of the strongly peaked angular distribution of neutrons the loss of neutron detection efficiency is about 20% but the gain in gamma-gamma coincidence rate is about 1.8.

The neutron wall might also be combined with light charged particle detection so that selection of reaction channels involving x neutrons can be done for a certain multiplicity of protons or alphas or combinations of these. Experimental data from the reaction ^{28}Si + ^{58}Ni at 110 MeV (Nyberg 1989) show strong selection of reaction channels like (^{28}Si,3p.n) ^{82}Y. The spectroscopy of this and a variety of other final nuclei are under evaluation. The 4π silicon detector used in these experiments will be presented in section 4.3.

4.2 HYSTRIX

Hystrix is a high efficiency detector for charged particles and has its name from the porcupine. (Lidén 1988, 1989). The detector is a sandwhich of NE102A and NE115 plastic scintillators coupled to a Hamamatsu 1924 photomultiplier Figure 4.

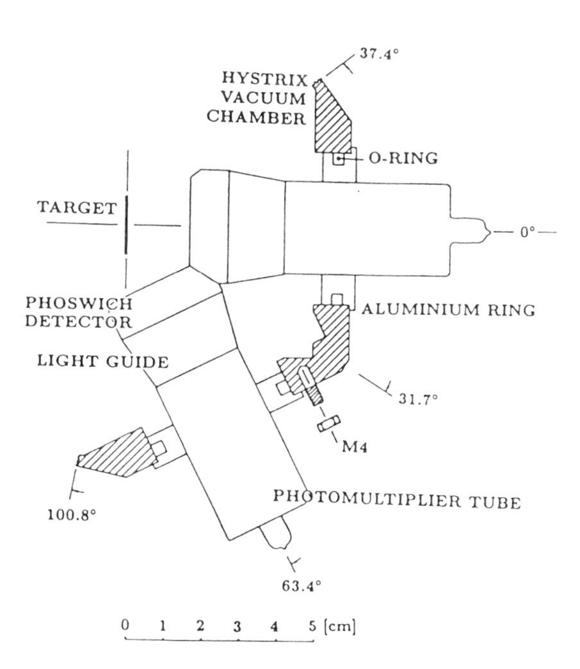

Figure 4. Cut through Hystrix showing details of detector and vacuum chamber Note that the amount of mass is minimized at angles of 37^0 and 101^0 where the Ge-spectrometers are positioned.

The geometrical solid angle of the detectors is about 94% of 4π. The device has its own vacuum chamber in the centre of the Ge-array and is producing well separated particle identification spectra of protons, alphas and combinations of these in such a way that f.ex. 1p,2p,3p and 4p events are uniquely separated from 1α,2α,1p1α,2p1α etc. Clearly the lower particle multiplicity spectra will be contaminated with the higher, but the higher will be essentially pure. Increasing the solid angle and the number of detectors would reduce the contamination but become a more delicate and complex device.

Measurements of the peak to total ratio for the Ge-spectrometers with HYSTRIX in situ gives essentially the same numbers as without. Note that the p.m. tubes are positioned at angles different from those of the Ge-spectrometers. Figure 4.

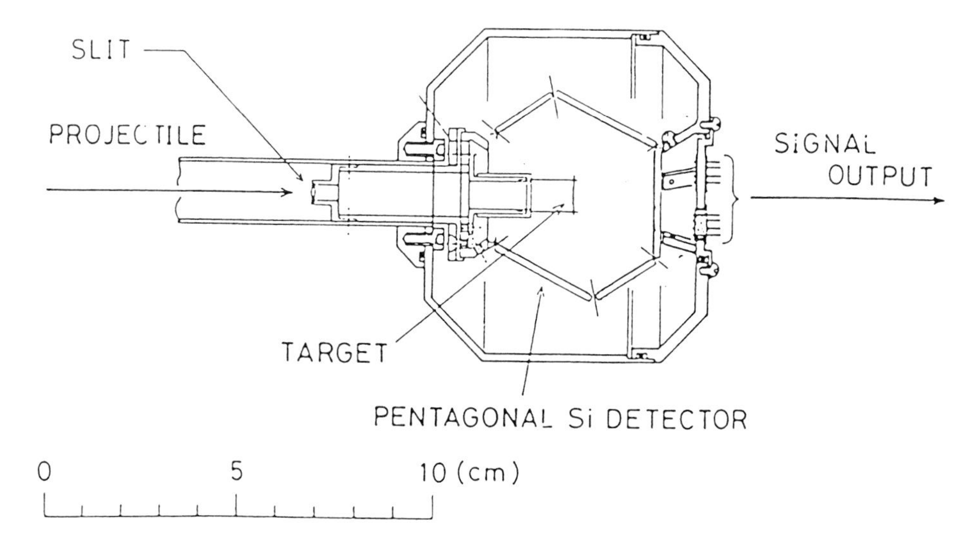

Figure 5. Cross-sectional view of the Si-ball and its vacuum chamber. Note the multiple pin feedthroughs for the high voltage and signals. This device fits inside the 60 element BaF_2 crystal ball see section 4.4.

4.3 The 4π Silicon ball.

The 4π silicon ball for detection of light charged particles shown in Figure 5 has been developped at The Kyushu University in Japan (Kuroyanagi 1989). In the first experiments the Si-ball consisted of 11 ion implanted detectors with a depletion thickness of 170μm. Each one of these were divided in two so that 22 could in principle be available. However, some of these were interconnected so that only 17 detectors were used during these experiments.

With its very small dimmension it fits with its own vacuum chamber inside the BaF_2 calorimeter, of the Nordball array (see section 4.4). The selectivity by the Si-ball is done by claiming coincidence between the gamma-ray spectrum and specific lines in the particle identified spectrum. Such a spectrum from a silicon induced reaction is shown in Figure 6. The total detection efficiencies for protons and α- particles were 57% and 47% respectively in the reaction ^{28}Si + ^{58}Ni at 100 MeV. Geometrically the solid angle should be more than 90% but the reduced numbers can be explained by failure of one of the detectors near 0^0 where the charged particle angular distribution peaks.

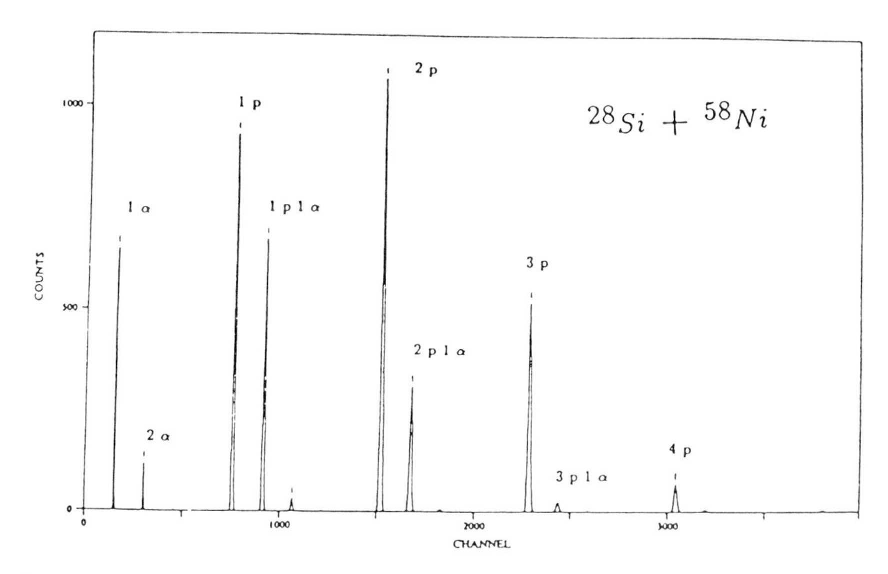

Figure 6. Particle identified spectrum from Silicon-ball used as a selective parameter for the gamma spectra.

A low statistics spectrum of the 4p-channel in the reaction ^{28}Si + ^{58}Ni is shown in Figure 7. Further data has been taken at higher bombarding energies. Experiments with the combination of 5 neutron detectors at 63^0 and the Si-ball as selective devices has been done at 110 MeV. Evaluation of data for ^{83}Zr and ^{82}Y is in progress.

4.4 The BaF_2 calorimeter.

A 60 element array of BaF_2 crystals 95 mm deep form a ball with an outer diameter of about 300 mm but leaving a central spherical hole with 100 mm diameter for the target and selective devices like the Si-ball. The solid angle for this detector is 96% of 4π. The individual detectors are mounted in quintets and their phototubes extend to the outer world through the pentagonal faces of the Nordball. Figure 8. The geometry is such that the Ge-spectrometers look to the target through a collimating hole formed by 3 adjacent BaF_2 crystals.

This instrument takes advantage of the very fast timing of the BaF_2 to give a gamma-neutron separation and an ultra-fast timing reference. The measured values of 490 ps FWHM and 1036 FWTM for pairs of these crystals will make neutron separation up to 10 MeV possible.

This inner ball of Nordball will be a very powerfull selective device. As mentioned earlier total gamma energy and multiplicity can be measured , hit pattern for spin orientation can be registered and isomer triggering will be available. These quantities will be used for event preprocessing. At the moment the ball is being assembled and the necessary software for the event handling under development. The data that are shown in this paper are taken with an array of ten 2" x 3" BaF_2 at a distance of 60 mm from the target as an interim ball with only 20% solid angle.

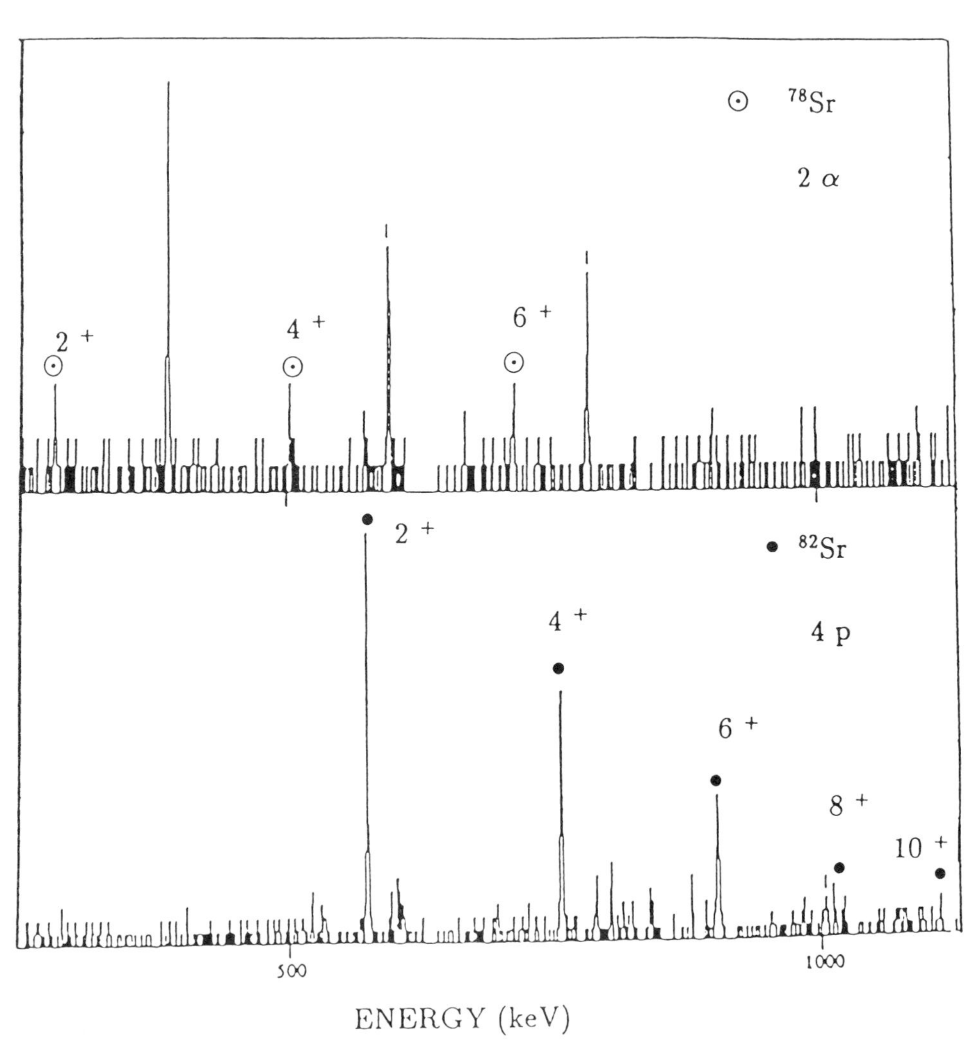

Figure 7. Gamma spectrum showing the 4 proton and 2 α exit channels respectively following the ^{28}Si + ^{58}Ni reaction at 100 MeV. The selection is done by gating from spectra of the type shown in Figure 6. The unassigned lines in the upper spectrum belong to exit channels with higher particle multiplicity.

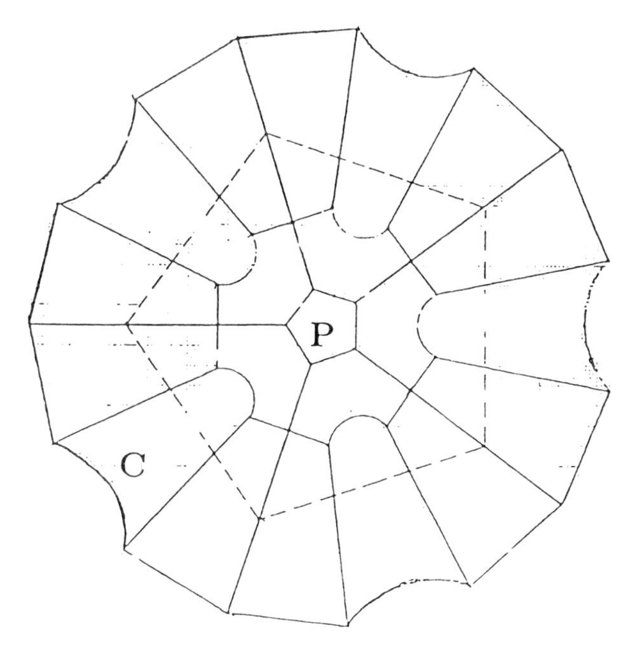

Figure 8. Quintet of BaF_2 crystals as seen from the target towards the outside of the ball. Each of the 5 crystals is coupled to a fast R1668 Hamamatsu pm-tube which extends through a pentagon of the Nordball frame. The conical structures of the crystals marked C, form the colimator for the anti-Compton shields together with 3 adjacent quintets. The small pentagon marked P is a hole separating the 5 individual BaF_2 detectors. Twelve of these quintets form the BaF_2 calorimeter of Nordball.

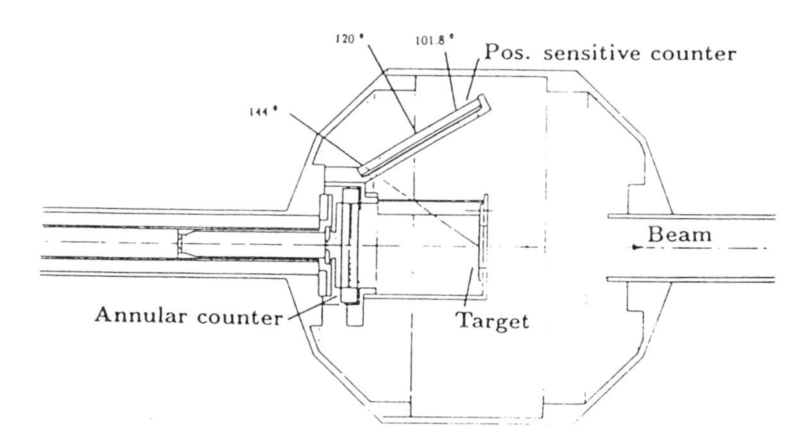

Figure 9. Detector for scattered beam projectiles in Coulomb Excitation experiments. The annular counter covers the angles from 163.9^0 to 176.4^0 and the position sensitive detectors from 101.8^0 to 144.0^0. Only one of the five position sensitive detectors is shown.

4.5 Position sensitive detectors for Coulomb Excitation.

A detector arrangement for measurement of inelastically scattered heavy ions have been developped to fit inside the BaF_2 inner ball and inside the Ge-spectrometer array (see Kato (1988). Figure 9). An anular counter covers angles from 163.9^0 to 176.4^0 and five arms extending from 144^0 to 101.8^0 carry position sensitive detectors.

Test experiments with this device has been performed with ^{32}S ions on a ^{154}Sm target and states up to 14^+ have been observed as well as well- known vibrational states.This detector is still under development.

5. Special Instrumentation.

Apart from instrumentation that primarily aims at higher degree of selectivity, 3 other instruments have been designed and tested in the Nordball. 1) A recoil distance (plunger) apparatus for measurement of short lifetimes. 2) A mini-orange electron spectrometer positioned at 180^0 for measurement of conversion coefficients. 3) A cryogenic target to make solid gas targets (f.ex. Xe) available. The high rates of gamma-gamma coincidences in the Ge-spectrometers will open new possibilities for studies with these instruments.

6. Experiments with Nordball

During the first year of experimental activity only beams from the FN tandem van de Graaff accelerator has been available. With a new ion source injecting completely resolved masses at 150 keV the tandem has been operating at 9.5 MV. This has put constraints on the spin and excitation energy region one has been able to access. Recently the booster accelerator has been completed and experiments involving ^{48}Ti beams at 225 MeV and intensity of 2-3 pnA is being performed, but a presentation of this research would be premature. Therefore the present report will present two experiments involving oxygen and fluorine beams.

6.1 Large Deformation Bands in ^{131}Ce.

Rotational bands with large moments of inertia and with large quadrupole moments (corresponding to $\beta_2 = 0.3 - 0.4$) have been observed by several groups in the mass $A \sim 130$ region during the last few years. The known cases are ^{130}La, ^{131,132}Ce, $^{133-137}$Nd and ^{135,137}Sm. Theoretical calculations suggest that the occupation of the $i_{13/2}$ neutron orbital is responsible for the appeareance of these bands. In most of the known cases, excluding ^{135}Sm and ^{137}Sm and maybe ^{135}Nd, the excitation energy and the spin of these bands relative to the known normally deformed bands is still unknown. The population of the well deformed bands relative to the normally deformed ones ranges from 1 to 20% in the cases studied so far. Projectiles with mass numbers 32 - 82 were used in the previous studies. In order to obtain more information about the feeding of the well deformed bands one can try to excite them by using projectiles of different masses, thus varying the entry state angular momentum and excitation energy.

We describe two different experiments. The first one is a study of ^{131}Ce, in which the well deformed band already was identified by Luo (1987). In a recent measurement by Nolan (1989b) it was shown that the band has a short lifetime consistent with a large quadrupole moment of about 7 eb.

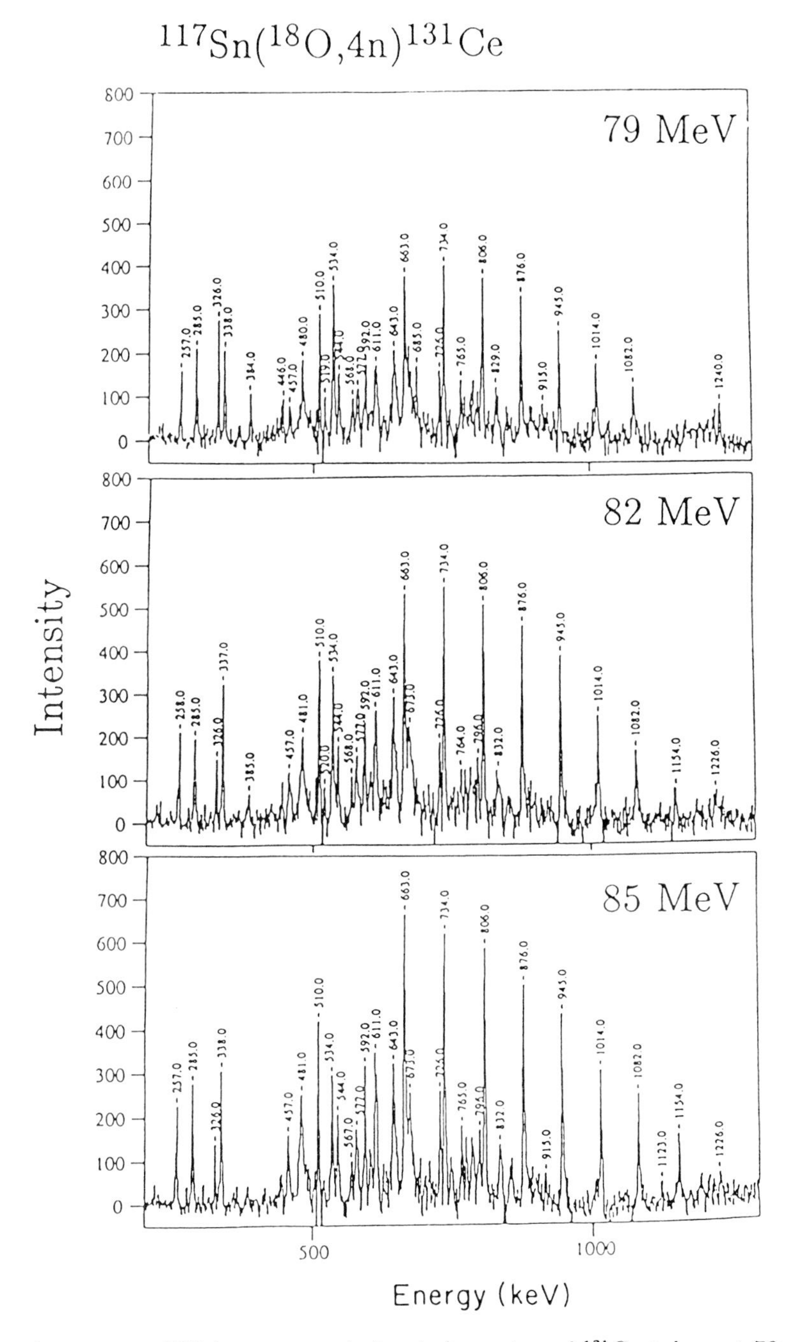

Figure 10. Efficiency corrected gated spectra of ^{131}Ce taken at 79, 82 and 85 MeV incident ^{18}O beam energy. Each spectrum is a sum of several clean gates.

The second experiment is a study of ^{133}Ce, in which case a search for a well deformed band was performed. Ma (1987) has earlier reported on a possible candidate for a $\nu i_{13/2}$ band in this nucleus, but no definite assignment of that band could be made. An extension of this band to higher spins and a search for the decay out of the band was done in the present study.

We have chosen to study odd-N nuclei since calculations of equilibrium deformations (Total Routhian Surfaces) indicate that the well deformed band becomes yrast at lower spin in odd-A than in even-even nuclei. This is an important factor due to the limited angular momentum input in the ^{18}O-induced reaction.

These experiments were performed with the NORDBALL equipped with 15-18 Ge-BGO spectrometers and a multiplicity filter of 10 BaF_2 detectors (total efficiency 20 %). At least two Ge detectors and one BaF_2 detector was required for an event to be stored. The times of each individual Ge and BaF_2 detector signal relative to the OR of all the BaF_2 detectors were also stored for each event.

The first experiment was done by using the reaction ^{117}Sn (^{18}O,4n)^{131}Ce. The target was a stack of two self supporting foils of ^{117}Sn, each with a thickness of about 400 $\mu g/cm^2$. An excitation function was measured using the projectile energies 79, 82 and 85 MeV. About 50 million $\gamma - \gamma$ coincidences were collected at each energy. In Figure 10, three γ-ray spectra measured at 79, 82 and 85 MeV are shown. The efficiency corrected spectra were obtained by summing several clean gates (the same gates were used for all three energies) in the well deformed band. Apart from the regular well deformed band, consisting of ten transitions, one can see γ-rays originating from the low spin part of known bands in ^{131}Ce. Because the spectra are a sum of many gates there are also some contaminant γ-rays, especially from ^{131}La. The three spectra shown in Figure 10 are very similar, except that there is an increase of the population of the well deformed band relative to the other γ-rays as a function of projectile energy.

A level scheme of ^{131}Ce is shown in Figure 11. Only the observed transitions in the well deformed band and some of the low lying normal deformed transitions are shown. The dashed arrows indicate the states to which the well deformed band decays. No discrete transitions connecting the two different structures were observed.

In the second experiment the nucleus ^{133}Ce was studied using the reaction ^{119}Sn(^{18}O,4n)^{133}Ce at 81 and 85 MeV. A stack of two self supporting foils of ^{119}Sn, each with a thickness of about 500 μg/cm^2, was used. About 66 and 115 million γ - γ coincidences were collected at 81 and 85 MeV, respectively. The two data sets at 81 and 85 MeV were added together in the present analysis.
In Figure 12 the measured relative γ-ray intensities in the two bands in ^{131}Ce (open symbols) and in ^{133}Ce (filled circles) are shown. The values were normalized to 100 % at the strongest transition. For ^{131}Ce, three different measurements are shown, corresponding to the beam energies 79 MeV (triangles), 82 MeV (squares) and 85 MeV (circles). The values for ^{133}Ce were extracted using the sum of the two measurements at 81 and 85 MeV. The population of the bands relative to the total population of the nucleus is about 4.0 %, 3.6 % and 2.4 % for ^{131}Ce at 79, 82 and 85 MeV, respectively, and about 9 % for ^{133}Ce (using the sum of the data sets taken at 82 and 85 MeV). In

Figure 12 also shown (as filled diamonds) the relative γ-ray intensities measured by Nolan (1988) using the reactions $^{98}Mo(^{36}S,3n)^{131}Ce$ at 155 MeV and $^{100}Mo(^{36}S,5n)^{131}Ce$ at 150 MeV.

Some comments can be given about the results shown in Figure 12. The three measurements for ^{131}Ce are very similar - that is, within the errors, the same feeding pattern of the well deformed band is observed at the different bombarding energies. There seems to be no "flat region" of the γ-ray intensity versus $\hbar\omega$, that is typical in many of the cases studied previously with heavier ions. The band is fed almost all the way down to the last observed state. As can be seen in the figure there is a striking difference between the light ion and the heavy ion reaction. The intensity pattern of the band in ^{133}Ce is strikingly similar to the one observed for the well deformed band in ^{131}Ce.

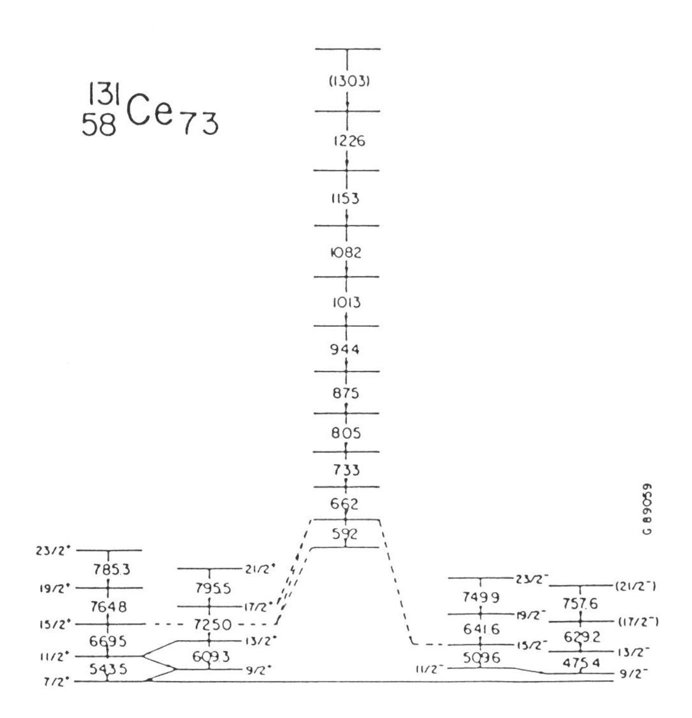

Figure 11. Partial level diagramme of ^{131}Ce. The connecting transitions (dashed) between the strongly deformed band and the low-lying structures are not observed.

The measured dynamical moments of inertia, $J^{(2)}$, for the well deformed band in ^{131}Ce (open squares) and three bands in ^{133}Ce (filled symbols) is shown versus $\hbar\omega$ in Figure 13. For ^{133}Ce the squares correspond to the "candidate" band, the down-triangles and up-triangles to positive and negative parity three quasi-particle bands, respectively. The behaviour of $J^{(2)}$ for ^{131}Ce is quite typical for a well deformed band in an odd-N nucleus in this mass region. It has ha broad bump at $\hbar\omega = 0.4 - 0.6$ MeV and it falls off at higher frequencies. The $J^{(2)}$ of the interesting band in ^{133}Ce, however, behaves quite differently.

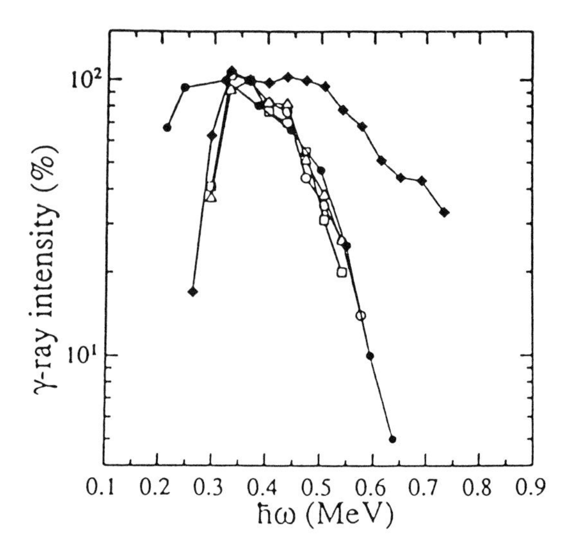

Figure 12. Relative γ-ray intensities in the strongly deformed band in ^{131}Ce and in ^{133}Ce. The values have been normalized to 100 % at the strongest transition. Filled diamonds are from (Luo 1987) ^{98}Mo(^{36}S,3n)^{131}Ce. The filled circles ^{133}Ce and open symboles ^{131}Ce from this work. Triangles, squares and circles are intensities at 79, 82 and 85 MeV respectively.

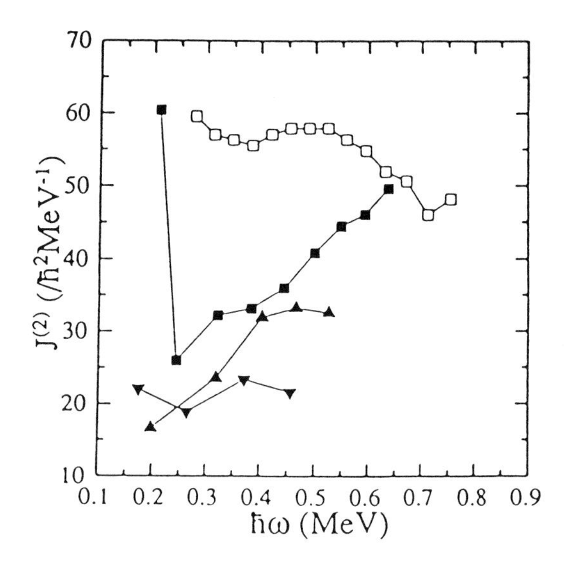

Figure 13. Dynamic moments of inertia J^2 for the well deformed band in ^{131}Ce (open squares) and 3 bands in ^{133}Ce filled symboles. The closed squares represent the band that might be the $\nu i_{13/2}$ band in ^{133}Ce. The down- and up-triangles correspond to positive and negative 3-q.p. bands respectively.

It is increasing rapidly over the whole high frequency region. At the highest frequencies it is much larger than a band with "normal" deformation in this mass region (compare the two other bands in ^{133}Ce shown in the figure).

From the behaviour of the $J^{(2)}$ shown in Figure 13 one cannot draw any definite conclusions about the deformation of this band, because quasiparticle alignments and crossing of bands whith large interactions can also give such effects. A measurement of the quadrupole moment of the band would show if this is a well deformed band or not. The effective life-time at the bottom of the band is probably rather large, because Ma (1987) were able to observe it, although they used a thick backed target.

6.2 Configuration-dependent Transition Rates in ^{157}Ho.

The quadrupole deformation observed in many nuclei is related to the occupation of orbitals with large orbital angular momentum j. This is understood from the deformation dependence of the single particle energies of nucleons in a Nilsson or Wood-Saxon potential. The effect of a specific configuration on the deformation can be studied in a nucleus with an unpaired particle and deformation dependent changes of shape by occupation of different single particle- or quasiparticle- states by the odd nucleon.

The relation between quadrupole deformation, quadrupole moment and transition rate is experimentally accessable by measurement of the latter quantity. It should be noted, however, that the information on deformation extracted from transition rates in odd nuclei may be somewhat obscured by the contribution of the unpaired particle spin to the total. This ambiguity decreases as the intrinsic spin represented by its projection K becomes small compared to the total spin I.

The experimental data reported here are obtained by application of the DSAM method to odd-even rare earth nuclei where a systematic investigation of the N=90 nuclei is carried out. We have chosen ^{157}Ho with its yrast states based on the $7/2^-$[523] proton and a strongly deformation driving yrare structure based on the $1/2^-$[541] proton, Radford (1988), for the transition rate measurements.

The bands were excited by the ^{142}Ce(^{19}F,4n) reaction at 85.5 MeV and recoils allowed to recoil out of the 350 μm^2 target into a 6 mg/cm^2 ^{208}Pb backing evaporated onto it. Fifteen Compton-suppressed Ge-spectrometers detected the γ-rays at angles of 37^0, 101^0 and 143^0 relative to the beam axis.

The experimental spectra show fully shifted peaks up to 59/2 and $61/2\hbar$ in the [523] and the [541] bands respectively. Figure 14 shows a part of the spectra for the [541]band. The upper spectrum is taken at 37^0 the middle at 101 ^{0} and the lower at 143^0. Also shown are the spectral shapes fitted to the peaks to account for the Doppler shift. Similar fits were performed for the [523] yrast band as well and form the experimental basis for the extracted transition rates.

A detailed discussion of the time dependence of the velocity distribution of recoils will be presented in a forthcoming paper (Gascon 1989). It should only be noted at this point that the "clock" to which nuclear lifetimes are compared is controlled by this distribution. The line-shape observed in the spectra is the result of a composite lifetime including several different feeding states. The obvious approach would therefore be to study spectra resulting from a unique feeding transition. Unfortunately this is not possible for statistical reasons in the present data set. Therefore a rather complex mode of analysis had to be performed involving the analytical solution of Bateman's equation for up to 15 states in a strongly coupled band ($\Delta I = 1$ and 2 transitions), each state being fed by a rotational cascade of up to 5 states. A detailed discription of this procedure and the extraction of the transition quadrupole moment for the states will be available in a forthcoming paper (Gascon 1989).

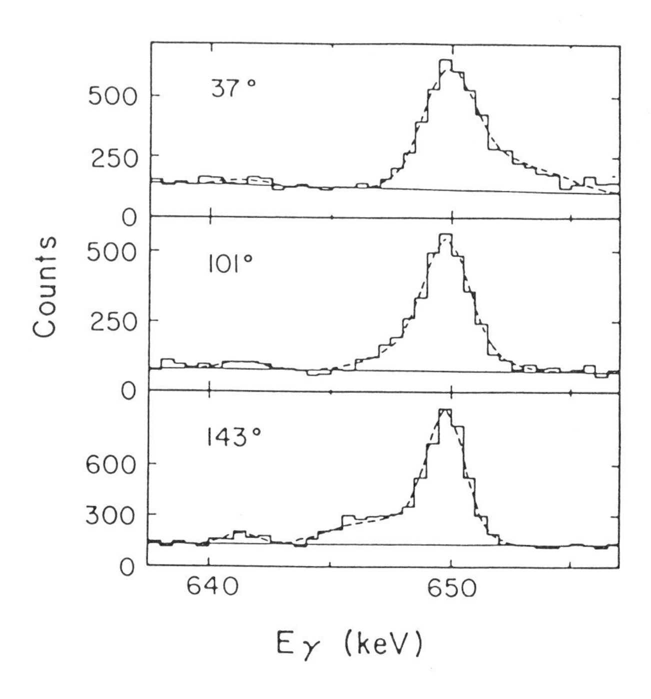

Figure 14. Peak shapes observed for the 37/2 to 33/2 transition in the $[541]1/2^-$ band of ^{157}Ho at 3 differnt angles. The histogrammes are data, the dashed line the least-squares fitted shape of the transition time analysis (Gascon 1989).

The transition quadrupole moments extracted for the $[523]7/2^-$ band are shown as weighted average values from the present Nordball experiment and the results from the ^{124}Sn(^{37}Cl,4n) reaction performed at Daresbury with TESSA 3 (Gascon 1989) in Figure 15. The weighted average of the electric quadrupole moment of the $\Delta I = 2$ transitions in the spin-range 45/2 to 55/2 in the $[523]7/2^-$ band is $4.96\ ^{+0.25}_{-0.19}$ eb. The value averaged in the spin range 33/2 to 49/2 states in the $[541]1/2^-$ band from the Nordball experiment is $5.62\ ^{+0.66}_{+0.5}$ eb.

The ratio of these two numbers ($1.13\ ^{0.14}_{0.12}$) is consistent with an expected increase of the quadrupole moment in the $[541]1/2^-$ configuration. The difference is even more significant if we compare the values based on data coming only from the fluorine experiment (i.e. Q_t=4.38 $^{+0.25}_{-0.19}$ eb; the resulting [541]1/2$^-$/[523]7/2$^-$ ratio, 1.28 $^{+0.17}_{-0.15}$, should be less tainted by the systematic error caused by the uncertainty in the stopping powers.

From the relationship:

$$Q_t = 0.0109\ Z\ A^{2/3}\beta(1 + 0.36\beta)$$

the quadrupole moment can be deduced. A condition for the validity of this relation is that there are no structural changes between states linked by the E2 transitions. This is not the case since the [523]7/2$^-$ band with its sharp backbend indicates a very small interaction (see Figure 16). For the [541]1/2$^-$ band, however, the Q_t value is obtained by averaging over the spin range from 33/2 to 49/2$\hbar$, in the middle of the region where it is interacting with the two-quasineutron structure, see Figure 16. The observed smoothness of the alignment implies a rather large interaction value, but a two band mixing calculation (as in Bengtson 1979) shows that this only leads to a reduction of a few percents of the apparent quadrupole moment. We thus conclude that the present data indicates that the quadrupole deformation of the nucleus ^{157}Ho is larger when the odd proton occupies the [541]1/2$^-$ configuration, rather than the [523]7/2$^-$ configuration, by about 10 to 40 %. This range corresponds to the size of the error bars obtained in a typical high-spin DSAM lifetime analysis, and to obtain a more precise result would require a significant improvement of the method. The experimental ratio is consistent with the value of 1.18 deduced from calculations of band-head quadrupole deformations. The calculated β values for the [523]7/2$^-$ and [541]1/2$^-$ configurations are 0.236 and 0.278, respectively. The corresponding experimental values from the present report, including the uncertainty in the stopping powers, are 0.217 $\pm$ 0.022 and 0.243 $^{+0.034}_{-0.030}$ (assuming γ = 0^0). In both bands, the deformation at high spin is comparable to that calculated at lower spins.

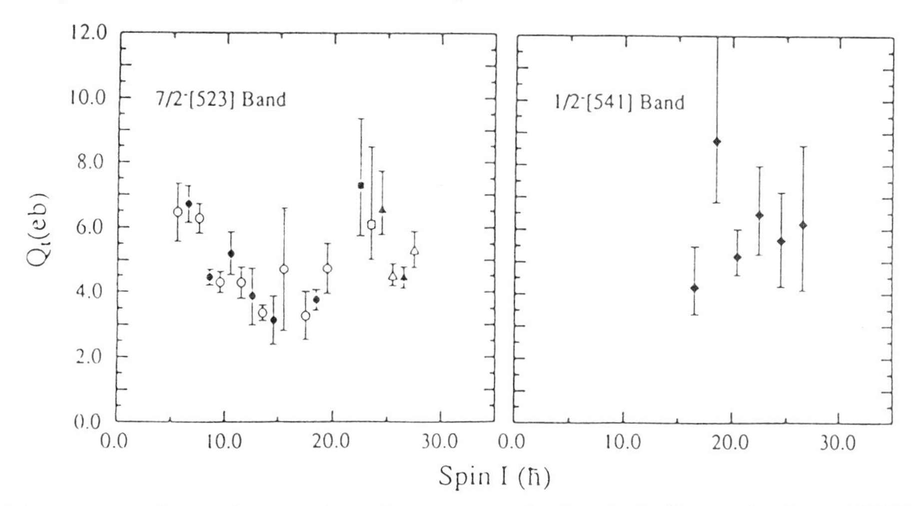

Figure 15. Transition quadrupole moments for bands built on the $h_{11/2}$ [523]7/2$^-$ and $h_{9/2}$ [541]1/2$^-$ configurations in ^{157}Ho. For the [523]7/2$^-$ band the values in the lower spin region are from RDM measurements by Hagemann (1984) re-evaluated with the branching ratios from Radford (1988). Beyond spin 20 the values are the average of measurements with the ^{124}Sn(^{37}Cl,4n) reaction (Gascon 1989) and the ^{142}Ce(^{19}F,4n) reaction in Nordball. The values for the [541]1/2$^-$ band are also from Nordball.

While the deformation of the [523]7/2$^-$ and [541]1/2$^-$ configuration bandheads that can be deduced from the present data lies within the expectation of calculations, the actual value for the [541]1/2$^-$ band is quite different from what would be deduced from a simple cranked-shell model interpretation of the measured crossing frequency of the first neutron crossing. Compared to the [523]7/2$^-$ band, the $i_{13/2}$ neutron crossing ($\hbar\omega_c$) is delayed by as much as 75 keV in the [541]1/2$^-$ band Figure 16.
The deformation dependence of this neutron crossing frequency, calculated from the cranked-shell model with a Woods-Saxon potential, is shown in Figure 17. Indeed, a significant increase of β_2 can cause a shift in $\hbar\omega_c$. It would also lower the $i_{13/2}$ orbital relative to the Fermi level, allowing for a larger interaction at the crossing, and could explain how a sharp backbend in the [523]7/2$^-$ band is transformed into a smooth upbend in the [541]1/2$^-$ band.

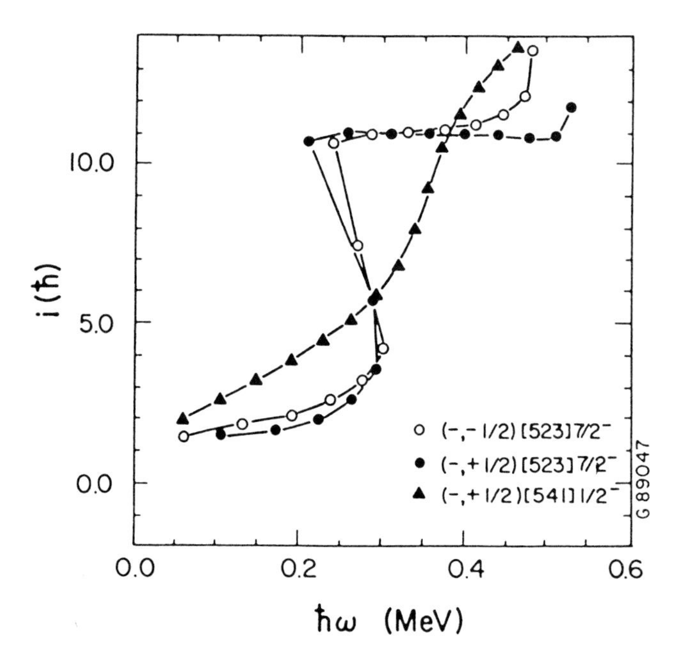

Figure 16. Rotational alignment of the two signatures of the $h_{11/2}$ [523]7/2$^-$ band and the $h_{9/2}$ [541]1/2$_-$ band as a function of rotational frequency.

The experimental values of $\hbar\omega_c$ and the deformations deduced from the present measurement are plotted for both bands in Figure 17. The crossing frequency in the [523] band is consistent with the measured deformation. In contrast, there is a large discrepancy for the [541]1/2$^-$ band. Even allowing for the possibility of a sizeable γ deformation, the crossing frequency corresponding to the β_2 deduced from this experiment cannot exceed 0.3 MeV, which is much lower than the measured value of $\hbar\omega_c$. The change in β_2 (and γ) cannot by itself explain the observed neutron crossing frequency.

It is surprising that the proton quasiparticle configuration can have such a dramatic effect on the neutron quasiparticle spectra beyond the part that can be explained by the polarisation of the core. The mechanism behind this interaction remains to be explained.

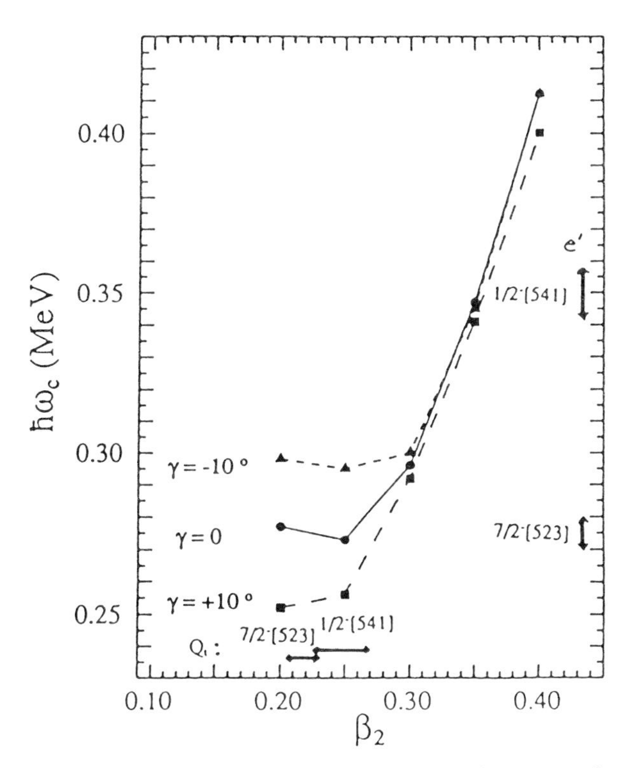

Figure 17. The relation between the rotational frequency for the alignment of the first pair of $i_{13/2}$ neutrons and the quadrupole deformation of the core for different values of γ-deformation. The calculated values are from a cranked shell model with a deformed Woods -Saxon potential and parameters from Nazarewicz (1985). Deduced values of β_2 and crossing frequency from the experiment is indicated by arrows.

7. References.

Arnell S E, Roth H A, Skeppstedt Ö, Nyberg J 1989, Contribution to "6th Nordic Meeting on Nuclear Physics", Univ. of Bergen Sci./Tech. no.208, july 4. pp 108-110.

Bengtsson R and Frauendorf S 1979 Nucl. Phys. **A314** p 27.

Bruce A M, et al. 1988 Phys. Lett.B **215** pp 237-241.

Gascon J et al. 1989 To be submitted to Nucl. Phys.A.

Hagemann G B et al. 1984 Nucl. Phys. **A424** p 365.

Herskind B 1988, Slide report from the Workshop on Nuclear Structure, NBI, Denmark. pp 65-69.

Jerrestam D et al. 1988, Accepted for publication in Nucl. Instr. Meth.

Kato N 1988, Development at the University of Kyusyu, Japan.

Kuroyanagi T and Mitarai S 1989, Development at the University of Kyusyu, Japan.

Lidén F, Johnson A, Kerek A, Dafni E and Sidi M, 1988 Nucl. Instr. Meth. **A273** pp 240-244.

Lidén F et al.,1989 Submitted to Nucl. Instr. Meth.

Luo Y-X et al. 1987 Z.Phys. **A329** p 125.

Ma R et al. 1987 Phys. Rev. **C36** p 2322.

Moszynski M, Bjerregaard J H, Gaardhøje J J, Herskind B, Knudsen P and Sletten G 1989 Nucl. Instr. Meth. **A280** pp 73-82.

Nazarewicz W et al. 1985, Nucl. Phys. **A435** pp 397-447.

Nolan P J and Simpson J 1989a, Report on : Gamic, The Gamma-Ray Microscope, Technical report University of Liverpool.

Nolan P J 1989b Talk presented at Bad Honnef March.

Nyberg J, Kuroyanagi T, Mitarai S,Arnell S E, Roth H A, Skeppstedt Ö 1989, Priv. com. of unpubl. data.

Radford D C et al. 1988, AECL Progress report, Physics and Health Sciences, Physics section, January-June 1988, (AECL-9758) pp 3-9.

Inst. Phys. Conf. Ser. No 105
Paper presented at Int. Conf. on Spectroscopy of Heavy Nuclei, Crete, Greece, 1989

Recent advances in very high resolution crystal gamma-ray spectroscopy

H.G. Börner, J. Jolie, S.J. Robinson and P. Schillebeeckx
Institut Laue Langevin, 156X, 38042 Grenoble Cedex, France

ABSTRACT: The ultra high resolution obtained with modern crystal spectrometers widens the field of possible applications in (n,γ) spectroscopy. Newly developed experimental techniques are discussed. In particular it is shown in some examples how the measurement of gamma ray induced Doppler broadening can be used as a new tool in nuclear structure studies.

INTRODUCTION

Energies, intensities and branching ratios of γ-transitions connecting nuclear states are important quantities which help us to understand the nature of the states involved. In addition the lifetimes of excited levels yield information about absolute transition rates. All together these are crucial quantities for the testing of nuclear models and are therefore vital for our understanding of nuclear structure.

The (n,γ) reaction is (in contrast to most other reactions) non selective in terms of nuclear structure and thus offers the possibility of studying very comprehensive sets of low lying nuclear states (irrespective of their underlying structure), as long as their spins are not too far away from those of the capture states. The unravelling of the resulting complex spectra demands the highest possible resolution. The instruments generally best adapted to this purpose are crystal spectrometers - in various geometries. Such instruments require very strong gamma sources, a condition best met at high flux reactors such as that at the ILL Grenoble.

High precision (n, γ) spectroscopy at the HFBR of the Institute Laue-Langevin now employs very sophisticated techniques using complementary crystal spectrometer geometries. The first generation of these spectrometers became operational in 1973 and they have since contributed considerably to our knowledge of nuclear structure. Just to cite a few examples of more than 100 nuclei now investigated:

i) ^{168}Er [1] has served as a famous testing ground for a huge variety of theoretical models due to its highly complete level scheme even beyond excitation energies of 2.5 MeV.

ii) ^{196}Pt [2] was the first example of a nucleus where the level scheme could be associated with the O(6) limit of the IBA.
iii) ^{195}Pt [3] and ^{195}Ir [4] were compared to supersymmetries which arise in the IBFA, as well as ^{76}As [5].
iv) Very complete level schemes obtained in the K and Cl region [6,7] have allowed the testing of statistical models.
v) Many vibrational states built on both groundstates and excited states have been discerned in actinide nuclei [8,9,10,11].

The flat crystal spectrometer GAMS4 has been operated since 1983 as a joint NIST/ILL facility. It was originally designed and successfully used for the determination of absolute gamma-ray energies in the region $\leq$ 7 MeV [13].

Though nuclear structure studies provide the main source of experiments carried out on these instruments, their versatility is demonstrated by applying the (n, γ)-reaction also to the study of problems in related fields: High precision measurements of standards and fundamental constants; studies of crystal properties (Debye Waller- and structure factor); atomic transitions; atomic collisions; and certain studies in fundamental physics.

We shall, in the following, restrict ourselves to the discussion of some new experimental possibilities which have recently been developed. Furthermore, we shall concentrate, essentially, on the feature of ultra high resolution rather than high precision.

I. SOME INSTRUMENTAL PROPERTIES

The crystal geometries used at the ILL are bent crystal spectrometers of Du Mond type, GAMS 1,2,3 [12] and a two axis flat crystal spectrometer (TACS), GAMS 4 [13]. GAMS 1,2,3 have a luminosity several orders of magnitude higher than that of GAMS 4 (Table 1) and are therefore used for studies where the measurement of an extended spectrum is required. GAMS 4 offers a resolving power which is superior by an order of magnitude (Table 1) to that of GAMS 1,2,3 and is used for high resolution measurements of, generally, a few specific transitions. These spectrometers are installed at both sides of the tangential through-tube H6-H7 of the Grenoble High Flux reactor. Targets are (in an in-pile arrangement) exposed to a neutron flux of $5.5 \times 10^{14} s^{-1} cm^{-2}$.

Table 1: Gross comparison of sensitivity and resolution of DuMond and TACS geometry at ILL

	Du Mond	Two Axis
Peak efficiency (approximate)	10^{-7}	10^{-12}
Resolution at 1 MeV (approximate, n=refelction order)	$\leq \frac{300}{n}$ eV	$\leq \frac{30}{n}$ eV

Further details concerning the instruments may be found elsewhere [12, 13, 14, 15].

II. SPECIFIC EXAMPLES FOR THE APPLICATION OF HIGH RESOLUTION (n,γ) SPECTROSCOPY

A The ^{176}Lu-Level Scheme

It had long been realized that ^{176}Lu might be a good candidate to study the age of stellar s-process material [16,17]. This is due to the fact that it is only produced in the s-process (being shielded from r-process production by stable ^{176}Yb) and that its 7^- ground state (details see [21]) decays to ^{176}Hf with a half-life of 3.6×10^{10} years. The ^{176}Lu to ^{176}Hf abundance ratio could then, in principle, give the key to the use of ^{176}Lu as a cosmic clock. However, there exists an isomeric 1^- state in ^{176}Lu (at approx. 120 keV) which is also strongly populated in n-capture and which also decays to ^{176}Hf, but with a halflife of 3.7 h. Now in a hot stellar environment excited states can be populated by thermal excitation and eventually decay back to both the ground and the isomeric state [18]. As such the initial population belance following s-process n-capture may be changed. The question as to whether such

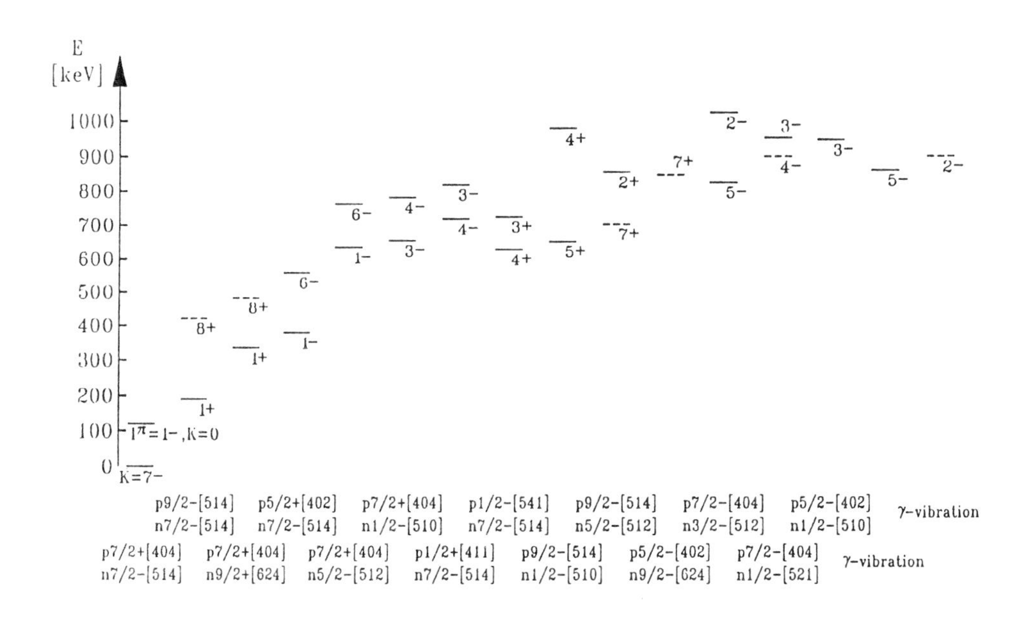

<u>Fig. 1:</u> Bandheads observed in ^{176}Lu. For the K^{π} = 0^- band the I^{π} = 1^- isomeric state is shown instead of the I^{π} = 0^- bandhead.

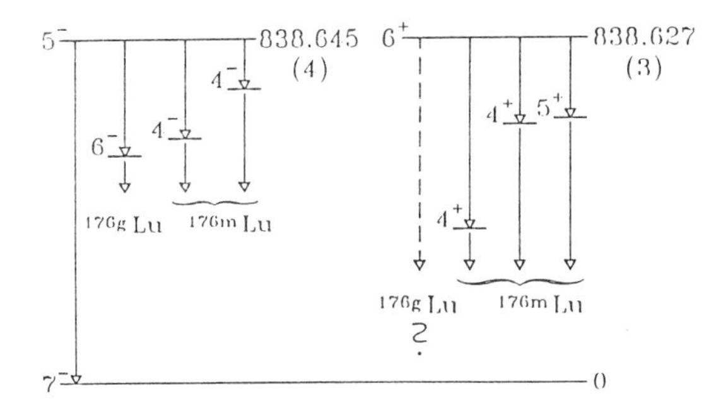

<u>Fig. 2:</u> The 838 keV level doublet and its decay properties

thermal excitation is important can only be answered via a very detailed knowledge of the level structure in ^{176}Lu.

Previous studies of ^{176}Lu [19,20] had yielded only two completely separate parts of the level scheme, one linked to the 7^- ground state and the other to the 1^- isomeric state. No intermediate γ-transition,

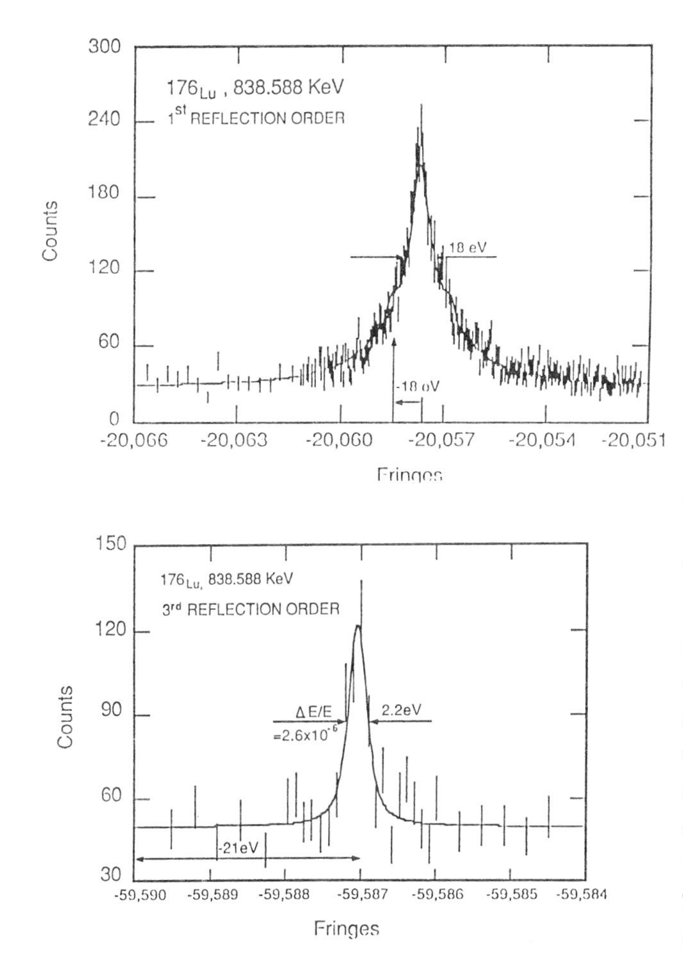

Fig. 3: Lineprofiles of the 838 keV transition in ^{176}Lu measured in n = 1st and 3rd order, respectively. The structure in the 1st order measurement is due to the diffraction behavior of the perfect crystals used. It emerges from interference effects which result from the abrupt boundaries created by surfaces of the crystals [13]. This structure also depends on the wavelength and reflection order and does not appear for the 3rd order here. The solid lines represent the theoretical lineshapes (and no fits) calculated for a 838 singlet. Only the peak height is normalized to the measurement.

linking these two parts (and thus establishing a possible path from gound state to isomeric state via thermal excitation) were known. In a recent investigation N. Klay et al. [21, 22] have used the Grenoble spectrometers as part of a very exhaustive study of the level scheme.

This has led to the assignment of more than 25 rotational bonds below 1 MeV (Figure 1). In addition levels have been observed which provide the link between the ground and isomeric states. Amongst these the most important astrophysically is the 5^- level at 838 keV which is the lowest lying state of this special group.

Yet there was still an ambiguity left: One can construct a second (6^+ or 5^+) 838 keV level via Ritz combination built on transitions which decay only into the isomeric group. This level is degenerate (Figure 2) within 18 ± 5 eV with the 5^- level. Does it also decay to the groundstate - and is hence important for the astrophysics - or not ? Coincidences do not give the answer.

We have measured the corresponding 838 keV transition (figure 3) using the flat crystal spectrometer and were able to clearly show that this transition is a singlet. The 5^- level remains the most important!

The influence of thermal excitations on the population equilibrium depends not only on the excitation energy but also on the halflives of the mediating levels involved. The fact that the 5^- level is directly fed by a primary transition allowed the application of the technique of Gamma Ray Induced Doppler broadening (GRID) to establish an important preliminary lower limit of $\tau \geq 1$ ps for the lifetime of this level (the method will be explained briefly in the next chapter). An upper limit of $\tau \leq 0.5$ ns has also been obtained by Andrejtschef et al [23] at ILL using the coincidence centroid shift method.

This new information now allows the more precise calculation of the s-process branching factor for the production of ^{176}Lu [22].

B Lifetime Measurements using GRID

It is well known that the lifetimes of γ-decaying nuclear states in the 10^{-14} - 10^{-12} s range are accessible by the Doppler shift attenuation (DSA) method whenever the initial recoil velocity v_R/c is comparable to the relative energy resolution $\delta E/E_\gamma$ of the γ-ray spectrometer. Typical recoil velocities from heavy ion reactions are in the order of 0.1 to 10% of the velocity of light. We have recently shown [24, 25] that ultra high resolution gamma spectroscopy can be used to extract lifetimes $<10^{-11}$ s of levels populated following neutron capture. Similarly to the

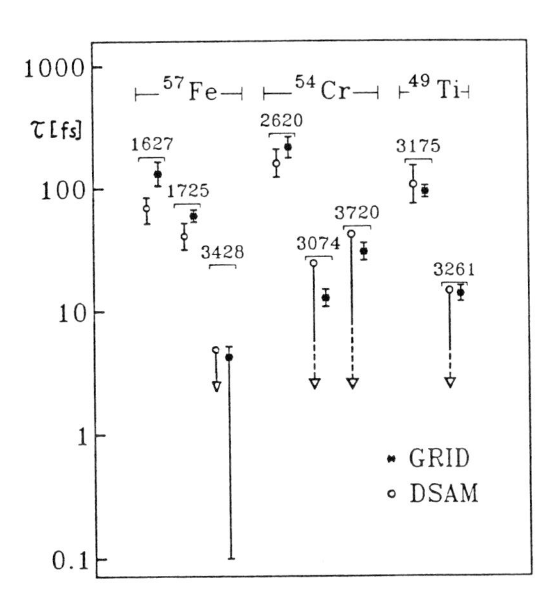

Fig. 4: Comparison of lifetimes obtained from GRID and DSAM.

DSA-method we use the fact that a recoiling nucleus is slowed down in the target material and that this slowing down time can be compared to the lifetime of an excited state. However, here the recoil is obtained from the emission of preceding γ-rays which leads to recoil velocities in the range 10^{-4} to 10^{-6} of the velocity of light. Since the initial recoil directions are isotropically distributed one measures a Doppler broadening. Details of the method are described elsewhere [25,31]. Here we merely summarize:

1) The initial recoil velocity is given by $v_R = E_0/Mc$ with E_0 the γ-ray energy causing the initial recoil and M the Mass of the recoiling nucleus.

2) The maximum Doppler broadening is $2\Delta E^D_{max} = 2\frac{E_1 E_0}{Mc^2}$ with E_1 the energy of the γ-transition depopulating the level of interest.

3) The observed lineshape is described by

$$I(E)=I_S(E)\exp(-\sum_{k=0}^{N-1}\frac{t_k}{\tau})+\sum_{k=1}^{N} I_D(E,v_{n-1})\left[\exp(-\sum_{k=0}^{n-2}\frac{t_k}{\tau})-\exp(-\sum_{k=0}^{n-1}\frac{t_k}{\tau})\right]$$

where $I_S(E)$ describes the stopping peak (taking into account thermal motion of the target nuclei), $I_D(E,v)$ describes the Doppler-broadening at a given velcity v, t_k is the averaged time between atomic collisions and is a function of a specific stopping theory (see ref[25]). The theoretical lineshape I(E) is folded with the instrumental response function (ref[25]) and then fitted to the measured lineshape.

Lifetimes have already been determined in a variety of nuclei. Fig. 41 shows a comparison with other data in the region around A=54. One can conclude that, especially in the region around 10 to 50 fs, the GRID-method is very attractive compared to other methods. Here we would like to discuss only one specific example showing that GRID can contribute to identify mixed symmetry states.

The lowest mixed symmetry state in vibrational like nuclei is predicted to be a 2^+ level corresponding to the antisymmetric combination of the collective proton and neutron components which form the 2^+_1 level. One of the basic signatures of such a state is thought to be a strong M1 transition to the 2^+_1 state. It has been proposed [26] that in ^{56}Fe the mixed symmetry strength is shared almost equally between the 2^+_2 (2658 keV) and 2^+_3 (2960 keV) levels with a summed M1 strength to the 2^+_1 level of 0.38 μ_N^2. Since the excitation energy of this mixed symmetry mode is predicted to depend on the number of valence nucleous [27] it was expected to lie higher in ^{54}Cr than in ^{56}Fe. Using GRID the line profiles of the 1785 keV (2^+_2 - 2^+_1) and 2239 keV (2^+_3 - 2^+_1) transitions in ^{54}Cr were measured [28] and gave lifetimes for the 2^+_2 (2620 keV) and 2^+_3 (3074 keV) states of 183 fs and 13 fs respectively. These give B(M1; 2^+_2 - 2^+_1) = 0.04 μ_N^2 and B(M1; 2^+_3 - 2^+_1) = 0.39 μ_N^2 which suggest that in ^{54}Cr the mixed symmetry strength is concentrated in the 2^+_3 level, leaving the 2^+_2 level as an almost pure 2 phonon state. These results have provided the first direct evidence for a pure mixed symmetry state and confirm the predicted energy dependence of such excitations.

C Measurement of the Electron Neutrino Helicity

A continuous effort has been made to further reduce background rates and even further increase the resolving power of GAMS4. The current high sensitivity to small recoil velocities now makes it possible to try a new approach to the determination of the neutrino helicity [29]. This is, in principle, a refinement of the Goldhaber experiment [30] employing neutron capture on ^{151}Eu followed by K-capture leading into ^{152}Sm. The decay scheme is shown in the insert Fig.5. The basic feature is that the angular momentum couplings between states leading to the 963 keV $1^- - 0^+$ transition imply a direct correlation between the helicity of the 963 keV transition and the neutrino helicity as a function of the angle between the linear momenta of the γ and neutrino. Regarding K-capture only, in the extreme limit γ's emitted antiparallel (parallel) to the neutrino will have the same (opposite) helicity as the neutrino itself, i.e. γ's which are emitted under 0° and 180° with respect to the recoil direction will have exactly opposite helicities. This recoil is caused by the emission of the neutrino. The helicity of the γ's can be measured by means of a polarimeter allowing the determination of the variation in photon helicity across the line profile. This is then equivalent to the determination of the helicity of the neutrino. The recoil direction is given by $\cos\Theta = \delta E/\Delta E^D_{max}$ where δE corresponds to the distance from the Center of the Doppler profile to a given energy cut within the profile with $0 < |\delta E| < \Delta E_{max}$. The neutrino induced recoil velocity is $v_R/c = 6.2 \times 10^{-6}$. This leads to a maximum Doppler broadening of 2 $\Delta E_{max} \simeq 12$ eV which is to be compared to the instrumental resolution of $\simeq 3$ eV (figure 5).

The discussion above is idealized. The correlation between neutrino helicity and the direction of neutrino emission can only be uniquely established for 963 keV γ-transitions which are emitted before any slowing down of the recoiling atom has occured. This is only fulfilled for some of the emitted photons. This part is a function of the lifetime of the 963 keV level (fortunately as short as ~ 40 fs) and the composition of the target. Initial measurements have been carried out to measure this quantity for targets of EuF_2, EuF_3 and Eu_2O_3. Essentially this means that only the border regions of the Doppler profile (close to 0° or 180°) will be useful for the measurements. This is due to the fact that these maximally Doppler shifted photons have undergone no slowing

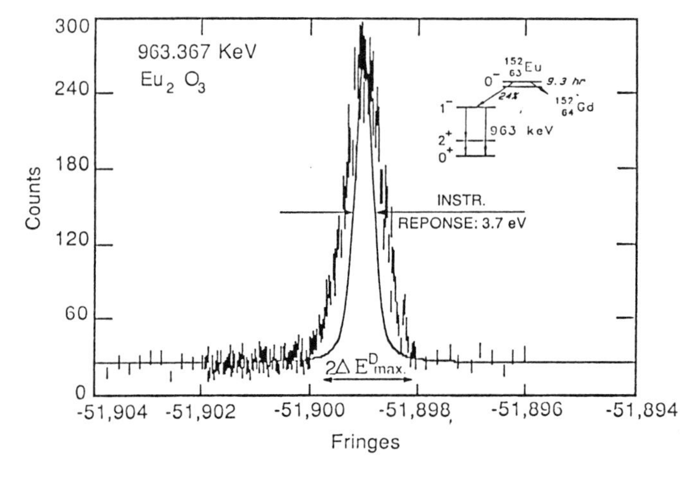

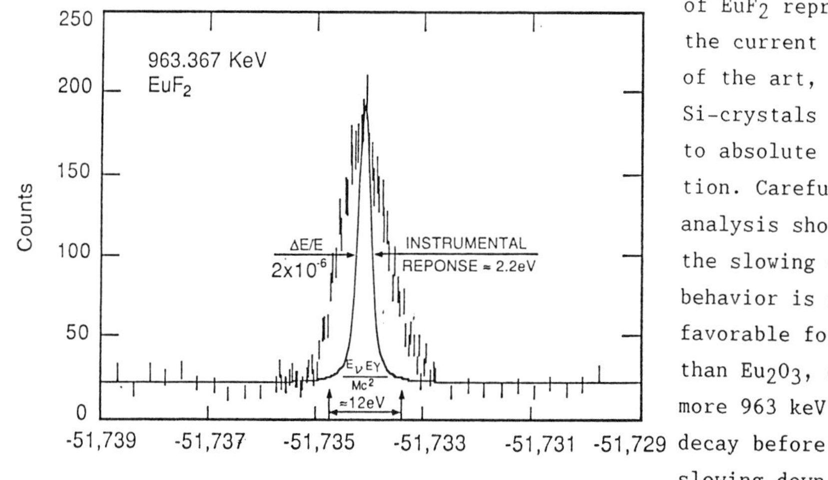

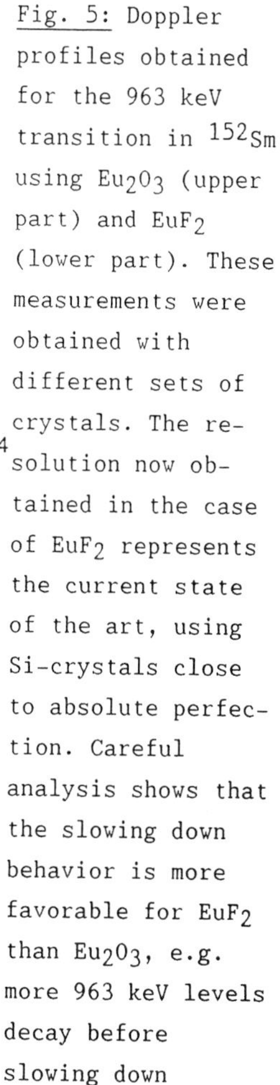
Fig. 5: Doppler profiles obtained for the 963 keV transition in ^{152}Sm using Eu_2O_3 (upper part) and EuF_2 (lower part). These measurements were obtained with different sets of crystals. The resolution now obtained in the case of EuF_2 represents the current state of the art, using Si-crystals close to absolute perfection. Careful analysis shows that the slowing down behavior is more favorable for EuF_2 than Eu_2O_3, e.g. more 963 keV levels decay before slowing down occurs.

down, thus keeping the opposite direction with regard to the neutrino, and are emitted at 0° and 180° with respect to the neutrino direction, thus having the same spin direction as the neutrino.

The aim of the experiment will be to put a limit on contributions of right handed currents in the left handed system of the neutrino helicity, with -hopefully- a precision somewhat better than current limits.

CONCLUSION

Neutron capture γ-ray spectroscopy is still a very powerful tool in probing the nucleus at low energy. In the last decades it has provided important tests of nuclear models via the construction of very complete level schemes. Now the extremely high resolution obtained allows the unravelling of spectra with even higher transition densities. This makes it possible to obtain detailed schemes of odd odd nuclei and/or to study the structure at higher excitation energies than in the past.

Ultra high resolution obtained with the flat crystal spectrometer opens a new field of applications for (n, γ) spectroscopy. The ability to observe Doppler broadening at low recoil velocities adds a new tool to nuclear structure study: GRID. This method not only allows the determination of lifetimes of nuclear excited states but leads also to the study of atomic stopping mechanisms in solids at low kinetic energies or to the study of fundamental properties like the neutrino helicity.

These new developments have all been obtained within the last two years and are still in their infancy. A lot more work, experimental and theoretical has still to be done. High resolution (n, γ) spectroscopy remains an exciting field.

REFERENCES

[1] W.F. Davidson, D.D. Warner, R.F. Casten, K. Schreckenbach, H.G. Börner, J. Simic, M. Stoyanovic, M. Bogdanovic, S. Koicki, W. Gelletly, G.B. Orr, M.L. Stelts
J. Phys. G7 (1981) 455

[2] J.A. Cizewski, R.F. Casten, G.J. Smith, M.S. Macphail, M.L. Stelts, W.R. Kane, H.G. Börner, W.F. Davidson
Nucl. Phys. A323 (1979) 349

[3] D.D. Warner, R.F. Casten, M. Stelts, H.G. Börner, G. Barreau
Phys. Rev. C26 (1982) 1921

[4] J.A. Cizewski, G.G. Colvin, H.G. Börner, F. Hoyler, S.A. Kerr, K. Schreckenbach
Phys. Rev. Lett. 58 (1987) 10

[5] J. Jolie, Poster to this conference
and
F. Hoyler, J. Jolie, G. Colvin, H.G. Börner, K. Schreckenbach, P. van Isacker, P. Fettweis, H. Göktürk, J.H. Dehaes, R.F. Casten, D.D. Warner, A.M. Bruce
to be submitted to Nucl. Phys.

[6] B. Krusche, C. Winter, K.P. Lieb, P. Hungerford, H.H. Schmidt, T. von Egidy, H.J. Scheerer, S.A. Kerr, H.G. Börner
Nucl. Phys. A 439 (1985) 219

[7] B. Krusche, K.P. Lieb, H. Daniel, T. von Egidy, G. Barreau, H.G. Börner, R. Brissot
Nucl. Phys. A 386 (1982) 245

[8] H.G. Börner, H.R. Koch, H. Seyfarth, T. von Egidy, W. Mampe, J.A. Pinston, K. Schreckenbach, D. Heck
Z. Phys. A286 (1978) 31

[9] J. Almeida, T. von Egidy, P.H.M. van Assche, J. Valentin, H.G. Börner, W.F. Davidson, K. Schreckenbach, A.I. Namenson
Nucl. Phys. A315 (1979) 71

[10] P. Jeuch, T. von Egidy, K. Schreckenbach, W. Mampe, H.G. Börner, W.F. Davidson, J.A. Pinston, R. Roussille, R.C. Greenwood, R.E. Chrien
Nucl. Phys. A 317 (1979) 363

[11] D.H. White, H.G. Börner, R.W. Hoff, K. Schreckenbach, W.F. Davidson, T. von Egidy, D.D. Warner, P. Jeuch, G. Barreau, W.R. Kane, M.L. Stelts, R.E. Chrien, R.F. Casten, R.G. Lanier, R.W. Lougheed, R.T. Kouzes, R.A. Naumann, R. Dewberry
Phys. Rev. C35 (1987) 81

[12] H.R. Koch, H.G. Börner, J.A. Pinston, W.F. Davidson, R. Roussille, J.C. Faudou, O.W.B. Schult
Nucl. Inst. Meth. 175 (1980) 401

[13] E. Kessler, G.L. Greene, M.S. Dewey, R.D. Deslattes, H.G. Börner, F. Hoyler
J. Phys. G 14 (1988) 167

[14] R.D. Deslattes, E.G. Kessler, W.C. Sander, A. Henins
Annals of Phys. 129 (1980) 378

[15] F. Hoyler, H.G. Börner, S. Robinson, G.L. Greene, E. Kessler, M.S. Dewey
J. Phys. G, 14 (1988) 161

[16] J. Andouze, W.A. Fowler, D.N. Schramm
Nature, 238 (1972) 8

[17] H. Beer, F. Käppeler
Phys. Rev. C21 (1980) 534

[18] H. Beer, F. Käppeler, K. Wisshak, R.A. Ward
Ap. J. Suppl. 46 (1981) 295

[19] M.K. Balodis et al.
Nucl. Phys. A194 (1972) 305

[20] R.A. Dewberry, R.K. Sheline, R.G. Lanier, L.G. Mann, G.L. Struble
Phys. Rev. C24 (1981) 1628

[21] N. Klay, H.G. Börner, F. Käppeler, H. Beer, G. Schatz, B. Krusche, S. Robinson, F. Hoyler, K. Schreckenbach, U. Mayerhofer, T. von Egidy, G. Hlawatsch
Report KfK 4508 (1989) 14

[22] N. Klay
Fifth Workshop on "Nuclear Astrophysics"
Schloß Ringberg, Feb. 1988,
in press

[23] W. Andrejtscheff et al.,
private communication

[24] H.G. Börner
J. Phys. G14 (1988) 143
and J. Jolie, H.G. Börner, F. Hoyler, S. Robinson, M.S. Dewey
Inst. Phys. Conf. Ser. 80 F 586

[25] H.G. Börner, J. Jolie, F. Hoyler, S. Robinson, M.S. Dewey, G. Greene, E. Kessler, R.D. Deslattes
Phys. Lett. B 215 (1988) 45

[26] S.A.A. Eid, W.D. Hamilton, J.P. Elliott
Phys. Lett. B 166 (1986) 267

[27] A.B. Balantekin, B.R. Barrett
Phys. Rev. C35 (1987) 878

[28] K.P. Lieb, H.G. Börner, M.S. Dewey, J. Jolie, S.J. Robinson, S. Ulbig, Ch. Winter
Phys. Lett. B 215 (1988) 50

[29] H.G. Börner et al. Research proposal to the Institut Laue Langevin (1988) (unpublished)

[30] M. Goldhaber, L. Grodzins, A.W. Sunyor
Phys. Rev. 109 (1958) 1015

[31] J. Jolie et al.
these proceedings

Inst. Phys. Conf. Ser. No 105
Paper presented at Int. Conf. on Spectroscopy of Heavy Nuclei, Crete, Greece, 1989

Excitation and decay of giant multipole resonances in intermediate energy heavy ion reactions

J. R. Beene and F. E. Bertrand

Oak Ridge National Laboratory,* Oak Ridge, Tennessee 37831, U.S.A.

ABSTRACT: The role of intermediate energy heavy ions in the study of giant multipole resonances is explored, with emphasis on gamma decay coincidence experiments. Experiments on ^{208}Pb bombarded by 84 MeV/nucleon ^{17}O are discussed and compared with earlier work at 22 MeV/nucleon. The role of Coulomb excitation in the 84 MeV/nucleon data is emphasized and some consequences for study of isovector resonance strength are explored. A comparison of the excitation and decay of the isovector giant dipole resonance in ^{208}Pb and ^{209}Bi excited with 84 MeV/nucleon ^{17}O scattering is presented.

1. INTRODUCTION

The study of simple nuclear excitations, such as single particle states or giant multipole resonances (GR), has taught us a great deal about the physics of nuclei. This talk is concerned with recent experiments carried out by our group at ORNL and GANIL on the excitation and decay of giant resonances.

The isovector giant dipole resonance has been studied almost since the beginning of quantitative experimental nuclear physics (Danos and Fuller 1965). The field was broadened considerably with the identification of non-dipole resonances in the early 1970's (Bertrand 1981). Since that time, a great deal of information on the gross properties (excitation energy, width, strength) has been accumulated. As an illustration, a sample of multipole vibrations which have been identified in ^{208}Pb (only those that can be ascribed to vibrations in ordinary space; spin and spin-isospin excitations, e.g., are omitted) are listed in Table 1. The prefix isoscalar (IS) or isovector (IV) refer to whether the oscillation of protons and neutrons is predominantly in phase (IS) or out of phase (IV). Of the resonances listed only the IVGDR, ISGQR and ISGMR are well established as compact peaks which exhaust a large fraction of the relevant energy weighted sum rule (EWSR) in a wide range of nuclei.

While much has been learned, there are still many open questions. For example there is good data on several isoscalar resonances, but except for the venerable IVGDR, there is little systematic evidence for strong compact IV states. Our knowledge of high multipolarity resonances ($L > 3$) is meager and sometimes contradictory, in part because the high L strength

*Operated by Martin Marietta Energy Systems, Inc. under contract DE-AC0521400 with the U.S. Department of Energy.

appears to be broadly fragmented. There is little information on concentrated collective strength between the IVGDR and the nucleon resonances (Δ, etc.). Multiphonon giant resonances or other exotic collective states might populate this "desert." Finally there is little data on the

Table 1

Resonance in ^{208}Pb	IS/IV	Excitation Energy	Designation in this Paper
Dipole	IV	13.6 MeV	IVGDR
Dipole	IS	17 ?	ISGDR
Quadrupole	IS	10.6	ISGQR
Quadrupole	IV	22 ?	IVGQR
Monopole	IS	13.9	ISGMR
Monopole	IV	32?	IVGMR
Octupole	IS	20	ISGOR
Hexadecapole	IS	12	ISGHR

microscopic structure of the GR states, or on the interaction of these simple, highly excited states with the large number of more complex states of the nuclear continuum at the same excitation (i.e., the damping process).

The experiments which we will discuss involve excitation of resonances (mostly in ^{208}Pb) using inelastic ^{17}O scattering, and the study of the subsequent photon decay of the resonance region in coincidence with the scattered particle. Most of the talk deals with data acquired at the French facility GANIL, using 84 MeV/nucleon ^{17}O beams. Some 22 MeV/nucleon data acquired at ORNL will also be discussed. These experiments represent first steps in an effort to address some of the questions mentioned above.

2. HEAVY ION EXCITATION OF GIANT RESONANCES

For the past decade heavy ions have been touted as potentially important tools in the study of GR. They are just beginning to live up to this promise. Generally speaking, heavy ions have not proved useful in conventional singles studies of GR by inelastic scattering because the angular distributions are not very sensitive to angular momentum transfer. The real virtue of heavy ions lies in the large excitation cross sections which can be achieved, making coincidence studies of resonance decay feasible, and perhaps making the identification of multiphonon GR states possible.

The cross section for GR excitation increases rapidly as the bombarding energy is increased. This is illustrated in Fig. 1, which shows the angular distribution for excitation of the ISGQR in ^{208}Pb by ^{17}O scattering at a range of energies, and in Fig. 2 which shows the peak differential cross sections (at an angle just inside the grazing angle) for excitation of various multipoles, all assumed to lie at an excitation energy of 14 MeV, and all assumed to exhaust 100% of the respective EWSR. The ratio of GR peak cross section to that of the underlying continuum also increases rapidly with bombarding energy. This is illustrated in Fig. 3, which shows the inelastic singles spectra observed with the ^{208}Pb(^{17}O,^{17}O') reaction at 22 MeV/nucleon and 84 MeV/nucleon. The latter spectrum shows a peak-to-continuum ratio at least three times better than the best observed with any other probe.

As the bombarding energy increases Coulomb excitation plays an increasingly important role. This has two important effects. (1) Unlike nuclear excitation, Coulomb excitation excites IS and IV states equally well. This makes study of IV states feasible in inelastic hadron scattering at high energy. (2) The excitation of quadrupole and especially dipole states, which benefit most from Coulomb excitation, increase much more rapidly than the cross section for excitation of the $L = 0$ or $L > 2$ GR. This is evident in Fig. 2, and is further illustrated in Fig. 4 which is a semiclassical Coulomb excitation calculation (Bertulani and Baur 1985) extending to much higher energy. Note that the excitation probability of the GDR continues to increase with bombarding energy eventually substantially exceeding the geometrical cross section. Consequently systematic study of multiphonon GDR excitation is a real possibility with relativistic heavy ions.

The probability of n-phonon resonance excitation scales roughly as Z^{2n} where Z is the charge of the exciting ion. Hence high Z ions will play an important role in future experiments. We will return to this topic briefly at the end of this talk. In the more modest experiments reported here, we chose ^{17}O as a projectile because it has only three particle-stable states. This simplifies the inelastic spectra, since if the ^{17}O projectile is excited above 3.8 MeV, it breaks up and is not detected as ^{17}O.

3. EXPERIMENTS

The experiments reported here involved inelastic scattering of ^{17}O by various targets at 22 MeV/nucleon (carried out at ORNL) and 84 MeV/nucleon (at GANIL). The reader interested in experimental details should consult Beene *et al.* (1989a and b) and (1988), Bertrand *et al.* (1988) and Barrette *et al.* (1988). A very brief outline will be given here.

In the 84 MeV/nucleon experiments at GANIL, the inelastically scattered ^{17}O particles were detected and identified in the SPEG energy loss magnetic spectrometer over a range of scattering angles from ~1° to 5°. Gamma rays in coincidence with inelastic scattering were detected in arrays of BaF_2 scintillators, and distinguished from neutrons by time of flight. Two gamma detector configurations have been used. In one case twenty-eight 10 cm x 14 cm long BaF_2 crystals with hexagonal cross section were arranged in four clusters of seven detectors. The second configuration was made up of six of these 7-pack arrays together with three groups of nineteen 6 cm x 20 cm long hexagonal cross section crystals.

The 22 MeV/nucleon experiments employed the ORNL Spin Spectrometer (a 70 detector, nearly 4π NaI detector crystal ball) for gamma detection, and a ring of six cooled surface barrier E-ΔE telescopes for ^{17}O detection and identification. The particle detectors were all positioned at $\theta = 13°$, with opening angles $\Delta\theta = 3°$ and $\Delta\phi = 9°$.

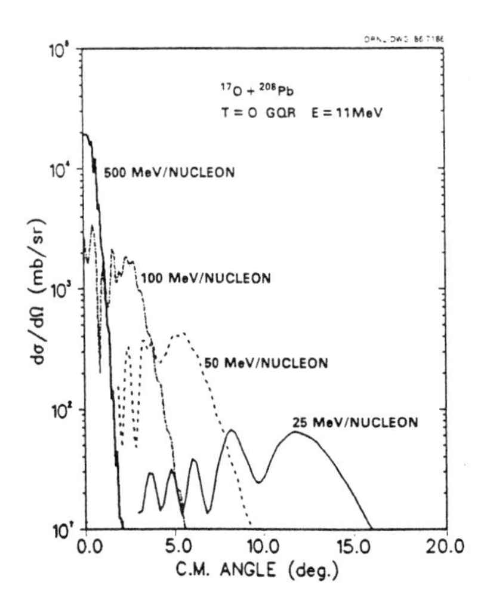

Fig. 1. Calculated angular distributions for the isoscalar giant quadrupole resonance excited by inelastic scattering of various energy ^{17}O ions.

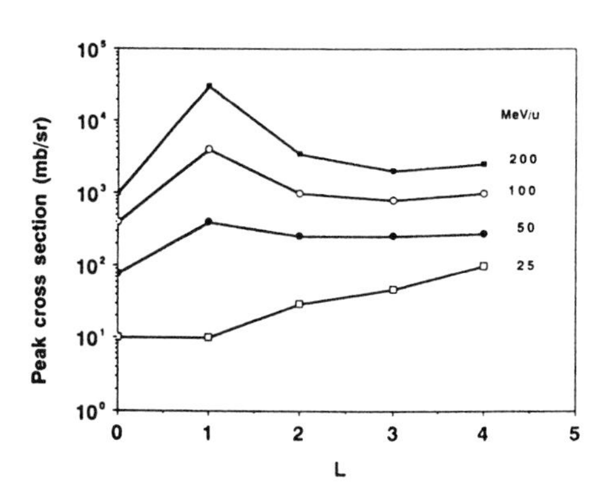

Fig. 2. The peak differential cross section as a function of multipolarity for fictitious states at 14 MeV having 100% of the corresponding EWSR, in the reaction $^{208}Pb(^{17}O,^{17}O')$ at the bombarding energies (MeV/nucleon) given. The L = 1 state is assumed to be isovector; all other isoscalar.

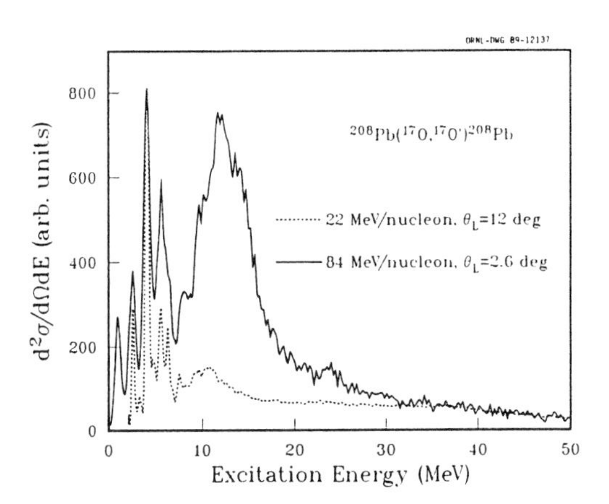

Fig. 3. Spectra from inelastic scattering of 84 (Barrette *et al.* 1988) and 22 (Beene *et al.* 1989b) MeV/nucleon ^{17}O from ^{208}Pb. The two spectra are normalized in the unstructured continuum near 40 MeV.

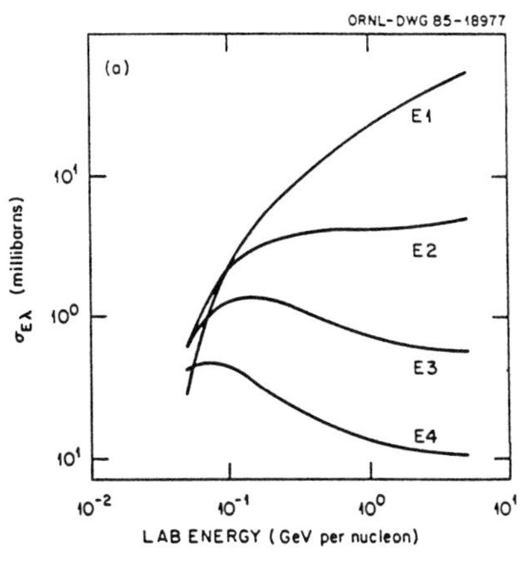

Fig. 4. Total cross section for excitation of a fictitious state at 22 MeV in ^{208}Pb which exhausts the full EWSR for the multipolarity given. The beam is ^{16}O.

4. EXCITATION AND DECAY OF THE IVGDR

In the $^{208}Pb(^{17}O,^{17}O')$ reaction at 22 MeV/nucleon, near the grazing angle, excitation of the ISGQR dominates the inelastic singles spectrum in the giant resonance region. This is illustrated in Fig. 5 which shows the decomposition of the GR region (Beene *et al.* 1989) (after an empirical background was subtracted). This decomposition is not a free fit, but was based on states observed in high resolution (p,p') (Bertrand *et al.* 1986) and photonuclear (Veyssiere *et al.* 1970) experiments. The main point here is that the IVGDR is very weakly excited (energy integrated cross sections are ~35 mb/sr for the GQR and ~5 mb/sr spread over a much larger energy range for the GDR). This was an important feature of the 22 MeV/nucleon experiment and made it possible to study the ground state gamma decay of the ISGQR, since if the IVGDR is strongly excited it will dominate the ground state gamma spectrum. The reason for this is illustrated in Fig. 6 which shows the relative ground state gamma decay width of a sharp giant resonance state of exhausting 100% of its EWSR, relative to the E1 GDR.

A similar decomposition (Barrette *et al.* 1988) of an 84 MeV/nucleon singles spectrum is shown in Fig. 7. In this case the IVGDR dominates the inelastic spectrum, with a cross section of almost 2.5 barns/sr, compared to about 700 mb/sr for the ISGQR. In Fig. 8, the spectrum resulting from requiring ground state gamma coincidences in the 84 MeV per nucleon data is shown. According to the spectral decomposition of Fig. 7, and the width ratios of Fig. 6, we would expect that the spectrum above ~9 MeV ought to be almost entirely due to the IVGDR. That this is indeed the case is demonstrated in Fig. 9 and Fig. 10 which show angular correlations between ground state gamma rays and the ^{17}O ejectile. In Fig. 9, this correlation is shown as a function of ^{17}O angle for fixed gamma detection angle. The solid line is a calculated result assuming pure Coulomb excitation of the GDR. The overall normalization of the calculation is adjusted to fit the data, giving an energy integrated ground state gamma branch for the IVGDR of $\Gamma_{\gamma 0}/\Gamma = 0.017 \pm .002$. Figure 10 shows the angular correlation for fixed ^{17}O angle ($\theta_{cm} = 2.7°$), as a function of gamma detection angle, compared with DWBA predictions for pure E1 excitation ($1^- \rightarrow 0^+$ decay). Clearly the E1 calculations describe the data in both Fig. 9 and Fig. 10 remarkably well.

The ground state gamma branch from the IVGDR can be accounted for quantitatively by applying the multistep compound emission (MSCE) formalism (Feshbach *et al.* 1980; Hussein and McVoy 1979; and Dias *et al.* 1986). Such calculations have been discussed extensively for both the 22 MeV/nucleon (Beene *et al.* 1989b and Beene *et al.* 1988), and the 84 MeV/nucleon data (Beene *et al.* 1988 and Beene *et al.* 1989a). In the MSCE calculations the total ground state gamma coincidence cross section is assumed to consist of an incoherent sum of terms which are interpreted as arising from successively more complicated stages of the reaction:

$$\sigma(X,X'\gamma_0) = \sum_{i=1}^{r} \sigma_i \ .$$

In the case of GR excitation and decay the first stage (i = 1) is the coherent 1p-1h resonance state which acts as a primary doorway state for the reaction process. This simple collective state eventually is damped into the continuum of complex compound states (r^{th} stage of the sum) which occupy the same region of excitation energy. The intermediate stages

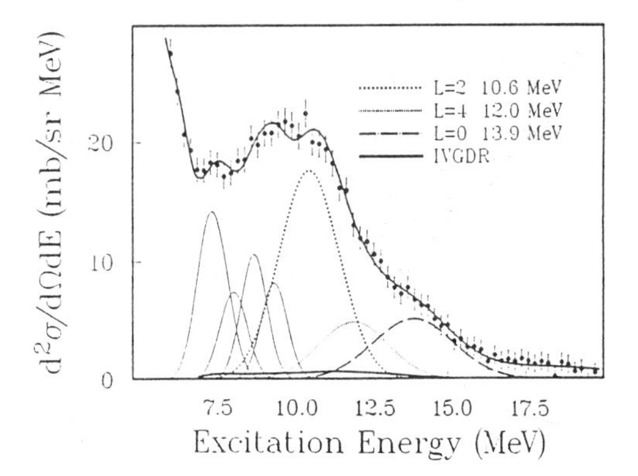

Fig. 5. Decomposition of a $^{208}Pb(^{17}O,^{17}O')$ inelastic singles spectrum taken at 22 MeV/nucleon and θ_{cm} = 14° (see Beene *et al.* 1989b).

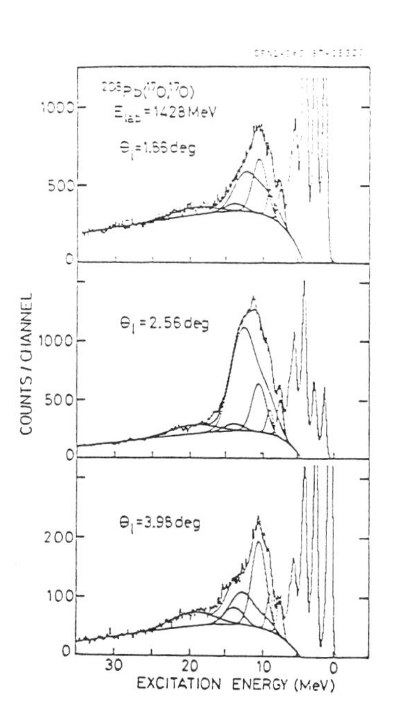

Fig. 7. Inelastic scattering spectra at θ_{lab} = 1.86°, 2.56°, and 3.98° from the $^{208}Pb(^{17}O,^{17}O')$ reaction at 1428 MeV. The solid curves show a decomposition of the spectra into resonance peaks at ~10.6 MeV (ISGQR), ~13.6 MeV (IVGDR), 14 MeV (ISGMR), 7.5 MeV, 9.1 MeV and a broad, undefined, peak centered at ~20 MeV and an underlying continuum (Barrette *et al.* 1988).

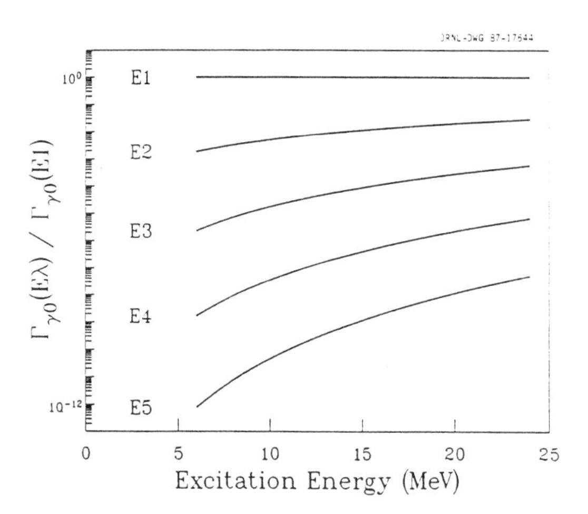

Fig. 6. Ground-state gamma widths of hypothetical sharp states fully exhausting the appropriate isovector or isoscalar energy weighted sum rule as a function of the excitation energy of the state, relative to the E1 width.

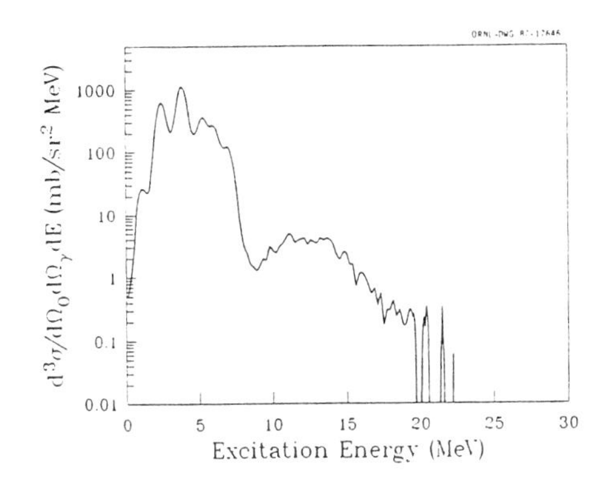

Fig. 8. Inelastic spectrum in coincidence with gamma rays to the ground state. (The ^{17}O angles are $\theta = 2.0°$-$3.5°$ in this case, the gamma detector angles are given in the text.)

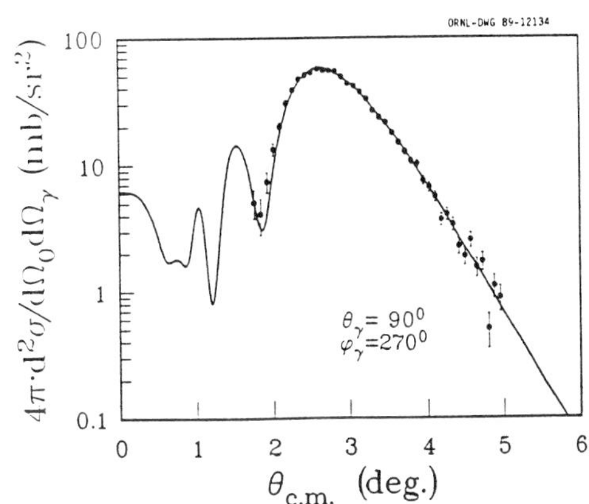

Fig. 9. ^{17}O-γ_0 angular correlation for the ^{208}Pb(^{17}O,^{17}O') reaction at 84 MeV/nucleon, for fixed γ angle $\theta_\gamma = 90°$, $\phi_\gamma = 270°$.

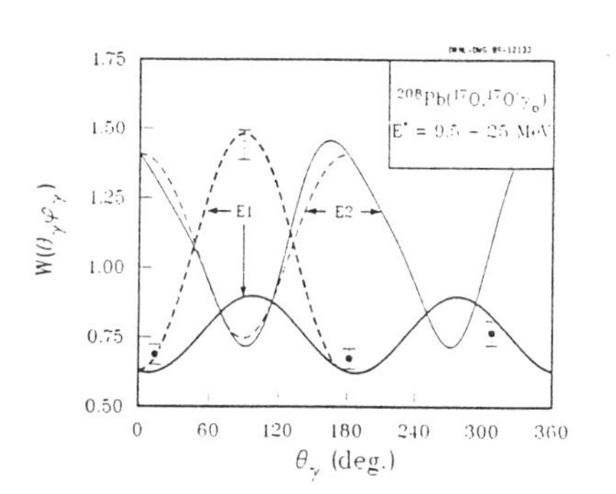

Fig. 10. The same correlation as fig. 15, but for fixed ^{17}O angle (θ - 2°-3°) and varying γ angle. The lines are from theoretical calculations assuming pure Coulomb excitation of the GDR. Filled data points and the solid curve lie in the reaction plane ($\phi = 0°$ and 180°). (For convenience the $\phi = 180°$ half plane is labeled by $\theta + 180°$.) The open point and dashed line refer to the $\phi = 270°$ half plane.

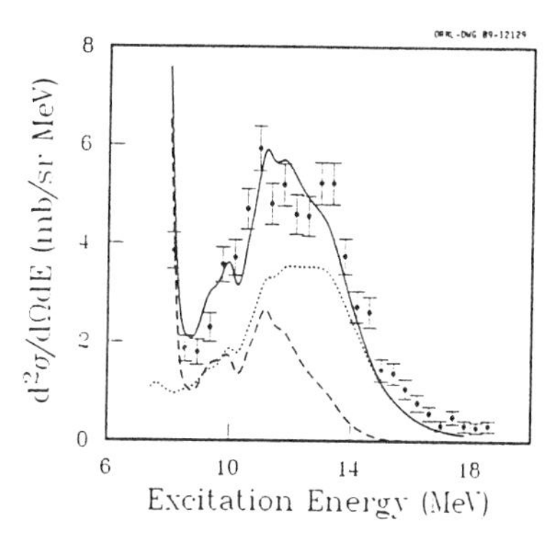

Fig. 11. The ground state gamma coincidence yield for 84 MeV/nucleon ^{17}O scattering on ^{208}Pb, compared with calculations as discussed in the text. The dashed line shows the compound contribution, and the dotted line the doorway contribution. The solid line is the sum.

$1 < i < r$ can in this case be taken to be the hierarchy of increasingly complex 2p-2h, 3p-3h etc. states through which the damping proceeds. Each stage in this damping process can contribute to the ground state gamma decay. In order to carry out quantitative calculations we have assumed that the most important contributions to the ground state gamma decay of giant resonances come from the first (1p-1h) stage and the rth or compound stage. An example of the application of this formalism is shown in Fig. 11 which shows the experimental distribution in excitation energy of the ground state gamma decay coincidences compared with a MSCE calculation considering only the IVGDR, with its strength distribution taken from the photonuclear data of Veyssiere *et al.* (1970). The calculation which involves no adjustable parameters agrees very well with both the distribution of γ_0 strength and the energy integrated branching ratio obtained from the fit in Fig. 9. The calculated branch is $\Gamma_{\gamma 0}/\Gamma = 0.016$ compared to the experimental value 0.017 ± 0.02.

The separate contribution of the "doorway" (dash-dot line) and compound (dotted line) contributions to the γ_0 coincidences are also shown in Fig. 11. The compound contribution is significant on the low energy side of the resonance but becomes increasingly less important at higher energy as the neutron decay branch of the compound states increases. The doorway contribution to the γ_0 decay depends only on the collective properties of the resonance (strength, spreading width, etc.) which change very slowly with A and Z from nucleus to nucleus. However, the compound contribution can be very dependent on the details of the nucleus being studied. Therefore a very sensitive test of the MSCE calculations can be made by comparing the γ_0 decay from pairs of judiciously chosen neighboring nuclei. A good case is the pair ^{209}Bi and ^{208}Pb. Figure 12 shows the singles inelastic spectrum at $\theta_{cm} = 2.5°$ for excitation of ^{208}Pb (solid) and ^{209}Bi (dashed). The distribution of excitation cross section (which is dominated by the IVGDR) is almost identical in the two cases. Figure 13 shows the ratio of ground state gamma decay cross sections for these two nuclei. Clearly the decay is not so similar as the excitation. The γ_0 yield in the region where compound emission is important for ^{208}Pb is much larger in ^{208}Pb than in ^{209}Bi. This is precisely what is predicted by the MSCE calculations, shown as a dotted line of Fig. 13. The compound γ_0 emission is smaller in ^{209}Bi since the density of available states in ^{208}Bi is much larger than in ^{207}Pb, resulting in larger relative neutron emission probability and consequently a smaller gamma decay branch. The agreement between the data and the MSCE calculations shown in Figs. 11 and 13 and for 22 MeV/nucleon data in Beene *et al.* (1989), is remarkable, especially in view of the approximations made. [A more rigorous application of MSCE ideas to this problem has been outlined by Dias *et al.* (1986).] The MSCE model has played a major role in the interpretation of charged particle and neutron decays of GR (e.g., Wagner 1980, Bracco *et al.* 1988 and 1989 and Brandenburg *et al.* 1987). We regard the success of the model in describing the relative simple electromagnetic decay case as an important demonstration of its applicability to GR decay.

A further interesting point can be made from the excitation cross section data in Fig. 12. It has been postulated (Nolte *et al.* 1986) that the total GDR strength in ^{209}Bi might be substantially larger than in ^{208}Pb as a result of meson exchange currents. An estimate of the size of the effect (Nolte *et al.* 1986) has been made using the renormalization of the ^{209}Bi ground state magnetic moment (again due to meson exchange effects). The near identity of the two spectra in Fig. 12 can only be understood if the GDR strength are the same in ^{209}Bi and ^{208}Pb within ~3%!

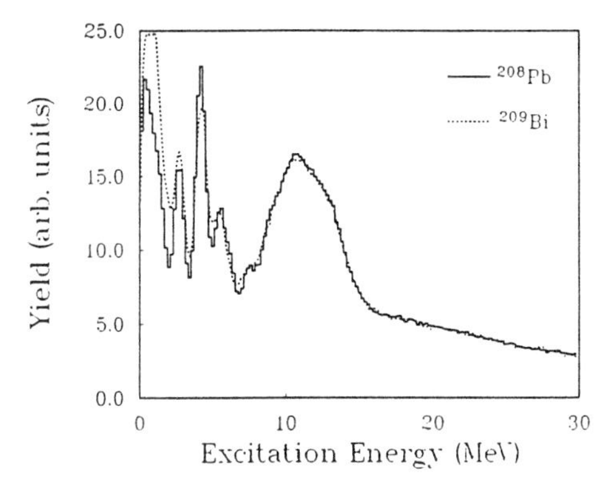

Fig. 12. Singles spectrum for 84 MeV/nucleon ^{17}O inelastically scattered by ^{208}Pb and ^{209}Bi at $\theta_{cm} \sim 2.7°$.

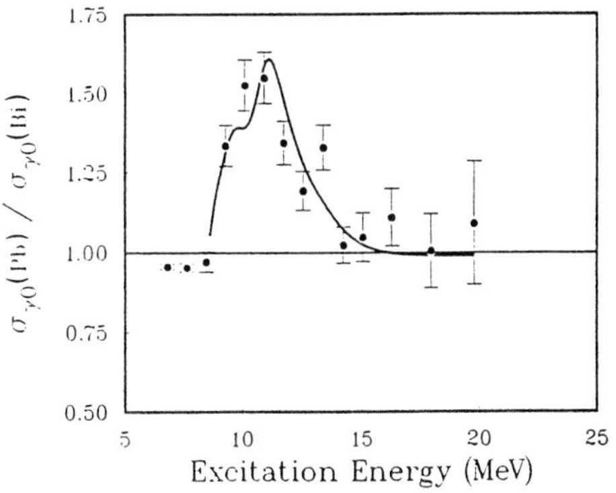

Fig. 13. Ratio of γ_0-coincident spectra for 84 MeV/nucleon ^{17}O scattered by ^{208}Pb and ^{209}Bi (points). The curve is the ratio predicted by the MSCE calculation (see text).

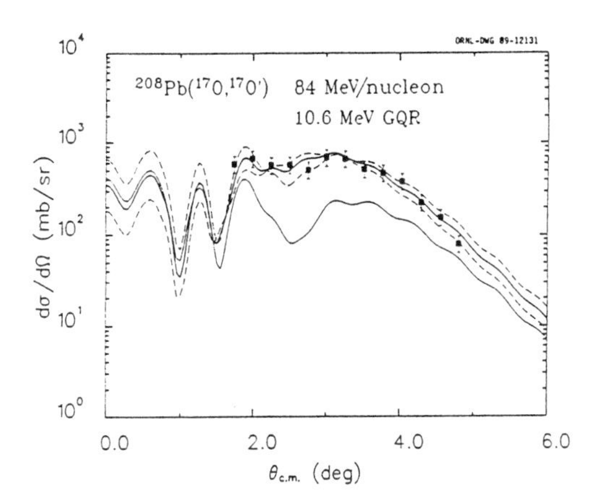

Fig. 14. The experimental angular distribution for the 10.6 MeV ISGQR in ^{208}Pb (data points). The lines are described in the text.

5. ISOSPIN CHARACTER OF THE 10.6 MeV GQR IN ^{207}Pb

Significant localized quadrupole strength (i.e., a GQR) was first identified in the early 1970's (Bertrand 1981) and has been studied systematically throughout the periodic table. It has been widely assumed that this is an isoscalar resonance, i.e., an in-phase vibration of neutrons and protons with approximately equal amplitude. This implies that neutrons and protons should contribute to the excitation matrix elements (see Satchler 1987 for details) in the ratio $M_n/M_p \sim N/Z$ (= 1.54 for ^{208}Pb). Recent pion scattering data have, however, been interpreted as implying that the 10.6 MeV GQR in ^{208}Pb is mostly a neutron excitation ($M_n/M_p \sim 3.8$) with an electromagnetic strength B(E2↑) ≈ 1100 e^2fm^2, which is almost 5 times smaller than would be expected from an isoscalar state exhausting ~60-100% of the isoscalar energy weighted sum rule (EWSR). In our work at 22 MeV/nucleon we have isolated the ground state gamma decay branch from the 10.6 MeV GQR in ^{208}Pb, and deduced a value B(E2↑) ~ 5300 e^2fm^4 for the L = 2 strength near 10.6 MeV in ^{208}Pb, corresponding to $M_n/M_p \sim 1.5 \pm 0.5$ and contradicting the pion scattering result (Beene *et al.* 1989b and Horen, Beene and Bertrand 1988).

Further support for the essentially isoscalar character of the 10.6 MeV resonance can be obtained from the 84 MeV/nucleon inelastic singles data. From Fig. 8 it can be seen that the IVGDR and the 10.6 MeV GQR account for a very large fraction of the excitation cross section in the GR region. From the evidence presented in Section 4 we believe that the IVGDR excitation can be accounted for extremely accurately using photonuclear strength distributions and Coulomb excitation calculations [it is important to take into account the excitation energy dependence of the Coulomb excitation process (Beene *et al.* 1989a and Bertrand *et al.* 1988)]. We can therefore extract the excitation cross section for the 10.6 MeV GQR very well. The resulting angular distribution is presented in Fig. 14.

The importance of Coulomb excitation for L = 2 states at 84 MeV/nucleon means that these data are extremely sensitive to the electromagnetic strength (i.e., proton matrix element) of the resonance. We have carried out an analysis of the data in which M_n/M_p is treated as a free parameter which is reported in more detail in Beene *et al.* (1989a). The basic idea of the analysis (see e.g. Satchler 1987, Bernstein *et al.* 1981, Rychel 1987 and Horen, Beene and Bertrand 1988) is that the nuclear excitation of the GQR depends on the sum of neutron and proton matrix elements [$\alpha(M_n + M_p)^2$] while the Coulomb excitation is proportional to B(E2↑) = $e^2M_p^2$. The Coulomb nuclear interference region just inside the grazing angle is very sensitive to $|M_n/M_p|$, while the total cross section, because of the strong Coulomb excitation is most sensitive to M_p. The best fit result is illustrated on Fig. 14 as a solid line which corresponds to $|M_n/M_p| = 1.7 \pm 0.4$ and B(E2↑) = 3980 ± 450. The dashed lines on either side of the solid line in Fig. 14 show the effect of changing M_n/M_p by one standard deviation, while the dotted line represents the cross section that would be expected for the transition potential with M_n/M_p = 3.8, deduced from pion scattering. If additional L = 2 strength at 8.9 and 9.1 MeV (which would not have been resolved from the 10.6 MeV peak in our 22 MeV/nucleon data, or the pion data) is included assuming the same M_n/M_p as deduced here for the 10.6 MeV peak, we obtain a total B(E2↑) = 5200 ± 800 for the GQR region. Clearly the results of pion scattering data grossly underestimate the electromagnetic strength (proton matrix element) of the 10.6 MeV GQR. These results for the electromagnetic strength of 10.6 MeV GQR are in remarkably good agreement with recent (e,e'n) data from Bolme *et al.*

(1988). This is illustrated in Fig. 15 which compares our B(EL)↑ distribution with the (e,e'n) data. This is especially gratifying in view of the long history of disagreement between electron and hadron scattering results for the ISGQR.

6. THE IVGQR IN ^{208}Pb

The principal motivation for our initial ^{208}Pb(^{17}O,^{17}O') experiments at 84 MeV/nucleon was to try to isolate the IVGQR In ^{208}Pb using a gamma-gamma coincidence technique. In our early work at 22 MeV/nucleon on the ISGQR, we found that the E1 decay branch from the ISGQR to the 3^- 2.6 MeV state in ^{208}Pb was many times smaller than would be expected from naive statistical arguments. This result was explained by two theoretical papers (Bortignon *et al.* 1984 and Speth *et al.* 1985) as partly due to cancellation between neutron and proton transition matrix elements which results from the isospin character (IS) of the ISGQR. These authors pointed out that the IVGQR should, on the other hand, have a very large gamma decay branch to the 3^- 2.6 MeV state. Our experiments (Bertrand *et al.* 1989) have mapped out the distribution of gamma-gamma coincidence yield resulting from this decay above an excitation energy of ~15 MeV (Fig. 16). If this triple coincidence yield is interpreted as arising form the IVGQR it implies the following properties of the strength distribution: $\langle E \rangle = 22.6 \pm 0.4$ MeV, $\sigma \sim 6 \pm 2$ MeV and strength ~50% of the EWSR.

7. HIGH LYING STRENGTH AND MULTIPHONON STATES

The search for collective strength in the large gap between the known nuclear giant resonances and the nucleon resonances is an exciting area of current research. Interesting but controversial results are already available from the Orsay group (e.g., Chomaz *et al.* 1986). The availability of good quality beams of higher energy high Z ions should make an important contribution to this search. Recent papers by Baur and his collaborators (Baur and Bertulani 1986 and 1989 and Baur and Baron 1989) and proposals by Braun-Munzinger *et al.*(1985) and Emling (1987) have emphasized the crucial role which Coulomb excitation using the intense field produced by high Z collision partners in intermediate energy and high energy heavy ion collisions. We believe the distinctive electromagnetic decay of states which involve coherent excitation of a GDR phonon will make it possible to isolate these particular two phonon states in the continuum, even in experiments carried out near 100 MeV per nucleon, using coincidences between inelastically scattered heavy ions and de-excitation photons. A combination of the observation of these distinctive particle gamma-gamma coincidences, and the strong excitor Z dependence of the excitation cross section (scales like Z^{2n}, where n is the phonon number) may make it possible to identify the 2 phonon strength even in a much more intense background of 1 phonon and more complex continuum excitations. This offers the near-term prospect of relatively high resolution spectroscopic studies of two phonon strength at facilities such as GANIL where high quality beams and magnetic spectrographs are available.

8. CONCLUSION

The results presented in this talk represent the early stages of quantitative use of intermediate energy heavy-ion scattering in the study of giant resonances. We have stressed the importance of Coulomb excitation of L = 1 and 2 states, and the very large excitation cross sections which make possible the quantitative study of weak decay branches. Our experiments

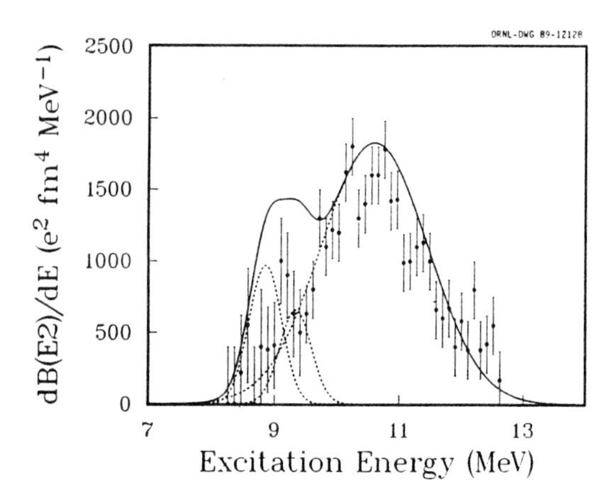

Fig. 15. The distribution of B(E2)↑ in the giant resonance region as obtained from the ^{208}Pb(^{17}O,^{17}O'γ) measurements of Beene *et al.* (1989) (solid curve) and the (e,e'n) movements of Bolme *et al.* (1988) (points with error bars). The contributions of the 8.8 and 9.3 MeV states and the 10.6 MeV state are shown separately as dashed curves.

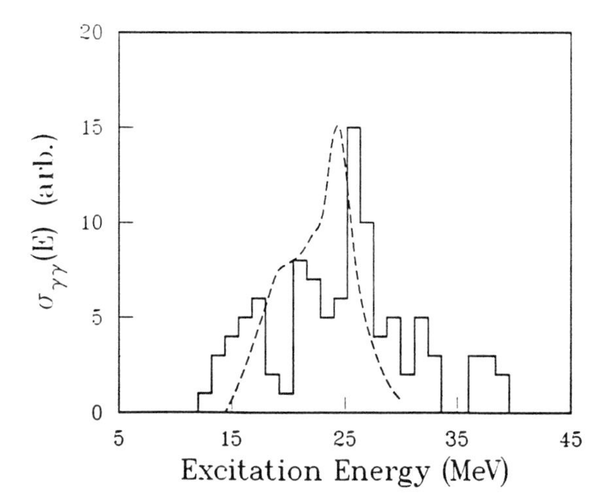

Fig. 16. The histogram is the triple coincidence data, $\gamma_1\gamma_2{}^{17}$O', (γ_1 > 10 MeV, γ_2 = 2.6 MeV). The curve is the predicted (Bertrand *et al.* 1989 distribution if IVGQR strength convoluted with the energy dependence of the probability of excitation by 84 MeV/nucleon ^{17}O on ^{208}Pb.

have already led to significant information on giant resonance gamma decay mechanisms and the isospin character of resonance states. We have begun to utilize the large Coulomb excitation cross sections to investigate isovector strength above the IVGDR. There is reason to hope that intermediate-energy experiments done with somewhat heavier projectiles will, along with gamma decay coincidences, enable us to investigate, in some detail, two phonon states with one or both the phonons being a GDR mode.

The experiments described here were carried out in collaboration with D. J. Horen, R. L. Auble, B. L. Burks, J. Gomez del Campo, M. L. Halbert, D. C. Hensley, J. E. Lisantti, R. L. Robinson, R. O. Sayer, and R. L. Varner from Oak Ridge; W. Mittig and Y. Schutz from GANIL; B. Haas and J.P. Vivien of Strasbourg, and J. Barrette, N. Alamanos, F. Auger, B. Fernandez, and A. Gillibert from Saclay, and A. Nathan from the University of Illinois.

REFERENCES

Danos M and Fuller E G 1965 *Annu. Rev. Nucl. Sci.* **15** 29 and references therein

Barrette J *et al.* 1988 *Phys. Lett.* **B209** 182

Baur G and Bertulani C A 1986 *Phys. Rev. C* **34** 1654

Baur G and Bertulani C A 1988 *Nucl. Phys.* **A482** 313c

Baur G and Baron N 1989 *J. Phys. G* **15** 661

Beene J R *et al.* 1989a to be published

Beene J R *et al.* 1989b *Phys. Rev. C* **39** 1307

Beene J R, Varner R L and Bertrand F E 1988 *Nucl. Phys.* **A482** 407c

Bernstein A M, Brown V R and Masden V A 1981 *Phys. Lett.* **103B** 255

Bertrand F E 1981 *Nucl. Phys.* **A354** 129c and references therein

Bertrand F E *et al.*, 1989 to be published

Bertrand F E *et al.* 1986 *Phys. C* **34** 45

Bertrand F E and Beene J R 1989 *Proceedings of International Nuclear Physics Conference Sao Paulo Brazil August 20-26* to be published

Bertrand F E, Beene J R and Horen D J 1988 *Nucl. Phys.* **A488** 163c

Bertulani C A and Baur G 1985 *Nucl. Phys.* **A442** 739

Bolme G O *et al.* 1988 *Phys. Rev. Lett.* **61** 1081

Bortignon P F, Broglia R A and Bertsch G F 1984 *Phys. Lett.* **148B** 20

Bowman J D *et al.* 1983 *Phys. Rev. Lett.* **50** 1195; Erell A *et al* 1984 *Phys. Rev. Lett.* **52** 2134

Bracco A *et al.* 1988 *Phys. Rev. Lett.* **60** 2603

Bracco A *et al.* 1989 *Phys. Rev. C* **39** 725

Brandenburg S *et al.* 1987 *Nucl. Phys.* **A466** 29

Braun-Munzinger P *et al.* 1985 Proposal 814 submitted to AGS Program Committee, accepted 1985

Chomaz et al., 1986 *Zent. Phys.* **A318** 41

Dias H, Hussein M S and Adhikari S K 1986 *Phys. Rev. Lett.* **57** 1998

Emling H 1987 *Proceedings of 8th High Energy Heavy-Ion Study, Berkeley* ed J W Harris and G J Wozniak p 503

Feshbach H, Kerman A and Koonin S 1980 *Annals of Physics* **125** 429

Horen D J, Beene J R and Bertrand F E 1988 *Phys. Rev. C* **37** 888

Hussein M and McVoy K 1979 *Phys. Rev. Lett.* **43** 1645

Nolte R *et al.* 1986 *Phys. Lett.* **B173** 388 but see also Dale D S *et al* 1989 *Phys. Lett.* **B214** 329

Rychel D *et al.* 1987 *Z. Phys.* **A326** 455

Satchler G R 1983 *Direct Nuclear Reactions* (Oxford University Press: Oxford)

Satchler G R 1987 *Nucl. Phys.* **A472** 215

Speth J *et al.* 1985 *Phys. Rev. C* **31** 2310

Veyssiere A *et al.* 1970 *Nucl. Phys.* **A159** 561

Wagner G 1980 *Proceedings of Conference on Giant Multipole Resonances* p 251

Inst. Phys. Conf. Ser. No 105
Paper presented at Int. Conf. on Spectroscopy of Heavy Nuclei, Crete, Greece, 1989

Angular momentum and temperature dependence of the GDR in Dy nuclei

J.P.S. van Schagen[1)], J.C. Bacelar[2)], M.N. Harakeh[1)], W.H.A. Hesselink[1)], H.J. Hofmann[2)], R.F. Noorman[2)], A.J.M. Plompen[1)], A. Stolk[1)], Z. Sujkowski[2)]

[1)] Nat. Lab. van de Vrije Universiteit, Amsterdam, The Netherlands
[2)] KVI, Groningen, The Netherlands

ABSTRACT: The statistical γ-ray decay of the GDR built on excited states in $^{151\text{-}156}$Dy has been investigated as a function of angular momentum and temperature. The selection on angular momentum has been made by using the total γ-ray energy and γ-ray multiplicity. The temperature dependence in the range T≈1.0 MeV to T≈2.0 MeV has been measured by taking data for two reactions, i.e. ^{114}Cd(^{40}Ar,xnγ)$^{154\text{-x}}$Dy* and ^{116}Cd(^{40}Ar,xnγ)$^{156\text{-x}}$Dy* at beam energies of 180 MeV and 200 MeV, respectively, populating the same final nuclei. A systematic shift of the strength to lower energies has been observed at increasing angular momentum, which can be attributed to a change of the nuclear shape from prolate to oblate.

1. INTRODUCTION

The existence of giant resonances built on excited states was already proposed in the fifties by Brink[1)], but the experimental verification of this conjecture was for the first time obtained by Newton et al.[2)]. Since then the study of the giant dipole resonance built on excited states has been at the forefront of nuclear research, both theoretically and experimentally[3)]. This is due to the fact that the statistical decay of the GDR in hot compound nuclei has opened up the way to study the properties of these nuclei under extreme conditions.

One of the features of the giant dipole resonance which has been established at zero temperature, i.e. the splitting of the GDR due to deformation[4)], could indeed be used as a means to investigate the shape of nuclei as a function of angular momentum and temperature. In axially symmetric deformed nuclei the GDR will split into two components of low and high frequency due to oscillations along the long and short symmetry axes, respectively. The strength distribution between the two components, S1 and S2, is then determined by the nuclear deformation; for prolate deformation S1/S2

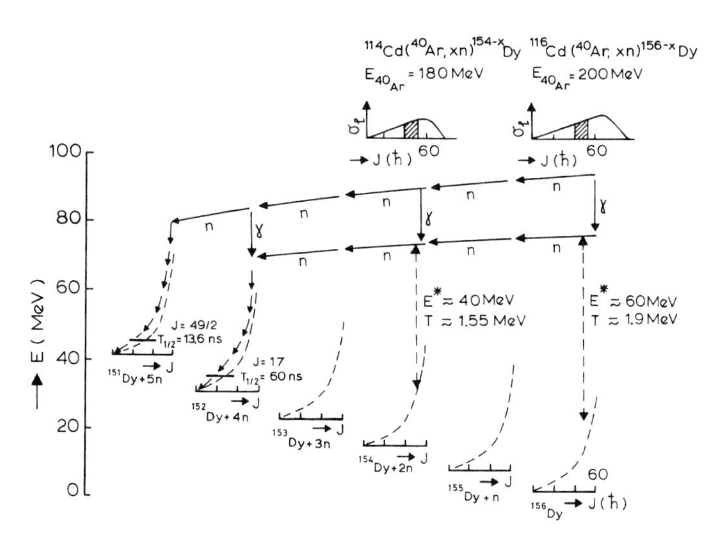

Fig 1. Schematic representation of the decay of the compound nuclei ^{154}Dy* and ^{156}Dy* formed by bombarding ^{114}Cd and ^{116}Cd targets with a ^{40}Ar beam of 180 MeV and 200 MeV, respectively

= 0.5 whereas for an oblate deformation S1/S2 = 2.0. Only at a relatively low temperature T $\leq$1 MeV one may expect a clean signature of a split GDR. At higher temperatures the GDR strength function will become rather structureless due to thermal fluctuations which cause a strong damping of the GDR [5,6].

The splitting of the GDR at finite temperature was observed for the first time in Er nuclei[7,8], in inclusive measurements in which no selection on angular momentum was made. Furthermore, the experiment by Gaardhøje et al.[8] indicated a shift from prolate to oblate shape with increasing temperature. The average angular momentum in both studies was relatively low. It is clear that a good selectivity on angular momentum is needed if one wants to extend these studies to higher angular momentum domains. This selectivity can be obtained by using information on the fold distribution and total energy of the γ-rays emitted in the reaction[9]. The dependence on the temperature can furthermore be investigated by performing two experiments leading to compound nuclei at different excitation energies and differing by one or two neutrons and which after emission of several neutrons end up in the same final nuclei.

We have combined both techniques in two experiments in which we investigated the statistical decay of the compound nuclei ^{154}Dy and ^{156}Dy, produced by bombarding ^{114}Cd and ^{116}Cd targets with ^{40}Ar beams of 180 MeV and 200 MeV, respectively. A schematic representation of the angular momentum and temperature domains covered in these experiments is given in Fig.1. The compound nucleus ^{156}Dy is formed at an average temperature of 1.9 MeV with an angular momentum distribution that peaks around 65 h. After emission of a high energy γ-ray and 4 or 5 neutrons, states in ^{151}Dy and ^{152}Dy are populated. The ^{154}Dy compound nucleus is formed at T ≈ 1.55 MeV with an angular momentum distribution that peaks around 55 h. In this case the populated final nuclei are the same as in the ^{156}Dy experiment. This enables us to compare both experiments to study the decay of the GDR in two temperature domains of T≈1-1.5 MeV and T≈1.5 -1.9 MeV as a function of angular momentum.

2. EXPERIMENTAL METHODS

The experiments described here are a continuation of the study performed by Stolk et al.[10]. The experiments were performed at the KVI in Groningen with ^{40}Ar beams from the AVF cyclotron. The target thicknesses were 1.5 mg/cm^2 for ^{114}Cd and 1.0 mg/cm^2 for ^{116}Cd. The beam energies were 180 and 200 MeV leading to initial excitation energies of the compound nuclei ^{154}Dy and ^{156}Dy of 75 MeV and 90 MeV, respectively. The beam particles and recoiling nuclei were stopped in a catcher foil 25 cm downstream of the target.

The high energy γ-rays were detected in a 10"x13" NaI detector, which was surrounded by a plastic anti-coincidence shield and a lead shield. The detector was placed at a distance of 66 cm from the target to be able to separate neutrons by time of flight.

The total γ-ray energy was measured with a 40 cm x 40 cm cylindrical sum-spectrometer, consisting of six segments. The two halves were separated by 14 cm. The multiplicity filter consisted of eight 5"x5" NaI detectors, placed in the gap at a distance of 20 cm from the target, and the six sum-spectrometer segments. Discrete γ-rays were detected with a Compton suppressed Ge-telescope which was also palced in the gap.

The catcher foil was surrounded by eight 3"x5" NaI detectors and one additional 5"x5" NaI detector to detect γ-rays decaying from the isomeric states in ^{151}Dy and ^{152}Dy. These detectors were shielded from the target by 10 cm of lead. Prompt high energy γ-rays detected with the large NaI detector were recorded if they were in coincidence with a prompt event in at least one of the 5"x5" NaI detectors and one of the sum-spectrometer segments. γ-rays detected in the detectors surrounding the catcher foil have been used as a selective trigger for events leading to the final nuclei ^{151}Dy and ^{152}Dy. Events due to γ-rays with energy ≤6 MeV were downscaled by a factor of 100.

Fig. 2 shows the time relation between delayed γ-rays from the catcher foil with respect to prompt γ-rays detected in the sum-spectrometer for different gate settings on the high energy γ-rays detected in the large NaI detector. In the time spectrum one recognizes a large bump corresponding to X-rays produced by beam particles in the Au foil, which might be in real or random coincidence with prompt γ-rays from the target, followed by a second broad bump due to delayed γ-rays from recoiling nuclei. The two slope components in the time spectrum are characteristic for the decay of the $J^{\pi}= 49/2^-$ isomer in ^{151}Dy and the $J^{\pi}=17^+$ isomer in ^{152}Dy.

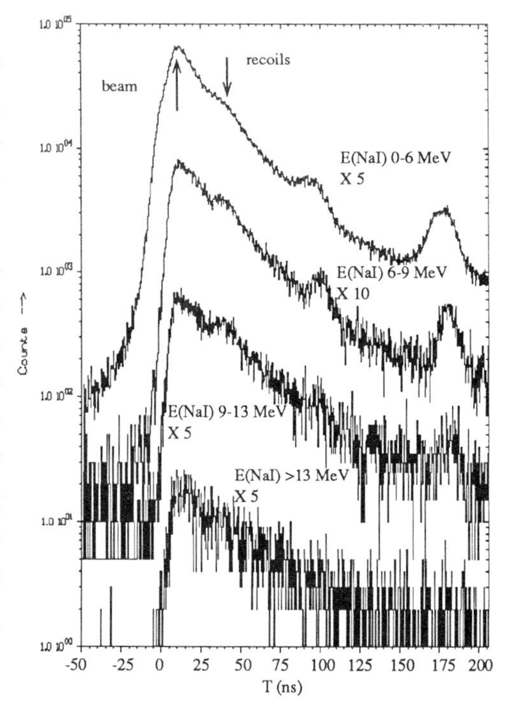

Fig 2. Time spectrum of γ-rays and X-rays from the catcher foil. The curves correspond to different gates in the high energy γ-ray spectrum.

3 RESULTS

The measured fold distribution is transformed into reaction multiplicity by Monte Carlo methods[10] taking into account effects like cross-talk, different efficiencies, the specific geometry of the setup and corrected for γ-rays detected by the array around the cathcher foil. The spectra gated on three different fold multiplicity and sum-energy combinations are shown in Fig. 3. The gaps in the spectra are due to the discriminator threshold for γ-ray energies above 6 MeV and the downscaling of the part below 6 MeV. The indicated mean angular momentum values are estimated from the deduced γ-ray multiplicities. The spectra can be compared directly because the data are obtained with the same setup and under the same conditions. Clearly, in both sets of spectra a shift of the strength to lower energies with increasing angular momentum can be observed. The same effect has been found by Hennerici et al.[9], Bruce et al.[11] and Stolk et al.[10]. Based on comparison with statistical model calculations both Bruce et al.[11] and Stolk et al.[10] concluded that this shift in strength can be interpreted as being due to a change in shape from prolate to oblate as a function of angular momentum. The spectra are being fitted by the computer code

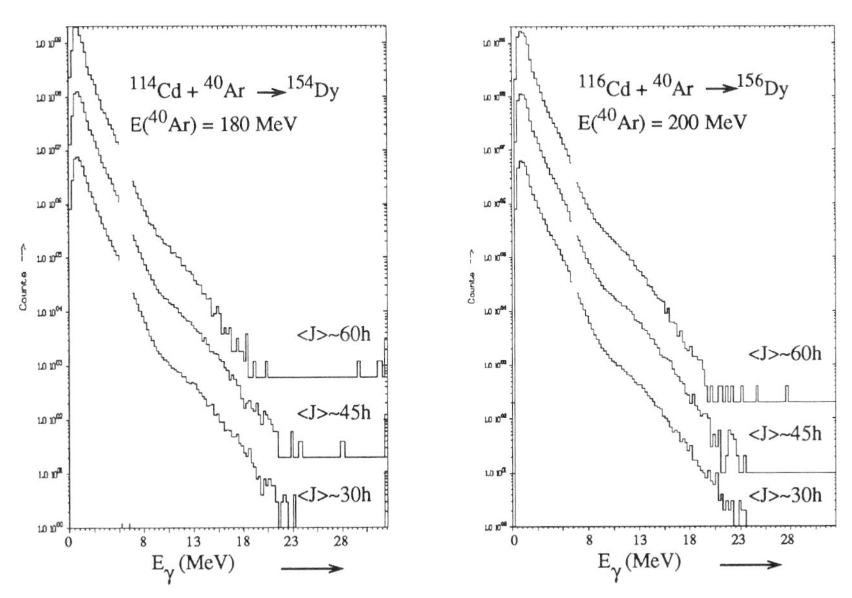

Fig. 3 Prompt high energy γ-ray spectra measured in the $^{114}Cd(^{40}Ar,xn)^{154-x}Dy$ reaction (left) and the $^{116}Cd(^{40}Ar,xn)^{156-x}Dy$ reaction (right). The low energy part has been multiplied with a factor of 100, i.e. the downscale factor in the experiment.

CASCADE[12)], in a similar procedure to that followed by Stolk et al.[10)]. Of these time consuming calculations only the preliminary results for the highest angular momentum bin at 200 MeV can be quoted here. The results are E_1=12.45 MeV, S_1=0.57, Γ_1=6.6 MeV, E_2=18.8 MeV, S_2=0.43 and Γ_2=15.1 MeV. These values are in reasonable agreement with the findings of Stolk et al..

In Fig. 4, the results of the statistical calculation are compared to the experimantal data on a linear scale after multiplying both by an exponential factor $\exp[E_\gamma/T]$ where an effective temperature of T=1.85 MeV was taken. The fit is excellent in the whole region between 8.5 and 22 MeV. The χ^2/N of the fit in this region is 0.75.

The spectra which were obtained with an explicit gate in the time spectra of the detector array around the catcher foil also show a shift of the strength to lower energies with increasing angular momentum. However, further analysis is needed before any conclusion can be drawn. Also no results can be given yet about the comparison of the experiments with E_{beam}=180 MeV and 200 MeV.

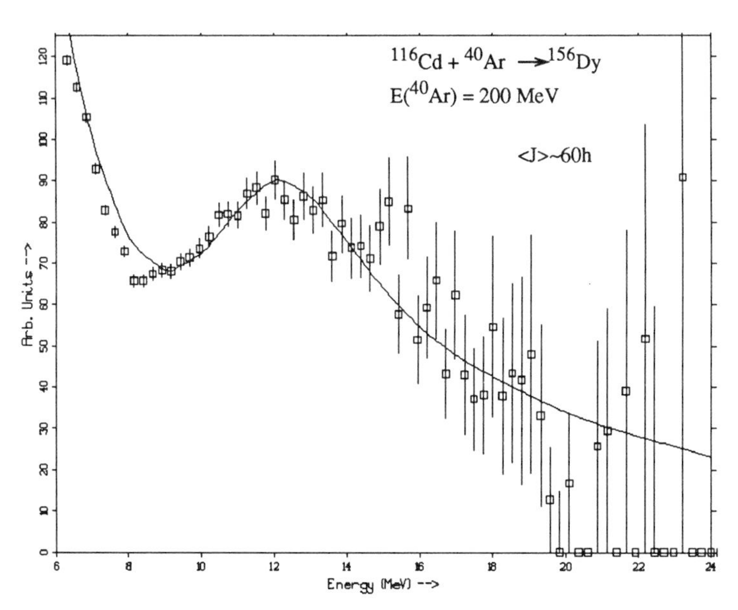

Fig. 4 A comparison of the data of the highest angular momentum bin with the result of the fit with a statistical model calculation using the computer code CASCADE. The data point and the curve of the fit are both multiplied with an exponential function exp[E_γ / T] taking an effective temperature of T=1.85 MeV.

REFERENCES

[1] D. Brink, thesis, *Oxford University* (1955) unpublished

[2] J.O. Newton et al., *Phys. Rev. Lett.* ***46*** (1981) 1383

[3] K.A. Snover, *Ann. Rep. Nucl. Phys.* ***36*** (1986)

[4] P. Carlos et al., *Nucl. Phys.* ***A219*** (1974) 61

[5] M. Gallardo et al., *Nucl. Phys.* ***A443*** (1985) 415

[6] Y. Alhassid et al., *Phys. Rev. Lett.* ***57*** (1986) 539

[7] C. Gossett et al., *Phys. Rev. Lett.* ***54*** (1985) 1486

[8] J.J. Gaardhøje et al., *Phys. Rev. Lett.* ***53*** (1984) 148

[9] W. Hennerici et al., *Nucl. Phys.* ***A396*** (1983) 329

[10] A. Stolk, Thesis, *Vrije Universiteit, Amsterdam* (1988); to be published

[11] A.M. Bruce et al., *Phys. Lett.* ***215B*** (1988) 237

[12] F. Puhlhofer, *Nucl. Phys.* ***A280*** (1977) 267; M.N. Harakeh, extended version

Inst. Phys. Conf. Ser. No 105
Paper presented at Int. Conf. on Spectroscopy of Heavy Nuclei, Crete, Greece, 1989

The analysis of gamma-ray induced Doppler broadening (GRID) measurements

J. JOLIE, S.J. ROBINSON, H.G. BÖRNER, P. SCHILLEBEECKX,
Institut Laue Langevin, 156X, 38042 Grenoble CEDEX, FRANCE.

ABSTRACT: The theory used to analyze Gamma-ray Induced Doppler broadening (GRID) and Neutrino Induced Doppler broadening (NID) measurements is reviewed. The analysis of GRID measurements is done using a collision model which can be used for both monoatomic and diatomic targets. The NID measurements are analysed using a phonon creation model.

1. INTRODUCTION.

Due to the wide range of lifetimes of excited nuclear states, several different experimental methods, each sensitive for a certain region, have been developed in the last decades. Here we discuss the GRID method which uses the (n,γ) reaction to produce the nucleus in an excited state. The lifetime is then determined by measuring the Doppler-broadened energy profile of gamma rays. The broadening is due to the emission of a gamma ray or a neutrino which gives a recoil to the atom. In contrast to the recoils used in DSAM the recoil velocity is very small and the atom has slows down within twenty Ångström.s It is clear that under such conditions no standard backing material can be used. Moreover, the known stopping theories become questionable at such slow velocities. To overcome these problems we have constructed, using simple assumptions, a description of the slowing down process. Here we will not attempt to give a description which is fully 'microscopic', e.g. including the crystal structure etc., but despite this is flexible enough to be used for many different targets. Since the experimental verification of our stopping theory is discussed by H. Börner et al. in these proceedings and has been published elsewhere [BÖR88,LIE88,JOL88,JOL89,ULB89] we only give the theory without applications to experiment here.

2. LINESHAPE OF A DOPPLER BROADENED LORENTZIAN.

When a photon with energy E_0 is emitted from a moving object at an angle θ with respect to the velocity **v**, the measured Doppler shifted energy is given by $E_0\,(1+(v/c)\cos(\theta))$ to first order in v/c ($v/c=10^{-4}$-10^{-6}). If the lifetime of the nuclear transition is very short, the observed distribution of the energy of the gamma ray is described by a Lorentzian shape with natural linewidth Γ. Inserting the Doppler shift into the Lorentzian shape one obtains:

$$I(E)dEd\theta = \frac{A\Gamma D(\theta)dEd\theta}{2\pi((E-E_0-E_0(v/c)\cos(\theta))^2+(\Gamma/2)^2)} \quad (1)$$

where D(θ) describes the distribution of the angles between the gamma ray and the direction of recoil. Assuming an isotropic distribution of the recoil directions, for the small solid angle in which the sample is observed, D(θ) = sin(θ)/2. Integrating (1) over θ yields the lineshape:

$$I_D(E,v)dE = \frac{A\,c}{2E0\pi v}\left[\arctan\left(\frac{2}{\Gamma}\left(E-E_0\left(1-\frac{v}{c}\right)\right)\right)-\arctan\left(\frac{2}{\Gamma}\left(E-E_0\left(1+\frac{v}{c}\right)\right)\right)\right] \quad (2)$$

Gamma-gamma correlation will lead to an anisotropic distribution D(θ).This anisotropy is lost when the atom scatters off a target atom. Thus only for very short lifetimes and for transitions of multipolarities L = 1 and 2, we have a distribution:

$$D(\theta)d\theta = \sin(\theta)(1 + a_2P_2(\cos(\theta)) + a_4P_4(\cos(\theta)))\ d\theta/2 \quad (3)$$

Then we obtain for the lineshape:

$$I_D(E,v) = (1-a_2/2+3a_4/8)I_0(E,v) + (3a_2/2-30a_4/8)I_1(E,v) +(35a_4/8)I_2(E,v) \quad (4)$$

with $I_0(E)$ given by (2) and (setting $\beta = v/c$ and $x = \beta E_0$):

$$I_1(E,v) = (A\Gamma/4\pi)\{2x^{-2} + (E-E_0)\, x^{-3}[\ \ln((E-E_0(1+\beta))^2 + \Gamma^2/4) \\ -\ln((E-E_0(1-\beta))^2+\Gamma^2/4)]\} + ((E-E_0)^2 - \Gamma^2/4)\, x^{-2}\, I_0(E,v) \quad (5)$$

$$I_2(E,v) = A\Gamma/(6\pi x^2) + 2A\Gamma\,(E - E_0)^2/(\pi x^4) - ((E-E_0)^2 + \Gamma^2/4)I_1(E,v)x^{-2} \\ + 2(E-E_0)^2((E-E_0)^2 - 3\Gamma^2/4)I_0(E,v)x^{-4} \\ + A\Gamma\,(E-E_0)(3(E-E_0)^2 - \Gamma^2/4)[\ \ln((E-E_0(1+\beta))^2 + \Gamma^2/4) \\ - \ln((E-E_0(1-\beta))^2+\Gamma^2/4)]\ /(4\pi\, x^5) \quad (6)$$

When incorporating gamma-gamma correlations formula (4) should be used instead of (2) for those decays occuring before a collision with any other atom can occur.

3.DETERMINATION OF THE LIFETIME WITH MONOATOMIC TARGETS.

After the emission of the first gamma ray γ_1 the nucleus will start to recoil with an initial velocity v_R given by $E\,\gamma_1/mc$, where m is the mass of the recoiling atom. Since the nucleus is moving in the sample it cannot recoil forever with the same velocity v_R, so after a while its velocity will have changed to v'. When it then emits a secondary gamma ray γ_2 the lineshape will be given by formula (2) but now with velocity v'. The total lineshape of γ_2 can then be used to determine the lifetime of the excited nuclear level, since the number of nuclei decayed with velocities v' and Γ depends on the lifetime τ of the level. To obtain the lifetime one needs a description of v(t), the velocity of the recoiling atom in the sample. We will describe this process in a discrete way by splitting it up into a set of averaged distances r(k) in which the atom will have a mean velocity v(k) (e.g. v_R=v(0)). First we consider monoatomic targets.

The atom moving in the target with velocity v(k) will encounter another atom after a certain mean distance r(k). Assuming that the atom undergoes an elastic collision with this atom at rest (v(k) >> $v_{thermal}$) and neglecting the mass difference between the recoiling nucleus and the atom at restand as the displacement energy, on average the recoiling nucleus will loose half of its kinetic energy in the collision. Thus after the collision the nucleus will recoil with a velocity $v(k+1) = v(k)/\sqrt{2}$. At the moment of the first collisionthe number of nuclei which have not decayed with γ_2 is $N_{nd} = N_0e^{-t(0)/\tau}$, with t(0) = r(0)/v(0) and the number that have decayed is $N_d = N_0(1-e^{-t(0)/\tau})$. Repeating the same procedure we obtain if the nucleus is stopped after the *N* th scatter the lineshape:

$$I(E)dE = N_0\{I_S(E)e^{-\sum_{k=0}^{N-1}t(k)/\tau} + \sum_{n=1}^{N} I_D(E,v(n-1))[\ e^{-\sum_{k=0}^{n-2}t(k)/\tau} - e^{-\sum_{k=0}^{n-1}t(k)/\tau}]\}. \quad (7)$$

Once v(k) and r(k) have been determined formula (7) contains two parameters: the lifetime τ and the total number of steps N. N will be determined such that after the last scatter v < $v_{thermal}$ and τ will be fitted. For the stopping peak $I_S(E)$ we use a Doppler broadened lineshape of a Maxwellian velocity distribution around a thermal velocity v_T:

$$I_S(E) = \int_0^{+\infty} 4\pi\,(3/2\pi\, v_T^2)^{3/2} \exp(-1.5(v/v_T)^2)\, v^2\, I_D(E,v)\,dv \qquad (8)$$

with $I_D(E,v)$ given by (2). Unfortunately this integral has no analytic solution and has to be solved numerically.

Now we still have to determine r(k). When r(k) is taken to be constant, this description is equivalent to the assumption of a constant kinetic energy loss $\Delta E_k/E_k$ as a function of the travelled distance r. Of course formula (3) is always a discrete description of such an exponential energy loss. This approach was followed in [JOL88]. Improvement of this description can be obtained by considering the repulsive interatomic potential. This potential takes into account the screening effect of the orbital electrons surrounding the nuclei of the two atoms which strongly reduces the Coulomb repulsion. To take account of the repulsive potential we proceed as follows. The approximate potential appropriate for the description of the interaction of identical atoms at separations between 0.5 to 2 Å is the Born-Mayer potential $V(r) = A \exp(-r/a)$. In order to obtain an estimate for r(k) an atomic radius d(E), depending on the kinetic energy E of the incident atom, is defined to be the distance of closest approach between the two atoms during the collision. Then we obtain $d(E) = a.\ln(2A/E)/2$ from $V(2d(E)) = E/2$. Now we define r(k), k referring now to the number of collisions the atom has undergone as the mean free path $r(k) = 1/(n\sigma)$, where n is the number of atoms per unit of volume and $\sigma = \pi\, d(E)^2$, and hence:

$$r_k = V/\pi(a \ln(4A/mv(k)^2))^2 \qquad (9)$$

where a and A are the parameters of the Born-Mayer potential and V is the volume occupied per atom. With the Born-Mayer potential, no fitting of a lineshape with a known lifetime is needed, since its parameters are tabulated for all atoms [ABR69]. This approach was successfully applied to determine lifetimes in ^{49}Ti [BÖR88] and ^{57}Fe[ULB89] and allowed the discovery of a mixed symmetry 2^+ state in ^{54}Cr[LIE88].

4. DETERMINATION OF THE LIFETIMES WITH DIATOMIC TARGETS

Since the target temperature at the in beam position is high (~600°C including self heating) the choice of monoatomic targets is limited to metals. In order to study other atomic nuclei the target will have to be made from salts, alloys, oxides etc. which contain the nucleus under study. Therefore we extended our model to diatomic targets NM [JOL89]. Since we are dealing with targets of the type NM in which the atom N is recoiling we need the Born-Mayer potentials, $V(r) = A_i e^{-r/a_i}$ (the index i now refers to the atom, e.g. A_n, a_n are the parameters for the potential between two N atoms) between the atoms N and N and also between the atoms N and M. The latter can be obtained from the formula for identical atoms using the combination rule of Smith [SMI72]:

$$(A_{nm}/a_{nm})^{2a_{nm}} = (A_n/a_n)^{a_n}\, (A_m/a_m)^{a_m} \qquad (10)$$

which gives the parameters A_{nm} and a_{nm} for the interatomic potential between the atoms N and M, as a function of the values A_i and a_i (i=n,m) for the interaction between the identical atoms i. Formula (10) contains two unknowns but we can take $a_{nm} = (a_n+a_m)/2$ since the variation of a_i with atomic number is small.

Following the same approach as in the preceding paragraph, the mean velocity v(k+1) which the recoiling atom with mass m_n and velocity v(k) has after hitting a target atom with mass m_m is given by $v(k+1) = (1-2m_n m_m/(m_n+m_m)^2)^{1/2}\, v(k)$ and the mean free path r(k) between the collisions of an atom n on an atom m is given as a function of the Born-Mayer parameters A_{nm} and a_{nm} by:

$$r_k = V/p(a_{nm} \ln(2A_{nm}(m_n+m_m)/m_n m_m v(k)^2))^2. \quad (11)$$

with V the volume occupied by a single atom. In the diatomic case we now have to define the collision sequence . As all nearest-neighbours of the N atoms are M atoms and vice versa, we assumed that a recoiling N atom first hits an M atom and use the formulas derived above. Now being near a M atom position it then hits preferentially a N atom and we take the mean free path and mean energy loss as derived in section 3. Thus the sequence of collisions is taken to be N on M, N on N and so on. This double layer approach is clearly an approximation and only holds when the scattering lengths r(k) are less than or equal to the interatomic distance. This approach has been applied to the slowing down of Ti in pure Ti and TiC [JOL89] and yielded consistent lifetimes allowing oneto discriminate between two sets of parameters for the Born-Mayer potential between C and C. It was also used to obtain an estimate of the lower limit of the lifetime of the 838keV level in ^{176}Lu.[BÖR89]

5. ANALYSIS OF NEUTRINO INDUCED DOPPLERBROADENING (NID).

Up to now we have been using the recoil originating from gamma-ray emission. It is however possible to use other nuclear reactions to produce the extra kinetic energy leading to Doppler broadened energy profiles of subsequently emitted gamma rays. In this section we will concentrate on a specific kind of beta-decay, namely K-electron capture.

$$^{A}{}_{Z}X_{N} + e^- \rightarrow {}^{A}{}_{Z-1}Y_{N+1} + \nu \quad (12)$$

The atom as a whole will, due to the emision of the neutrino, obtain a recoil velocity given by $v_R = Q/m(^AY)c$, with Q the Q-value of the reaction and $m(^AY)$ the mass of the final atom (neglecting secondary effects such as X-ray emission or emission of Auger electrons). This particular beta decay reaction gives, as in the case of gamma emission, a unique value for the initial recoil energy, provided the Q value is less than 1.022 MeV. In contrast to gamma emission it induces a very small recoil energy of the order of a few eV to the final nucleus.

We now focus our attention on K-electron capture from the 0^- isomeric state of ^{152}Eu to the 1^- level at 963 keV in ^{152}Sm. The neutrino emission gives the final nucleus a recoil energy of 3.0 eV [KAL70]. Initially this reaction was choosen for reasons which are more connected to fundamental physics, namely the measurement of the helicity of the neutrino[BÖR89]. However, the electron capture reactions can also be useful for nuclear physics since they allow to redo with better precision the measurements done using nuclear resonance fluorescence with the ultra centrifuge technique[MET65], and thus will yield lifetimes of nuclear excited levels fed by β^- decay. The very low initial recoil velocity completely changes the slowing down process due to the fact that the recoil energy does not allow the atom to leave its lattice position. This is most drastically illustrated by the case of Eu_2O_3 and EuF_3. Although for EuF_3 the volume occupied per atom is smaller the observed Doppler broadening is larger, in contrast to what one expects from a collision model. This is a clear indication that instead of loosing its energy due to collisions, the recoiling atom remains bound in the lattice and looses its energy through the creation of phonons.

Using the nuclear resonance fluorence technique Langhoff et al. have studied the slowing down of atoms with very small kinetic energies in solids. They concluded that the phonon model (PM) gave the best description for these small recoils[LAN69]. Here we use the simplest version of the phonon model which is the Debye approximation [LAN69]. This is motivated by the fact that the frequency spectrum of the solids are normally determined using inelastic neutron scattering, which is unfortunately not possible for the targets one can use for our measurements which require large cross-sections for neutron capture. Then the recoil velocity is given by [LAN69]:

$$v(t) = 3v_0\{(2/\omega_D^2 t^2)\cos(\omega_D t) + ((1/\omega_D t) - (2/\omega_D^3 t^3))\sin(\omega_D t)\} \quad (13)$$

with ω_D the Debye frequency. The Doppler broadened lineshape is now given by:

$$I_D(E,\tau) = \int I_D(E,\tau,v(t,\omega_D))(N_0/\tau)\, e^{-t/\tau}\, dt \qquad (14)$$

where $v(t,\omega_D)$ and $I_D(E,\tau,v)$ are given by formula (13) and (3). The integration of (14) has to be done numerically. For practical reasons, e.g. the way our computercode is written we do not integrate (14) but proceed as follows. First a constant time interval is defined. Then we consider N subsequent interval t(k) for which we calculate the average velocity v(n) by numerical integration of (13). These quantities are used as input in formula (7).

Since the recoil velocities are not much above the thermal velocity we take the thermal broadening into account as a Gaussian which is folded over I(E)dE. The thermal width σ_T is determined from a measurement of a transition decaying from a level with a known long lifetime (in the case of ^{152}Sm the transition from the 121.78 keV level with $\tau = 1.4$ ns to the ground state). In order to illustrate the slowing down we give in Fig. 1 the obtained v(t) for Sm recoiling in four different target compositions. From the figure it is clear that EuF_2 is, of the four, the best suited for the helicity measurement since the recoiling atom is less slowed down in the target.

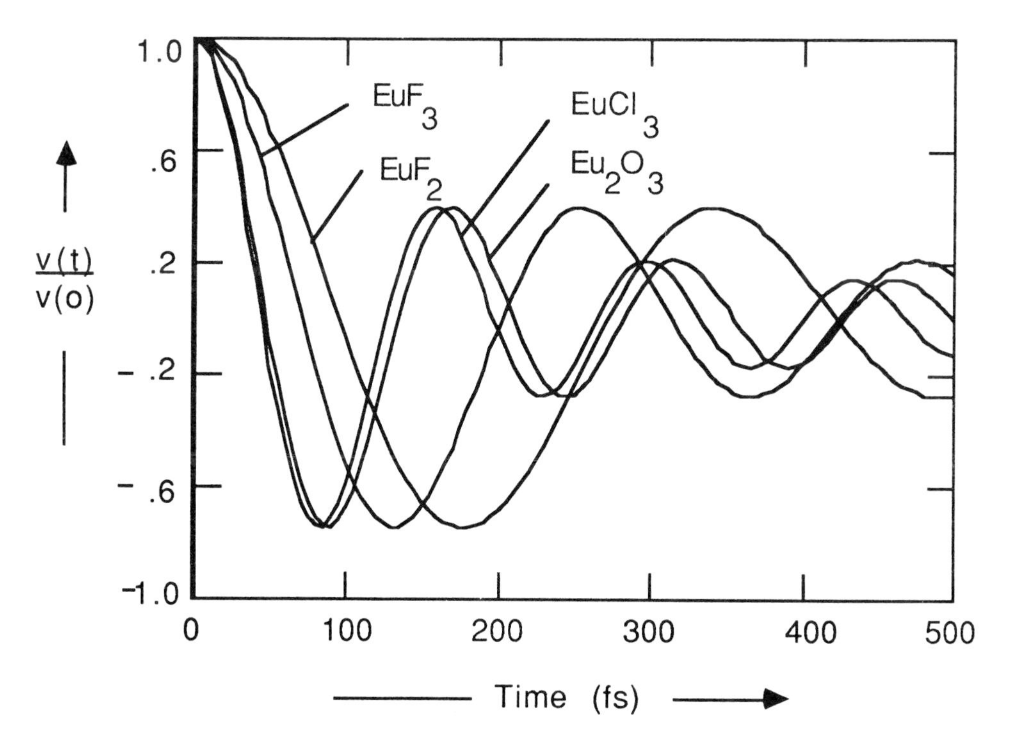

Fig 1: The obtained slowing down for Sm recoiling in Eu_2O_3, EuF_3, EuF_2 and $EuCl_3$ using the phonon creation model.

6. CONCLUSION

We have reviewed the stopping theory elaborated for the analysis of GRID measurements. Since the initial recoil is of the order of a few hunderds eV/atom or lower, the stopping is dominated by atomic collisions (nuclear stopping) and the exchange effects between the electrons (electronic stopping) are neglected. The first description [JOL88] has been substantially improved by taking into account the interatomic potential. Since we use tabulated parameters for the Born-Mayer potential no fitting of the slowing-down process for each target is needed. This approach was successfully applied to ^{49}Ti, ^{54}Cr and ^{57}Fe. In order to describe the slowing down in more complex targets we extended the theory to diatomic targets [JOL89], this allows the study of the interatomic potential in different alloys and to test the consistency of the stopping theory. In this contribution we showed how very low energy recoils, as induced by neutrino emission after electron capture, lead to a different slowing down process. Due to the fact that the atom does not receive enough kinetic energy to break the bonds with the other atoms the slowing down is described by the oscillations of the atom around its equilibrium positions. These can be described using a phonon creation model based on the Debye approximation. In this case an 'effective' Debye frequency has to be fitted to the observed lineshape.

Finally, we want to add some comments.
1) The collision model is based on statistical averaging of different quantities, and its success may be due to positive interference between poor approximations, e.g. defining the atomic radius from the minimum distance in a head-on collision using the calculated Born-Mayer potential parameters.
2) The fact that we don't have to consider the displacement energy can be explained by the energy dependence of the atomic radius and the mean free paths r(k). When the kinetic energy of the recoiling atom is low the path becomes very small and reflects the bouncing of the atom in between the target atoms without displacing them.
3) In the diatomic case one should be careful when applying the double layer approach. The physical picture we used only holds for targets where the interatomic distance is nearly equal to r(0). On the other hand this approach can be seen as a first approximation of the real probability distribution for the collision sequence.
4) It is very questionable that the Debye approach holds at such high temperatures. The fitted Debye frequency has to be seen as an effective frequency or a parameterization of the slowing down process. The next step would consist of using two Debye frequencies one for the acoustical and one for the optical vibrations. Work is in progress to test this.

In conclusion, we have developed a stopping theory to describe the slowing down of recoiling atoms as occuring in GRID and NID measurements. Notwithstanding the approximations the theory yields good results [BÖR89] and is almost parameter free.

REFERENCES.

[ABR69] A.A. Abrahamson, Phys. Rev. 178 (1969)76.
[BÖR88] H.G. Börner, J. Jolie, F. Hoyler, S.J. Robinson and M. S. Dewey, G.L. Greene, E.G. Kessler Jr., R.D. Deslattes, Phys. Lett. B. 215 (1988) 45.
[BOR89] H.G. Börner, J. Jolie, S.J. Robinson, P. Schillebeeckx, these proceedings.
[JOL88] J. Jolie, H.G. Börner, F. Hoyler, S. Robinson and M. S. Dewey, in Inst. of Phys. Conf. Ser. No. 88, p. 586.
[JOL89] J. Jolie, S. Ulbig, H.G. Börner, P.K. Lieb, S.J. Robinson, P. Schillebeeckx, E.G. Kessler Jr., M. S. Dewey, and G.L. Greene, acc. for publ. in Europhys. Lett.
[KAL70] J. Kalus, Phys. Rev. B (1970) 3569.
[LAN69] H. Langhoff, J. Weiss, and M. Schumacher, Z. Physik 226 (1969) 59.
[LIE88] K.P. Lieb, H.G. Börner, M. S. Dewey, J. Jolie, S.J. Robinson, S. Ulbig, and Ch. Winter, Phys. Lett. B. 215 (1988) 50.
[MET65] F. R. Metzger, Phys. Rev 137 (1965)1415.
[SMI72] F.T. Smith, Phys.Rev. A 5 (1972) 1708.
[ULB89] S. Ulbig et al. subm. to Nucl. Phys. A.

Paper presented at Int. Conf. on Spectroscopy of Heavy Nuclei, Crete, Greece, 1989

Shape isomerism at no spin

P. Bonche [1,2], S.J. Krieger [1], P. Quentin [1,3], M.S. Weiss [1], J. Meyer [4], M. Meyer [4], N. Redon [4], H. Flocard [5] and P.-H. Heenen [6].

[1] Department of Physics, Lawrence Livermore National Laboratory, Livermore, CA 94550, USA
[2] Service de Physique Théorique, DPhG, CEN Saclay, F - 91191 Gif sur Yvette Cedex, France
[3] Laboratoire de Physique Théorique, Unité associée au C.N.R.S., Université Bordeaux 1, F - 33170 Gradignan, France
[4] Institut de Physique Nucléaire (et IN2P3), Université Lyon 1, F - 69622 Villeurbanne Cedex, France
[5] Division de Physique Théorique, Unité de Recherche des Universités Paris XI et Paris VI associée au C.N.R.S, Institut de Physique Nucléaire, F - 91406 Orsay Cedex, France
[6] Maître de Recherches F.N.R.S., Physique Nucléaire Théorique, Université Libre de Bruxelles, CP 229, B - 1050 Brussels, Belgium

ABSTRACT: Lattice constrained Hartree-Fock plus BCS calculations are performed to obtain potential energy surfaces in order to investigate the occurence of shape isomerism in heavy Osmium, Platinum and Mercury isotopes.

[*] Work supported in part by US Department of Energy under Engineering Contract NW 7405-ENG-48, in part by SDIO/IST administered by NRL and by NATO under Grant number RG 0195/85.

1. Introduction

The concept of superdeformation was first introduced some twenty years ago (Strutinsky 1967) to explain the fission isomers observed in the actinides (Polikanov *et al* 1962). For the first time, there were nuclei with, in the potential energy surfaces, a secondary minimum which corresponds to very elongated shapes with an axis ratio of 2 to 1. It has reappeared as an explanation of the rare earth nuclei gamma ray spectra observed at high spin (Nyako *et al* 1984).

There is no reason to believe that shape isomerism is restricted to the actinide region or high spin states. Secondary minima in the potential energy curves have previously been found for Mercury isotopes within Hartree-Fock calculations (Cailliau *et al* 1973). However no definite prediction concerning isomeric states were made at that time. More recently, predictions for the existence of these isomers in non fissile even-even nuclei have been made by Girod *et al* (1989) and Bonche *et al* (1989) using microscopic calculations and by Chasman (1989) up to I = 40 within cranked Strutinsky method.

The region of heavy Osmium, Platinum and Mercury isotopes have been considered in this paper. In section 2, the procedure of our static Hartree-Fock plus BCS calculations is briefly described. The collective potential energy surfaces are presented in section 3 in terms of usual deformation parameters for these nuclei as the characteristics of the calculated secondary minima. The stability of our results is also investigated with respect to the choice of the effective force. Discussion and conclusions are presented in section 4.

2. Procedure description

Static constrained Hartree-Fock plus BCS calculations have been performed according to the discretization procedure described by Bonche *et al* (1985). The S III parametrization of the Skyrme effective force has mainly been used. Pairing correlations are introduced by means of the usual constant G pairing matrix. The collective quadrupole deformation space is mapped through doubly constrained Hartree-Fock quantities : Q_0 the mass quadrupole moment in barns and γ the asymmetry angle in degrees. Computations have generally been carried out at three values of the asymmetry angle γ : 0, 30 and 60 degrees. In some cases, calculations for 15 and 45 degrees have also been performed. The relevant sextant has been mapped by using a Q_0 mesh size of 200 fm^2.

3. Results

In this region, fourteen even Osmium isotopes, from ^{186}Os to ^{212}Os, eleven Platinum even isotopes, from ^{186}Pt to ^{206}Pt, and seventeen Mercury even isotopes, from ^{186}Hg to ^{218}Hg, have been calculated with the Skyrme S III effective force.

An example of potential energy surfaces is presented in Figure 1 for some Platinum isotopes. We retrieve for the absolute minima the well-known prolate - oblate transition (Sauvage-Letessier *et al* 1981) between ^{188}Pt and ^{196}Pt. The prolate - oblate energy difference is small and there exists a possibility of having triaxial equilibrium shapes. The relation between triaxiality and shell effects has been discussed explicitly by Redon *et al* (1986).

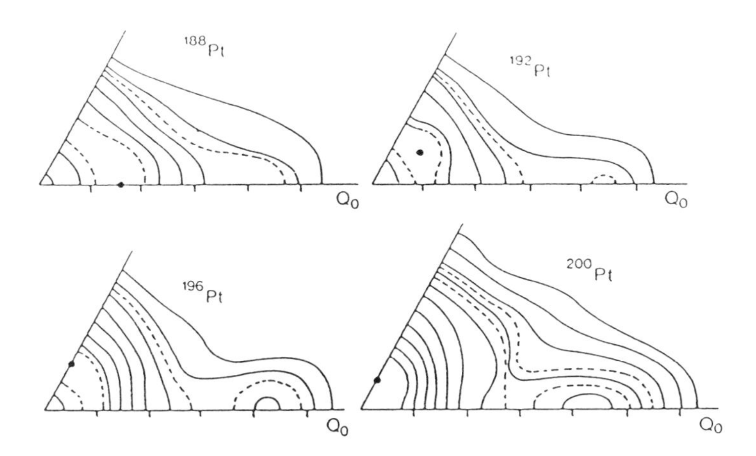

Fig. 1. Hartree-Fock energy surfaces for Platinum isotopes. The black dots stand for the absolute energy minima. From these minima, contour lines are drawn every 2 MeV. Dashed lines are sometimes plotted every 1 MeV. Deformation is measured by the quadrupole moment Q_0 (in barn) and the distance between two tickmarks on the axis corresponds to a 10 barn increase. The γ angle varies from 0 to 60 degrees.

A characterization of possible isomeric states is given in Figures 2,3. For each series of isotopes, excitation energies defined as the energy differences between the ground and isomeric states are given, as well as the depths of the second wells. The sequences of nuclei which exhibit a second minimum extend across the $N = 126$ neutron magic number for both Osmium, from ^{188}Os to ^{212}Os, and Mercury isotopes, from ^{192}Hg to ^{218}Hg. As for the Platinum isotopes, the isomeric region starts at $N = 110$ and extends up to at least $N = 128$ which is the last isotope considered here. A well marked maximum of the excitation energy for the isotopes corresponds to the magic neutron number $N = 126$.

The appearance of second minima occurs at values of the mass quadrupole moment Q_0 of approximately 45 barns. Q_0 can be related to the commonly published β_2 values of the deformation by : $\beta_2 = \sqrt{\frac{\pi}{5}} \cdot \frac{Q_0}{A\,r^2}$, where A is the nucleon number and r is the r.m.s. radius of the mass distribution. This prescription yields typical values of β of 0.5, as listed in Table 1 for Mercury isotopes. Another estimate is provided of an axis ratio q defined from the expectation values of the x^2 (or y^2) and z^2 operators : $q = \sqrt{\frac{\langle z^2 \rangle}{\langle x^2 \rangle}}$.

Strictly speaking, superdeformation is generally assigned to an axis ratio of 2 to 1, whereas our secondary minima, while quite deformed, have a ratio of 1.6 to 1. This deformation is stable beside the asymmetry angle γ. The stability with respect to the octupole deformations has also been checked.

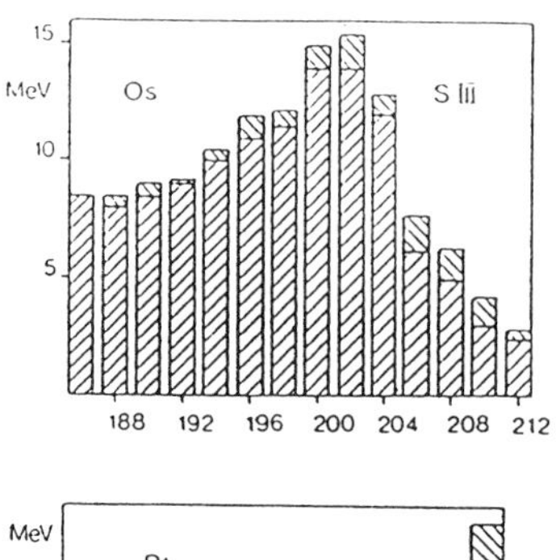

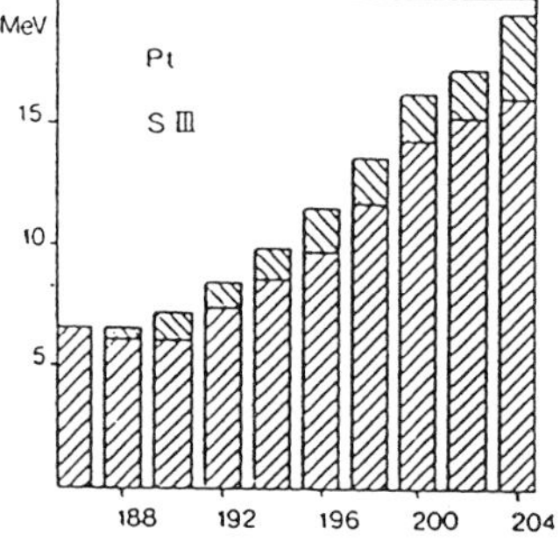

Fig. 2. Excitation energies and well depths for secondary minima in Os and Pt isotopes. The lower (upper resp.) part of each bar diagram corresponds to the excitation energy (well depth resp.).

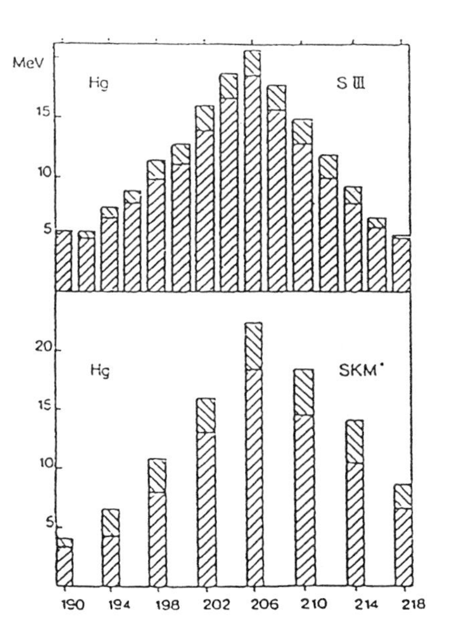

Fig. 3. Same as Fig. 2 for Hg isotopes calculated for the S III Skyrme effective effective force (upper part) and for the SkM* one (lower part).

A	Q_0 (*barns*)	β	q
192	42.5	0.53	1.58
194	43.9	0.54	1.59
196	43.0	0.52	1.57
198	42.6	0.51	1.55
200	43.3	0.51	1.55
202	43.5	0.50	1.55
204	43.6	0.50	1.54
206	43.6	0.49	1.53
208	43.7	0.48	1.52
210	44.0	0.48	1.52
212	45.2	0.48	1.52
214	46.7	0.49	1.53
216	50.0	0.51	1.56
218	52.4	0.53	1.58

Table 1 Deformation of the Hg isotopes at the secondary minima

The stability of our results to changes in the Skyrme interaction has been studied. Calculations of some Mercury isotopes have been performed with the SkM* interaction of Bartel *et al* (1982). The surface tension associated with the SkM* force is more realistic than the corresponding SIII quantity but the shell structure yielded by the SIII force is more compatible with spectroscopic data. As seen from Figure 3, SkM* force yields results in excellent qualitative agreement with those obtained with the Skyrme S III force with the exception of a secondary minimum, yet very shallow, appearing with the SkM* force in ^{190}Hg.

4. Discussion and conclusions

We have investigated the occurence of shape isomerism in Osmium, Platinum and Mercury isotopes within the microscopic constrained Hartree-Fock plus BCS framework. We are mostly interested by isomeric states decaying electromagnetically to the ground state. For that, we impose an arbitrary upper limit of 8 MeV for the excitation energy in order to rule out neutron emission as much as possible. A no less arbitrary minimum of 0.5 MeV is fixed for the well depth so that the local minimum will be quasi-static. With these conditions, only some isotopes are proposed for experimental verification : ^{190,192}Pt, ^{206,208,210}Os and ^{194,196,214}Hg. This does not imply that shape isomerism may not been found in neighbouring nuclei, our prescription was only meant to delineate the most promising candidates. Moreover, a superdeformed band has been observed at low spin in ^{191}Hg by Moore *et al* (1989). We have also carried out an experiment with the Château de Cristal in Strasbourg to observe such superdeformed states in ^{189}Pt.

Historically, the concept of superdeformation has been introduced in the context of shape isomerism in the actinides, characterized by an axis ratio of 2 to 1. It was later extended to shapes inferred for region of high spin states in rare earth nuclei, as they presented similar deformations. Our candidates for shape isomerism in the Osmium-Platinum-Mercury region have a smaller axis ratio, namely 1.6 to 1. However they present all the known characteristics of shape isomerism at no spin other than fission. It is therefore quite tempting to extend the concept of superdeformation to these new states, eventhough their ratio of major to minor axis is smaller. Furthermore, quasi static states with a 2 to 1 axis ratio are not expected at low excitation energy in nuclei ligther than actinides.

REFERENCES

Bartel J, Quentin P, Brack M, Guet C and Håkanson H -B 1982 Nucl. Phys. **A385** 269

Beiner M, Flocard H, Van Giai Nguyen and Quentin P 1975 Nucl. Phys. **A238** 29

Bonche P, Flocard H, Heenen P -H, Krieger S J and Weiss M S 1985 Nucl. Phys. **A443** 39

Bonche P, Krieger S J, Quentin P, Weiss M S, Meyer J, Meyer M, Redon N, Flocard H and Heenen P -H 1989 Nucl. Phys. A in press

Cailliau M, Letessier J, Flocard H and Quentin P 1973 Phys. Lett. **B46** 11

Chasman R R 1989 Phys. Lett. **B219** 227

Girod M, Delaroche J P, Gogny D and Berger J F 1989 Phys. Rev. Lett. **62** 2452
Moore E F, Janssens R V F, Chasman R R, Ahmad I, Khoo T L, Wolfs F L H, Ye D, Beard K B, Garg U, Driegert M W, Benet Ph, Grabowski Z W and Cizewski J A 1989 in press
Nyako B M, Cresswell J R, Forsyth P D, Howe D, Nolan P J, Riley M A, Sharpey-Schafer J F, Simpson J, Wand N J and Twin P J 1984 Phys. Rev. Lett. **52** 507
Polikanov S M, Druin V A, Karnaukhov V A, Mikheev V L, Pleve A A, Skobelev N K, Subbotin V G, Ter-Akop'yan G M and Fomichev V A 1962 Sov. Phys. JETP **15** 1016
Redon N, Meyer J, Meyer M, Quentin P, Weiss M S, Bonche P, Flocard H and Heenen P -H 1986 Phys. Lett. **B181** 185
Sauvage-Letessier J, Quentin P and Flocard H 1981 Nucl. Phys. **A370** 231
Strutinsky V M 1967 Nucl. Phys. **A95** 420

Inst. Phys. Conf. Ser. No 105
Paper presented at Int. Conf. on Spectroscopy of Heavy Nuclei, Crete, Greece, 1989

A detection system for proton radioactivity with an electrostatic deflector

A. Gillitzer, T.Faestermann, F.Heine, M.Köpf, J.Siegl
Technische Universität München, D-8046 Garching, F.R.Germany

ABSTRACT: In order to investigate particle decays of shortlived nuclei, in particular proton radioactivity, an electrostatic deflector, which separates fusion evaporation residues from the primary beam, has been set up. It is designed for electric rigidities of $E \cdot \varrho$=10MV and accepts a solid angle of 9msr. The large area gas detection system 1.3m behind the target consists of two parts. The first part measures time, energy, energy loss, range and position of the evaporation residues and the second time, energy, range and position for the emitted protons or alpha particles. For evaporation residues from the fusion of ^{58}Ni and ^{54}Fe an efficiency of 40% has been observed and background from beam particles has been reduced by a factor of 10^5. As an alternative for the gas detector system a 10x10cm^2 Si detector array is being developed.

MOTIVATION

An important part of the experimental work in nuclear physics is devoted to the effort to extend the region of known isotopes and to study the newly synthesized exotic nuclei by analysing their decay modes. For neutron deficient nuclei the borderline of stability is determined by the proton binding energy and proton decay from the ground state will be the main decay mode of nuclei beyond that line. Therefore, the study of proton radioactivity is the natural method to obtain information about ground state configurations and nuclear masses in this region of nuclei.
Until now, five ground state proton emitters are known [1]; two of them, ^{113}Cs and ^{109}I, have been found at the Munich tandem accelerator postaccelerator system. A third isotope in the T_z=3/2 series above Z=50, ^{105}Sb, could be shown to be

proton unstable by combining proton and α decay energies, but with a proton decay branch probably too small to be observable [2,3]. From the measured decay energy and halflife and from shell model systematics we consider a $d_{5/2}$ ground state to be the most probable configuration for the two proton emitters. The decay, however, is hindered by factors of 17 for ^{113}Cs and 7 for ^{109}I, compared to halflife calculations using a WKB approximation. It is conceivable that the increasing hindrance is due to the transition from the more spherical nuclei near Z=50 to nuclei in the deformed region with collective degrees of freedom.

For a better understanding of the observed spectroscopic factors it would be helpful to obtain also information on neighbouring nuclei. In particular, it would be interesting to observe proton decay of a very neutron deficient Sb isotope, because it has only one proton above the closed shell, and no spectroscopic hindrance is expected.

Unfortunately the cross sections for the production of these neutron deficient nuclei are very small. Both known proton emitters in this region were produced by a fusion reaction using a ^{58}Ni beam via the evaporation of one proton and two neutrons with a cross section of about 30μb. A drop of more than one order of magnitude is expected for each additional neutron being evaporated. Therefore still more sensitive detection techniques have to be developed.

EXPERIMENTAL TECHNIQUE

To continue the search for proton radioactivity - and more generally, to study particle decays of shortlived nuclei - an electrostatic recoil separator has been built (see Fig.1). This separator has three relevant features :

i) It is very efficient due to its large acceptance and good background suppression.

ii) It is short (1.3m from the target to the detection system).

iii) It is rather simple and could be set up within a comparatively short time.

The separator is designed for typical electric rigidities of E·ϱ=10MV. It consists of two pairs of electrodes, each 12cm high and 25cm long. The first pair has a distance of 4cm, the second opens from 4cm to 6cm and its anode has a horizontal slit to let the primary ion beam pass through to the Faraday cup. The geometrical acceptance of the separator is ±2° horizontally, defined by a slit at the entrance, and ±3.7° vertically, defined by the height of the deflector plates, corresponding to a solid angle of 9msr.

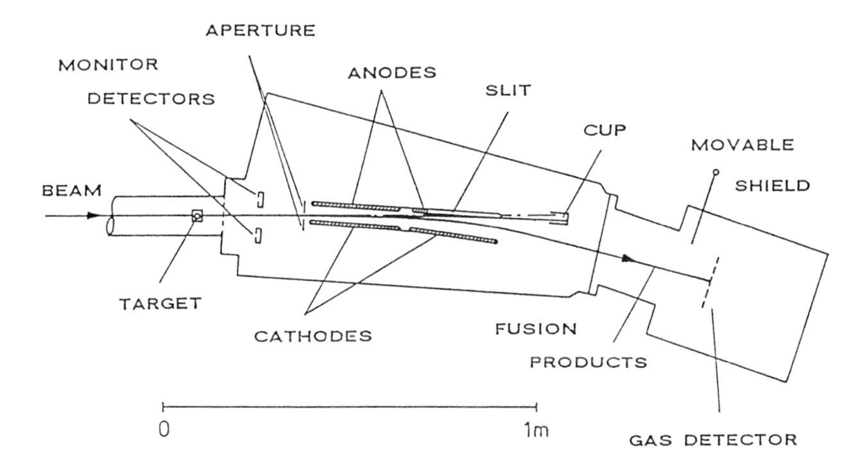

Fig.1
Schematic view of the electrostatic recoil separator

A maximum voltage of 160kV has been applied to the deflector, however a voltage of 100kV is sufficient to deflect evaporation residues from the fusion of 235MeV ^{58}Ni with ^{54}Fe by an angle of 8°, twice as much as the primary beam. For this reaction, leading to the proton emitter ^{109}I, nearly 40% of the evaporation residues were observed in the detector system at a ratio of scattered beam particles to residues of 5:1. If the horizontal acceptance was reduced, this background ratio could be improved to 1:1 at a reduced efficiency of 10%.
With the short length of the apparatus of little more than 1m flight times of the fusion products from the target to the detector system are typically 100ns. Thus also shortlived nuclei can be studied. Because of the dramatic decrease of the halflife of a proton unstable nucleus with increasing decay energy, the sensitivity of the experimental method for short halflives is essential for the search for new proton emitters.
As there are no focusing elements, a large area detector system is required to cover the full solid angle of the separator.The present gas detector system (see Fig.2) is composed of two main parts:
The fusion products are stopped and identified in the first part, particles from their decay are analysed in a separated second gas volume.
The 'heavy ion' part consists of a twodimensionally position sensitive avalanche counter measuring the time of flight with respect to the pulsed beam, and an ionization chamber measuring the energy and the range in the detector gas.

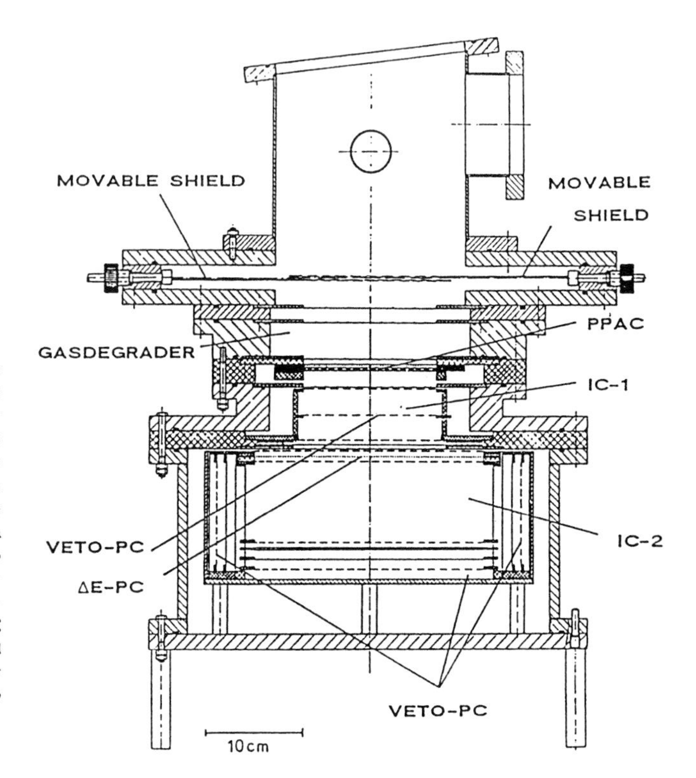

Fig.2
Schematic view of the gas detector system. The detector system is composed of two separate units, to identify the evaporation residues (a parallel plate avalanche counter and a ionization chamber IC-1) and to measure the decay particles emitted in forward direction (a ΔE proportional counter and an ionization chamber IC-2, surrounded by veto proportional counters).

The second part consists of a twodimensionally position sensitive proportional counter and a second ionization chamber. Decay particles emitted in forward direction are identified by their energy loss in the proportional counter, their range in the ionization chamber, and their energy.
The second ionization chamber is surrounded by veto proportional counters, to suppress background signals from particles not stopped in the detector, especially β-delayed protons. If the halflife is short enough, the signals of the decay particle and the mother nucleus can be correlated. As both parts of the detector system are position sensitive, the track of the mother nucleus and the angle of emission of the decay particle are fully determined, and its measured energy can be corrected for unobserved energy loss. In addition, the background from the decay of longlived evaporation residues is considerably reduced. The separator and the detector system were tested with the isomeric proton decay of ^{53m}Co, produced with a ^{24}Mg(^{32}S,p2n) reaction (E_p=1.56MeV, σ_p≈8μb, $T_{1/2}$=247ms) [4] and the ground state proton decay of ^{109}I (E_p=0.81MeV, σ_p≈30μb, $T_{1/2}$=109μs). Spectra of the latter test are shown in

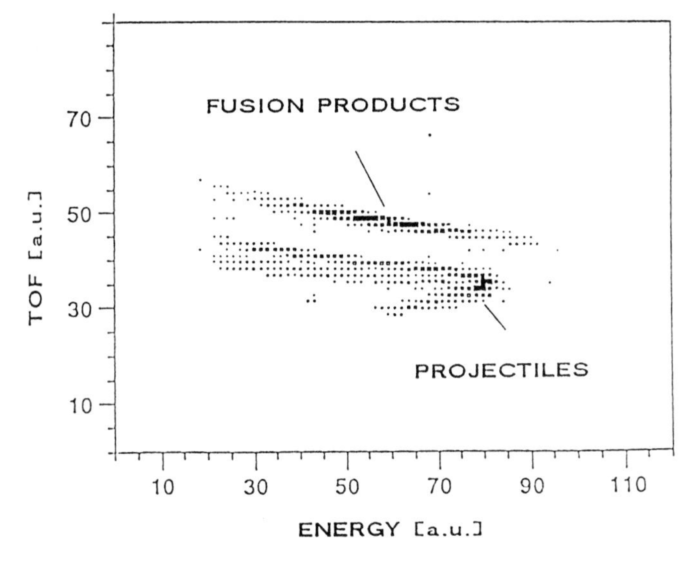

Fig.3
Twodimensional time of flight vs. energy spectrum of heavy ions observed in the first part of the gas detector system with a 235MeV ^{58}Ni beam and a 0.5mg/cm^2 ^{54}Fe target. Fusion evaporation residues are clearly separated from scattered beam particles.

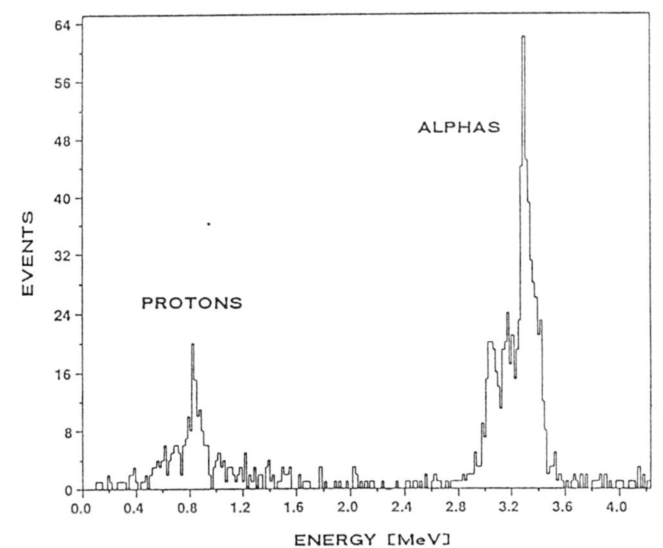

Fig.4
Energy spectrum of protons and α-particles observed in the second part of the detector system during the time interval between the beam pulses.

Figs.3 and 4. The proton emitter was produced by the fusion of 235MeV ^{58}Ni with ^{54}Fe. The fusion evaporation residues are identified by their time of flight and energy and well separated from scattered beam particles (Fig.3). The second spectrum (Fig.4) shows a projection on the total energy of decay particles, observed during the time intervals between the beam pulses. Protons and α-particles from the decay of neighbouring isotopes are distinguished by their energy loss and range signals. As can be seen in the spectrum there is still some background left from protons which are not stopped

in the detector system and not fully suppressed by the veto detectors. During this test, however, correlations of protons and mother nuclei have not yet been used. A total efficiency of 5% for the observation of the proton decay was achieved.
A much better energy resolution for the decay particles can be obtained, if the mother nuclei are implanted directly into silicon detectors. Therefore, as an alternative, a large area array, consisting of 10×10 PIN photodiodes of $1cm^2$ each, is being developed. These silicon detectors are industrially manufactured and have an energy resolution and timing characteristics equivalent to those of surface barrier detectors. The array is read out in rows and columns, to obtain an independent energy and timing signal for each detector. The thickness of 220μm is sufficient to stop protons up to an energy of 4.5MeV and to separate low energetic protons due to ground state decays from β-delayed protons, which are much higher in energy and mostly not stopped in the detector. Because of the good timing characteristics, the measurement of a decay process is possible a few microseconds after the implantation of a fusion product. Thus also the decay of shortlived nuclei can be analysed with good energy resolution.

REFERENCES

1. S.Hofmann, 'Particle Emission from Nuclei', M.Ivascu, D.N.Poenaru eds., (CRC Press, Boca Raton, USA, 1987), Vol.2, p.25
2. T.Faestermann, A.Gillitzer, K.Hartel, P.Kienle, E.Nolte, Phys.Lett. 137B, 23 (1984)
3. A.Gillitzer, T.Faestermann, K.Hartel, P.Kienle, E.Nolte, Z.Phys. A326, 107 (1987) and T.Faestermann, A.Gillitzer, K.Hartel, P.Kienle, E.Nolte, Proc. Fifth Int. Conf. on Nuclei far from Stability, Rosseau Lake, Canada, 1987, p.739
4. K.P.Jackson, C.U.Cardinal, H.C.Evans, N.A.Jelley, J.Cerny, Phys.Lett. 33B, 281 (1970)

Inst. Phys. Conf. Ser. No 105
Paper presented at Int. Conf. on Spectroscopy of Heavy Nuclei, Crete, Greece, 1989

Evidence for fractional occupancy of shell orbits from electron scattering

C. N. Papanicolas and S. E. Williamson

Nuclear Physics Laboratory and Department of Physics, University of Illinois, Champaign, IL 61820

ABSTRACT: Single arm electron scattering provides one of the best probes for the measurement of occupation numbers of shell orbitals in nuclei. We review in this paper the available experimental evidence and extract from it occupation numbers for heavy nuclei and for, in particular, ^{208}Pb. We conclude that occupation probabilities of single particle states just below and above the Fermi energy of 0.7 ± 0.1 and $0.1^{+0.05}_{-0.10}$ respectively are consistent with the available experimental evidence.

1 INTRODUCTION

During the last decade it became increasingly obvious that nuclear mean field theory could not provide a consistent description of the very diverse and precise data from single arm electron scattering on heavy nuclei (Frois and Papanicolas (1987)). As experiments focused on the details of the phenomena through more accurate and more exclusive measurements and with increasing emphasis on separated (transverse-longitudinal) responses, the discrepancy between theory and experiment grew larger and patterns began to emerge. In the mid-eighties it was suggested that the discrepancy between mean field predictions and data was primarily the manifestation of the influence of correlations – effects that could not be accounted for within the framework of mean field theory with phenomenological interactions. Substantial theoretical and experimental progress has been achieved since then in elucidating many of the issues, and a consistent picture has emerged lending further credence to this suggestion. Theory and experiment have converged on occupation numbers for the last filled orbitals in heavy nuclei (typically lead) ranging from 0.6 to 0.9. Results from other probes, as diverse as (e, e′ p) (Lapikas (1989)), (d,^{3}He) reactions (Wagner (1989)), Gamow-Teller resonances (McFarlane (1986)), and low energy neutron scattering (Johnson *et al* (1987) and Das and Finlay (1989)) agree with the results derived from single arm electron scattering, thus providing a consistent overall picture indicative of an emerging consensus on this issue.

2 ORBITALS AND OCCUPATION NUMBERS

Empirical results concerning fractional occupancy of orbitals inevitably must face two difficult issues.

1. The model dependence involved in calculating mean field orbitals inevitably propagates to the derived occupation numbers; and
2. the ambiguities involved in extracting information about orbitals from specific physical measurements are substantial and probe dependent.

The first issue is of general concern; it affects all experimental results, and it inevitably enters as an overall systematic uncertainty to all experiments. The second point is more closely related to uncertainties in reaction theory. Different probes are subject to different systematic uncertainties. The consistency of results obtained from these probes will therefore be limited to the sum of these uncertainties.

The issue concerning the model dependence of orbitals has been pursued recently by the Liege (C. Mahaux *et. al*) and the Urbana (V. R. Pandharipande *et. al*) groups. We provide below a summary of the key definitions and conclusions, limiting ourselves to those necessary for a meaningful discussion of the data. A detailed exposition of the underlying theoretical issues can be found in references by Jaminon *et al* (1985), Mahaux and Ngô (1984), and Lewart *et al* (1988).

Given a model choice for the single particle wave function basis $[\phi_\alpha]$ which has the appropriate completeness and orthonormality properties (e.g. Hartree-Fock or oscillator wave functions), it is always possible to write the one-body density distribution as:

$$\rho(\mathrm{r}) = \frac{1}{4\pi \mathrm{r}^2} \sum_{\alpha,\alpha'} (2\mathrm{j}+1)\, \rho_{\alpha\alpha'}\, \phi_\alpha(\mathrm{r})\, \phi^*_{\alpha'}(\mathrm{r}) \tag{1}$$

where α is an all encompassing quantum number and $\rho_{\alpha\alpha'}$ is the appropriate one-body density matrix for that basis. A unitary transformation U can be found that will render ρ diagonal. This transformation defines (see Lowdin (1955) and Schafer and Weidemuller (1971)) then the "natural orbitals" ζ_α :

$$\zeta_\alpha \equiv \mathrm{U}_{\alpha\beta}\, \phi_\beta$$

In the unique basis of natural orbitals the density matrix is by definition diagonal and the one-body density distribution takes its simplest form:

$$\rho(\mathrm{r}) = \frac{1}{4\pi \mathrm{r}^2} \sum_{\alpha} (2\mathrm{j}+1)\, \mathrm{n}_\alpha\, \zeta^*_\alpha(\mathrm{r})\, \zeta_\alpha(\mathrm{r}) \tag{2}$$

where the diagonal elements $\rho_{\alpha\alpha'}$ have been expressed more economically (only one index needed) as "occupation probabilities" n_α .

In addition to the natural orbitals two other types of orbitals can be defined uniquely: the mean field orbitals and the quasi particle (or overlap) orbitals. The mean field orbitals, $\chi(r)$ can be defined as the wave functions that allow the one body density to be written as:

$$\rho(r) = \frac{1}{4\pi r^2} \sum_{\varepsilon_i < \varepsilon_f} (2j+1)\, \chi_i(r)\, \chi_i^*(r) \tag{3}$$

while the quasiparticle or overlap orbitals are defined by:

$$\psi_\alpha(r) = \langle \Psi^{N-1)} | \Psi^N \rangle \tag{4}$$

where $|\Psi^N\rangle$ and $|\Psi^{N-1}\rangle$ are the many-body wave functions describing the systems of N and N − 1 constituents.

The above definitions are chosen in such a way that in the absence of correlations, the natural orbital, the quasiparticle orbital, and the mean field orbital coincide.

The basis of natural orbitals provides an excellent theoretical concept and a very useful language for the discussion of occupation probabilities (Lowdin (1955)). In practice, to the best of our knowledge, it is not known how to calculate natural orbitals for finite nuclei exactly. A prescription for the approximation of natural orbitals in the nuclear many-body problem can be found in the work of Jaminon *et al* (1985). Particularly illuminating is the study of single particle orbitals in liquid helium drops by Lewart *et al* (1988), where these issues are extensively studied in ^{3}He drops of 20 to 240 atoms. These fermionic systems can be subjected to rigorous theoretical investigation because of the simple interaction between their constituents. They exhibit many of the features found in atomic nuclei (e.g. shell structure), and therefore they provide and obvious modeling environment for the investigation of these issues. These authors, through analytical calculations and by variational Monté Carlo estimation, have calculated mean field, quasiparticle, and natural orbitals in such droplets. In Figure 1 the 3s mean field, quasiparticle, and natural orbits for a ^{3}He N = 70 droplet are shown.

Several approximations to the one-body density are also of considerable value as they are more easily realizable in model calculations. One can define a Fermi gas density $\rho^{FG}(r)$:

$$\rho^{FG}(r) = \frac{1}{4\pi r^2} \sum_\alpha (2j+1)\, \phi_\alpha^*(r)\, \phi_\alpha(r) \tag{5}$$

which is obviously defined within the given model space. The suppressed occupation probability has the known Fermi-Dirac distribution at zero temperature:

$$n_\alpha = \begin{cases} 1 & \varepsilon < \varepsilon_f \\ 0 & \varepsilon > \varepsilon_f \end{cases} \tag{6}$$

A subset of this class of one-body densities are the various types of Hartree-Fock distributions, $\rho^{HF}(r)$.

In Figure 2 we show the distribution of occupation numbers one might expect for a finite nucleus in the limit of vanishing correlations. The occupation numbers have the familiar step-wise distribution given by equation 6, which is indicated in the figure by the solid curve. Correlations induce virtual scattering, which leads to occupation numbers less than 1 below the Fermi energy and greater than 0 above it. A discontinuity at the Fermi energy persists and its magnitude, Z, is a measure of the influence of correlations in that system Migdal (1967).

$$Z = n_- - n_+ \tag{7}$$

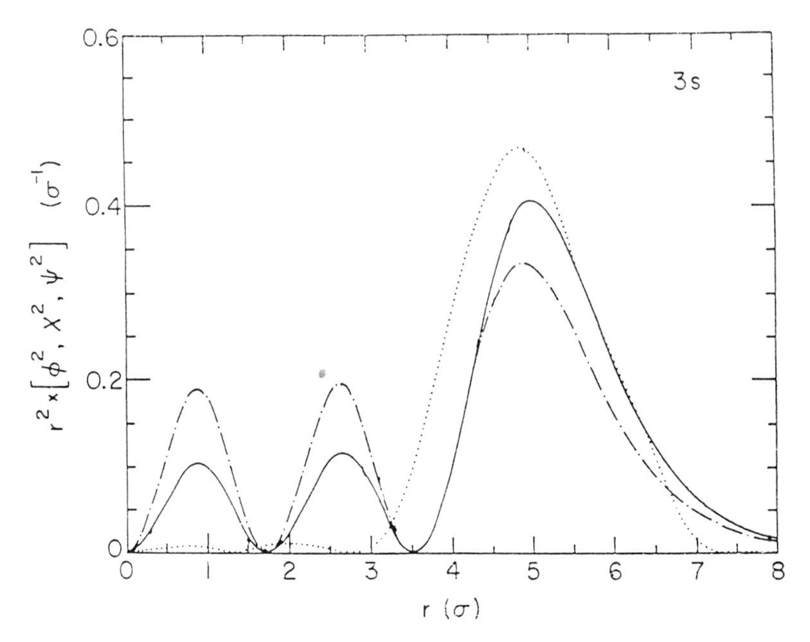

Figure 1: The shape of mean field (dash-dot), quasiparticle (solid curve), and natural (dotted curve) orbitals for a ^{3}He droplet (Lewart *et al* (1988)).

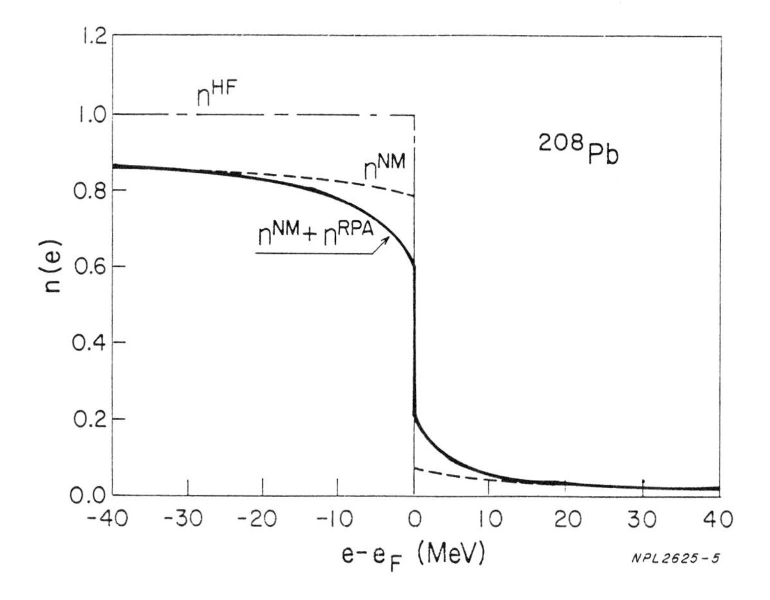

Figure 2: An estimate of the occupation numbers for ^{208}Pb (thick line) by Pandharipande *et al* (1984). n^{HF} represents the mean field distribution defined by equation 6.

where n_+ and n_- denote the occupation probabilities of orbitals lying immediately below and above the Fermi energy. In the language of the Landau theory, Z is the pole strength of the quasiparticle term (Speth *et al* (1977)).

Partial occupancy of shell orbits is required by our very basic notions about the nuclear force and medium. Its manifestations are numerous, perhaps the most irrefutable being the existence of the imaginary part of the nuclear optical potential. This has been recognized from the beginning of nuclear mean field theory. It is thoroughly dealt with, for instance, in the early work of Brueckner and coworkers.

Finally we have found that a semantic confusion permeates the literature leading to possible misunderstandings. The term "occupation probability" is used to mean the probability of a given orbital being occupied. This is a well defined concept. However, in an arbitrary nucleus this probability is influenced by a number of effects. Three distinct mechanisms are often invoked to explain deviations from the mean field stepwise distribution:

1. configuration mixing,
2. pairing,
3. correlations in the many-body system.

Clearly, of the three mechanisms, only the last is not expected to vary significantly among neighboring nuclei. It provides the underlying basis to which effects (1) and (2) must be added in order to describe the structure of any heavy nucleus. Historically, this third effect has been mostly ignored with pairing and configuration mixing being calculated starting from a mean field basis. The confusion arises because various authors discuss the depletion of occupancy resulting from different combinations of these three effects. In our discussion we deal solely with the fractional occupancy resulting from the correlations of the many-body system.

3 INFLUENCE OF CORRELATIONS ON (e, e') MEASUREMENTS

It is remarkable that, forty years after the birth of the shell model, the occupancy of shell orbits and the limits of validity of nuclear mean field theory is a subject of active research. The imprecise knowledge of occupation probabilities at this point can be attributed to many factors, some of the most important being the following:

1. Most of the available data concern phenomena dominated by the nuclear surface where it is particularly difficult to test the concept of the mean field due to the rapid variation in nuclear density.
2. Of the measurements that possess sensitivity in the saturated nuclear interior only a small fraction can be interpreted reliably because of the large uncertainties in the reaction mechanism.
3. It is only for a very few select cases that the effects of pairing and configuration mixing can be either controlled or ignored, permitting fractional occupancy due to many-body correlations to be isolated and experimentally studied.

Single arm electron scattering possesses many of the desired features for the study of correlations and occupation numbers. It is a weak probe, and it samples the nuclear interior easily. It is also well understood. The reaction theory allows us, in favorable cases, to relate cross sections to currents and densities unambiguously with an uncertainty limited only by effects arising from dispersion corrections. However, in typical cases, contributions arising from meson exchange currents and core polarization contribute to the cross section at the 10-20% level. These effects must be taken into account if this reaction is to be used in the search for the influence of correlations.

The third consideration listed above is very restrictive. In practical terms it means that one must investigate doubly closed shell nuclei and their isotopes/isotones to minimize configuration mixing. If, in addition, we require that pairing be inconsequential, we find only one candidate: ^{208}Pb and its neighboring nuclei.

In the remainder of this article we will consider a number of phenomena that manifest the influence of correlations in heavy nuclei. After once establishing that a phenomenon is generally observed for a number of nuclear systems, we will always seek to quantify it in terms of fractional occupancy in the lead region for the reasons we discussed above.

Three distinct areas of electron scattering have provided evidence for the influence of correlations in heavy nuclei:

1. elastic scattering (charge and magnetic),
2. inelastic scattering to states of "pure" single particle nature, and
3. the longitudinal quasielastic response.

The charge distributions of doubly closed shell nuclei, reconstructed from elastic electron scattering, provided the first evidence that something "anomalous" was going on. Data, gathered from practically every electron scattering facility in the last 35 years combined with the very precise measurements of muonic x-ray transitions yield the charge densities shown in Figure 3. The empirical densities are compared with the Hartree-Fock-Bogoliubov calculations of Decharge and Gogny (1980) performed with a finite-range density dependent effective force. The disagreement is most pronounced in ^{208}Pb, which *ab initio* was expected to be the most favorable case for a mean field description. This comparison shows that modern mean field calculations with effective interactions can account rather accurately for the sizes and the surface properties of heavy nuclei but systematically overpredict the density fluctuation in the nuclear interior. This discrepancy has been haunting every mean field calculation that attempts to describe properties of medium and heavy spherical nuclei consistently.

Mean field theories proved to be inadequate in describing the separated longitudinal/transverse responses of quasielastic electron scattering. Figure 4 shows the longitudinal response function of ^{40}Ca at a momentum transfer of 410 MeV/c (Meziani *et al* (1985)). It has proven to be the case for all medium and heavy nuclei that the longitudinal quasielastic response (which is expected to be the "cleanest", showing little sensitivity to the presence of exchange currents) is significantly overestimated by mean field theoretical calculations.

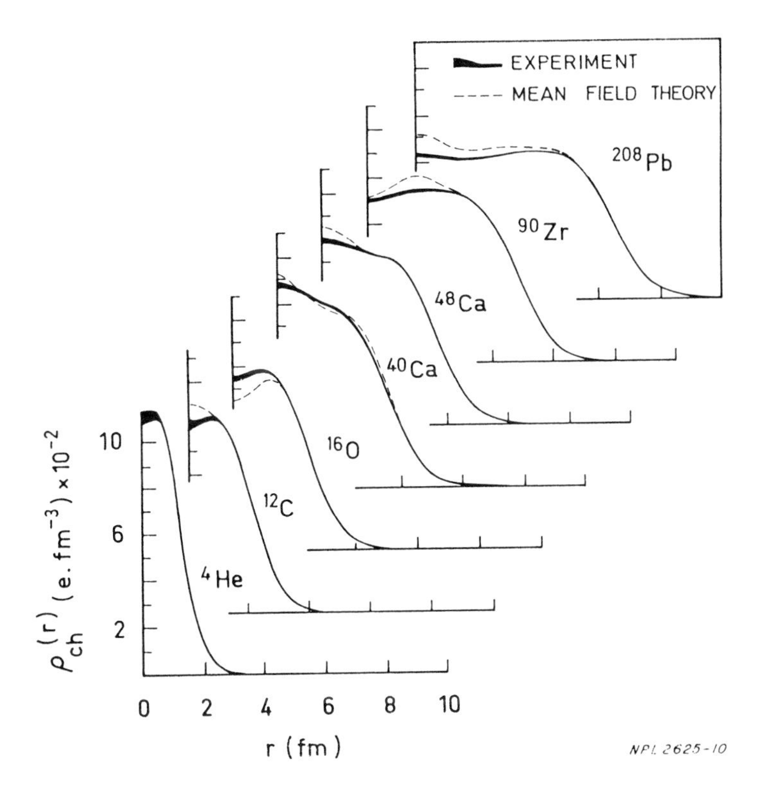

Figure 3: Very high precision characterizes the empirical charge distribution. Mean field calculations (Hartree-Fock) systematically overestimate the charge density in the interior of heavy nuclei.

Finally a third area of significant systematic discrepancies between empirical data and mean field predictions appears in the study of states of "pure" 1p-1h character through inelastic electron scattering. In the most favorable cases (typically closed shell nuclei) the transition amplitude approaches 50% of the mean field amplitude. In typical cases of open shell nuclei the discrepancy is much bigger. Typical is the case of stretched configurations.

Stretched configurations involve 1 $\hbar\omega$ transitions that yield maximum realignment of the total angular momentum, e.g. $\nu(i_{13/2} \rightarrow j_{15/2})_{14^-}$ in ^{208}Pb. It is nearly impossible to construct excitations of such high angular momentum in any other way at nearby energies. This implies absence of mixing and thus insures the relative purity of the single particle character of the transitions. These states have been extensively studied throughout the periodic table with a variety of probes. In Figure 5 the observed strength for such states are shown and compared to the limit predicted by the independent particle model.

The three cases discussed above are perhaps the most dramatic and best known cases of the inadequacy of mean field theory to describe single arm scattering data. They are not

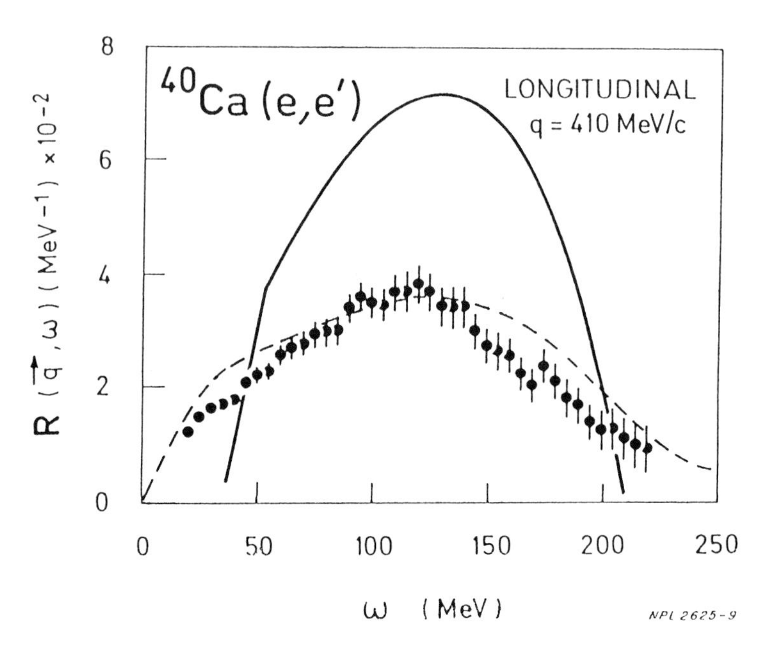

Figure 4: Longitudinal response function of ^{40}Ca at a momentum transfer of 410 MeV/c. The solid curve is a Fermi gas prediction. The dashed curve is the result of a calculation for nuclear matter that treats correlations (Fantoni and Pandharipande (1987))

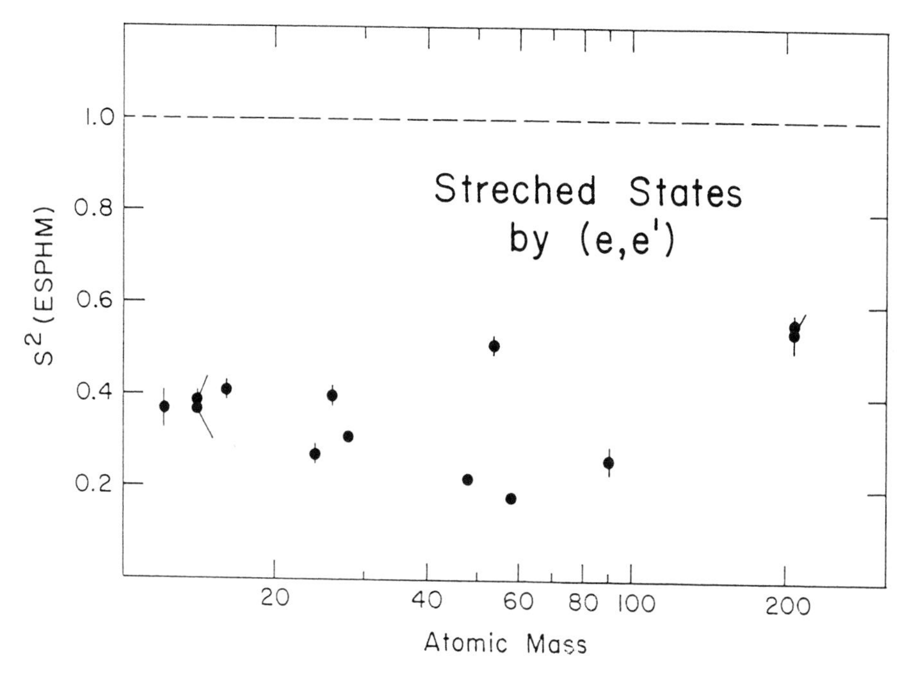

Figure 5: The strength of stretched states throughout the periodic table.

the only ones. As a result, a consistent picture has emerged – nuclear mean field theory consistently overpredicts single arm electron scattering data. The discrepancy is most dramatic in cases where single particle amplitudes dominate. Given the generality of this phenomenon, we should view with skepticism any theory that attempts to resolve the discrepancy at any one of the areas discussed earlier while ignoring the situation in the others. We will attempt to show in the remainder of this paper that many-body correlations and the resulting fractional occupation of mean field orbits provide an explanation capable of resolving a significant part of this discrepancy in all of these areas.

4 QUENCHING FACTORS AND OCCUPATION NUMBERS

In electron scattering one measures form factors, F(q), which can be mapped as a function of the momentum transfer q. The form factor that characterizes a transition between initial state $|i\rangle$ and final state $|f\rangle$ can be separated into two parts

$$F(q) = Q\, F_{sp}(q) + F_{bg}(q) \tag{8}$$

where $F_{sp}(q)$ is the "single particle" form factor and will have a shape very similar to that predicted by an independent particle model. Its amplitude will be different from that of mean field theory since the occupation probabilities will reduce the probability of finding the particle in the particle orbit and will block the transition by the partial occupation of the hole. The momentum transfer independent amplitude Q is solely a function of the occupation numbers. In heavy nuclei (in the limit of $A \rightarrow \infty$ and where $|f\rangle \simeq |i\rangle$) the quenching factor Q is given by (Pandharipande *et al* (1984)):

$$Q \simeq n(p)[1 - n(h)] \tag{9}$$

In the limit where mean field theory accurately describes reality then $n(p) = 1$, $n(h) = 0$ and $Q = 1$, as expected.

The presence of the background form factor $F_{bg}(q)$ is a necessary consequence of correlations. Correlations create complicated multiparticle-multihole configurations which cause partial occupancy and quenching ($Q < 1$). Scattering from the complicated multiparticle-multihole configurations leads to a response $F_{bg}(q)$ entirely different from that of $F_{sp}(q)$, the background response. The situation here is quite analogous to that encountered in the familiar case of scattering from a state with configuration mixing. The observed form factor again can be separated into the piece associated with the single particle state and the term arising from configuration mixing. In Figure 6 a diagrammatic expansion up to the 2p-2h term is shown. The identification of the terms that bring about the quenching and contribute to the "background" form factor $F_{bg}(q)$ is indicated.

Because the multiparticle-multihole background term is quite complicated no characteristic behavior in momentum space should be observed. In heavy nuclei one expects the background form factor to be featureless and peaked at low momentum transfers as all collective form factors do. We should expect then that at high momentum transfer ($q \simeq k_f$) that the measurements will be primarily sensitive to $QF_{sp}(q)$ while at low momentum transfers will be sensitive to $F(q) = QF_{sp}(q) + F_{bg}(q)$. This underlying assumption is crucial and its validity permeates the analysis of the experimental data that follows.

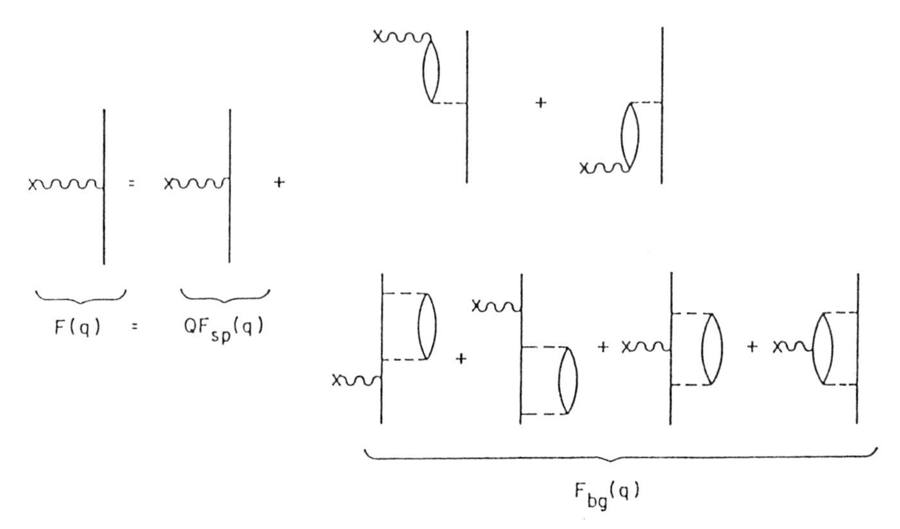

Figure 6: The separation of the observed form factor F(q) into a quenched single particle piece $QF_{sp}(q)$, and a background piece $F_{bg}(q)$ can be easily understood in a diagrammatic expansion. Terms up to second order in the interaction are shown.

The presence of the background term is also necessary for the restoration of a number of conservation laws at the long wavelength limit, implied by the appropriate Ward identities. For instance at low momentum transfers we must recover the Mott limit as conservation of charge requires. On the other hand its high momentum behavior is not constrained by any rigorous theorems, it can only be revealed through a microscopic calculation.

The above considerations have led us to define (Papanicolas (1985)) the following experimental procedure for the measurement of Q (quenching factor) in a heavy nucleus:

1. Measure the form factor of a bound single particle transition in a heavy nucleus at high q. High momentum transfers are taken to mean $q \simeq k_f$, the Fermi momentum, typically the region up to which form factors of a single particle state stay flat before taking an exponential dive. At these high momentum transfers the background term is assumed to be insignificant.
2. Assure that extraneous contributions (such as meson exchange currents, configuration mixing and pairing) are not significant or that they have been rendered so through subtraction.
3. Calculate within the framework of a successful mean field theory the expected single particle form factor and its excitation energy.
4. If the excitation energy and shape of the calculated form factor closely resemble the measured one, a simple scaling of the theoretical F(q) will yield the quenching factor Q.

"Heavy nucleus" in this paper will always be taken to be ^{208}Pb, its isotopes and isotones. The requirement of a bound state provides easy access to model wave functions of a

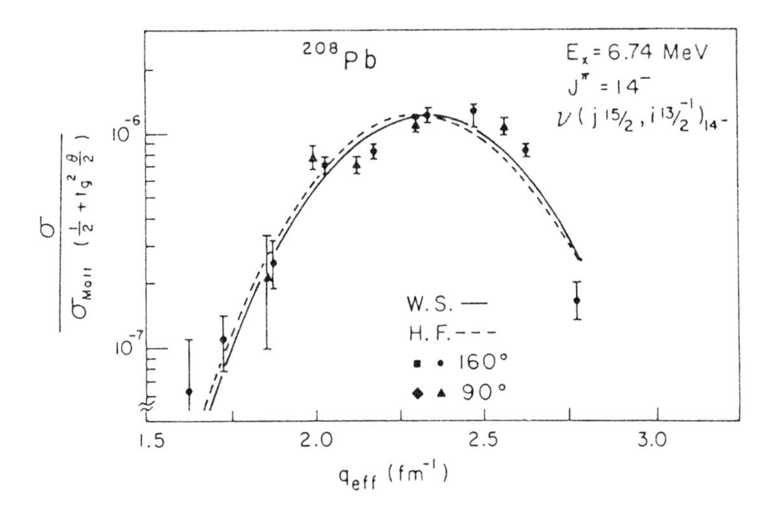

Figure 7: Extracting quenching factors for states of single particle character. The high spin states of ^{208}Pb satisfy all criteria (see text). The choice of Hartree-Fock or Woods-Saxon basis has little effect on the deduced quenching factor (see also Table 2).

"successful" mean field theory (such as those of density dependant Hartree-Fock) without having to face the difficult issues of the continuum wave functions.

The above procedure is best demonstrated through an example. As such we choose the M14 excitation of ^{208}Pb (see Lichtenstadt *et al* (1979)). Lead, the heaviest doubly-closed shell system available, constitutes the best testing ground for the principles outlined above. The $1\hbar\omega$ stretched states are well isolated and therefore relatively immune to mixing. As such the $J^\pi = 14^-$ state satisfies the criterion of a single particle transition being widely recognized as resulting from the $\nu(i_{13/2}^{-1}, j_{15/2})$ configuration. The measured form factor at high q (where it naturally peaks) exhibits the same shape as the one calculated using Hartree-Fock (DME) (Negele and Vautherin (1972)) wave functions. As shown in Figure 7 when the F_{sp} form factor is scaled by $Q = 0.71 \pm 0.5$ reasonable agreement with the data is achieved.

In the following sections we proceed to examine the available elastic (magnetic and charge) data in the lead region and then certain types of inelastic transitions that conform to our criteria. The suggested procedure may involve substantial model error which we have no way of evaluating at this time. Its reliability can be judged by the consistency of the results it yields and their compatability with related quantities deduced by alternative methods.

5 ELASTIC MAGNETIC SCATTERING

Unpaired nucleons in odd-even nuclei give rise to elastic magnetic scattering. In the case of systems neighboring doubly closed shells elastic magnetic form factors can be taken as an excellent example of single particle excitations.

Elastic magnetic scattering has been systematically studied during the last decade and it was recently reviewed by Donnelly and Sick (1984). Mean field theory is found to predict the shape of the observed form factors rather accurately throughout the periodic table. In most cases the calculated form factors overestimate the measured values. This is especially true for nuclei neighboring magic shells.

For the reasons we have extensively discussed in the previous section, the lead region is of particular interest. The elastic form factors of ^{209}Bi ($\pi h_{9/2} \otimes |0^+\rangle_{208}$) and ^{207}Pb ($\nu 3p^{-1}_{1/2} \otimes |0^+\rangle_{208}$) have been measured (deWitt Huberts (1978), Platchkov *et al* (1982), and Papanicolas *et al* (1987)). ^{209}Pb ($\nu 2g_{9/2} \otimes |0^+\rangle_{208}$) and ^{207}Tl ($\pi 3s^{-1}_{1/2} \otimes |0^+\rangle_{208}$) are unstable and therefore inaccessible to electron scattering experiments. Due to the instability of ^{207}Tl, ^{205}Tl ($3s^{-1}_{1/2} \otimes |0^+\rangle_{206}$) is of special interest and it has been the subject of many investigations recently.

Meson exchange current contributions and core polarization corrections play an important role in the measurements on all three nuclei (^{209}Bi, ^{207}Pb, and ^{205}Tl). A number of theoretical estimates indicate that meson exchange currents enhance the cross section in the momentum transfer region of interest ($1.5 < q < 3.0$ fm^{-1}) while core polarization depletes it. We will assume here that these two effects cancel each other. This assumption is compatible with the best available calculations and certainly very convenient. However, it introduces additional model error in the analysis. Configuration mixing is of no consequence when dealing with ^{207}Pb and ^{209}Bi. It is important for the case of ^{205}Tl, and it needs to be accounted for before inferences about partial occupancy can be derived.

Transfer reactions and shell model calculations indicate that the ground state of ^{205}Tl is best described as:

$$|1/2^+\rangle_{205} = \alpha|3s^{(-)}_{1/2}\rangle \otimes |0^+\rangle_{206} + \beta|2d^{(-)}_{5/2}\rangle \otimes |2^+\rangle_{206} + \gamma|2d^{(-)}_{3/2}\rangle \otimes |2^+\rangle_{206} \qquad (10)$$

Recent (d,^{3}He) measurements (Grabmayr *et al* (1985)) find $\alpha^2 = 0.89 \pm 0.08$; this value is in reasonable agreement with a recent ^{206}Pb(e, e$'$p) measurement (Quint *et al* (1986)), which gives $\alpha^2 = 0.79 \pm 0.02$. The perturbation calculation of Zamick *et al* (1975) yields $\alpha^2 = 0.74$. Such a description of the ^{205}Tl ground state will produce at high momentum transfers ($q_{\text{eff}} \simeq 2.0$ fm^{-1}) a form factor nearly identical to the one labeled "single particle" but quenched by a factor of α^2. The form factor for the $|3s^{(-)}_{1/2}\rangle \otimes |0^+\rangle_{206}$ component of the ground stat wave function dominates in the region of the last maximum ($q_{\text{eff}} \simeq 2.0$ fm^{-1}). This is due to the known (Papanicolas *et al* (1984)) rapid fall-off of the ^{206}Pb quadrupole form factor and to the fact that the $|3s^{(-)}_{1/2}\rangle$ single particle form factor extends to substantially higher momentum transfers than do the $|2d^{(-)}\rangle$ single particle form factors because of its more rapid oscillation in coordinate space. Detailed calculations indicate that the $(2d^{(-)}_{5/2}, 2d_{5/2})_{M1}$ and $(2d^{(-)}_{3/2}, 2d_{3/2})_{M1}$ contributions to the cross sections were an order of magnitude smaller than the $(3s^{(-)}_{1/2}, 3s_{1/2})_{M1}$ contribution. The influence of these terms is further diminished when they are multiplied by their corresponding small spectroscopic factors, β^2 and γ^2. Inclusion of this extra quenching will improve the agreement between existing calculations and the measured form factor for ^{205}Tl in the region of the last maximum.

The form factors of ^{205}Tl, ^{207}Pb, and ^{209}Bi at the region of the last measured maximum allow us to test whether the observed reduction can be explained by partial occupancy

A	h	Z	Reference
^{205}Tl	$\pi 3s_{1/2}$	$0.7 \pm .07$	Papanicolas *et al* (1987)
^{209}Bi	$\pi 1h_{1/2}$	$0.7 \pm .08$	Platchkov *et al* (1982) and deWitt Huberts (1978)
^{207}Pb	$\pi 3p_{1/2}$	$0.7 \pm .07$	Papanicolas *et al* (1987)

Table 1: Z factors derived from elastic magnetic scattering. The character of the unpaired nucleon and the nucleus in which it was studied are also given. The stated uncertainty reflects only the experimental error.

of ^{208}Pb orbitals. Partial occupancy in ^{208}Pb implies (see Pandharipande *et al* (1984)) that the single particle magnetic form factors of neighboring nuclei are quenched with respect to the independent-particle model prediction by a factor of $Z = n_+ - n_-$, where n_+ and n_- denote the occupation probabilities of orbitals lying immediately above and below the Fermi energy. The quenching factor $Q = 0.7 \pm 0.07$ obtained by comparing the independent-particle model prediction to our measured ^{207}Pb and ^{209}Bi form factor determines the Z factor of the neutron shell directly. The quenching factor $\tilde{Q} = 0.55 \pm 0.07$ observed in ^{205}Tl is due to both partial occupancy and configuration mixing. For this nucleus we can write $\tilde{Q} = \alpha Z$. Taking $\alpha^2 = 0.79 \pm 0.02$ from recent $(e, e' p)$ measurements (Quint *et al* (1986)) we find $Z = 0.7 \pm 0.07$ for the proton shell, as well. The experimentally extracted quenching and Z factors are tabulated in Table 1. Our value, $Z = 0.7 \pm 0.07$, for the discontinuity of occupation numbers at the Fermi energy, differs significantly from the value used in mean field calculations ($Z = 1.0$).

6 INELASTIC SCATTERING

Inelastic scattering to states of known single particle character provides a sensitive test of mean field theory predictions and naturally leads to the extraction of quenching factors. Complementary data for occupation probabilities to those derived from elastic scattering can be obtained from such transitions. Inelastic excitations actually constitute a richer source of information because many more orbitals can be tested besides the valence orbitals to which elastic magnetic scattering and isotonic charge differences are necessarily limited. We shall examine here form factors derived from inelastic scattering to the two most obvious classes of single particle (hole) transitions: High spin transitions in ^{208}Pb and neutron hole transitions in ^{207}Pb.

It is generally the case that mean field theory predicts the excitation energy and the shape of the form factor correctly but invariably it overestimates the strength. Our prescription for the extraction of occupation probabilities is directly applicable. Actually we have already used the case of the M14 transitions of ^{208}Pb to demonstrate this prescription (see Figure 7). The high resolution inelastic electron scattering from ^{208}Pb has mapped not only the form factor of the 14^- but also other high spin states such as 12^+, 12^-, 10^+, etc.. Their form factors are nicely predicted by Hartree-Fock wave functions. In Table 2 we summarize the quenching factors obtained from the most recent very high resolution ($\Delta E \simeq 15$ keV) studies at Bates (Connelly (1989)). The new results, richer and more precise than the earlier studies of ten years ago, reconfirm the conclusions reached based on the earlier data.

E (MeV)	J^π	h	p	t	Q
^{208}Pb(e, e′) (from Connelly (1989))					
5.29	11^+	$1i_{13/2}$	$2g_{9/2}$	N	$0.63 \pm .06$
5.86	11^+	$1i_{13/2}$	$1i_{11/2}$	N	$0.78 \pm ..04$
6.11	12^+	$1i_{13/2}$	$2g_{9/2}$	N	$0.64 \pm .03$
6.44	12^-	$1i_{13/2}$	$1j_{15/2}$	N	$0.68 \pm .04$
6.75	14^-	$1i_{13/2}$	$1j_{15/2}$	N	$0.73 \pm .02$
6.86	9^-	$1h_{11/2}$	$1i_{13/2}$	P	$0.74 \pm .03$
7.07	12^-	$1h_{11/2}$	$1i_{13/2}$	P	$0.56 \pm .03$
7.07 + 7.09	12^-	$1h_{11/2}$	$1i_{13/2}$	P	$0.71 \pm .02$
^{207}Pb(e, e′) (from Papanicolas *et al* (1980))					
0.57	$5/2^-$	$2f_{5/2}$	$3p_{1/2}$	N	$.65 \pm .05$
0.90	$3/2^-$	$3p_{3/2}$	$3p_{1/2}$	N	$.65 \pm .05$
1.63	$13/2^+$	$1i_{13/2}$	$3p_{1/2}$	N	$.47 \pm .05$
2.34	$7/2^-$	$2f_{7/2}$	$3p_{1/2}$	N	$.55 \pm .05$
2.73	$9/2^+$	$3p_{1/2}$	$2g_{9/2}$	N	$.50 \pm .05$
3.51	$11/2^+$	$3p_{1/2}$	$1i_{11/2}$	N	$.65 \pm .05$

Table 2: Quenching factors Q derived from inelastic scattering to high spin states of ^{208}Pb and neutron hole states of ^{207}Pb. The spin-parity of the transition and its character are also given.

Single particle (hole) transitions in odd-even nuclei neighboring doubly closed shells share certain features with the stretched states examined above. They have a clearly identified single particle (hole) character and in general they are well isolated which prevents any mixing. They are found at very low excitation energies where resolution problems do not arise. Once again these states have been studied with inelastic electron scattering and they are found to be rather accurately described in shape but overestimated in amplitude by mean field theory. Such cases are known throughout the periodic table from ^{15}N to ^{207}Pb.

In ^{207}Pb these states have been studied but no corresponding measurements exist for ^{209}Bi. The low energy excitation spectrum for ^{207}Pb is dominated by single neutron hole states. These transitions result by pushing the $3p_{1/2}$ neutron hole deeper and deeper into more bound orbitals. It can be seen in Figure 8 that the form factors of these states can be adequately described by Hartree-Fock transition densities after the transition amplitudes have been adjusted. Our recent measurements at Bates, now under analysis, have improved the statistical accuracy of the data shown. The scaling of the transition amplitudes yields the quenching factors in Table 2, along with those of the high spin states of ^{208}Pb.

7 THEORETICAL ESTIMATES

A theoretical estimate of the occupation probabilities characterizing the shell orbits of ^{208}Pb is offered by Pandharipande *et al* (1984). It is assumed that they can be reasonably approximated by the occupation numbers of nuclear matter, n_{NM}, suitably modified to account for the effects of surface vibrations. The occupation probabilities for nuclear

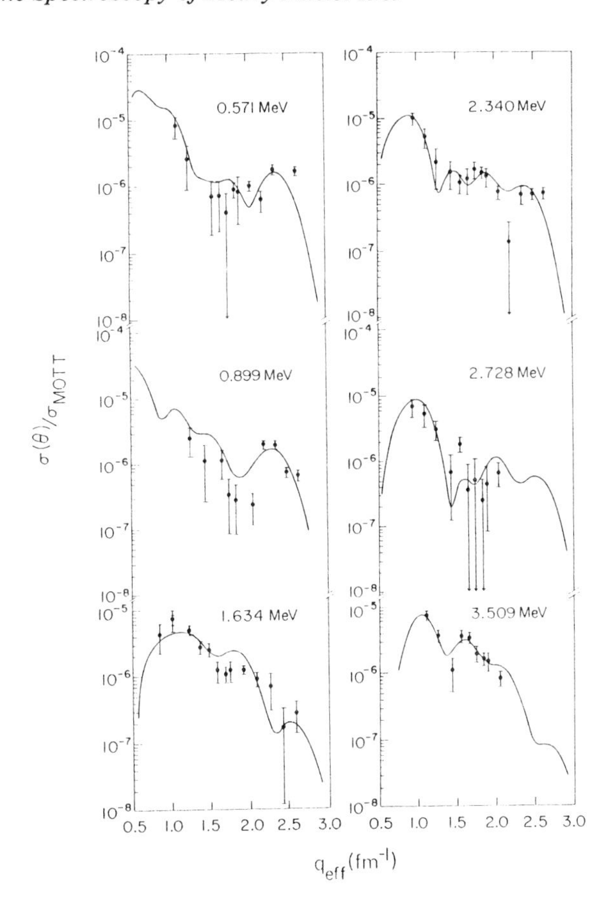

Figure 8: Form factors of ^{207}Pb neutron hole states and Hartree-Fock predictions scaled by the quenching factors listed in Table 2.

matter have been calculated by Fantoni and Pandharipande (1984) using a realistic nucleon-nucleon interaction. In this calculation $n_{NM}(K)$ is found to have little density dependence. The strong energy dependence is assumed to also apply in the case of ^{208}Pb where the following prescription is adopted

$$n_\alpha(\varepsilon_\alpha - \varepsilon_F) = n_{NM}(\varepsilon_\alpha - \varepsilon_F) + \delta n_\alpha^{RPA} \tag{11}$$

where n_α is the occupation number characterizing the shell orbital ϕ_α of ^{208}Pb whose single particle energy is ε_α. The nuclear matter $n_{NM}(\varepsilon - \varepsilon_F)$ occupation number is further reduced (enhanced) for the particle (hole) orbits of ^{208}Pb by δn_α^{RPA} due to the new degrees of freedom associated with the nuclear surface. δn_α^{RPA} has been taken from the calculations of Decharge *et al* (1981) and Decharge and Sips (1983). Most of the nuclear matter effect is found to be either due to short range correlations or due to interactions induced by the tensor force. The RPA result is derived using a central density dependent interaction (no tensor part). It is thus assumed that double counting effects are small.

The resulting occupation numbers are those we have shown in Figure 2. This distribution indicates that substantial amount of correlations are present in the ground state of ^{208}Pb. Most of the effect, as can be seen in the figure, is due to mechanisms already at work in nuclear matter. It is true however that the surface vibrations contribute substantially to the depletion of the valence orbitals.

In this estimate the surface orbitals are characterized by an occupation of $\sim$0.65 for the particles and of $\sim$.08 for hole states lying immediately below the Fermi sea. This according to equation 9 will yield a quenching of the single particle transitions associated with them (such as the data we have examined in the previous section) of 0.6 ± 0.1 in reasonable agreement with the data discussed earlier.

Fantoni, Pandharipande, and coworkers using the same interaction that leads to the occupation probabilities displayed in Figure 2 have shown that they can explain most of the depletion that systematically appears in the quasielastic response of heavy nuclei (see discussion in section 3 and Figure 4). This interaction induces correlation among the nucleons, which results in the quenching of the amplitude of the quasielastic peak by shifting a significant fraction of its strength to higher excitation energies.

Mahaux and Sartor (1989), using a dispersion relation approach, constrain the real part of the mean field at negative energies from the well known (from scattering experiments) field at positive energies. Knowledge of the energy dependence of full mean field between +40 MeV and −60 MeV allows them to derive among other properties, the spectroscopic factors and occupation probabilities for the neutron shell of ^{208}Pb. Right below the Fermi energy, they calculate an occupation probability of 0.85 and a Z factor of 0.7 in excellent agreement with our results.

8 ON THE SHAPE OF WAVE FUNCTIONS

Our discussion and analysis of the empirical data has been entirely focused on occupation numbers. Actually this has been the dominant trend in the field in the recent years. However, the appropriate wave functions to use are not the mean field theoretical ones, which we have employed in our analysis. As discussed in our introductory sections, they are appropriate only in the limit of vanishing correlations. The quenching factors we have derived indicate that the influence of correlations is strong and therefore not only should we expect reduction of occupation probabilities for orbitals below the Fermi energy, we should also expect significant modification of their shape.

Lewart *et al* (1988), in their work on ^{3}He droplets suggest that as correlations set in, one can approximate the quasiparticle (or overlap) wave function $\psi_\alpha(\mathrm{r})$ in terms of the mean field wavefunction $\chi_\alpha(\mathrm{r})$ by:

$$\psi_\alpha(\mathrm{r}) = \frac{1}{\sqrt{\mathrm{Z}}}\left[1 - \mathrm{c}\frac{\rho(\mathrm{r})}{\rho(\mathrm{r}=0)}\right]\chi_\alpha(\mathrm{r}) \tag{12}$$

where c is a constant and $\rho(\mathrm{r}=0)$ is the density in the nuclear interior. This local density approximation reproduces rather accurately their Monté Carlo estimation of $\psi_\alpha(\mathrm{r})$.

Ma and Wambach (1983) arrive at a similar conclusion by exploring the influence of correlations within the framework of the Landau theory of Fermi liquids. Their derivations based on the work of Migdal, parallels to some extent the approach of Khodel and

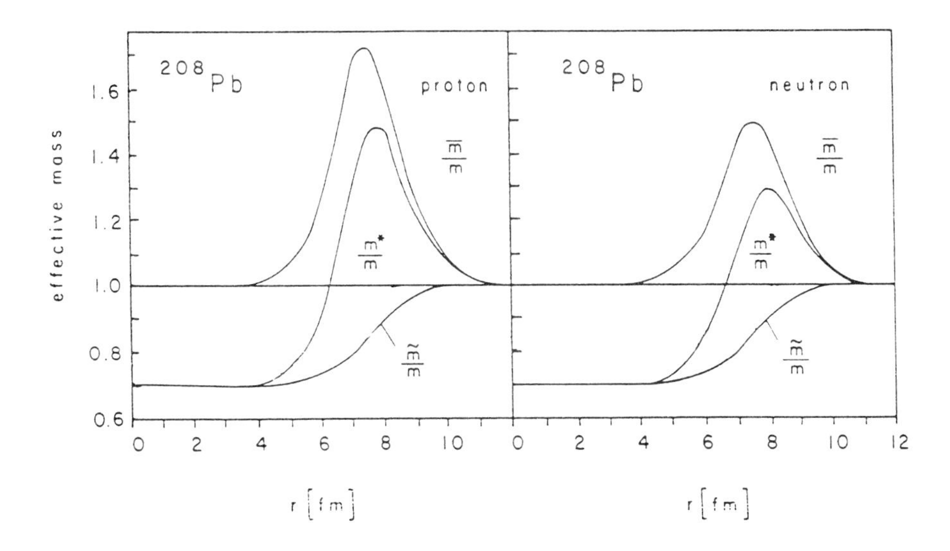

Figure 9: Radial dependence of the effective mass, m*(r), for the neutron and proton shell of ^{208}Pb (Ma and Wambach (1983)).

Saperstein (1982). They find that

$$\psi_\alpha(\mathrm{r}) \propto \sqrt{\frac{\mathrm{m}^*(\mathrm{r})}{\mathrm{m}}}\chi_\alpha(\mathrm{r}) \tag{13}$$

where m*(r) is the effective mass of the nucleon in the nuclear medium. They estimate m*(r)/m for ^{208}Pb, shown in Figure 9, by requiring that the modified mean field describes the level spacing near the Fermi energy. The radial dependence exhibited by m*(r)/m will result in a quasiparticle wave function that is quenched relative to the mean field wave function in the interior in a fashion similar to that predicted by Lewart *et al* (1988). Ma and Wambach (1983) actually predict an enhancement in the nuclear surface since m*(r)/m becomes greater than unity. This enhancement arises from the frequency dependence of the optical potential (Jeukenne *et al* (1976)).

We attempt here to derive experimentally the shape of a quasiparticle orbit by reexamining the ^{206}Pb – ^{205}Tl charge density difference. The charge density difference arises from two processes: from the charge that is missing due to the absence of the $3s_{1/2}$ proton wave function and the charge density redistribution arising from core polarization. Core polarization corrections can reliably be calculated for the lead isotopes in the framework of density dependent Hartree-Fock and such calculations have been successfully tested by a number of experiments. The most convincing test is provided by their successful prediction of the charge density difference of ^{208}Pb – ^{207}Pb and ^{207}Pb – ^{206}Pb (Cavedon *et al* (1987)). In order to derive the shape of the quasiparticle orbit we have subtracted from the experimentally obtained charge density difference of ^{206}Pb – ^{205}Tl (Cavedon *et al* (1982) and Frois *et al* (1983)) the theoretically calculated core polarization contribution. The derived result is shown in Figure 10 together with the square of the mean field $3s_{1/2}$ proton wave function. The result bears striking similarity to the Monté Carlo

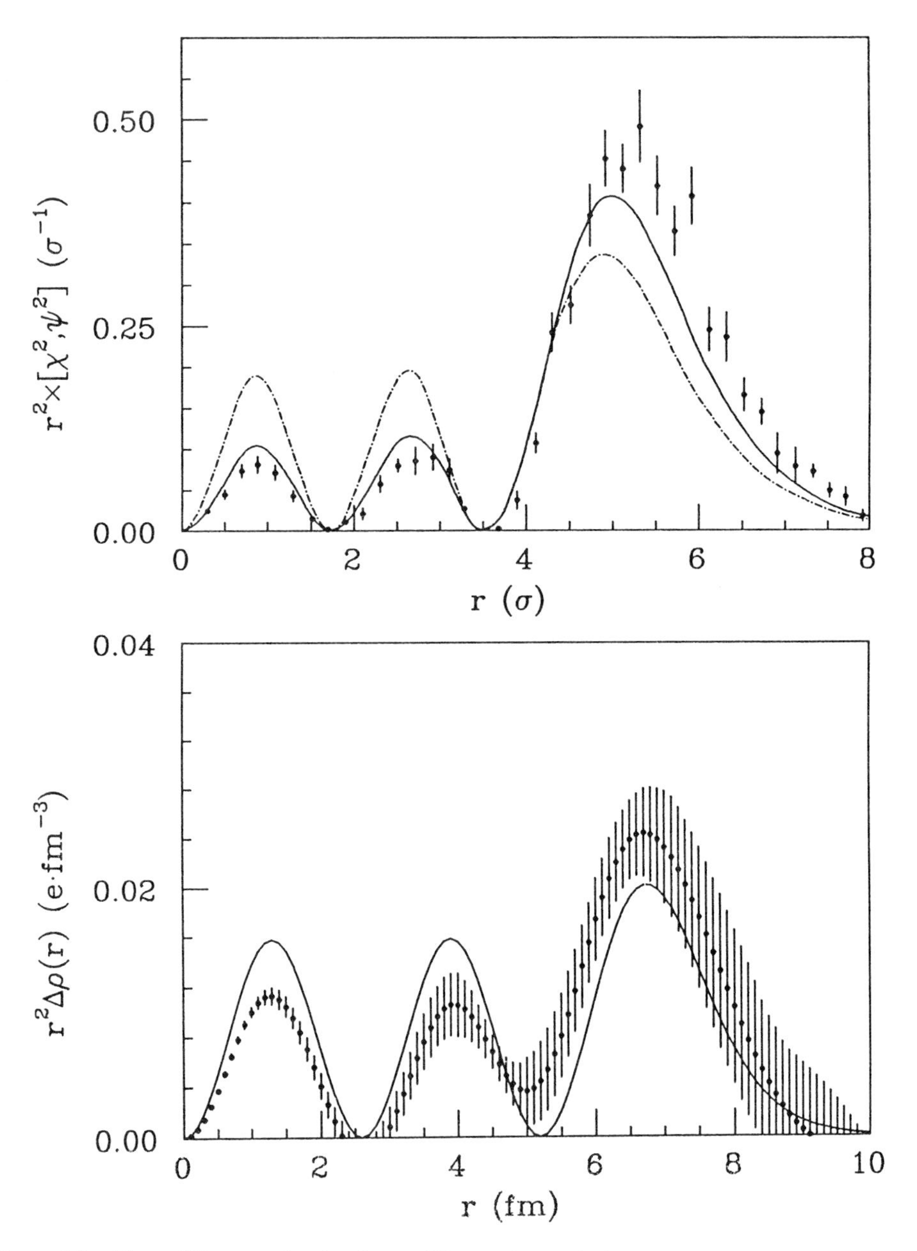

Figure 10: Top: ^{3}He atomic droplets. The estimated density difference $\rho_{N=70}(r) - \rho_{N=69}(r)$ (data points) compared with the squares of the mean field (dashed curve) and quasiparticle (solid curve) orbitals. Bottom: Nuclei. The ^{206}Pb – ^{205}Tl charge density difference with core polarization removed, compared to the square of the mean field wave function.

result obtained by Lewart *et al* (1988) for the density difference for ^{3}He droplets of 70 and 69 particles (these droplets resemble ^{205}Tl and ^{206}Pb in that they differ by a constituent occupying the 3s orbit). The experimental results, like the Monté Carlo results, show that the quasiparticle orbital, as compared to the mean field orbital, is suppressed in the nuclear interior and enhanced at the nuclear surface. As we have already pointed out, this is in excellent agreement with the predictions of Ma and Wambach (1983).

This first visualization of the shape of a quasiparticle orbit is gratifying in many respects. First, the interest in such a quantity transcends nuclear physics. It is of fundamental importance to all condensed matter physics. However, it cannot be obtained from the study of extended (infinite) systems, because in infinite systems, the shape of orbits is constrained by translational invariance. The ability of many-body theory to anticipate the shape correctly is impressive, and it is another reminder of how productive the interplay between theory and experiment has been. Armed with this new information we believe that it is now appropriate to attempt a reanalysis of the available data using the appropriate wave functions (not mean field model wave functions as has been the case up to now) in order to achieve a better understanding of the influence of correlations in nuclei.

ACKNOWLEDGEMENTS

We would like to thank B. Frois, V. R. Pandharipande and J. Wambach for many illuminating discussions. This work is supported by the U. S. National Science Foundation under grant NSF PHY 86-10493.

REFERENCES

Cavedon J M, Frois B, Goutte D, Huet M, Leconte P, Papanicolas C N, Phan X H, Platchkov S K, Williamson S E, Boeglin W and Sick I 1982 *Phys. Rev. Lett.* **49** 978

Cavedon J M, Frois B, Goutte D, Huet M, Leconte P, Phan X H, Platchkov S K, Papanicolas C N, Williamson S E, Boeglin W, Sick I and Heisenberg J 1987 *Phys. Rev. Lett.* **58** 195

Connelly J P 1989 Ph.D. Thesis U New Hampshire and private communication

Das R K and Finlay R W (1989) private communication and to be published

Decharge J and Gogny D 1980 *Phys. Rev.* **C21** 1568

Decharge J, Sips L and Gogny D 1981 *Phys. Lett.* **98B**, 229

Decharge J and Sips L 1983 *Nuc. Phys.* **A407**, 1

deWitt Huberts P K A 1978 in Proc. Conference on Modern Trends in Elastic Electron Scattering, Amsterdam edited by C. de Vries (Instituut voor Kernphysisch Onderzoek, Amsterdam, The Netherlands, 1978).

Donnelly T W and Sick I 1984 *Rev. Mod. Phys.* **56** 461

Fantoni S and Pandharipande V R 1984 *Nuc. Phys.* **A427** 473

Fantoni S and Pandharipande V R 1987 *Nuc. Phys.* **A473** 234

Frois B, Cavedon J M, Goutte D, Huet M, Leconte P, Papanicolas C N, Phan X H, Platchkov S K, Williamson S E, Boeglin W, and Sick I 1983 *Nuc. Phys.* **A396** 409c

Frois B and Papanicolas C N 1987 *Ann. Rev. Nuc. Part. Sci.* **37** 133

Grabmayr P *et al* 1985 *Phys. Lett.* **164B** 15

Jaminon M, Mahaux C and Ngô H 1985 *Nuc. Phys* **A440** 228

Jeukenne J P, Lejeure A and Mahaux C 1976 *Phys. Rep.* **C25** 83

Johnson C H, Horen D J and Mahaux C 1987 *Phys. Rev.* **C36** 2252

Khodel V A and Saperstein E E 1982 *Phys. Rep.* **92** 183

Lapikas L 1989 "The Quasielastic (e, e′ p) Reaction on Heavy Nuclei" this volume and references therein

Lewart D, Pandharipande V R and Pieper S C 1988 *Phys. Rev.* **B37** 4950

Lichtenstadt J, Heisenberg J, Papanicolas C N, Sargent C P, Courtemanche A N and McCarthy J S 1979 *Phys. Rev.* **C20** 497

Lowdin P O 1955 *Phys. Rev.* **97** 1474

Ma Z Y and Wambach J 1983 *Nuc. Phys* **A402** 402

Mahaux C and Ngô H 1984 *Nuc. Phys.* **A431** 486

Mahaux C and Sartor R 1989 *Nuc. Phys.* **A493** 157

McFarlane M H 1986 *Nuc. Phys* **B182** 265

Mezziani Z E *et al* 1985 *Phys. Rev. Lett.* **54** 1233

Migdal A B 1967 Theory of Finite Fermi Systems and Applications to Atomic Nuclei (John Wiley & Sons, NY 1967)

Pandharipande V R, Papanicolas C N and Wambach J 1984 *Phys. Rev. Lett.* **53** 1133

Papanicolas C N, Lichtenstadt J, Sargent C P, Heisenberg J and McCarthy J S 1980 *Phys. Rev. Lett.* **45** 106

Papanicolas C N, Heisenberg J, Lichtenstadt J, McCarthy J S, Goutte D, Cavedon J M, Frois B, Huet M, Leconte P, Phan X H, Platchkov S K and Sick I 1984 *Phys. Rev. Lett.* **52** 247

Papanicolas C N 1985 in Proc. IUCF Workshop on Nuclear Structure at High Spin, Excitation and Momentum Transfer, Bloomington edited by H. Nann, AIP Conf. Proc. 142 (American Institute of Physics, New York, 1986) 110

Papanicolas C N, Cardman L S, Heisenberg J, Schwentker O, Milliman T, Hersman B, Hicks R, Peterson G, McCarthy J S, Wise J and Frois B 1987 *Phys. Rev. Lett* **58** 2296

Platchkov S K, Bellicard J B, Cavedon J M, Frois B, Goutte D, Huet M, Leconte P, Phan X H, deWitt Huberts P K A, Lapikas L and Sick I 1982 *Phys. Rev.* **C25** 2318

Quint E N M *et al* 1986 *Phys. Rev. Lett.* **57** 186

Schafer L and Weidemuller H A 1971 *Nuc. Phys.* **A174** 1

Speth J, Werner E and Wild W 1977, *Phys. Rep.* **33C** 127

Negele J W and Vautherin P 1972 *Phys. Rev.* **C5** 1472

Wagner JW 1989 in Progress in Particle and Nuclear Physics **24** edited by A. Foessler in press and references therein

Zamick L, Klemt V and Speth J 1975 *Nuc. Phys.* **A245** 365

Inst. Phys. Conf. Ser. No 105
Paper presented at Int. Conf. on Spectroscopy of Heavy Nuclei, Crete, Greece, 1989

Collective properties of ^{208}Pb and ^{226}Ra

H.J. Wollersheim
Gesellschaft für Schwerionenforschung, Darmstadt, West Germany

ABSTRACT: The nuclei ^{208}Pb and ^{226}Ra have been Coulomb excited using ^{208}Pb beams. In case of ^{208}Pb a possible candicate for the 2-octupole-phonon multiplet has been found. For ^{226}Ra level energies and reduced transition probabilities of high-spin states have been determined. The results are discussed in terms of a stable octupole deformation.

1. INTRODCUTION

The description of nuclear systems in terms of elementary degrees of freedom has been of great importance for the understanding of the dynamics of nuclei [1]. Surface and pairing vibrations have been specially useful due to the high degree of harmonicity found in the spectra of normal spherical nuclei when described in terms of those vibrations [1-3]. The first excited states of most doubly magic nuclei have spin-parity 3^- and are interpreted as one-phonon vibrations of octupole character. The collective nature of these states is inferred from the observed enhancement of the transition probabilities to the ground state. In the particular case of ^{208}Pb the first excited state is at 2.6 MeV and the enhancement over the single-particle values is of the order of 30. This picture of the lowest-lying 3^- levels in magic nuclei seems unimpeachable although the multi-phonon states, whose existence is implied by this description, remain unobserved. The present search for such excitations is therefore of great interest as their observation would provide a direct check on the degree of harmonicity of the associated octupole-vibrational motion.

When very low-lying negative-parity states were also discovered in Rn, Ra and in the neutron-deficient isotopes of Th, the possibility of strong octupole correlation effects were suggested. Recently theoretical calculations [4-7] of the nuclear shape have predicted stable octupole deformations in the region $130 \leq N \leq 146$, $84 \leq Z \leq 92$. The reflection asymmetry of a stable octupole shape gives rise to a staggered sequence of levels of positive and negative parity. These states are connected by strong E1 transitions competing with the stretched E2 transitions. The present Coulomb-excitation experiment on ^{226}Ra was performed at GSI in order to extend the experimental information up to high-spin states. This technique allows the possibility to determine most of the electric dipole, quadrupole and, for the first time, octupole transition moments up to spin 18 for yrast states.

2. EXPERIMENTAL PROCEDURES

A ^{226}Ra target was bombarded with a 4.7 MeV/A ^{208}Pb beam obtained from the UNILAC at GSI. The target was a 2 mm deposit of 400 $\mu g/cm^2$ $^{226}RaBr_2$ on a 50 $\mu g/cm^2$ C-backing and covered by a protective layer of 40 $\mu g/cm^2$ Be. The recoiling target nuclei and scattered projectiles were detected in an arrangement of position-sensitive parallel-plate avalanche gas counters: an annular detector ($15° \leq \vartheta_L \leq 45°$, $0° \leq \phi_L \leq 360°$) and four particle detectors, symmetrically positioned around the beam axis, each covering an angular range of $53° \leq \vartheta_L \leq 90°$ and $\Delta\phi_L = 84°$. The particle detectors were made position sensitive in the scattering angle direction (ϑ_L) by means of a delay line readout incorporated into the multi-strip detector anode. The information about the azimuthal angle (ϕ_L) was obtained by dividing the thin metalized cathode foil of the annular detector into 20 radial segments of 18° each. By measuring the kinematical correlation between the scattering angles of both reaction partners, the recoiling ^{226}Ra nuclei could be uniquely identified. The coincident emission of γ-rays was observed in seven Ge detectors (three of them Compton suppressed) positioned at angles of 30° (Ge1-Ge4) and 150° (Ge5-Ge7) with respect to the beam direction. In addition a multiplicity filter of six NaI detectors located at 90° was used to select events of high γ-multiplicity. The information about the scattering angle of the excited Ra nuclei allowed to correct for the large Doppler shifts of the emitted γ-rays due to the high recoil velocities (up to ~10% c). Moreover, the measurement of the scattering angles provided the impact parameter dependence of the γ-ray yields which allowed the determination of individual B(E1), B(E2) and B(E3) values.

3. RESULTS AND DISCUSSIONS

3.1 Coulomb Excitation of ^{226}Ra

Fig.1 shows the summed γ-spectrum of all Ge detectors, after Doppler-shift correction and suppression of the Compton background, accumulated under the additional requirement that one of the collision partners was found at $15° \leq \vartheta_L \leq 45°$ and at least one γ-ray was detected by the NaI array. Transitions up to $18^+ \rightarrow 16^+$ and $17^- \rightarrow 15^-$ can be assigned in the ground-state and in the low-lying negative parity band, respectively. The level scheme[8] for ^{226}Ra shows the characteristic interleaving even-odd spin sequence down to spin 7. Such a sequence is expected in nuclei with stable octupole deformation.

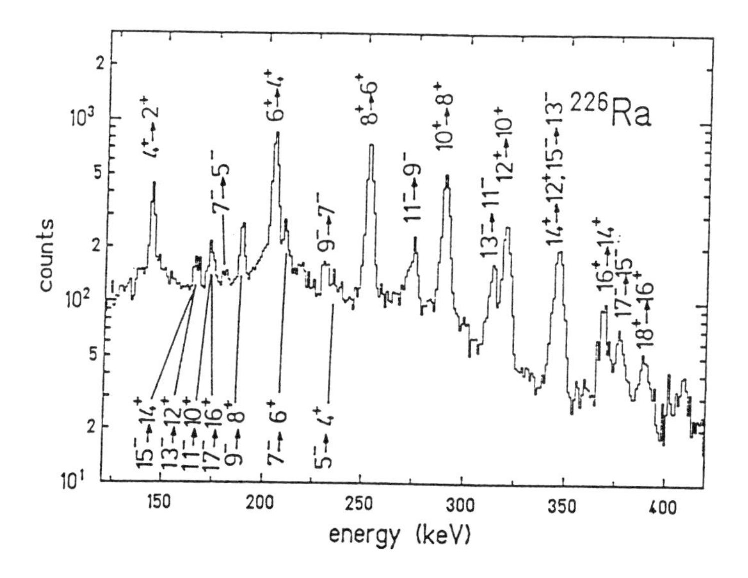

Fig. 1: Doppler-shift corrected γ-ray spectrum following the Coulomb excitation of ^{226}Ra by a 4.7 MeV/A ^{208}Pb beam.

Further support for a stable octupole deformation comes from the enhanced E1 and E3 transition moments connecting the opposite-parity states. The method we use to determine the electric transition moments by multiple Coulomb excitation is based on the comparison of experimental to calculated γ-yields for a large range of scattering angles. The experimental data on the intrinsic electric dipole D_o, quadrupole Q_t and, for the first time, octupole O_t transition moments are summarized in fig. 2. The octupole deformation is determined by the E3 transition moments which are less sensitive to single-particle effects than the dipole moments.

The average O_t values at high spins are well described by the rotational model using the transition strength of $B(E3,0^+ \rightarrow 3^-) = 1.1e^2b^3$. The collectivity of these B(E3) values in ^{226}Ra is about three times larger than the one encountered in octupole-vibration type heavier actinide nuclei. The transition quadrupole moments Q_t for the positive and negative parity bands are quite different for low spin states: in the ground state band we find an average value of 7.2 b and in the octupole band an average value of 11.3 b. However, above spin $I = 10\hbar$ a change of the nuclear shape occurs. The quadrupole deformation of both, positive and negative parity states, is the same ($Q_t = 9.8$ b) as expected for levels in a single rotational band. The electric dipole moments D_o could be determined from the measured E1/E2 branching ratios. They show a dramatic increase with increasing angular momentum. The calculated electric dipole moment - using the liquid - drop formula and the theoretical deformation parameters[5] - is indicated in fig. 2. The experimental data seem to approach this theoretical value at high angular momenta, just as the quadrupole moments for the different parity states are merging to a common value.

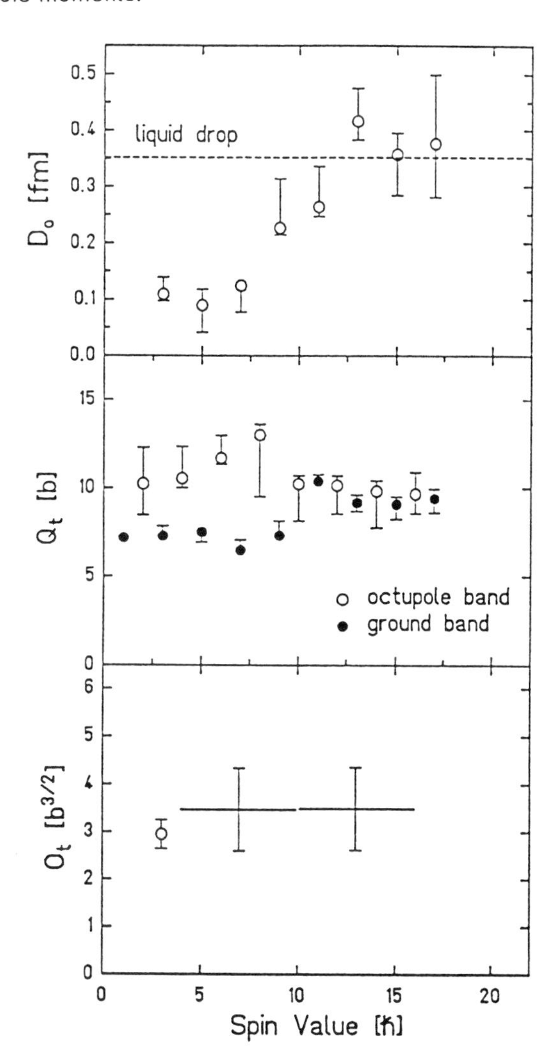

Fig. 2: Electric transition dipole D_o (top), quadrupole Q_t (center) and octupole O_t (bottom) moments for the yrast band in ^{226}Ra.

3.2 Search for Two-Octupole-Phonon States in ^{208}Pb

The purpose of the present experiment was a new attempt to locate some of these two-octupole-phonon states in ^{208}Pb using electromagnetic excitation with heavy ions. For a harmonic vibrator one expects a quadruplet of two-phonon ($3^- \otimes 3^-$) states with spins and parities 0^+, 2^+, 4^+ and 6^+ at about twice the energy of the 3^- state, i.e. at 5.23 MeV. Only a few states with positive parity are known below 6 MeV excitation energy, but in spite of several attempts[9-11] it has not been possible to identify any good candidates for the two-phonon multiplet. The energy of the two-phonon states have been estimated by a number of authors[2,12,13]. In the first place, Blomquist[12] pointed out that the observed large quadrupole moment Q=-0.34(15) of the collective 3^- state in ^{208}Pb[14] implied a departure from harmonicity due to the coupling between the quadrupole and octupole motions which would give rise to a substantial splitting of the two-octupole phonon quartet. Fig. 3 shows the effect of the quadrupole moment on the energy splitting. Other calculations[2,13] showed that the effect of including the interaction with the two-phonon pairing vibration modes are also not negligible but amounts to about the same energy shift in the opposite direction. Those discrepancies illustrate the importance attached to the identification of these two-phonon octupole vibrational states.

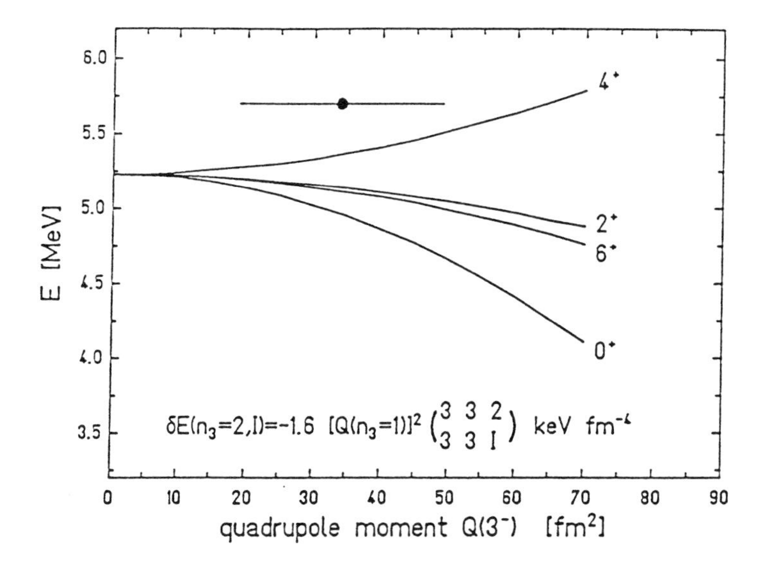

Fig. 3: Energy splitting of the 2-octupole-phonon vibrational states as a function of the quadrupole moment Q of the first excited 3^- state in ^{208}Pb. The measured quadrupole moment[14] is indicated.

A possible way to excite the two-phonon octupole vibration is to use electromagnetic excitation with ^{208}Pb projectiles. We have calculated the inelastic cross section leading to the $(3^- \otimes 3^-)_I$ states and have found that at least the 6^+ state should be observed in a suitable arranged particle-γ experiment. The search for the 2-phonon states was carried out with the same experimental set-up as described in sect. 2. Fig. 4 (top) shows a Doppler-corrected γ-ray spectrum observed by bombarding a 1 mg/cm^2 ^{208}Pb target with 6.2 MeV/A ^{208}Pb projectiles. The peak at 2.485 MeV (marked with an asterisk) just below the $3^- \rightarrow 0^+$ transition is identified as a transition between higher spin states because of its γ-multiplicity measurement. The most probable decay branch of the 2-phonon state can be assigned to a $6^+ \rightarrow 5^-$ transition. Fig. 4 (bottom) shows the result of a $\gamma\gamma$-measurement between Ge and NaI detectors with a gate set on the $5^- \rightarrow 3^-$ transition (E_γ = 583 keV) in the NaI spectrum. The unidentified transition is clearly seen in the Ge-spectrum and it also appears in a γ-spectrum gated on the $3^- \rightarrow 0^+$ transition. However, the bad energy resolution of the NaI detector, mainly due to the large Doppler broadening, did not allow to separate the $5/2^- \rightarrow 1/2^-$ transition in ^{207}Pb (E_γ = 570 keV) so that also γ-transitions in ^{207}Pb and ^{209}Pb, produced by transfer reactions, are identified. The intensity of the γ-transition at 2.485 MeV is approximately a factor of 2 larger than the cross section calculated in the harmonic vibrator limit. The effects of nuclear excitation and simultaneous excitation of projectile and target nucleus are not included in the semi-classical Coulomb excitation calcuclations. With an improved $\gamma\gamma$-coincidence experiment we hope to place the observed γ-transition in the level scheme of ^{208}Pb.

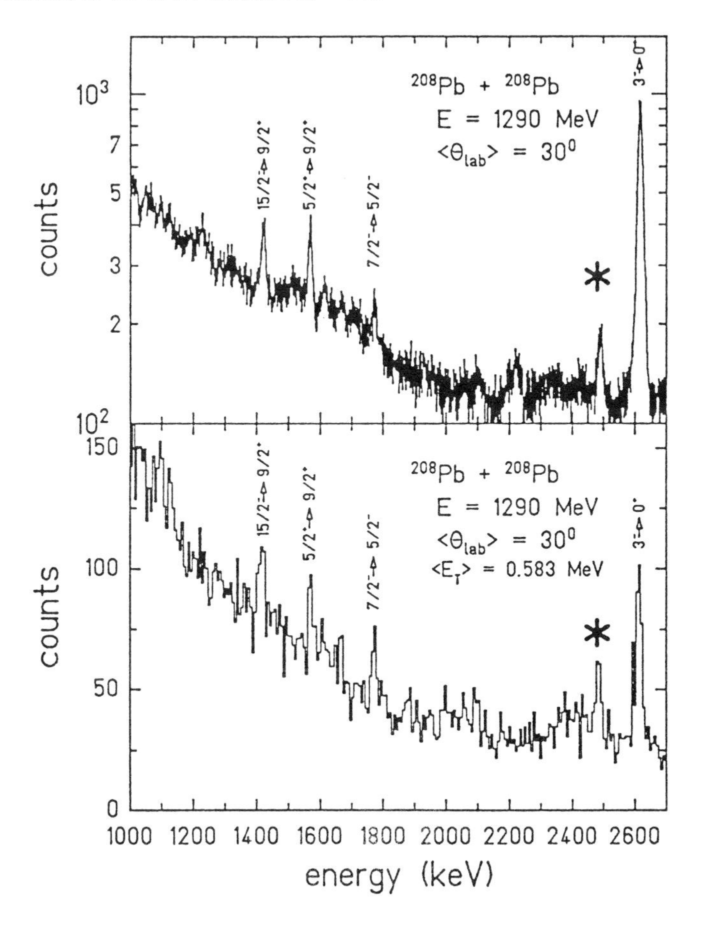

Fig. 4: Doppler-corrected γ-spectrum obtained for the ^{208}Pb+^{208}Pb system at 6.2 MeV/A. Transitions in ^{207}Pb and ^{209}Pb resulting from transfer are also indicated (top). Doppler-corrected $\gamma\gamma$-spectrum between Ge and NaI detectors with a gate set on the $5^- \rightarrow 3^-$ transition in the NaI spectrum (bottom).

4. SUMMARY

The results presented in this paper demonstrate that Coulomb excitation by heavy ion now provides a sophisticated method for detailed nuclear-structure studies of nuclei with octupole deformed (^{226}Ra) and octupole vibrational (^{208}Pb) character. The octupole effects in ^{226}Ra show up consistently in the level structure, the electric transition moments, and in the high-spin behaviour. In case of ^{208}Pb we have found a possible candidate for the 2-octupole-phonon multiplet.

ACKNOWLEDGEMENT

The experiments described in this article have been performed together with my colleagues P. von Brentano, J. de Boer, P. Butler, P. Egelhof, H. Emling, C. Fahlander, C. Fleischmann, E. Hauber, W. Henning, R. Holzmann, R. Julin, K. Kaiser, R. Kulessa, C. Lauterbach, T. Matulewicz, R. Reinhardt, C. Schandera, R. Simon, and R. Wirowski, whom I thank for the very good collaboration.

REFERENCES

1. A. Bohr and B.R. Mottelson, Nuclear Structure, Vol. II (Benjamin, New York, 1975)
2. I. Hamamoto, Phys. Rep. 10 (1974), 63
3. D.R. Bes et al., Phys. Rep. 34 (1977), 1
4. A. Gyurkovich et al., Phys. Lett. 105B (1981), 95
5. G. A. Leander et al., Nucl. Phys. A388 (1982), 452
6. S. Frauendorf et al., Phys. Lett. 141B (1984), 23
7. W. Nazarewicz et al., Nucl. Phys. A429 (1984), 269
8. C. Lauterbach et al., International Conference on the Spectroscopy of Heavy Nuclei, Crete, 1987
9. M.A.J. Mariscotti et al., Nucl. Phys. A407 (1983), 98
10. P.R. Christensen et al., unpublished, quoted in Phys. Lett. 70B (1977) 292
11. R. Julin et al., private communication
12. J. Blomquist, Phys. Lett. 33B (1970), 541, and Phys. Lett. 208B (1988), 331
13. R.A. Broglia et al., Phys. Lett. 37B (1971), 159 and 257
14. R.H. Spear et al., Phys. Lett. 128B (1983), 29

Inst. Phys. Conf. Ser. No 105
Paper presented at Int. Conf. on Spectroscopy of Heavy Nuclei, Crete, Greece, 1989

The quasi elastic (e, e′p) reaction on heavy nuclei

L. Lapikás

National Institute for Nuclear Physics and High Energy Physics, section K (NIKHEF-K),
P.O. Box 4395, 1009 AJ Amsterdam, The Netherlands

ABSTRACT : Measurements are discussed of the spectral function for the (e,e'p) reaction on $^{40,48}Ca$, ^{90}Zr, $^{142,146}Nd$ and ^{208}Pb. In the experiment on ^{40}Ca we studied the influence of various kinematical conditions on the deduced momentum distributions, as well as the dependence on the virtual photon momentum. For ^{90}Zr the outgoing proton energy was varied between 61 and 165 MeV in order to investigate the role of the final-state interaction. In the case of the Nd isotopes we examined the effect of the deformation on the spectral function. For ^{208}Pb the spectroscopic strengths have been obtained as a function of the binding energy of the single particle orbital. In all cases the resulting spectroscopic strengths and occupations are compared with various theories that include correlations and with results from the (d,^{3}He) reaction. Finally the mass dependence of the quasi-particle strength for valence states is discussed in relation to nuclear matter theory.

1. INTRODUCTION

From the first high resolution measurements of the (e,e'p) reaction that were carried out at NIKHEF-K on ^{12}C, ^{51}V and ^{90}Zr [1,2] spectroscopic strengths for valence states have been extracted of about 50% of the Independent Particle Shell Model (IPSM) sum-rule limit. Within the frame work of the quasi-particle theory this value would imply an occupation n=55-75% of states just below the Fermi surface. The large uncertainty that is introduced when going from quasi-particle strength to occupancy arises from two sources: i) the predictions are mostly based on nuclear-matter calculations in which the character of the employed short- and long-range nucleon-nucleon interaction plays a crucial role, ii) the transition from (infinite) nuclear matter to a finite nucleus involves the complicated estimate of the coupling to surface vibrations or RPA correlations. In spite of these difficulties theory is making progress in the calculation of correlations in nuclei along two different roads. Benhar et al. [3] perform a nuclear matter calculation with correlated wave functions and then make the transition to the finite nucleus ^{208}Pb by a modification of the imaginary part of the mass operator. Mahaux et al. [4] employ a dispersion-relation approach, in which the correlations are effectively embedded in the imaginary part of the mass operator. In both calculations appreciable depletions arise for states near the Fermi level.

In order to enable a meaningful comparison between theory and experiment the pureness of the (e,e'p) reaction as a probe of spectroscopic strength has to be carefully investigated. Among the most important ingredients that are being examined are: i) the validity of the Impulse Approximation [5,6,7,8,9,10], ii) the description of the Final State Interaction [11,12], iii) the role of two-step processes [13] and charge-exchange contributions [14].

In the present paper I will present a number of experimental results for $^{40,48}Ca$,

^{142,146}Nd and ^{208}Pb, in addition to some tests of the description of the reaction mechanism for ^{40}Ca and ^{90}Zr. Finally I will present a comparison of our experimental data with theoretical calculations and with results obtained from the (d,^{3}He) reaction.

2. PRINCIPLES OF THE (e,e'p) REACTION

The formalism of the (e,e'p) reaction has been described in detail elsewhere [15], so only some major features are presented here. The energy ω and momentum $\mathbf{q}$ lost by the electron are transferred to a proton with binding (missing) energy E_m and (missing) momentum $-\mathbf{p}_m$, which then leaves the nucleus with momentum $\mathbf{p} = \mathbf{q} - \mathbf{p}_m$ and kinetic energy T_p. In the plane-wave Impulse Approximation (PWIA) the coincidence cross section can be written as

$$d^6\sigma / de'd\mathbf{p} = K\sigma_{ep}S(\mathbf{p}_m, E_m), \tag{2.1}$$

where K is a kinematical (phase-space) factor, σ_{ep} is the (off-shell) electron-proton cross section, e.g. the current-conserving σ_{ep}^{cc1} from de Forest [16], and $S(\mathbf{p}_m,E_m)$ is the spectral function, which is the probability to find in the parent nucleus a proton with binding energy E_m and momentum $\mathbf{p}_m$. For the transition to a discrete state with quantum numbers α (α=[nlj]) one can write

$$S(\mathbf{p}_m,E_m) = \rho_\alpha(\mathbf{p}_m)\delta(E_m-E_\alpha)\ , \tag{2.2}$$

where the momentum density

$$\rho_\alpha(\mathbf{p}_m) = |\Phi^\alpha_{if}(\mathbf{p}_m)|^2 \tag{2.3}$$

is the Fourier transform squared of the overlap between initial and final nuclear states, which is often called the bound-state wave function (BSWF). In the independent-particle shell model (IPSM) this BSWF is (proportional to) the single-particle or mean-field wave function. More generally it can be identified with a quasi-particle wave function.

Including the final-state interaction, i.e., after knock-out the proton is not in a free state but in a scattering state, described for instance by an optical-model (OM) wave function. Then the cross section can be written as

$$\sigma^{DWIA} \approx K\sigma_{ep}S^D(\mathbf{p}_m, E_m, \mathbf{p}), \tag{2.4}$$

where now the distorted spectral function and accompanying momentum density also depend on $\mathbf{p}$. Because of (2.4) all measured (and calculated) coincidence cross sections are given in terms of the reduced cross section $\sigma/K\sigma_{ep}$ which in factorised DWIA is the distorted momentum-density distribution

$$\rho_\alpha{}^D(\mathbf{p}_m,\mathbf{p}) \equiv \sigma/K\sigma_{ep}. \tag{2.5}$$

The factorised form (2.4) is an approximation; all actual calculations are done in unfactorised DWIA. The accompanying distorted momentum distribution is

$$\rho_\alpha{}^D(\mathbf{p}_m,\mathbf{p}) = \int \chi_\mathbf{p}^{(-)*}(\mathbf{r})\ |\Phi^\alpha_{if}(\mathbf{p}_m)|^2 \exp(i\ \frac{A-1}{A}\ \mathbf{q}.\mathbf{r})\ d\mathbf{r} \tag{2.6}$$

where $\chi_\mathbf{p}^{(-)*}(\mathbf{r})$ is the distorted proton wave, for which usually an OM wave function is taken. As can be seen the main assumptions in this formalism are that the virtual photon couples to a bound proton in the same way as to a free proton (Impulse Approximation) and that the final-state interaction effects between the outgoing proton and the residual nucleus can be described by an OM wave function. Implicit in the latter is that the photon has coupled to the knocked-out proton (no two-step knock-out).

3. TESTS OF THE (e,e'p) REACTION MECHANISM

In this section I shall discuss some experiments that were designed to obtain information on the mechanism of the (e,e'p) reaction. For that purpose we write the coincidence cross section (eq. 2.1) in the more general form [17]

$$d^6\sigma/de'd\mathbf{p} = K\sigma_{Mott}\{W_L + \varepsilon^{-1} W_T + W_{TT}\cos 2\alpha + \varepsilon' W_{TL}\cos\alpha\}, \qquad (3.1)$$

where $\varepsilon^{-1} = 1 + 2(\mathbf{q}^2/Q^2)\tan^2(\theta_e/2)$ is the virtual photon polarization, α the angle between the reaction plane and the plane through $\mathbf{q}$ and $\mathbf{p}$, and ε' a further kinematical parameter. The W_i's (i=L, T, TT, TL) are generalized structure functions that depend in a complicated way on $\mathbf{p}_m$, E_m, $\mathbf{q}$ and $\mathbf{p}$, and on the nucleon elastic form factors $F_L(Q^2)$ and $F_T(Q^2)$. In parallel kinematics the structure functions W_{TT} and W_{TL} become identical zero.

In PWIA one can show that the spectral function $S(\mathbf{p}_m, E_m)$ can be factorized from these structure functions and one is left with the expression (2.1). However, distortions have a different effect on each of the structure functions and hence measuring the same part of the spectral function under different kinematical circumstances will provide a powerful check on the understanding of the final-state interaction (see sect 3.1 and 3.3). Similarly, a variation of the four-momentum transfer $Q^2 = \mathbf{q}^2 - \omega^2$ in the reaction, while keeping other variables ($|\mathbf{p}_m|$, E_m, T_p) constant, will provide insight into the validity of the Impulse Approximation, i.e. the use of the free proton charge and magnetic form factors $G_E(Q^2)$ and $G_M(Q^2)$ in the expressions for W_i. If one assumes the Impulse Approximation a further reduction of the four structure functions in eq. (3.1) can be carried out with the result that the coincidence cross section can be written as

$$d^6\sigma/de'd\mathbf{p} = K\sigma_{Mott}\frac{\varepsilon_x(1-\varepsilon)}{\varepsilon(1-\varepsilon_x)}\{F_L^2(Q^2) + \varepsilon_x^{-1} F_T^2(Q^2)\}, \qquad (3.2)$$

where ε_x is a generalized form of ε which reduces to ε in parallel kinematics. When one neglects off-shell effects one also has $F_L(Q^2) = G_E(Q^2)$ and $F_T(Q^2) = (Q/2M_p)\, G_M(Q^2)$.

3.1 The (e,e'p) reaction on ^{40}Ca in parallel and (q,ω) constant kinematics

In this experiment we measured the momentum distributions in the range $-200 < p_m < 300$ MeV/c for the $1d_{3/2}$ ground-state transition and the $2s_{1/2}$ transition to the $1/2^+$ state at 2.552 MeV in the reaction $^{40}Ca(e,e'p)^{39}K$. The momentum distributions were obtained for a constant proton energy T_p = 100 MeV (p=450 MeV/c) under two different kinematical conditions :

I. *(anti) parallel kinematics*, i.e. $\mathbf{q} // \mathbf{p} // \mathbf{p}_m$, where we take the sign convention $p_m = |\mathbf{p}| - |\mathbf{q}|$, and hence q varies from 650 MeV/c at p_m = - 200 MeV/c to 150 MeV/c at p_m = + 300 MeV/c.

II. *(q,ω) constant kinematics*, (also called perpendicular kinematics since in the case $|\mathbf{p}| = |\mathbf{q}|$ one has $\mathbf{p}_m \perp \mathbf{q}$ for small p_m). Here q was kept constant at 450 MeV/c and the range in p_m was achieved by varying the angle between $\mathbf{q}$ and $\mathbf{p}$.

Fig. 1 shows the measured momentum distributions for the $1d_{3/2}$ and $2s_{1/2}$ transitions in (anti) parallel kinematics. The curves are fits to the data where the free parameters are the spectroscopic factor S and the radius r_0 of the Woods-Saxon well used to generate the BSWF. The CDWIA calculations included proton distortion (optical model potential from Schwandt [18]) and first-order eikonal approximation for the electron [19]. The left/right asymmetry in the data is rather well described especially so for the minima in the $2s_{1/2}$ momentum distribution.

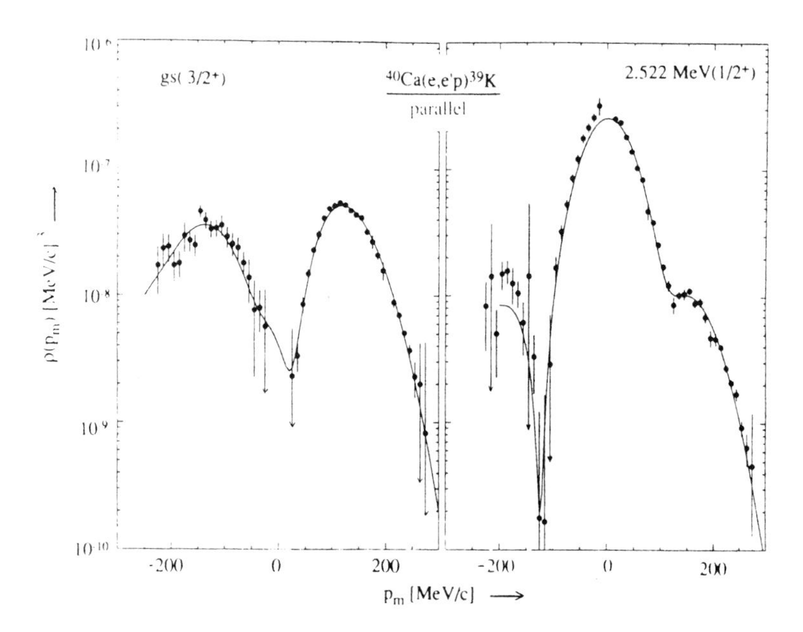

Fig. 1. *Momentum distributions for the $2s_{1/2}$ and $1d_{3/2}$ transitions in the reaction $^{40}Ca(e,e'p)^{39}K$ measured in parallel (p_m >0) and antiparallel (p_m<0) kinematics. The proton kinetic energy is 100 MeV. The curves represent CDWIA calculations for bound-state wave functions with r_0 (radius of the WS well) and S (spectroscopic factor) fitted. The optical model potential is from ref. [18].*

Fig. 2 shows the momentum distributions for the same transitions as in fig. 1, but now measured in (q,ω) constant kinematics. The curves are calculated in CDWIA with the parameters r_0 and S as deduced from the fits to the parallel data (see fig. 1). The curves reproduce the data rather well, considering the large differences in kinematical circumstances. When the data of fig. 2 are fitted with r_0 and S as free parameters we obtain differences of ≤ 4 % in r_0 (i.e. ≤ 2 % in the rms radius of the BSWF) and ≤ 15 % in S. These differences might be due to: i) the uncertainty in the choice of the optical model potential, or ii) the inadequacy of the Impulse Approximation. In order to study the first possibility we will investigate the freedom in the choice of the optical model potential by performing a reanalysis of existing (p,p) data [18]. As to the second possibility we note that the virtual-photon parts of the kinematics I and II are quite different. In parallel kinematics (I) the momentum transfer q varies from 150 MeV/c at p_m=300 MeV/c to 650 MeV/c at p_m=-200 MeV/c. Hence θ_e and thus ε changes. In the (q,ω) constant kinematics (II) we have q=450 MeV/c for all data. Hence, a possible enhancement of the ratio of transverse to longitudinal nucleon structure functions as found in refs. [5,6,9,10] will show up predominantly at negative p_m (where ε^{-1} is large) in kinematics I, but as a nearly overall enhancement in kinematics II (where ε^{-1} is constant for all p_m). A similar observation holds for a possible modification of the Q-dependence of the bound proton form factors, which will have a constant effect in kinematics II, but a p_m-dependent effect in kinematics I.

From the data of figs. 1 and 2 we deduced the spectroscopic factors for the $1d_{3/2}$ and $2s_{1/2}$ transitions to be 2.3±0.3 and 1.05±0.05, where the error reflects the uncertainty due to the different values obtained in each kinematics. The low values (i.e. 58±5% and 52±3%, of the IPSM sum-rule limit, respectively) are also found from (e,e'p) experiments for valence orbitals in other (medium) heavy nuclei (see sect. 7).

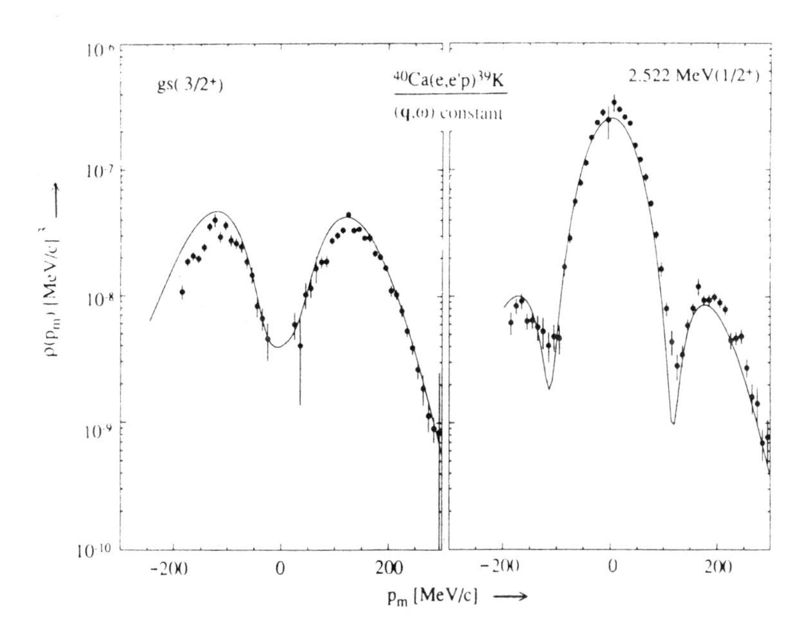

*Fig. 2. Momentum distributions for the $2s_{1/2}$ and $1d_{3/2}$ transitions in the reaction $^{40}Ca(e,e'p)^{39}K$ measured in (q,ω) constant kinematics. The sign of p_m is such that $p_m > 0$ for **p** right of **q** and $p_m < 0$ for **p** left of **q**. The proton kinetic energy is 100 MeV. The curves represent CDWIA calculations for bound-state wave functions with r_o and S fixed to the values of fig. 1. The optical model potential is the same as in fig. 1.*

3.2 The momentum-transfer dependence of the (e,e'p) reaction

As shown in eq. (3.1) and (3.2) it is in general possible to separate the contributions to the cross section of the longitudinal and transverse nucleon structure functions G_E and G_M. In parallel kinematics the relative weight of these contributions is given by the virtual photon polarization ε, in non-parallel kinematics by a similar parameter ε_x. Since we wanted to study the dependence of the structure functions on Q^2 we chose to measure the reaction ^{40}Ca(e,e'p)^{39}K for a constant ε_x, a constant p_m = 140 MeV/c and a constant T_p = 100 MeV in kinematics with different Q^2. We measured both the parallel and antiparallel kinematics and two momentum transfers in between (q = 305, 585 , 480 and 380 MeV/c, respectively). Since we kept ε_x fixed we needed four different beam energies for these measurements.

Fig. 3 shows the preliminary results of these measurements with constant $\varepsilon_x = 0.44$. In the figure we plot the ratio $\sigma^{exp} / \sigma^{ref}$ as a function of Q^2 for the $1d_{3/2}$ transition. The reference cross section σ^{ref} was calculated in CDWIA with r_0 and S as determined in parallel kinematics (see sect. 3.1). In PWIA we expect R = 1. For the $1d_{3/2}$ transition we have an indication that R differs from unity. This effect might be explained by an enhanced magnetic or reduced electric nucleon structure function. The dashed line in fig. 3 indicates the effect of using a ratio $R_G = [G_M(Q^2)/G_E(Q^2)](Q/2M_p) = 1.2\mu_p$ instead of the PWIA value $R_G = \mu_p$. Similar values of this ratio were found in experiments on ^{6}Li [6] and ^{12}C [5] at NIKHEF, and on ^{4}He [7] and ^{40}Ca at Saclay [9].

Clearly from the present data one may not conclude that there is evidence for a Q-dependence of the coincidence cross section other than that predicted by the Impulse Approximation. A similar observation was made in recent (e,e'p) experiments on ^{4}He [7],

^{12}C [10, 8] and ^{40}Ca [9]. In a further test we also have performed a longitudinal-transverse separation in the same way as carried out in refs. [5, 6]. The analysis of these data is underway.

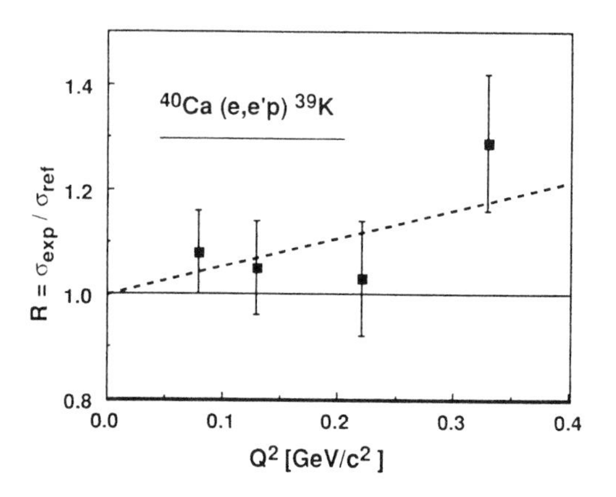

Fig. 3. Ratio $\sigma^{exp}/\sigma^{ref}$ of experimental cross sections σ^{exp} for the $1d_{3/2}$ transition in the reaction $^{40}Ca(e,e'p)^{39}K$ measured in mixed kinematics (see text) relative to the cross section σ^{ref} calculated with the parameters of fig. 1. The mixed kinematics were chosen such that only Q^2 varied, whereas the kinematical variables ε_x, T_p and p_m were kept constant. In the Impulse Approximation one would expect R=1.

3.3 The proton energy dependence of the (e,e'p) reaction

In eq. (2.5) it was shown that the final-state interaction between the proton and the residual nucleus distorts the momentum distribution. The amount of distortion is related to the kinetic energy of the outgoing proton and to the energy-dependence of the optical model potential that is used to generate the proton waves in the scattering state. In order to test whether this part of the reaction description is well understood, we performed an (e,e'p) experiment in which the outgoing proton energy was varied between 61 and 165 MeV.The measurements were carried out on ^{90}Zr in the missing momentum region $120 \leq p_m \leq 160$ MeV/c. The resulting momentum distributions for the $2p_{1/2}$, $2p_{3/2}$, $1f_{5/2}$ and $1g_{9/2}$

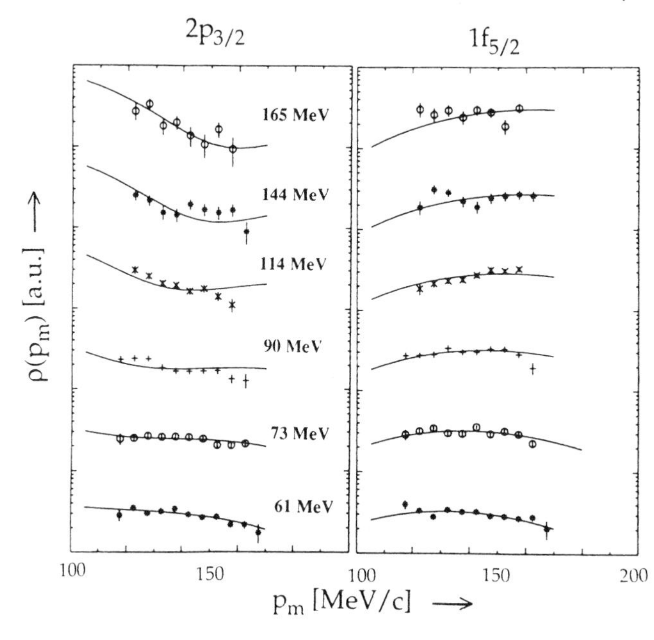

Fig. 4. Momentum distributions $\rho(p_m)$ for the $2p_{3/2}$ and $1f_{5/2}$ transitions in the reaction $^{90}Zr(e,e'p)^{89}Y$ at six different proton energies. The curves represent CDWIA calculations with the BSWF's of ref. [2] and the optical model potential of ref. [18]. The amplitudes of the curves (i.e. the spectroscopic factors) have been scaled to fit the data.

transitions have been compared with CDWIA predictions (i.e., both proton and electron distortions are included, see fig. 4). For the BSWF's we used Woods-Saxon wave functions as determined in an earlier (e,e'p) experiment [2], in which the complete momentum distributions were measured between 0 and 300 MeV/c at outgoing proton energies of 70 and 100 MeV.

In an initial analysis the optical-model parameters were obtained from a global parametrisation of Schwandt et al. [18]. The corresponding curves are shown in fig. 4. They represent fits to the data where the amplitude (i.e. the spectroscopic factor S) was treated as a free
parameter. The deduced $S(T_p)$ showed a systematic dependence on T_p, the largest being dS/dT_p = 0.3 %/MeV for the $1f_{5/2}$ transition. In order to investigate whether this trend depends on the chosen optical-model parametrisation we performed optical-model fits [11] to a set of elastic proton scattering data (T_p=65-160 MeV) on ^{90}Zr, including both volume (W) and surface (W_d) absorption. These fits show a rather large freedom in the value of W. If we take a dependence of W on T_p with a slope dW/dT_p=0.03 we obtain spectroscopic factors that are independent of T_p within ±10% for all four transitions (see fig. 5).

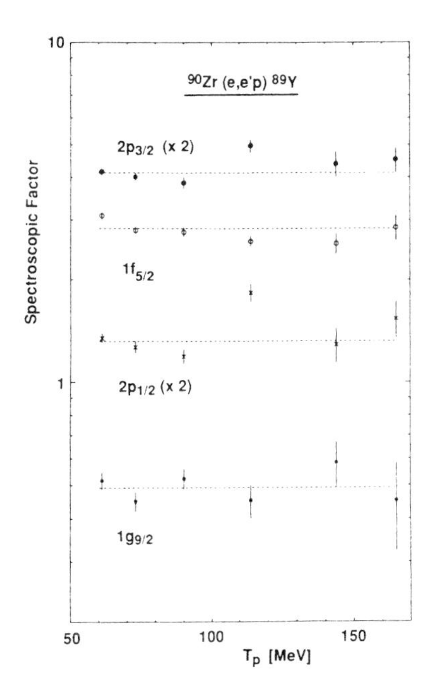

Fig. 5. Spectroscopic factors obtained for the $2p_{3/2}$, $2p_{1/2}$, $1g_{9/2}$ and $1f_{5/2}$ transitions in the reaction $^{90}Zr(e,e'p)^{89}Y$ at six different proton energies T_p. The energy dependence of the optical model potential used to extract these values is discussed in the text. The dashed curve indicates the average value for each transition.

The results presented here are not independent of other assumptions on the description of the reaction, most notably on the Impulse Approximation. If one assumes that e.g. the ratio $R_G=[G_M(Q^2)/G_E(Q^2)](Q/2M_p)$ is different from the free proton ration $R_G=\mu_p$, this will introduce a dependence of the momentum distributions on the kinematical conditions, which varied from a forward electron scattering angle θ_e=33° at 61 MeV proton energy to 83° at 165 MeV. Hence the relative contribution of the transverse and longitudinal coupling would vary with kinematics and for R_G= 1.2μ_p (see sect. 3.2) one would obtain a 15 % differential effect when going from the kinematical situation at 61 MeV to 165 MeV.

The obtained average values for the spectroscopic factors of the four transitions (see fig. 5) agree within the errors with the previously determined results [2]. The statistical spread in the values of $S(T_p)$ obtained from the present analysis indicates that the dependency of the spectroscopic factors on the choice of the optical-model potential is ≤ 10 %.

4. THE (e,e'p) REACTION ON Nd ISOTOPES

Momentum distributions for proton knockout have been measured at NIKHEF for a number of spherical nuclei (^{16}O, $^{40,48}Ca$, ^{90}Zr, ^{208}Pb). However, information on knockout from deformed nuclei is hardly available. Here we present the first results of (e,e'p) measurements on ^{142}Nd and ^{146}Nd. The latter nucleus is calculated [20] to be permanently deformed (β=0.22,γ=0), whereas the calculated shape of ^{142}Nd is spherical. The aim of the present experiment is twofold: i) a comparison of the knockout from the $2d_{5/2}$ and $1g_{7/2}$ valence proton shells, in order to establish a possible effect of the deformation on the momentum distributions; ii) a study of the fragmentation of the deeper lying $1g_{9/2}$ shell, in relation to the question whether inner shells are deformed as well.

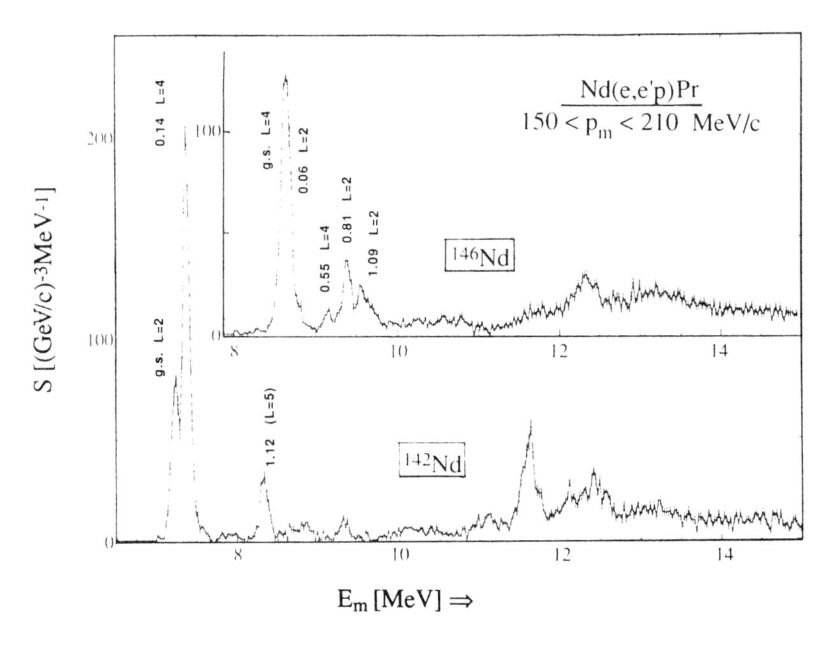

Fig. 6. *Missing-energy (E_m) spectra for the (e,e'p) reaction on ^{142}Nd and ^{146}Nd (inset) for missing momentum (p_m) values corresponding to the maximum of the $1g_{9/2}$ distribution. The different structure of the $1g_{9/2}$ energy distribution (expected between 10 and 16 MeV) is clearly visible.*

In fig. 6 we show missing-energy spectra for both isotopes in the missing-momentum range $150 < p_m < 210$ MeV/c, i.e. the region where the $1g_{7/2}$ momentum distribution is at its maximum. One expects this 1g strength to reside between 11 and 15 MeV missing energy. For the spherical nucleus ^{142}Nd the strength is dominated by a peak at E_x=11.6 MeV, whereas for ^{146}Nd the strength is spread over a larger energy domain. This observation is in agreement with theoretical calculations of the spreading mechanism in the framework of the quasi-particle phonon model (QPM) [21] (see fig. 7) and with similar measurements of the (d,^{3}He) reaction on the Nd isotopes [22].

The extremely good energy resolution (70 keV, see fig. 6) obtained in the present experiment has enabled to extract the momentum distributions of a number of discrete transitions to states below 2 MeV excitation energy in the residual ^{141}Pr and ^{145}Pr nuclei. Fig. 8 shows $\rho(p_m)$ for the lowest $2d_{5/2}$ transition in both nuclei. A surprising result is that within error bars the experimental data show no difference in shape. Indeed, when fitting these data with CDWIA momentum distributions we obtain equal rms radii for the $2d_{5/2}$ bound state wave functions calculated in a Woods Saxon well (5.16±0.10 fm for ^{142}Nd, 5.21± 0.10 fm for ^{146}Nd). A similar conclusion holds for the $1g_{7/2}$ transitions (5.20±0.10 fm for ^{142}Nd, 5.10±0.10 fm for ^{146}Nd). Moreover the radii agree with the values 5.13 fm and 5.21 fm predicted for the $2d_{5/2}$ and the $1g_{7/2}$ orbit, respectively, in a spherical Hartree-Fock calculation employing the G_σ interaction [36].

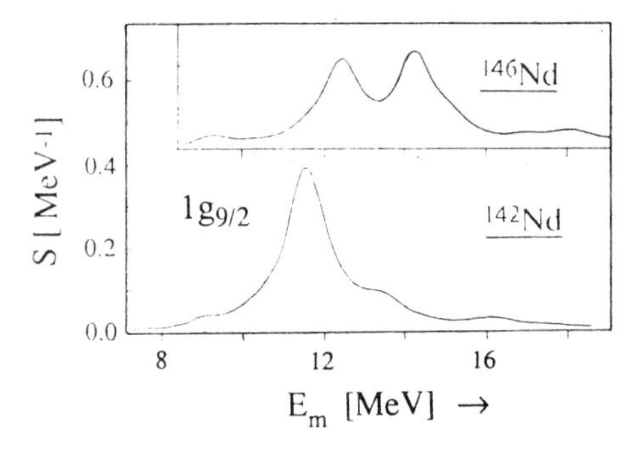

Fig. 7. Spreading of the $1g_{9/2}$ proton hole strength as calculated in the quasi-particle phonon model [21] for ^{142}Nd and ^{146}Nd (inset). The theoretical energy distributions have been folded with a 1 MeV wide (FWHM) Lorentzian shape.

Another interesting result derives from the total $2d_{5/2}$ and $1g_{7/2}$ spectroscopic strength observed in states up to 2 MeV in both nuclei. In ^{142}Nd we find summed spectroscopic factors 1.47 and 3.15 for the $2d_{5/2}$ and $1g_{7/2}$ transitions, respectively, whereas the corresponding numbers for ^{146}Nd are 1.70 and 2.71. The statistical errors in these numbers are smaller than 2%. The relative systematic error due to differences between the optical potentials of ^{142}Nd and ^{146}Nd is estimated to be 5 %. Apparently when going from the spherical nucleus ^{142}Nd to the deformed ^{146}Nd strength moves from the $1g_{7/2}$ to the $2d_{5/2}$ orbit. This is in agreement with the expectation that in a deformed nucleus the state with the higher spin (i.e. the $1g_{7/2}$ orbital) is pushed up in binding energy and hence will get a lower occupation.

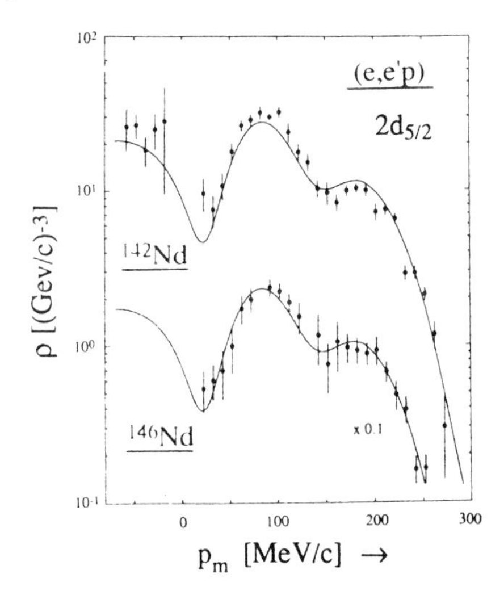

Fig. 8. Momentum distributions for the lowest $2d_{5/2}$ transitions observed in the (e,e'p) reaction on ^{142}Nd and ^{146}Nd. The curves represent CDWIA calculations for bound-state wave functions with r_0 (radius of the WS well) and S (spectroscopic factor) fitted. The optical model potential is from ref. [18].

Finally we note that the total strength in the $2d_{5/2}$ and $1g_{7/2}$ orbits is practically equal in both isotopes: 4.62 and 4.41, corresponding to 46 ±5 % and 44±5%, respectively of the 10 valence protons outside the closed ^{132}Sn core. This low strength agrees with similar findings from the (e,e'p) reaction on other heavy nuclei like ^{90}Zr [2] and ^{208}Pb [23] (see sect. 7).

5. PROTON KNOCKOUT FROM ^{208}Pb

5.1 Experiment

The spectral function of ^{208}Pb was measured [23] with the high resolution two-spectrometer set-up at NIKHEF-K [24] at missing momentum values between -50 and 300 MeV/c in parallel kinematics (**p//q**). The kinetic energy of the outgoing proton was kept constant at 100 MeV. Most data have been taken at an incoming electron energy of 412 MeV, but for experimental reasons the data at $p_m > 200$ MeV/c were taken at lower beam energies. The experimental momentum distributions presented in this paper have been corrected for this difference by transforming all data to the highest beam energy. The total systematic error resulting from a standard analysis of the data [2, 23] is 3 % on S and hence also on the momentum distributions.

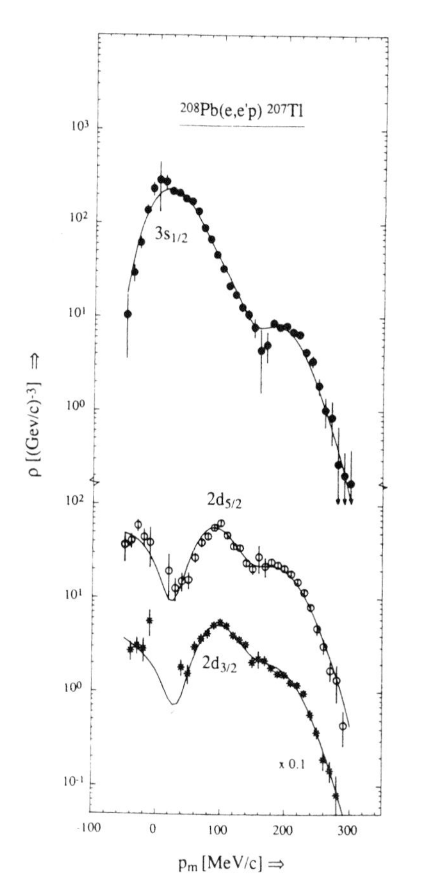

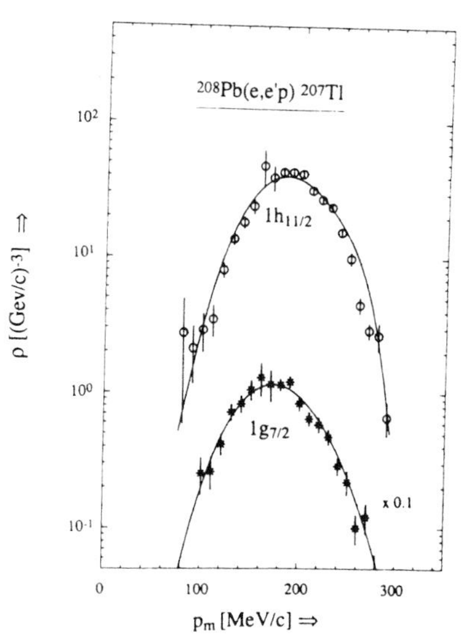

Fig. 9. Momentum distributions for $3s_{1/2}$, $2d_{3/2}$, $2d_{5/2}$, $1h_{11/2}$ and $1g_{7/2}$ proton knockout in the reaction ^{208}Pb(e,e'p)^{207}Tl. The curves represent CDWIA calculations employing standard WS wave functions, of which the normalization and radius have been adjusted to fit the data.

5.2 Momentum Distributions

Experimental momentum distributions for some discrete transitions to the low-lying states in ^{207}Tl are shown in fig. 9. They correspond to knock-out of $3s_{1/2}$, $2d_{3/2}$, $2d_{5/2}$, $1h_{11/2}$ and $1g_{7/2}$ protons from the ground state of ^{208}Pb. Also shown in these figures are fits to the data employing Woods-Saxon (WS) bound-state wave functions. In the fits the radius r_0 of the WS well and the normalisation (i.e. the spectroscopic factors S) were treated as free parameters. The curves were calculated with the code DWEEPY [19], which takes both the proton and the electron distortions into account. For the calculation of the proton distortion we determined optical model parameters from a fit to 98 MeV proton scattering data [18] from ^{208}Pb. The freedom in the choice of the OM parameters results in a 5% uncertainty in the amplitude of the calculated $\rho^D(p_m)$. The electron distortion was calculated in second-order eikonal approximation. Straightforward application of this method resulted in momentum distributions that were unable to describe the data at low p_m, which means, in our kinematics, at high momentum transfer q. We therefore modified the eikonal approximation by introducing a slightly different effective momentum transfer. This resulted in 5% different spectroscopic strengths but appreciably better fits. The obtained spectroscopic factors, expressed as a fraction of the IPSM sum rule value (2j+1), are between 40 and 50 % for the $3s_{1/2}$, $2d_{3/2}$, $2d_{5/2}$ and $1h_{11/2}$ transitions. For the $1g_{7/2}$ transition the spectroscopic factor is even lower, but this strength is fragmented into a number of transitions as shown in sect. 3.3.

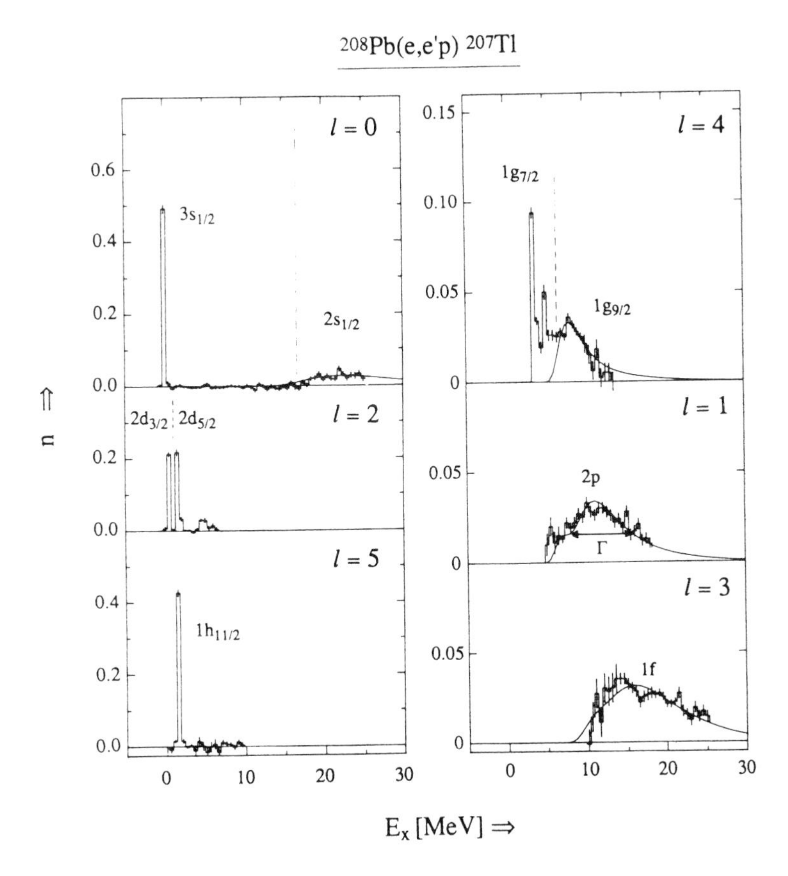

Fig. 10. Results of the l-decomposition of the proton spectral function of ^{208}Pb in an excitation energy region from 0 to 25 MeV showing the spectroscopic strength in each bin as a fraction S/(2j+1) of the IPSM sum-rule limit. The curves shown are Lorentzians with an energy dependent width Γ according to ref. [25]. The dashed lines separate excitation energy regions where different orbits were employed in the l-decomposition.

5.3 Integrated Strength

Once the description of the discrete transitions is satisfactory, one can determine the strength residing in the continuum by a so called *l*-decomposition. For this purpose the measured spectral function in every 0.5 MeV interval was fitted with an incoherent sum of calculated momentum distributions in order to determine the contributions of each orbital angular momentum (*l*) to the total observed strength. The resulting strength distributions are shown in fig. 3. Here one can observe that going to deeper-lying shells the low-lying discrete structure develops into broader distributions characterized by a spreading width Γ that is in reasonable agreement [26] with an empirical description deduced by Brown and Rho [25].

In fig. 11 we present the integrated strengths for the various single particle orbitals as obtained from the distributions shown in fig. 10. Here each distribution was integrated over the excitation energy range from 0 to E_{peak} +2Γ, with -if necessary- a correction for the strength above the experimental upper limit of 25 MeV excitation energy. Fig. 11 clearly shows that the summed spectroscopic strength is about 50% of the IPSM sum rule value for valence states and rises to about 75% for the deeply bound $2s_{1/2}$ and 1f states. In sect. 6 we shall compare these values with various theoretical predictions.

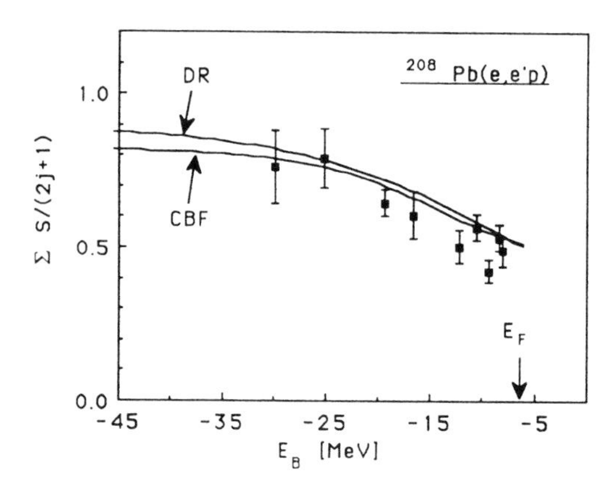

Fig. 11. Integrated strength $\Sigma S/(2j+1)$ as a fraction of the sum-rule limit observed in the reaction $^{208}Pb(e,e'p)$ for various single particle orbits. The $1g_{9/2}$, 2p, 1f and 2s occupations have been corrected for unobserved strength in the interval 25 MeV < E_B -E_F<E_{peak}+2Γ assuming the Lorentzian shapes of fig. 10. The curves are predictions based on correlated basis functions (CBF) [3] and on the dispersion relation (DR) approach [4].

6. THEORETICAL APPROACHES TO THE SPECTRAL FUNCTION

In an infinite system of interacting nucleons (nuclear matter) the spectral function is related [4] to the imaginary part of the Green's function:

$$G(k,E) = [E - k^2/2m - M(k,E)]^{-1} \tag{6.1}$$

where M(k,E) = V(k,E) + iW(k,E) is the mass operator for a nucleon with momentum k and energy E. The spectral function is

$$S(k,E) = \pi^{-1}\, \mathrm{Im}G(k,E) = \frac{\pi^{-1}W(k,E)}{[E - k^2/2m - V(k,E)]^2 + W(k,E)^2} \tag{6.2}$$

and the quasi-particle peak is located at an energy e(k), which is the solution of $e(k) = k^2/2m + V(k,e(k))$. Expanding V in the vicinity of E=e(k) then yields the quasi-particle spectral function

$$S_{QP}(k,E) = \frac{\pi^{-1}[s(k)]^2W(k,E)}{[E - e(k)]^2 + [s(k)\, W(k,E)]^2} \tag{6.3}$$

where $s(k) = [1 - \partial V/\partial E|_{E = e(k)}]^{-1}$ is the quasi-particle strength (also called effective mass m^*/m), which for $k \to k_F$ is equal to the discontinuity $Z(k_F)$ across the Fermi surface. The occupation probability is

$$n(k) = \int_{-\infty}^{\varepsilon_F} S(k,E)dE \tag{6.4}$$

Various authors have calculated S and n, mainly along two different lines. In the dispersion relation approach [28,19,29], one relates the dispersive part ΔV of V to W via

$$\Delta V(k,E) = \pi^{-1} \int_{-\infty}^{\infty} W(k,E)(E' - E)^{-1} dE' \tag{6.5}$$

and uses W from e.g. proton scattering. This yields the occupations as shown in fig. 12. The corresponding energy distributions (eq. 6.2) are shown in fig. 13 with $W = \Gamma / 2$ for the nuclear matter (NM) prediction [37] of Γ and for the empirical parametrisation by Brown and Rho [25].

In order to compare these predictions with our observables i.e. strength integrated over the finite experimental energy region $e(k) - 2\Gamma < E < \varepsilon_F$, we have calculated the ratio R of the integral in eq. (6.4) with lower limits $e(k) - 2\Gamma$ and $-\infty$, respectively. This yielded R = 0.83 practically independent of e(k). In fig. 11 we compare the experimental integrated strength with R times the quasi-particle strengths calculated in the dispersion relation (DR) approach [4].

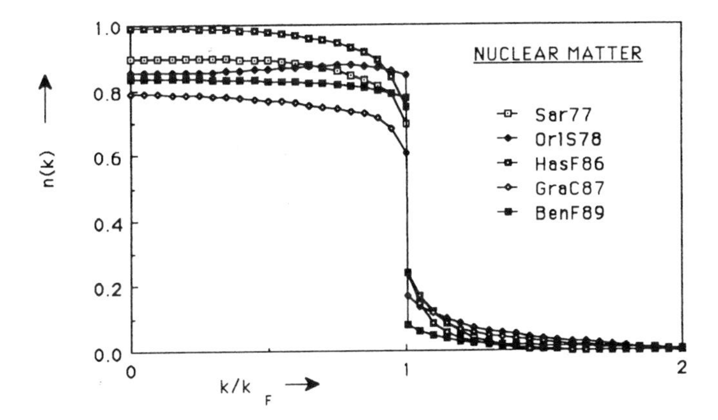

Fig. 12. Occupations n(k) in nuclear matter as a function of the momentum k (in units of the Fermi momentum k_F) for various types of nucleon-nucleon interactions and calculational approaches (Sar77 : ref. [27], OrlS78 : ref. [28], HasF86 ref. [29], GraC87 : ref. [30], BenF89 : ref. [3], see text).

In another approach S is evaluated in nuclear matter by explicitly computing all the diagrams up to 2p2h contributing to M, starting from correlated basis functions (CBF)[3, 31] or in Brueckner Hartree-Fock (BHF)[30]. This leads to the occupations as shown in fig. 12 for nuclear matter. Benhar *et al.* [3] then make the transition to a finite system by inclusion of the surface effects through a modification of W in such a way as to reproduce our experimental data for the spreading width [26]. This yields the curve labelled CBF in fig. 11.

From fig. 11 we conclude that the agreement between experimentally observed strength and the DR and CBF predictions is reasonable. Both theories predict the correct energy dependence. The remaining difference (≈10 %) between theory and experiment may be caused by uncertainties in the theoretical calculations (see. e.g. fig. 12) or in the uncertainty in the interpretation of the (e,e'p) reaction mechanism. The main uncertainties in the latter are related to the Final-State Interaction (≈ 5%) or a possible modification of the structure functions for bound protons [5, 6]. However, recent (e,e'p) experiments on ^{4}He [7], ^{12}C [10, 8] and ^{40}Ca [9] indicate that the q-dependence of the electron-proton coupling is consistent with that of free nucleon form factors.

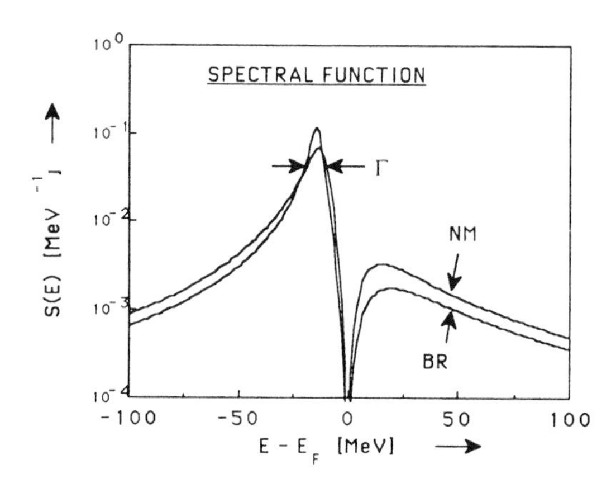

Fig. 13. Energy dependence of the spectral function (see eq. (6.3)) with the empirical form of Brown and Rho (BR) (ref. [25]) for the spreading width Γ and with the nuclear matter (NM) (ref. [37]) dependence for Γ. The curves were calculated for a hole state located at E_B=15 MeV below the Fermi level E_F. The integral from - ∞ to E_F represents the hole occupation (n), the integral from $-E_B$ -2Γ to E_F represents the quasi-hole strength (S). The depletion (1-n) which represents strength in the particle states is given by the area above E_F. The total area under both curves is unity.

7. MASS DEPENDENCE OF THE QUASI-PARTICLE STRENGTH

For a large number of nuclei the quasi-particle strength S for valence states has been measured at NIKHEF-K. The strengths were obtained by summing the spectroscopic factors for the (mostly few) transitions corresponding to the valence orbital. In most cases *l*-decompositions were carried out to verify that these indeed represented all valence strength up to 20 MeV excitation energy. Fig. 14 shows that ΣS/(2j + 1) is clearly mass dependent, going from ~ 0.8 for the few body systems to ~ 0.5 for heavy nuclei. A tentative explanation of this dependence is the following. In nuclear matter there is no energy gap between the states below and above the Fermi level, but for finite nuclei this gap behaves approximately as an inverse power *p* of A. Hence one expects a relation between the quasi-particle strength

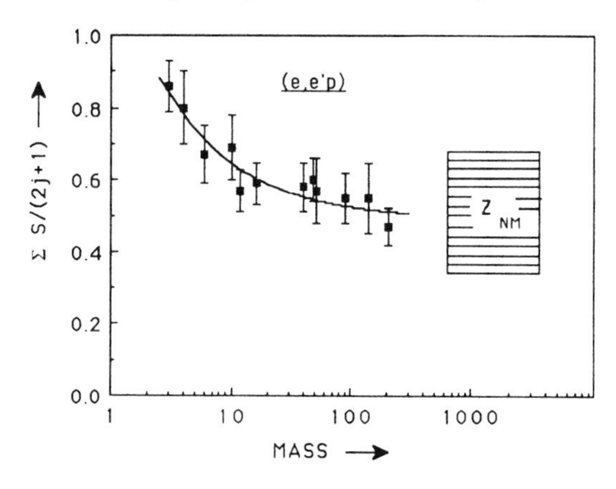

Fig. 14. Quasi-particle strength ΣS/(2j+1) for valence orbitals observed in the reaction (e,e'p) as a function of the mass of the target nucleus. The curve represents a fit to the data according to eq. (7.1). The hatched area indicates the range of predictions for Z_{NM} as calculated in various nuclear matter theories (see fig. 12).

in nuclear matter $Z(k_F)$ and that for valence states in finite nuclei S(A), which is approximately

$$S(A) = Z(k_F) + c.A^p. \tag{7.1}$$

The curve shown in fig. 14 was obtained by fitting Z and c to the data for p=-2/3. The fitted value Z = 0.50 ± 0.05 is well within the range for Z_{NM} = 0.35 - 0.70 for the various nuclear matter calculations (see hatched area in fig. 14). It would be worthwhile to investigate what the relation is of the described effect with e.g. the nuclear mass dependence of RPA correlations or of the effective mass.

8. COMPARISON BETWEEN (e,e'p) AND (d,^{3}He) RESULTS.

As has been illustrated in sect. 7 the spectroscopic strength for valence orbitals as deduced from (e,e'p) reactions is about 50%. In contrast to this, results derived from the (d,^{3}He) reaction have often been found to be close to the IPSM sum rule limit. However, the sensitivity of the extracted spectroscopic factors to the assumed parameters of the BSWF is much larger in the case of the hadronic reaction [32]. Moreover if one wants to compare the (e,e'p) and (d,^{3}He) results one should use the same (non-local) BSWF's. Below I will discuss two comparative studies on the (e,e'p) and the (d,^{3}He) reaction: i) spectroscopic strengths for a large number of transitions in ^{48}Ca, ii) spectroscopic strengths for valence transitions in a range of nuclei.

8.1 Proton spectroscopic factors in ^{48}Ca.

Momentum distributions for proton knockout from ^{48}Ca have been deduced for $2s_{1/2}$, $1d_{3/2}$ and $1d_{5/2}$ transitions in a recent experiment at NIKHEF. From these distributions we have deduced the spectroscopic factors S and the radii r_0 of the WS well which was used to calculate the BSWF. With these BSWF's we have carried out a reanalysis of polarised (d,^{3}He) data [33] at a deuteron energy of 79 MeV and of (d,^{3}He) data at 56 MeV [34] in a non-local finite-range DWBA calculation. We observe differences of up to 20% for the spectroscopic factors deduced in this way from the two (d,^{3}He) experiments. This finding is in agreement with the estimated [35] size of the systematic error to be associated with spectroscopic factors deduced from the (d,^{3}He) reaction.

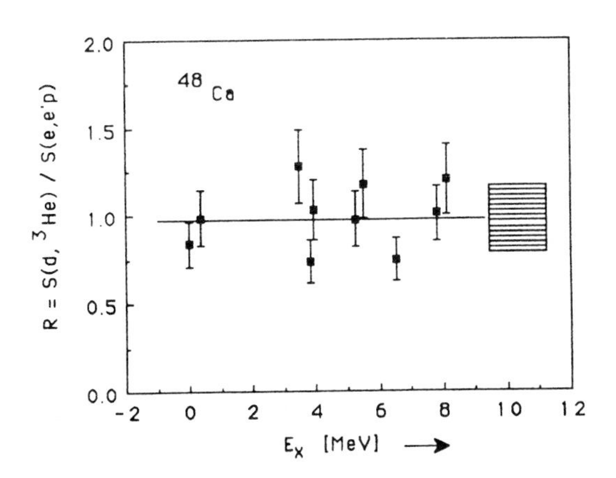

Fig. 15. Ratio R = S(d,^{3}He)/S(e,e'p) of spectroscopic factors for various transitions involving proton transfer with the (d,^{3}He) and the (e,e'p) reaction on ^{48}Ca as a function of excitation energy in the residual nucleus. The solid line represents the average of the data, the hatched area its standard deviation.

In fig. 15 we plot the ratio of the (d,^{3}He) spectroscopic factors as obtained from the reanalysis of data of ref. [33] to our (e,e'p) results. The average of these values is R=0.98±0.19, where the error represents the statistical spreading in the ratio's. Since the systematic uncertainties for the spectroscopic factors are of the order of 20% for the (d,^{3}He) reaction [35] and 5-10% for the (e,e'p) reaction (see sect. 3) it is concluded that for ^{48}Ca the results of both reactions are in excellent agreement.

8.2 Proton spectroscopic factors for valence transitions.

A similar reanalysis as described above for ^{48}Ca(d,^{3}He) data has been carried out for a number of other nuclei where good (e,e'p) and (d,^{3}He) data are available. The results of this comparison are shown in fig. 16 for proton knockout from valence orbitals in nuclei in the mass range from ^{12}C to ^{208}Pb.

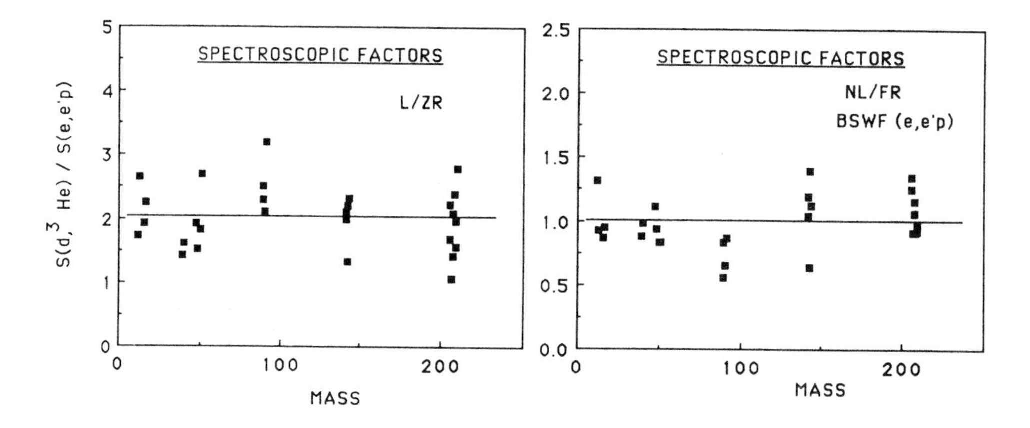

Fig. 16. Ratio R = S(d,^{3}He) / S(e,e'p) of spectroscopic factors for transfer of protons with the (e,e'p) and the (d,^{3}He) reaction on various nuclei. Left panel with original (d,^{3}He) spectroscopic factors extracted from a local zero-range (L/ZR) DWBA analysis, involving a variety of Ansatzes for the BSWF. Right panel with (d,^{3}He) values obtained after a reanalysis of the (d,^{3}He) data in non-local finite-range (NL/FR) DWBA using the BSWF as deduced from the corresponding (e,e'p) experiment. The solid lines indicate the average ratio.

In the left panel we plot the ratio of the original (d,^{3}He) spectroscopic factors to our (e,e'p) values. All published (d,^{3}He) data used in fig. 16 were extracted from a local zero-range (L/ZR) DWBA analysis, involving a variety of Ansatzes for the BSWF. The solid line shows the average ratio R=2.03±0.49 (standard deviation). In the right panel we have replaced the (d,^{3}He) values by the spectroscopic factors we obtain after a reanalysis of the (d,^{3}He) data in non-local finite-range (NL/FR) DWBA using the BSWF as deduced from the corresponding (e,e'p) experiment. From the indicated average value R=1.01±0.25 (standard deviation) we conclude that within the systematic uncertainties associated with the analysis of either reaction their results basically agree. This corroborates the finding in sect. 7, that for (medium) heavy nuclei the spectroscopic strength for proton knockout from valence orbitals is about 50% of the IPSM sum-rule limit.

9. CONCLUSION

From measurements of the spectral function of a number of (medium) heavy nuclei (^{40}Ca to ^{208}Pb) by the (e,e'p) reaction, we have obtained spectroscopic strengths for valence orbitals of about ΣS = 0.4-0.6 times the IPSM sum rule value. The (e,e'p) results agree with the results from the (d,^{3}He) reaction, provided that the same BSWF's are used in a non-local finite-range DWBA analysis. New tests of the reliability of the description of the (e,e'p) reaction mechanism show a small deviation from the Impulse Approximation, while the uncertainty due to the calculation of the Final-State Interaction was found to be $\leq$ 10 %. Given these uncertainties the observed fractional strength $\Sigma S/(2j+1)$ = 0.4-0.6 for valence orbitals agrees with predictions from nuclear matter calculations that include sizable short-range correlations and to which RPA correlations have been added in order to correct for finite size effects. For the purpose of the comparison between the (e,e'p) results and the nuclear-matter predictions it is essential that both the coupling to surface vibrations and the role of the quasi-particle tail be properly included. Evidently, for a more quantitative understanding of the structure of the spectral function there is a need for an approach in which these aspects are treated together in a consistent way for finite nuclei.

Acknowledgements

I like to thank Drs. C. Mahaux and H.P. Blok for stimulating discussions and Drs. E.N.M. Quint, G.J. Kramer, H.J. Bulten, J.B.J.M. Lanen for providing data and calculations. This work is part of the research program of the National Institute for Nuclear Physics and High-Energy Physics (NIKHEF) made possible by the financial support from the Foundation for Fundamental Research of Matter (FOM) and the Netherlands' Organization for the Advancement of Research (NWO).

References.

1. G. van der Steenhoven *et al.*, Nucl. Phys. **A480** (1988) 547.
2. J.W.A. den Herder *et al.*, Nucl. Phys. **A490** (1988) 507.
3. O. Benhar *et al.*, preprint INFN-ISS 89/2, and Proc. 4th Workshop on Perspectives in Nuclear Physics at Intermediate Energies, Trieste, 1989.
4. C. Mahaux and R. Sartor, Nucl. Phys. **A481** (1988) 381 and **A484** (1988) 205
5. G. van der Steenhoven *et al.*, Phys. Rev. Lett. **57** (1986) 182.
6. G. van der Steenhoven *et al.*, Phys. Rev. Lett. **58** (1987) 1727.
7. A. Magnon, Proc. 4th Workshop on Perspectives in Nuclear Physics at Intermediate Energies, Trieste, 1989.
8. L. Weinstein, Proc. 4th Workshop on Perspectives in Nuclear Physics at Intermediate Energies, Trieste, 1989.
9. D. Reffay-Pikeroen *et al.*, Phys. Rev. Lett. **60** (1988) 776.
10. R.W. Lourie, AIP Conference Proceedings **176**, p103, 1988.
11. H.P. Blok *et al.*, Phys. Lett. **B198** (1987) 4.
12. M. Bernheim *et al.,* Nucl. Phys. **A375** (1982)381
13. G. van der Steenhoven *et al.*, Phys. Rev. **C32** (1985) 1787 and Nucl. Phys. **A484** (1988) 445.
14. G. van der Steenhoven *et al.*, Phys. Lett. **B191** (1987) 227.
15. S. Frullani and J. Mougey, Adv. Nucl. Phys. **14** (1984) 1.
16. T. de Forest Jr, Nucl. Phys. **A329** (1983) 232.
17. P.J. Mulders, Nucl. Phys. **A459** (1986) 525 and Phys. Rep. in press.
18. P. Schwandt *et al.*, Phys. Rev. **C26** (1982) 55.
19. C. Giusti and F. Pacati, Nucl. Phys. **A473** (1987) 717.
20. L. von Bernus *et al.*, Proc. Top. Conf. on Problems of Vibrational Nuclei, Zagreb, 1974.

21. V.G. Soloviev, J. de Physique **45** (1984) C4, and private communication.
22. M.H. Harakeh, private communication.
23. E.N.M. Quint, PhD thesis, University of Amsterdam, 1988.
24. C. de Vries *et al.*, Nucl. Instr. & Meth. **223** (1984) 1.
25. G.E. Brown and M. Rho, Nucl. Phys. **A372** (1981) 397.
26. L. Lapikás, Proc. 4th Workshop on Perspectives in Nuclear Physics at Intermediate Energies, Trieste, 1989.
27. R. Sartor, Nucl. Phys. **A289** (1977) 329.
28. H. Orland and R. Schaeffer, Nucl. Phys. **A299** (1978) 442
29. R.W. Hasse *et al.*, Phys. Lett. **B181** (1986)5
30. P. Grange *et al.*, Nucl. Phys. **A473** (1987) 365.
31. V.R. Pandharipande *et al.*, Phys. Rev. Lett. **53** (1984) 1133
32. H.P. Blok, Proc. Int. Conf. on Nuclear Reaction Mechanism, Calcutta, 1989.
33. S.M. Banks *et al.*, Nucl. Phys. **A437** (1985) 381
34. N. Matsuoka *et al.*, Ann. Rep. RCNP Osaka, 1985, and private communication.
35. G.J. Kramer *et al.*, Nucl. Phys. **A477** (1988) 55.
36. A.J.C. Burghardt, PhD thesis, University of Amsterdam, 1989.
37. J.P. Jeukenne, A. Lejeune and C. Mahaux, Phys. Rep. **25C** (1976) 83.

Inst. Phys. Conf. Ser. No 105
Paper presented at Int. Conf. on Spectroscopy of Heavy Nuclei, Crete, Greece, 1989

200 GeV/A heavy ions at CERN: what has been learnt?

R. SANTO +

INSTITUT FÜR KERNPHYSIK, UNIVERSITÄT MÜNSTER,
4400 MÜNSTER, WILHELM-KLEMM-STR. 9, W.-GERMANY

ABSTRACT: Ultrarelativistic heavy ions became available at the end of 1986, when a ^{16}O beam was accelerated at the CERN SPS up to an energy of 200 GeV per nucleon. Three multipurpose and a number of specialized experiments have been performed in two runs in 1986/87 with ^{16}O and ^{32}S beams, respectively. Measured data on transverse energies and particle multiplicities provide evidence for the formation of a state of high energy density close to values predicted by lattice QCD calculations for a quark gluon phase transition. Different characteristic signals from this deconfined state are being investigated by different collaborations and are discussed.

+ WA80-collaboration, supported in part by BMFT

I. Introduction

At the end of 1986 the first heavy ion beam was accelerated in the CERN accelerator complex up to an energy of 200 GeV·A. This marked the opening of a new area of experimental research called "ultrarelativistic heavy ion reactions", related both to high energy physics as well as heavy ion reactions. The main aim of these experiments is the study of nuclear matter under extreme conditions in particular the search for a possible transition of normal hadronic matter to a deconfined state of quarks and gluons, sometimes called "quark gluon plasma"[1][2][3][4][5]. Unfortunately, this transition cannot be calculated by standard Quantum Chromodynamics (QCD) since the magnitude of the strong interaction within nuclear matter prevents a perturbative treatment in this regime. At present, the physics in this domain can only be treated by the so called "Lattice QCD", a statistical method invented by K. Wilson. Various calculations within this recipe have been performed in the last years and have, indeed, shown that a phase transition from ordinary nuclear matter to quark matter is expected to occur at critical energy densities of ε_{crit} = 2-4 GeV/fm^3 and critical temperatures of T_{crit} = 200-300 MeV. Although these energy densities are more than an order of magnitude higher than the energy density of groundstate nuclear matter, they are not completely out of the range accessible to energetic heavy ion reactions. Apart from the motivation in studying nuclear matter in its extremes and testing QCD in the strong interaction regime there is also a cosmological interest in the quark gluon plasma. According to current models, the Universe may, in the very first moments, have consisted of free quarks and gluons, which subsequently have condensed to the normal hadronic matter of today. Attempts to create the quark gluon plasma in the laboratory by a "little bang" will therefore shed some light on the "Big Bang".

When the heavy ion experiments started in 1986 with an ^{16}O beam of 3200 GeV total energy, nothing was known about the reaction mechanism at these energies apart from a few cosmic ray events. The experiments had therefore the difficult task of covering the expected physics as complete as possible, whilst at the same time remaining open and sensitive to the unknown.

II. Experimental Schemes and Setups

The relevant quantities used in the analysis and representation of the data are collected in table 1.

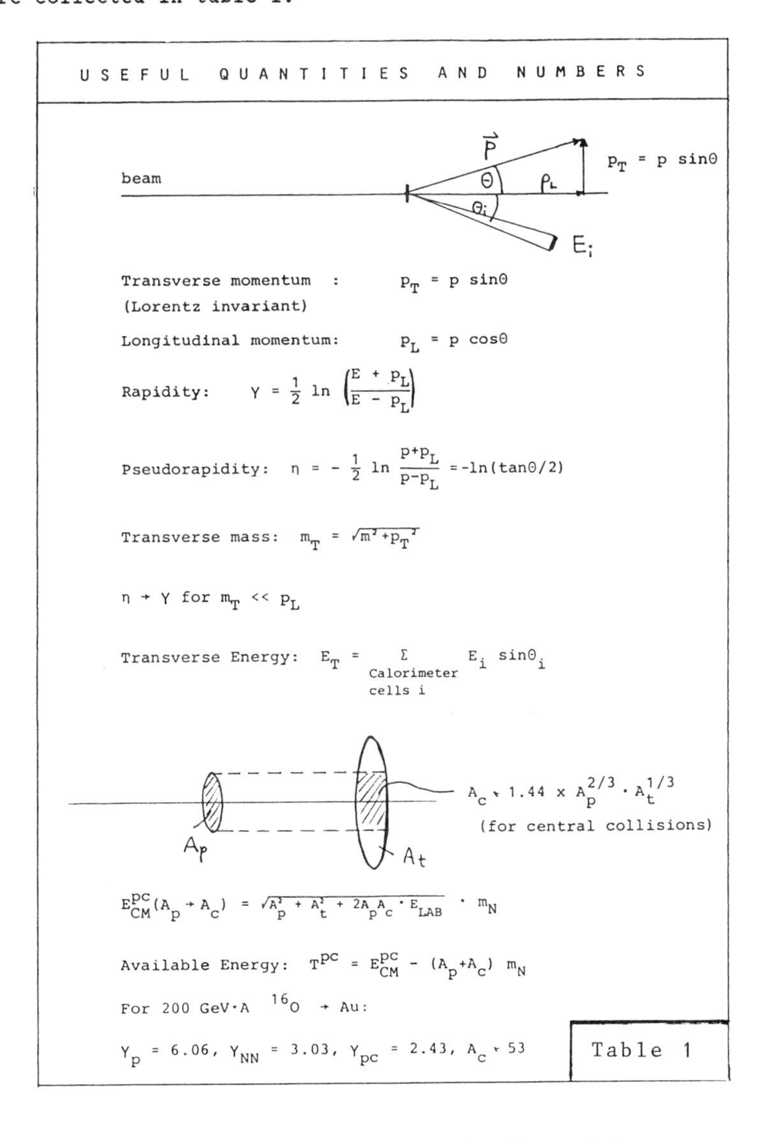

USEFUL QUANTITIES AND NUMBERS

beam; $\vec{P}$; Θ; p_L; Θ_i; E_i; $p_T = p \sin\Theta$

Transverse momentum (Lorentz invariant): $p_T = p \sin\Theta$

Longitudinal momentum: $p_L = p \cos\Theta$

Rapidity: $Y = \frac{1}{2} \ln \left(\frac{E + p_L}{E - p_L}\right)$

Pseudorapidity: $\eta = -\frac{1}{2} \ln \frac{p+p_L}{p-p_L} = -\ln(\tan\Theta/2)$

Transverse mass: $m_T = \sqrt{m^2 + p_T^2}$

$\eta \rightarrow Y$ for $m_T \ll p_L$

Transverse Energy: $E_T = \sum_{\text{Calorimeter cells } i} E_i \sin\Theta_i$

A_p; A_t; $A_c \approx 1.44 \times A_p^{2/3} \cdot A_t^{1/3}$ (for central collisions)

$E_{CM}^{pc}(A_p \rightarrow A_c) = \sqrt{A_p^2 + A_t^2 + 2A_pA_c \cdot E_{LAB}} \cdot m_N$

Available Energy: $T^{pc} = E_{CM}^{pc} - (A_p + A_c)\, m_N$

For 200 GeV·A $^{16}O \rightarrow Au$:

$Y_p = 6.06,\ Y_{NN} = 3.03,\ Y_{pc} = 2.43,\ A_c \approx 53$

Table 1

The scheme of the experimental arrangement for determining energy densities is shown in fig 1a. For the situation depicted in fig. 1b, the projectile suffers little interaction in the target, so E_{ZDC} is close to E_{beam} and, at the same time E_T and N_{ch} are small. This situation is expected to apply to peripheral reactions with large impact parameter.

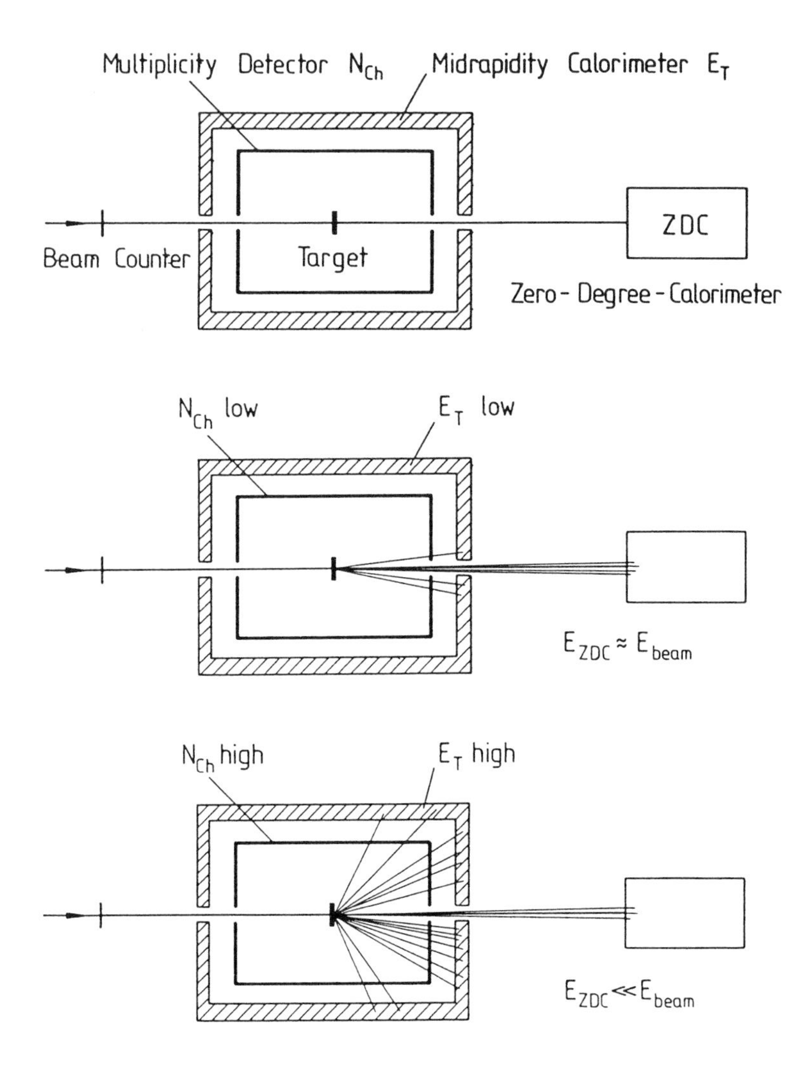

Fig. 1. a) Schematic sketch of experimental arrangement
b) Typical pattern for a peripheral reaction or high nuclear transparency
c) Typical pattern for a central collision with large stopping

It alternatively may, however, also apply to a central collision with large transparency. Fig. 1c describes the situation for a central collision with strong stopping of the projectile within nuclear matter. In this case E_T and N_{ch} are expected to be large but E_{ZDC} will be small compared to E_{beam}.

Fig. 2 shows the measured relation between N_{ch} and E_{ZDC} observed in the WA80 experiment and Fig. 3 gives a visual impression of a central event from the streamer chamber experiment NA35.

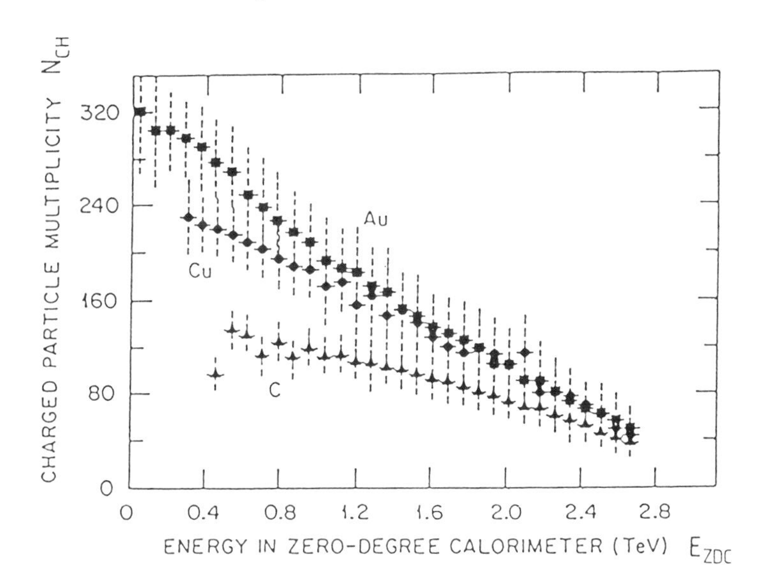

Fig. 2. Charged particle multiplicity N_{ch} versus energy in the zero-degree-calorimeter E_{ZDC} for ^{16}O on different targets at 200 GeV·A (measured by WA80)

Fig. 3. NA35 Streamer chamber picture of a ^{16}O + Pb collision

It has to be emphasized that the huge multiplicities emerging from heavy ion reactions constitute a tremendous experimental problem much different from standard elementary particle experiments, and require particular techniques. Fig. 4 gives a sketch of the three large multi-purpose experiments of the first generation. The experimental details and coverages of the detectors are listed in table 2.

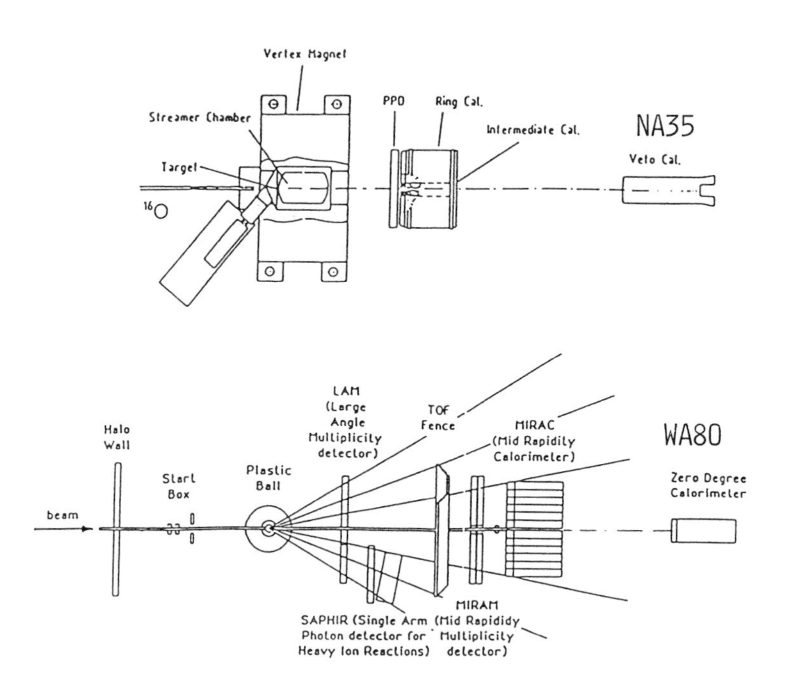

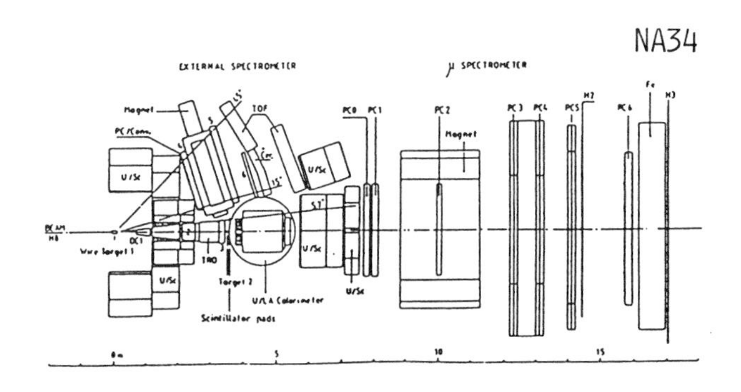

Fig. 4. Sketch of experimental setup for the multi-purpose experiments NA34, NA35 and WA80

EXPERIMENTS	D E T E C T O R S	ACCEPTANCES (FOR 200 GeV·A)
NA34 HELIOS	Uranium Calorimeter	$-0.1 < \eta < 5.5$
	Liquid Argon Calorimeter	
	Silicon-Multiplicity Detector	$0.9 < \eta < 4.9$
	External Magnetic Spectrometer	$0.9 < \eta < 2.0$
		$\Delta\Phi < 5°$
NA35	Ring Calorimeter	$2.08 < \eta < 3.72$
	Photon Position Detector (PDP)	$1.95 < \eta < 3.37$
	Zero Degree Calorimeter	$0 < \Theta < 0.3°$
		$5.9 < \eta$
	Streamer Chamber	$0.6 < Y < 4.6$
WA80	Midrapidity Calorimeter (MIRAC)	$2.4 < \eta < 5.5$
	Zero Degree Calorimeter (ZDC)	$0 < \Theta < 0.5$
		$5.9 < \eta$
	Midrapidity Multiplicity D. (MIRAM)	$2.17 < \eta < 4.74$
	Large Angle Multiplicity D. (LAM)	$1.3 < \eta < 2.4$
	Plastic Ball	$< \eta$ 1.31
		$30 < \Theta < 160°$
	Photon Leadglass Detector (SAPHIR)	$1.5 < \eta < 2.1$
		$\Delta\Phi < 20°$
NA38	EM-Calorimeter	$2 < \eta < 4.2$
	Dimuon-Spectrometer	$2.8 < \eta < 4.0$
		$(2 < M < 4$ GeV/$c^2)$

Table 2. Detector arrangements and acceptances for the NA34, NA35, NA38 and WA80 experiments

III. Energy Densities

Fig. 2 and 3 already indicate that in central heavy ion collisions a large amount of the longitudinal energy is converted into transverse energy associated with a large number of created particles. In all experiments it is found that N_{ch}, E_T and E_{ZDC} are strongly correlated and can individually be used for event characterization. This is also seen from fig. 5 and fig. 6 where transverse energy spectra[6] and multiplicity spectra[7] are shown, respectively. In both figures a strong A dependence is observed indicating appreciable - but not complete - stopping. For the most central events a scaling of E_T and N_{ch} is found with approximately $A_t^{1/3}$. This suggests a simple geometrical picture of the reaction where the projectile interacts only with the target nucleons in the geometrical overlap (see also table 1). With this simple assumption the maximum transferred energy and the energy density can be estimated. It is found

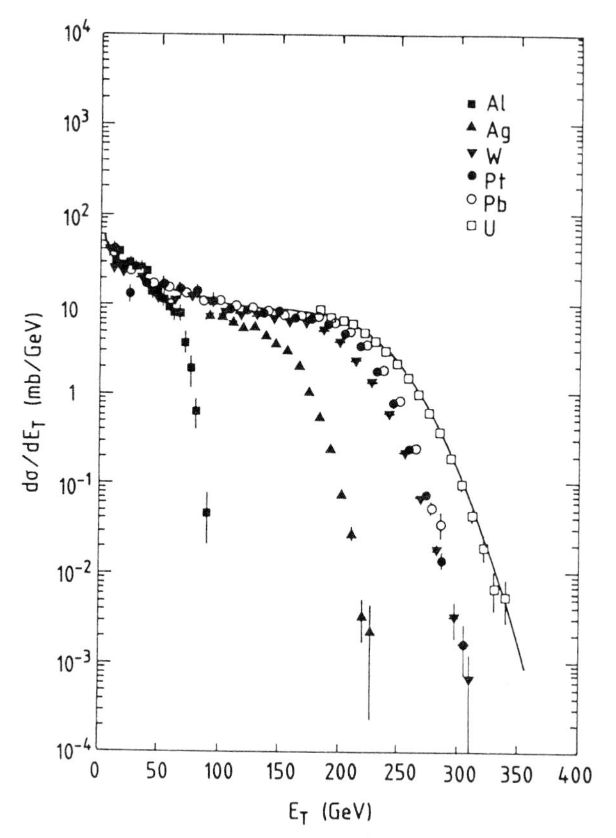

Fig. 5. Transverse energy spectra for ^{32}S on different targets at 200 GeV·A and $-0.1<\eta<2.9$ (measured by NA34)

that more than 50% of the projectile energy is converted into transverse energy. There are a number of other prescriptions for calculating the energy density from measured E_T and N_{ch}. Among these the Bjørken formula has widely been used, although it strictly applies to even higher energies. According to Bjørken[8], the energy density ε is given by

$$\varepsilon = \frac{1}{\tau_0 \pi R_p{}^2} \left(\frac{dE_T}{dy}\right)_{max}$$

where for (dE_T/dy) the measured value of $(dE_T/d\eta)$ is taken at midrapidity, πRp^2 is the cross section area of the projectile-target interaction and τ_0 is the formation time estimated to about 1 fm/c. For central collisions of ^{16}O projectiles with heavy targets all experiments (NA34, NA35, WA80) arrive at energy density estimates of ε = 2-3 GeV/fm^3.

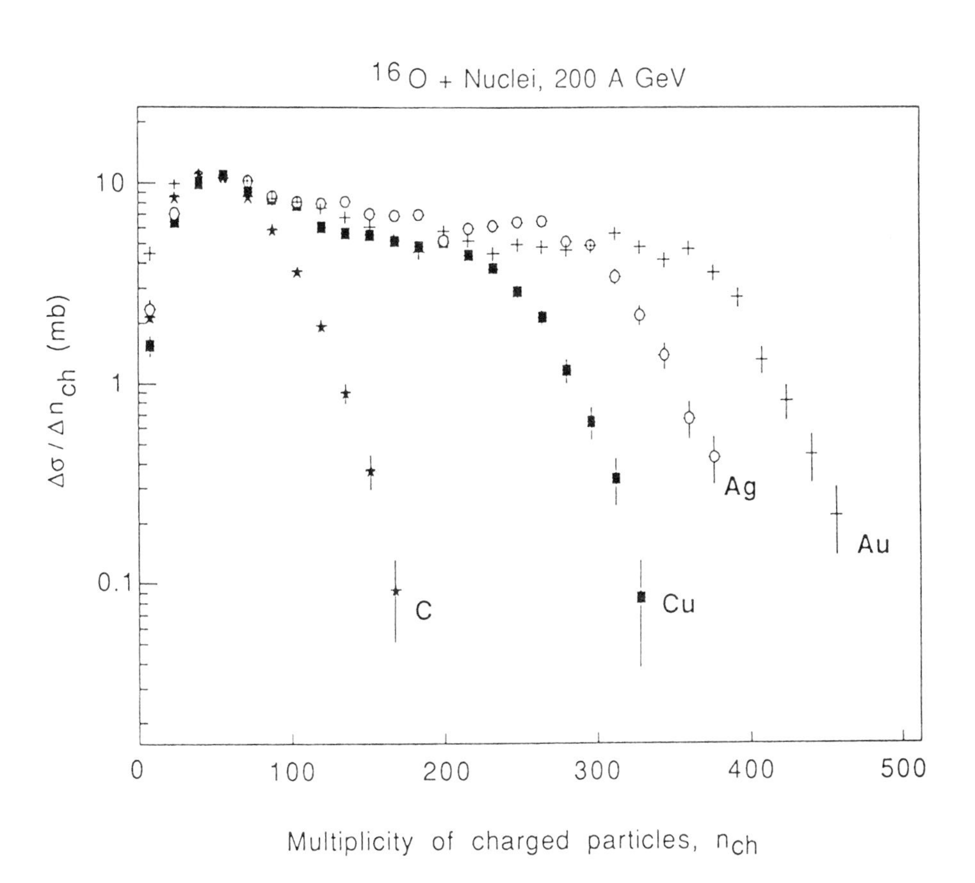

Fig. 6. Charged particle multiplicities from ^{16}O on different targets at 200 GeV·A and $-1.7<\eta<4.2$ (measured by WA80)

Fig. 7 shows an appreciable increase of the transverse energy[6] by increasing the projectile mass from ^{16}O to ^{32}S. Because of the larger cross section area the energy density is, however, only slightly increased.

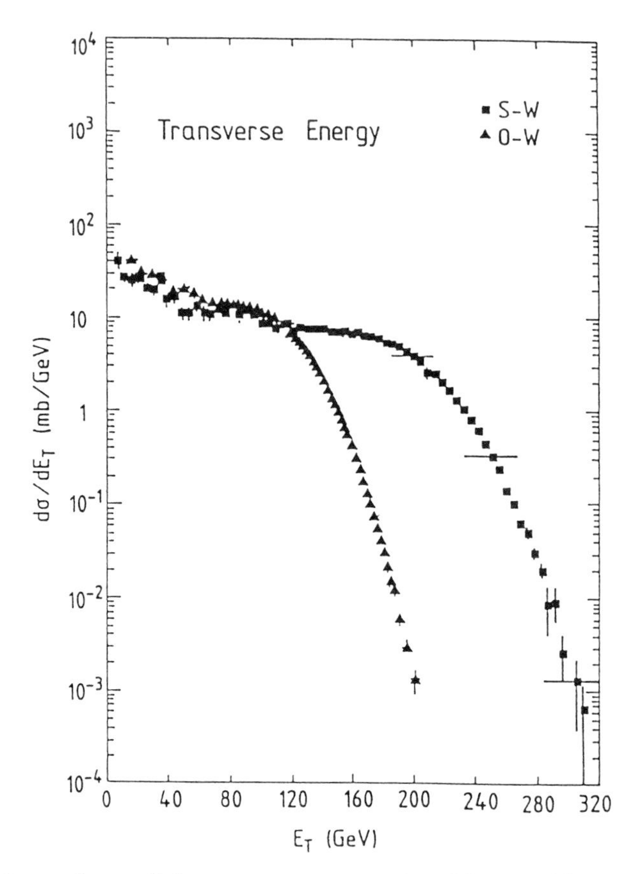

Fig. 7. Comparison of transverse energy for ^{16}O and ^{32}S projectile on a W target at 200 GeV·A (NA34)

The multiplicity and transverse energy measurements described above lead to the remarkable result, that the energy densities achieved in the first ultrarelativistic heavy ion experiments with 200 GeV·A ^{16}O and ^{32}S projectiles are close to the values predicted by lattice QCD for a quark gluon phase transition. This implies that a search for characteristic signals from this possible new phase is meaningful, in particular with heavier projectiles.

IV. Study of the Reaction Zone

After the existence of a region of high energy density has been established, the investigations now concentrate on determining the size and the thermodynamic state of the hot and compressed reaction zone. The NA35 collaboration has studied correlations[4] between charged pions in order to determine the size of the emitting source through the Hanbury-Brown-Twiss Effect. Fig. 8 shows correlation functions at different rapidities. Standard analysis leads to radii of the emitting source increasing from $R_T \approx 4.3$ fm (at $1<y<2$) to $R_T \approx 8.1$ fm (at $2<y<3$).

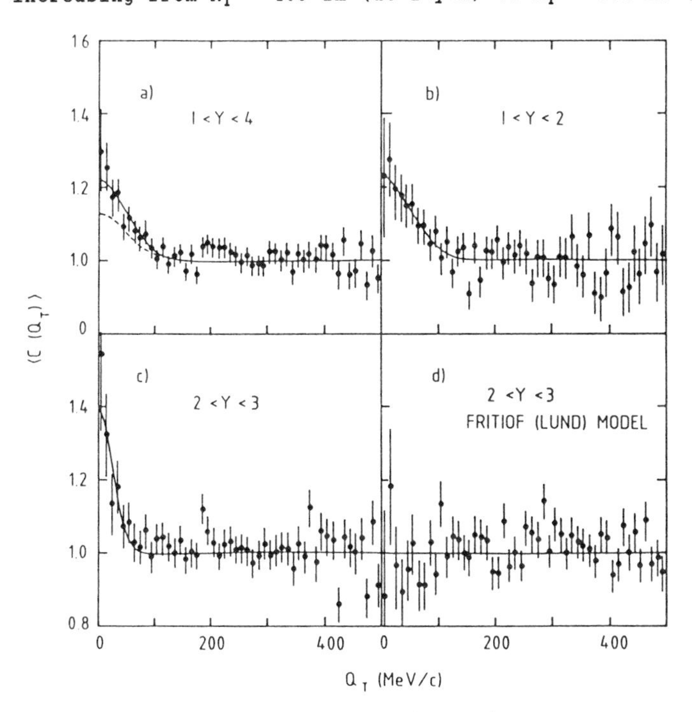

Fig. 8. Correlation function for two pions with transverse momentum difference Q_T together with projected Gaussian fits.

Transverse momentum spectra of charged and neutral pions have been measured by the NA34, NA35 and WA80 collaboration. The spectra display a roughly exponential behaviour with differences in the slopes between peripheral and central reactions. In thermodynamical models, where the particles in addition to their thermal motion gain momentum from the expanding source, this difference can be described. The detailed peripheral and central spectra of fig. 9, however, display a more complicated behaviour. While the peripheral data are in nearly perfect agreement with pion spectra from pp-collisions at comparable energies including the flattening of the slope for $p>1.8$ GeV/c, the central data

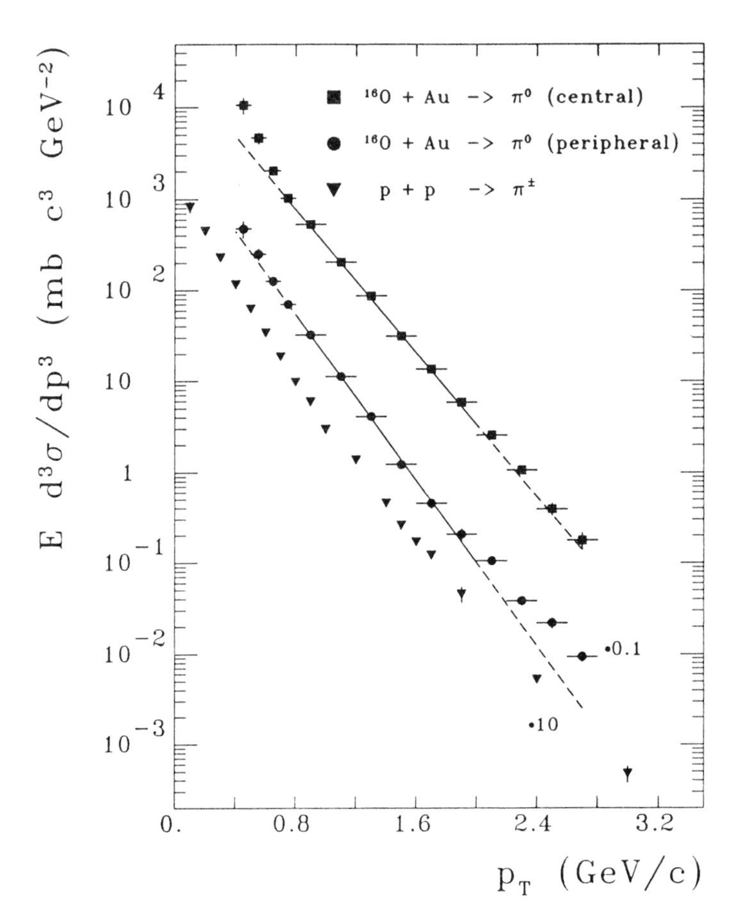

Fig. 9. Transverse momentum p_T spectra of π^0 from ^{16}O + Au collisions at 200 GeV·A (measured with the SAPHIR detector of WA80) compared with charged pion data from pp collisions at comparable energy of $\sqrt{S}$ = 23 GeV. The lines are exponential fits to the intermediate p_T range.

show characteristic deviations for low and high p_T not readily understood in current models or extrapolations of pp-collisions. It is hoped, that careful theoretical investigations of this behaviour may give information on the pecularities of the heavy ion compared to pp-reactions.

V. Direct Photons, Strangeness Production, J/ψ Suppression

Direct Photons and Leptons

A number of characteristic signals have been proposed as evidence for the formation of a quark gluon plasma. Unfortunately they all consist in an enhancement or suppression of signals already present in normal hadronic environments. A very accurate absolute determination of yields and a reliable comparison with the normal hadronic case is therefore required. Among the various signals the direct photons and leptons are the cleanest ones (perhaps also the most difficult to measure) since they leave the interaction zone without distortion by the hadronic shell. The WA80 has started the measurement of direct photons using a finely granulated lead glass detector (SAPHIR). The main problem here are the huge particle and photon multiplicities from the produced pions. By reconstructing the π^0 and η the photons from meson decays can be identified and compared with the measured total photon yield. Differences between peripheral and central reactions and to p induced data have been observed, but in the present state of the analysis the errors are still too large to draw definite conclusions. Similar findings have been made by NA34 with a different experimental setup and somewhat restricted range.

Strangeness Production

The idea here is that the s, $\bar{s}$ population may reach chemical equilibrium with the u, $\bar{u}$, d, $\bar{d}$ quarks in the plasma state, whereas no such

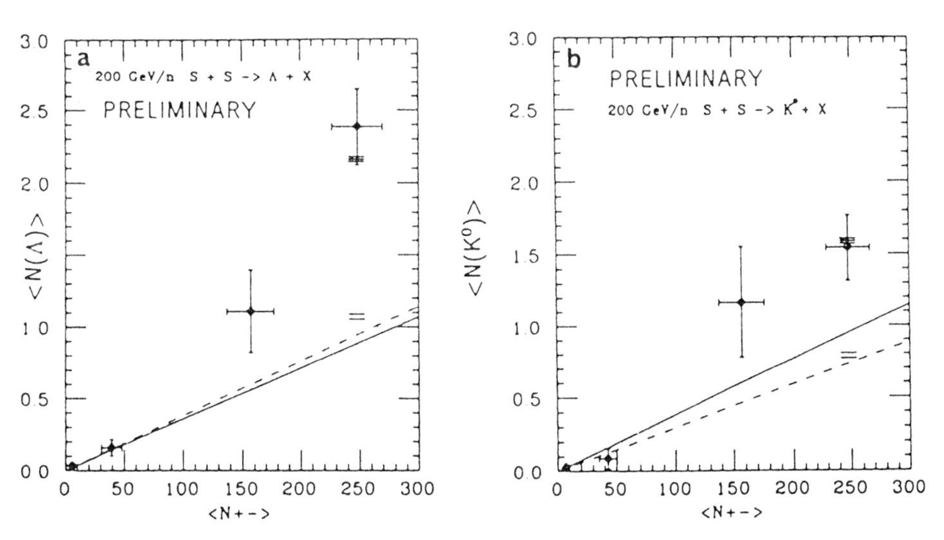

Fig. 10. Mean multiplicities of and K^0 particles as a function of the mean charged particle multiplicity for 200 GeV·A ^{32}S collisions (NA35). The solid lines are predictions of the Fritiof model and the dashed lines those of an independent nucleon-nucleon model.

equilibrium is known to exist in hadron-hadron interaction. The reason is the faster relaxation constant for the low mass $S\bar{S}$ pair in comparison to $K\bar{K}$ and pair creation time scales at typical temperatures around 200 MeV. With their streamer chamber the NA35 collaboration has concentrated on determining the strange particle yields[9] as a function of the total charged particle multiplicity. The preliminary data are shown in fig. 10. It seems that within the experimental acceptances an enhancement over trivial extrapolations from pp scattering is observed for the more central collisions.

J/ψ Production Suppression

This meson is supposed to be created from $c\bar{c}$ formation in the high energy density colour-magnetic "string" field. Initially close together in momentum space, the $c\bar{c}$ pair may stick together to form a J/ψ, or get rescattered into "open charm" channels D, D. Matsui and Satz have argued that the former channel is suppressed if the single string is embedded within closely packed adjacent strings, the typical configuration of a nuclear collision. If the bundle of strings (a pre-equilibrium state) develops into a quark-gluon-plasma, Debeye-screening should decrease the effective quark-quark interaction range to a degree where no hadrons of 1 fm typical size can be formed: the c, $\bar{c}$ quarks migrate uncaptured into bound J/ψ states, separating and ending up mostly in D, D mesons at hadronization. The result would be a suppression of the J/ψ yield in plasma creating reactions.

The CERN NA38 experiment has measured J/ψ production[5] in ^{16}O and ^{32}S collisions with U target nuclei. The experiment adapted a similar approach to the strangeness measurement: they compared peripheral with central nuclear collisions, observing the change in the ratio of J/ψ resonance to continuum Drell-Yan muon pair production. Ensembles of peripheral and central collisions were defined by mid-rapidity total transverse electromagnetic shower energy. Fig. 11 shows the NA38 result for 200 GeV·A ^{16}O + U peripheral collisions. The J/ψ peak in the di-muon invariant mass spectrum resides above the exponential Drell-Yan continuum. In comparing central to peripheral collisions, NA38 reports a suppression factor of 0.6 in the J/ψ to continuum signal. Unlike the direct photon and lepton pairs the signals from strangeness and charm production are strongly influenced

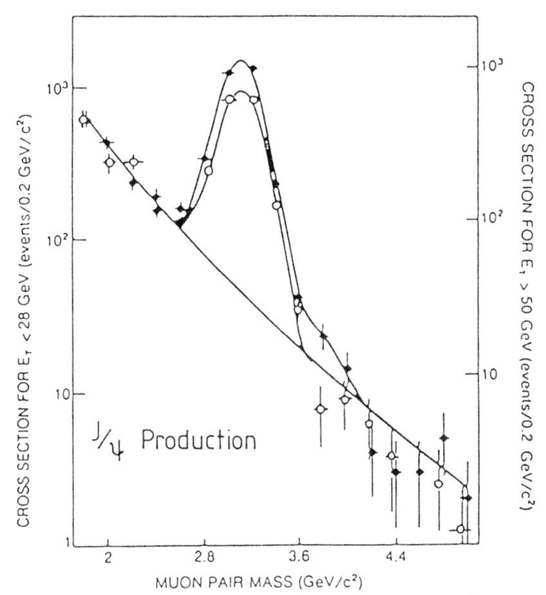

Fig. 11. Variation of J/ψ production as seen in the $\mu^+\mu^-$ invariant mass spectrum measured by the NA38 collaboration. A suppression of the more central collisions (open circles) is observed relative to the more peripheral ones (black dots). Note that the background is different in both cases and the data are normalized to the background.

by details of the hadronization phase of preequilibrium processes. Definite conclusions can therefore only be drawn when more detailed calculations of the dynamics are available.

VI. Summary and Outlook

The first series of ultrarelativistic heavy ion experiments at CERN have shown, that high energy densities are created in these collisions, which are close to values predicted by lattice QCD for a phase transition to quark matter. Some of the data - in particular p_T spectra - show distinct deviation from pp scattering which seem specific for the heavy ion systems.

Characteristic signals proposed for a quark gluon plasma phase transition have been investigated, but in the present state they all are not conclusive either by too large systematic and statistical errors, or by uncertainty in the hadronic reaction dynamics.

References

Most of the work on quark matter is published or cited in the proceedings of the Quark Matter Conferences. The experimenatl data from the CERN runs are contained in the 1987 and 1988 Conference proceedings.

1) Quark Matter 1983 (Brookhaven) Necl. Phys. A 418 (1984)
2) Quark Matter 1984 (Helsinki) Lecture Notes in Physics 221 (1984), Springer
3) Quark Matter 1986 (Asilomar) Nucl. Phys. A 461 (1987)
4) Quark Matter 1987 (Nordkirchen) Z. Phys. C 38 (1988)
5) Quark Matter 1988 (Lenox) Nucl. Phys. A (1989)
6) NA34-Helios collaboration, CERN-EP/88 - 121 preprint
7) WA80-collaboration, Phys. Lett. B 202 (1988) 596
8) I.D. Bjørken, Phys. Rev. D 27 (1983) 140
9) NA35-collaboration, Z. Phys. C 43 (1989) 25

Inst. Phys. Conf. Ser. No 105
Paper presented at Int. Conf. on Spectroscopy of Heavy Nuclei, Crete, Greece, 1989

Theoretical problems in quark–gluon plasma physics

N.G. ANTONIOU
Department of Physcis, University of Athens, GR-157 71,
Athens, Greece.

ABSTRACT

The standard theoretical proposals for the study of the formation, evolution and hadronization of the quark-gluon plasma in relativistic ion collisions are presented. The problems associated with the nature of the quark-hadron phase transition are emphasized and recent ideas about the dynamical origin of large fluctuations in the rapidity distributions are discussed in connection with the hypothesis that the hadronization mechanism of the quark-gluon plasma is a second-order phase transition.

I. Introduction

Quark-Gluon plasma physics is a new, rapidly developing field, in the interface on nuclear and particle physics, dealing with the macroscopic properties of complicated, dense systems of strongly interacting particles expected to get created in relativistic ion collisions. One of the main objectives of these experiments and the related theoretical studies is to reveal the mechanism of the quark-hadron phase transition which not only plays a fundamental role in the structure of quantum chromodynamics (QCD) at large distances but also it has been one of the building elements in the process of matter formation in the early cosmology (Alock 1989)

Philosophically, the attempt to discover the fundamental laws of nature by studying complicated structures (quark matter) in experiments with complicated systems (relativistic heavy-ion collisions) must be regarded as a proposal to abandon the conventional reductionism in physics, a method which, for example, cannot be applied if one wants to study the properties of the vacuum, the most fundamental but also one of the most complicated physical systems. In fact, the studies with relativistic ion collisions probe the structure of the non-pertarbative QCD vacuum and this shows the significance of these experiments since the two outstanding fundamental problems in particle physics (a) the missing symmetry and (b) the colour confinement, are related to the properties of the vacuum. We believe that there is a new challenge in the field of quark-gluon plasma physics, not only because in the experiments with relativistic ions we expect to discover a new state of matter but also because we may establish the remarkable fact that the microscopic system of elementary particles is not a self-contained unity but it depends strongly on the coherent properties of the macroscopic world (Lee 1981).

II. Quantum Chromodynamics

The elementary constituents of the hadronic world (quarks, gluons) interact according to the fundamental QCD Lagrangian which is invariant under local SU(3) gauge transformations, specifying the colour properties of the theory, and has the following structure:

$$\mathcal{L} = -\frac{1}{4} F^{a}_{\mu\nu} F^{\mu\nu}_{a} - \sum_{f} \bar{\psi}^{f}_{a} (i\not{\partial} - g\not{A})^{ab} \psi^{f}_{b} \tag{1}$$

where

$$F^{a}_{\mu\nu} = \partial_\mu A^{a}_{\nu} - \partial_\nu A^{a}_{\mu} - g f^{a}_{bc} A^{b}_{\mu} A^{c}_{\nu} \tag{2}$$

In eqs. (1) and (2) the Dirac spinors $\psi^{f}{}_{a}$ are associated with the quarks and the gauge fields $A^{a}{}_{\mu}$ with the gluons. There is a summation in the flavour index f and a contraction in the colour indices (a,b..). The quantities $f^{a}{}_{bc}$ are the structure constants of the gauge group. There are are two important properties of the system governed by the Lagragian (1) which are relevant to the relativistic ion physics: (1) At small distances the colour interaction becomes very weak (asymptotic freedom) and (2) at large distances the QCD force becomes very strong and it prohibits the liberation of the coloured particles (colour confinement). In the experiments with relativistic ions one may exploit the property of asymptotic freedom in order to create quark matter in a hadronic environment of very high density and temperature. On the order hand studying the behaviour of hadrons at the end of the hadronization period, near the critical temperature ($T<200$ MeV), one may trace the characteristics of the quark-hadron phase transition which is the macroscopic realization of the colour confinement mechanism. For this purpose one has to examine the thermal properties of QCD introducing the partition function for the thermodynamic quark-gluon system. We have:

$$Z = \mathrm{Tr}\left[\exp\left(\frac{H-\mu N}{T}\right)\right] \tag{3}$$

where H(L) is the Hamiltonian of the system as determined by the QCD Lagrangian density L. The energy and net baryon number densities (ε, ρ_b) are calculated with the help of eq. (3) as follows:

$$\varepsilon = \frac{T^2}{V}\left(\frac{\partial \ln Z}{\partial T}\right)_{\mu,V} \qquad \rho_b = \frac{T}{V}\left(\frac{\partial \ln Z}{\partial \mu}\right)_{T,V} \tag{4}$$

At high temperatures the energy density has a Stefan-Boltzmann behavior (asymptotic freedom) which in a theory without quarks corresponds to the density of a relativistic gluon gas:

$$\varepsilon_{SB} = \frac{37\pi^2}{30} T^4 \tag{5}$$

At lower temperatures and especially in the region T=100-200 MeV we are facing with the difficulties of the non-perturbative sector of the theory (QCD) and we have to solve three outstanding problems in connection with the quark-gluon plasma and its relation to the ordinary hadronic matter:

(1) The existence and the nature of the quark-hadron phase transition,

(2) the existence and the nature of the chiral symmetry breaking phase transition and (3) the relation of the critical temperatures (T_c, T_{ch}) of these critical phenomena.

For this purpose an extensive program of QCD studies on a lattice has been developed according to the Wilson's proposal (Wilson 1974). In the case of a pure gauge theory, the partition function Z_G defined on a space-time lattice is given by the form:

$$Z_G = \int \prod_{\ell} \mathcal{D}U_\ell \; \exp\left[\frac{1}{g^2}\sum_P \mathrm{Tr}\,(U_P + U_P^\dagger)\right] \tag{6}$$

where the summation runs over the space-time plaquettes P of the lattice and U_p denotes the product of the group elements along the four links of the plaquette as follows:

$$U_P = U_\mu(x)\, U_\nu(x+\mu)\, U_\mu^\dagger(x+\nu)\, U_\nu^\dagger(x) \tag{7}$$

For the confinement-deconfinement phase transition, the order parameter of the system is the Wilson loop average <W(B)>, associated with the contour (B) and given by the equations: $W_B = \Pi_B\, U_{1B}$

$$\langle W_B \rangle = \frac{1}{Z} \int \prod_{\ell} [dU_\ell]\; W_B \exp\left[\frac{1}{g^2}\sum_P \mathrm{Tr}\,(U_P + U_P^\dagger)\right] \tag{8}$$

The quantity $\langle W_B \rangle$ is related with the potential energy V(R) of two static quarks separated by a large distance R. We have:

$$V(R) \sim \ln \langle W_B \rangle \tag{9}$$

and the condition for confinement is $\langle W_B \rangle = 0$. In fact the confinement-deconfinement phase transition is established by the following

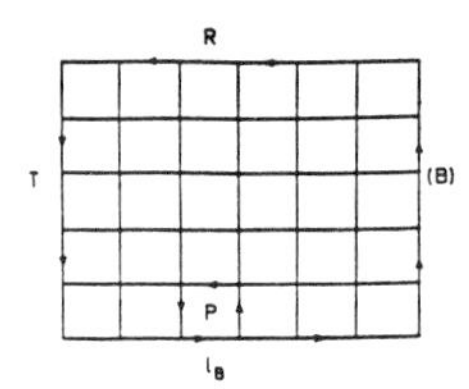

Fig. 1

behaviour of the order parameter:

$$\langle W_B \rangle = 0 \quad , \quad T < T_c$$
$$\langle W_B \rangle \neq 0 \quad , \quad T > T_c \qquad (10)$$

On the other hand the complete QCD Lagrangian (with quarks and gluons) in γ_5-invariant ($\Psi \to \gamma_5 \Psi$) and therefore the quark mass must be generated by a spontaneous chiral symmetry breaking mechanism which leads to on additional phase transition from the quark-gluon plasma with massless quarks to a system with massive quarks at approximately the same critical temperature, T_{ch} T_c. The order parameter for this critical phenomenon is the chiral condensate $<\psi\psi>$ which at $T=T_{ch}$ is expected to have the following behaviour.

$$\langle \bar{\psi}\psi \rangle = 0 \quad , \quad T > T_{ch}$$
$$\langle \bar{\psi}\psi \rangle \neq 0 \quad , \quad T < T_{ch} \qquad (11)$$

Finally, in order to study the nature of the confinement-deconfinement phase transition, the behaviour of the correlation length near the critical temperature is needed. For this purpose it is suitable to use the Polyakov loop $P(x,y,z)$ which, like the Wilson loop, is a gauge invariant operator, defined on the lattice by imposing periodic boundary conditions along the time direction instead of closing the the contour (fig. 1). With the help of this operator, the correlation legnth z may be defined through the equation:

$$C(z) \equiv \sum_{x,y} \left\langle \frac{1}{L_x L_y} P(x,y,z) \right\rangle = A e^{-z/\xi} + B \qquad (12)$$

where A,B are constants. In the actual computations (Monte Carlo) the size of the lattice is chosen $L_x \otimes L_y \otimes L_z \otimes L_t$ with $L_t << L_x = L_y << L_z$ in order to be able to study the asymptotic behaviour (z) of C(z) near the critical temperature.

Given this theoretical background, an extensive computational program with Monte Carlo simulation has been performed during the last few years and the main results of these studies which are relevant to quark-gluon plasma physics can be summarized, for the periods 1983-1984 and 1988-1989 as follows. The early Monte Carlo computations (1983-1984) suggest that both the confinement-deconfinement and the chiral breaking phase transitions are of first-order (Satz 1985) with approximately equal critical temperatures at a value T_c . On the other hand, recent Monte Carlo computations (1988-1989) with somewhat larger lattices show that the confinement-deconfinement critical behaviour, in a pure SU(3) gauge theory, is in agreement with a

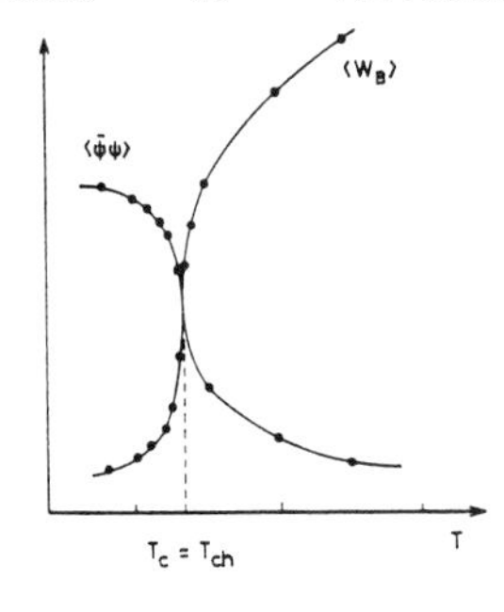

Fig. 2a

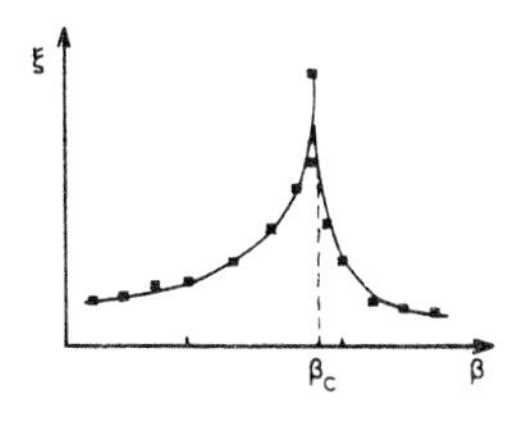

Fig. 2b

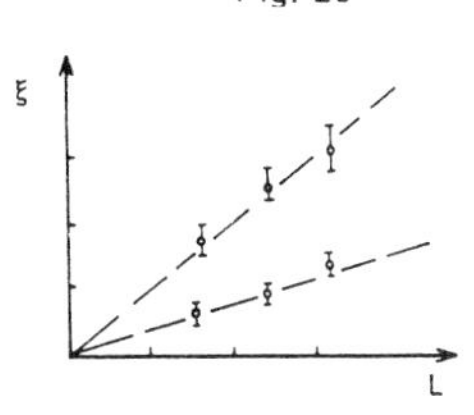

Fig. 2c

continuous transition (Bacilieri et al. 1988) where as the inclusion of fermions in a four flavour theory (Gavai et al. 1989) leads to a smaller transition temperature, T_C 100MeV. In fig. (2) we show a typical behaviour of the order parameters $\langle W_B \rangle$ and $\langle \bar{\psi}\psi \rangle$ corresponding to a first-order phase transition (fig. 2a) and the behaviour of the correlation length near the critical temperature suggesting a second-order phase transition (fig. 2b,2c).

III. Ultra-relativistic nuclear collisions

In order to study the thermodynamic properties of the system produced in a highly relativistic nuclear collision, the following correspondence of the basic "macroscopic" quantities with the observables of the experiments is a necessary, useful tool:

(1) The temperature corresponds to the slope of P_t-distributions

(2) The energy density is the total transverse energy of the produced particles per volume

(3) The entropy density is the particle multiplicity per volume.

(4) The chemical potentials correspond to the flavour content of the produced system (baryon number, strangeness) per volume.

The energy density created in a relativistic nuclear collision is a characteristic quantity for the onset of the deconfinement mechanism and it can be easily expressed in terms of the multiplicities produced localy in the process (Bjorken 1983).

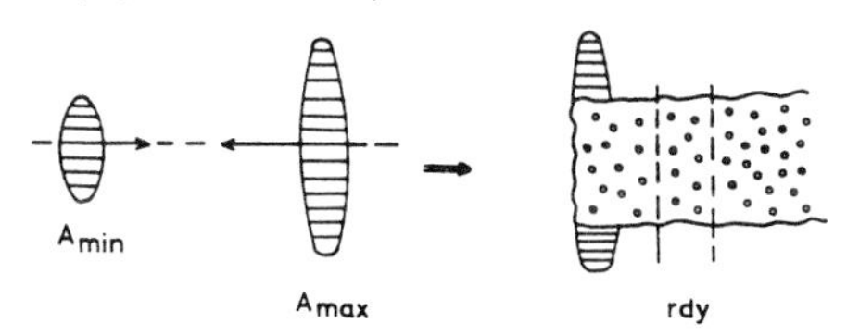

Fig. 3

In fig. (3) the collision of two nuclei ($R_{min}=R_0 A_{min}^{1/3}, R_{max}=R_0 A_{max}^{1/3}$) is shown schematically and the energy density corresponding to the volume element $dV_t=\pi R_{min}^2 \tau dy$ (dy is the rapidity element and τ is the proper time of the corresponding matter element) is given by the Bjorken formula:

$$\varepsilon_\tau = \frac{3}{2} \frac{\langle m_T \rangle}{\pi R_0^2 A_{min}^{2/3}} \frac{1}{\tau} \left(\frac{dn}{dy}\right)_{ch} \qquad (13)$$

where the rapidity density of charged particles is involved and

$m_T=(p_T^2+m^2)^{1/2}$. The energy density (13) created in a collision with relativistic ions has to be compared with the normal nuclear (ε_A) and hadronic (ε_N) densities given by the equations:

$$\varepsilon_A = \frac{m_A}{\frac{4}{3}\pi R_A^3} \simeq 0.13 \text{ GeV fm}^{-3}$$
$$\varepsilon_N = \frac{m_N}{\frac{4}{3}\pi R_N^3} \simeq 0.45 \text{ GeV fm}^{-3} \qquad (14)$$

From this comparison one estimates that the quark matter regime may be easily reached if the energy density produced is sufficiently higher than the value corresponding to the normal hadronic matter, namely if ε_T>2-3GeV fm^{-3} for τ=1fm.

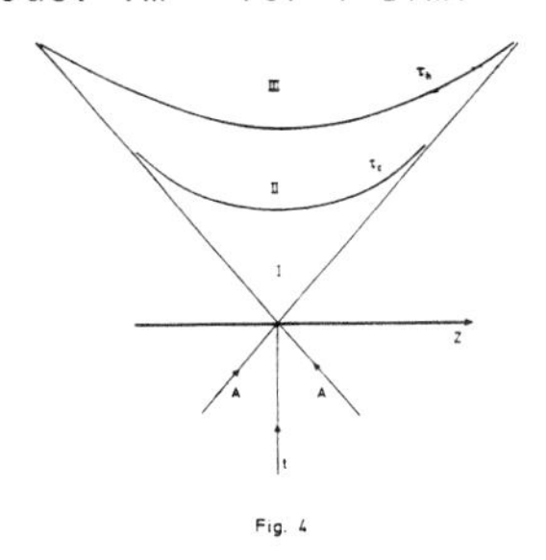

Fig. 4

The new phenomena which are expected to emerge, due to the extreme conditions imposed on the hadronic matter in the experiments with relativistic ions, depend strongly on the space-time evolution of the system which is assumed to follow the geometry of the space-time hyperbolas, $c^2t^2-z^2=c^2\tau^2$, during the longitudinal expansion of the quark-gluon plasma and the critical period related to the hadronization process (fig. 4). In the region (I) of the space-time ($0<\tau<\tau_c$) the temperature is high and the quark-gluon system behaves most of the time as a perfect relativistic fluid with energy density $\varepsilon\sim T^4$. When the temperature of the system approaches the critical value $T=T_c$, the confining forces start to operate and the hadronization process takes place during the period $\tau_c<\tau<\tau_h$ (region II). Finally for $\tau>>\tau_h$ (region III) the temperature is relatively low and a fully hadronized system appears which is organized in the central region as a perfect pion gas with low energy density. We also note that if the quark-hadron phase transition is of first-order, the temperature remains constant, $T=T_c$, during the hadronization period whereas if the transition is smooth (second-order) the temperature decreases continuously for $\tau_c<\tau<\tau_h$.

On the basis of the above discussion we distinguish two kinds of phenomena and related observables in connection with the new physics expected in these processes: (a) The effects which have their origin in the space-time region (I) and provide us with signals for the formation of quark-gluon plasma and (b) the effects which are related to the hadronization region (II) and are expected to reveal the nature and the parameters of the quark-hadron phase transition. In the first case the study of the following phenomena has been proposed:

(1) The production of single photons and dileptons
(2) The production of strangeness
(3) The J/ψ suppression

In the second case, in order to study the quark-hadron phase transition, the following studies, related to particle production, are of particular interest:

(1) The dependence of the transverse momentum on the multiplicity
(2) The pion interferometry

(3) Fluctuations in the rapidity space and intermittency.

IV. Signals of the quark-gluon plasma

4.1 The production of single photons and dileptons

In the region I (fig. 4) due to the high temperature of the quark-gluon plasma, the perturbation theory in QCD is, in principle, applicable and the production of direct photons may be studied with the help of the diagrams in fig. (5) where one has to add also the crossed terms. The production rate of the single photons computed within the perturbative approach has the following temperature dependence (Sinha 1983, Hwa et al. 1985):

$$\Gamma(T) \sim \alpha\alpha_s T^4 \ln\left(\frac{T}{m_q}\right) \tag{15}$$

Fig. 5

where a and a_S are the QED and QCD coupling constants respectively. Theoretically, the appearence of a mass singularity in the expression (15) requires a non-perturbative approach in order to obtain a more adequate estimate. Experimentally, on the other hand, although the single photons produced may convey the signal of the quark-gluon plasma formation, since they are not affected by the hadronic environment, their detection is somewhat problematic because of the considerable noise expected to be present mainly due to the high multiplicity of π^0 in the hadronic gas.

For this reason one has to study also the virtual photons which lead to dileptons with a production rate given in terms of the thermal distribution of the quarks-antiquarks in the plasma. On the contrary the dilepton production at zero temperature depends on the quark-antiquark structure functions inside the interacting nuclei (Drell-Van mechanism). The cross section of the Drell-Van process is given by the following expression based on the QCD parton model (fig. 6):

$$\frac{d\sigma_{DY}}{dM^2 dy} = \frac{4\pi\alpha^2}{9M^4}\sum_q e_q^2\left[G_{q/A}(x_1,M^2)\,G_{\bar{q}/B}(x_2,M^2)\right] + (q\leftrightarrow\bar{q}) \tag{16}$$

where x_1, x_2 are the longitudinal kinematical variables of the partons (quarks) and M is the invariant mass of the dilepton pair. The thermal emission rate of dileptons from the quark phase is given by the following simplified expression:

$$\frac{dN}{d^4x\, dM^2\, dy\, d^2p_T} = \frac{\alpha^2}{8\pi^4}\sum_q e_q^2 \exp\left(-\frac{E}{T}\right) \tag{17}$$

Finally, one has to include also the emission of dileptons from the hadronic phase through the conventional mechanism of p-dominance in the process $\pi^+\pi^-\rightarrow l^+l^-$ which takes place in the pion gas. One easily finds the corresponding emission rate:

$$\frac{dN_h}{d^4x\, dM^2\, dy\, d^2p_T} = \frac{\alpha^2}{8\pi^4}\,\frac{1}{12}\,F^2(M)\, e^{-E/T} \tag{18}$$

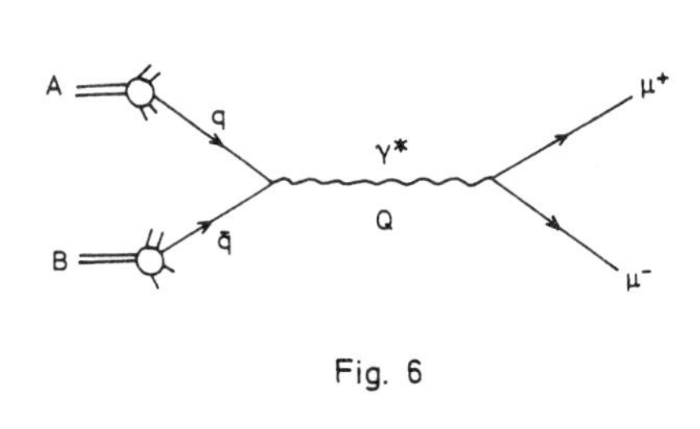

Fig. 6

where $F(M^2)$ is the pion form factor dominated by the ρ-pole.
Although the extensive study of the above mechanisms (eqs.16,17,18) has shown (Kajentie et al.) that there are kinematical regions in which the thermal emission of dileptons from the quark phase may be singled out, one must admit that the complications due to the hadronic environment, create difficulties for the detection of the dilepton signal of the quark-gluon plasma.

4.2 The production of Strangeness

The proposal that the increase of strange particle production in relativistic ion collisions is a signal for the formation of quark-gluon plasma (Koch et al. 1986)is based on the following two observations:

(a) In the quark phase there are six (6) degrees of freedom associated with the strange quark (two degrees correspond to the spin and three to the colour) whereas in the hadronic phase there are partically two (2) degrees of freedom (K^- and K^0) since the heavier strange particles are thermodynamically suppressed.

(b) The mass of the strange quark (m_s=180MeV) is considerably smaller compared to the mass of the K-meson (m_s=500MeV).

Whith these qualitative arguments one expects that, at a given temperature, the density of strange quarks in the quark phase is larger than the density of K-mesons in the hadronic phase. Moreover if we consider the baryonic content of the quark matter produced, we may estimate the chemical potential for the light quarks (u,d) in a baryon rich region of the phase space, according to the condition:

$$\mu_u = \mu_d \equiv \mu \sim 300 \text{ MeV} \qquad \left(\mu \simeq \frac{m_N}{3}\right) \tag{19a}$$

On the other hand for the antiquarks we have the relations:

$$\mu_{\bar{u}} = \mu_{\bar{d}} = -\mu \qquad \mu_s = \mu_{\bar{s}} = 0 \tag{19b}$$

With these conditions the strange quark (s) and light antiquark (u,d), densities are given by the following simplified forms:

$$\frac{n_s}{V} = \frac{n_{\bar{s}}}{V} = 6\int \frac{d^3p}{(2\pi)^3} \exp\left[-\frac{(p^2+m_s^2)^{1/2}}{T}\right] \approx 3\,\frac{T m_s^2}{\pi^2} K_2\left(\frac{m_s}{T}\right) \tag{20a}$$

$$\frac{n_{\bar{u}}}{V} = \frac{n_{\bar{d}}}{V} = 6\int \frac{d^3p}{(2\pi)^3} \exp\left(-\frac{p+\mu}{T}\right) \approx \frac{6T^3}{\pi^2} \exp\left(-\frac{\mu}{T}\right) \tag{20b}$$

where $K_2(z) \cong (\pi/2z)^{1/2} e^{-z}$ is a modified Bessel function. From eqs. (20) we finally obtain, for $T = m_s$

$$\frac{n_{\bar{s}}}{n_{\bar{q}}} \simeq \frac{1}{4} \left(\frac{\pi}{2}\right)^{1/2} \exp\left(\frac{\mu - m_s}{T}\right) \qquad (21)$$

where $n_{\bar{q}} = n_{\bar{u}} + n_{\bar{d}}$.

In the two extreme cases $\mu=0$ and $\mu=300MeV$ we obtain $n_{\bar{s}}/n_{\bar{q}}=1/9$ and 1/2 respectively, whereas strangeness production in the ordinary hadronic process gives $1/20 < n_{\bar{s}}/n_{\bar{q}} < 1/4$. It is therefore suggestive that the search of strangeness in the baryon rich region of the phase space is most appropriate for the study of the quark-gluon plasma but experimentally the realization of this proposal is very difficult. For this reason a detailed study, beyond the simple estimate (21), of the strangeness abundance in the central region (μ=0) is needed taking into account the evolution of the entropy during the longitudinal expansion of the system.

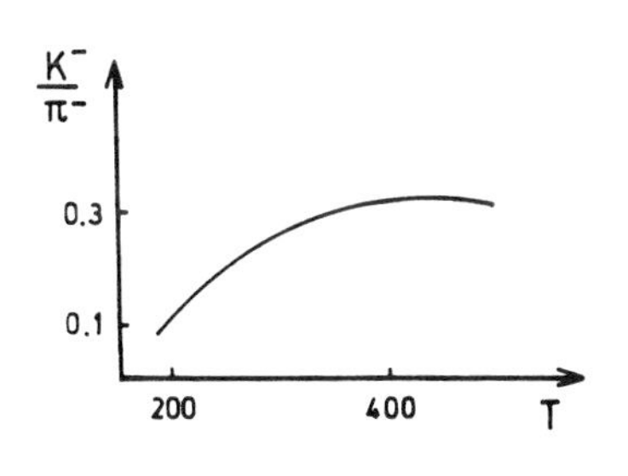

Fig. 7

Adopting the Bjorken model for the evolution of the quark-gluon plasma one may show (Kapusta et al.) that the ratio K^-/π^- in the central region is higher than the expected level ($K^-/\pi^-=0.08$) from the production of strangeness in the ordinary hadronic process (fig. 7). This study suggests that the strangeness abundance in the central region may continue to offer an indication for quark-gluon plasma formation despite the fact that the chemical potential is very low in this region.

4.3 The J/ψ suppression

The formation of heavy quark bound states in the environment of a hot quark-gluon plasma depends on the behaviour of the interquark potential at high temperatures. The phenomenological, non relativistic potential at zero temperature for the J/ψ bound state, has the form:

$$V(r) = -\frac{\alpha}{r} + \sigma r \quad (\alpha \simeq 0.5) \qquad (22)$$

At high temperatures ($T>T_C$) the linear term does not surrive and the Coulomb potential is modified by a screening mechanism which introduces a Debye length $r_D(T)$ in the interaction as follows:

$$V(r) = -\frac{\tilde{\alpha}}{r} \exp\left(-\frac{r}{r_D}\right) \qquad (23)$$

The study of the bound state problem in the quark system cc with the potential (23) suggests that the J/ψ particle formation is prohibited if the Debye length $r_D(T)$ becomes smaller than a typical radius r_ψ characteristic of the size of the bound state (Matsui et al. 1986). On the other hand, using perturbation theory of lowest order in QCD one obtains (Karch 1989):

$$\frac{1}{r_D^2} = 4\pi\alpha_s T^2 \left(1 + \frac{N_f}{6}\right) \tag{24}$$

where N_f is the number of flavours. For typical values of the parameters involved we find that for T>200MeV the formation of quark-gluon plasma leads to a suppression mechanism for J/ψ production in relativistic heavy ion collisions.

It is very encouraging that this phenomenon has been observed in one of the CERN experiments (NA 38) but a critical study has shown that, due to the complexity of the hadronic environment, there exist other competing mechanisms which lead to the same effect and therefore the J/ψ suppression in relativistic ion collisions cannot be alone a decisive signal for quark-gluon plasma creation in these reactions.

V. Quark-Hadron phase transition

Altough the study of the phenomena related to the early period of the quark-gluon plasma evolution ($T>T_c$) in relativistic ion collisions is of great importance, the theoretical charification and the experimental verification of the quark-hadron phase transition in these processes has a fundamental significance not only because it is related with the confinement mechanism but also because the nature of this phase transition affects in a critical manner our understanding of the process of nucleosynthesis in the early history of the universe. On the other hand our theoretical ideas about the connection of the quark-hadron phase transition with the multiparticle production pattern in relativistic ion collisions depend strongly on our basic assumption about the order of the transition. In what follows we discuss some of these ideas and the related phenomena either in the case of a first or in the case of a second-order phase transition.

5.1 Transverse momentum and multiplicity

In a first-order phase transition, the entropy density (s) of the system decreases while the temperature remains constant at the critical value $T=T_c$ and the system changes from the quark phase ($T>T_c$) to the hadronic phase ($T<T_c$) as shown in fig. (8). The characteristic entropy-temperature relation near the critical point implies (Van Hove 1982) a similar dependence of the multiplicity density on the average transverse momentum in a hadronic experiment, due to the analogy $dn/dy \sim s$, $\langle p_T\rangle \sim T$. The proposal now is that such a behaviour (fig. 9) is a signal of a first-order quark-hadron phase transition and moreover the plateau value $\langle p_T\rangle_c$ of the transverse momentum is a

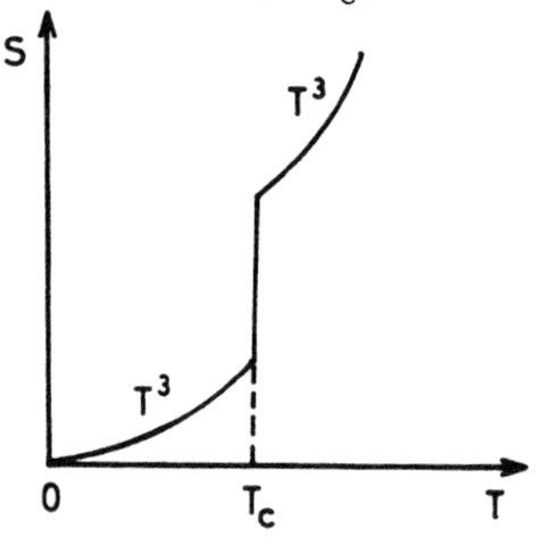

Fig. 8

measure of the critical temperature (T_c).

Altough there exist some interesting indications for a second rise of $\langle p_T \rangle$ as a function of multiplicity in experiments with cosmic rays (Yamashita et al. 1984) and also in a recent collider experiment (Alexopoulos et al. 1988) one has to take seriously into account the occurence of specific QCD mechanisms (minijets, high multiplicity process in QCD) which give rise to the same effect and which may also be considered as a possible explanation of this behaviour (Wang et al. 1989, Ryskin 1988). Moreover if the quark-hadron phase transition is of second order, the above dependence of $\langle p_T \rangle$ on the multiplicity density is no longer valid and one has to establish new criteria for the detection of a higher order quark-hadron phase transition in the experiments with relativistic ions.

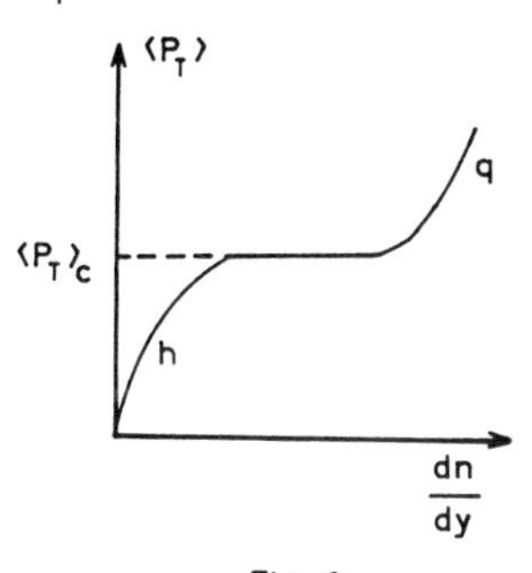

Fig. 9

5.2 Pion interferometry and quark-hadron phase transition

Pions emitted from random sources are correlated in their two-particle momentum distribution due to Bose-Einstein interference. The correlation function is defined as follows:

$$R(k_1, k_2) = \frac{P(k_1, k_2)}{P(k_1)\,P(k_2)} \qquad (25)$$

where $P(k_1, k_2)$ and $P(k)$ are the two-particle and one-particle inclusive densities, respectively. In a simple model with point-like sources distributed according to the density $\rho(x,t)$, the correlation function (25) measures the Fourier transform of the source density distribution:

$$R(k_1, k_2) = 1 + |\rho(\vec{k}_1 - \vec{k}_2, \omega_1 - \omega_2)|^2 \qquad (26)$$

One important question regarding the quark-hadron phase transition is to study its consequences on the pion interferometry. This problem has been examined by comparing the source of a pion gas without phase transition and with energy density (Pratt 1986):

$$\rho_0(r) = \rho_0 \left(1 - \frac{r^2}{R^2}\right) \quad (R \simeq 8\,\mathrm{fms}) \qquad (27)$$

with the bag model representing a first-order phase transition with energy densities for the two phases:

$$\rho_h = \frac{\pi^2}{10} T^4 \quad , \quad \rho_q = \frac{37\pi^2}{30} T^4 + B \qquad (28)$$

The results illustrated in fig. (10) correspond to the kinematical constraint $k_1+k_2=0$ and

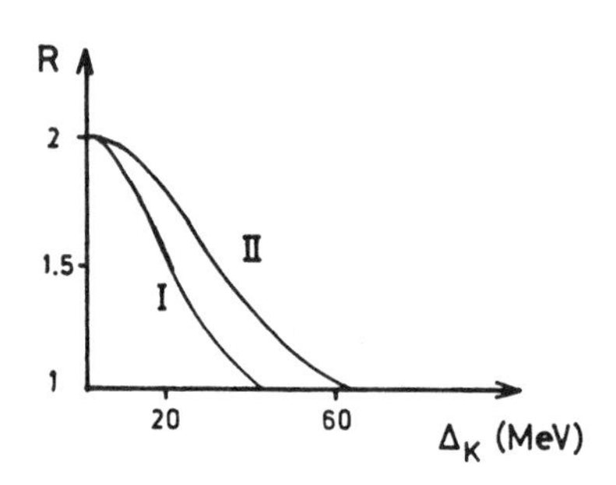

Fig. 10

show that the effect of the phase transition is appreciable leading to a larger correlation (curve II) than the expanding free gas of pions (curve I) at least in a restricted domain of the phase space. This study suggests that an extensive experimental program on pion interferometry in relativistic ion collisions is needed in order to measure systematically these effects, whereas new theoretical ideas about the consequences of a second-order phase transition on the pion interferometry must also be developed.

5.3 Fluctuations-Intermittency

Motivated by a series of recent QCD investigations on a lattice, which show that the confinement-deconfinement phase transition is smooth (Bacilieri et al. 1988) and guided by several recent phenomenological studies on the moments of the multiplicity distributions (Kittel 1989) which reveal a remarkable pattern of power laws in decreasing domains of the longitudinal phase space (intermittency) one is oriented towards the following proposal for the hadronization mechanism of the quark-gluon plasma:

(a) The quark-hadron phase transition is of second-order

(b) A realization of the quark-hadron phase transition is obtained through the critical behaviour of the one-dimensional Feynmann-Wilson fluid in the rapidity space.

With these assumptions one may study the rapidity distribution of hadrons at the final stage of the hadronization process by exploring the universal properties of the Feynmann-Wilson fluid near the critical point ($z=1, T \gtrsim T_c$ or $z=1, T=T_c$). For this purpose we may use two different but complementary approaches in order to charify the physical content of the method:

(a) The Ginzburg-Landau approach in the limit $z=1$, $T=T_c$

(b) The Kadanoff approach in the limit $z \gtrsim 1$, $T=T_c$

In the Ginzburg-Landau model, the effective action has the following form:

$$\Gamma[\sigma] = \int_{-\Delta/2}^{+\Delta/2} dy \left[\alpha \left(\frac{d\sigma}{dy}\right)^2 + \beta\sigma^2 + \gamma\sigma^4 \right] \qquad (29)$$

where $\sigma(y)$ is a stochastic field (Scalapino et al. 1973) associated with the hadron density $\rho(y)$ in the rapidity space, Δ is the size of the system and α, β, γ are temperature dependent coefficients with the condition $\beta(T_c)=0$ imposed by the criticality of the system. The inclusive density is given by the equation:

$$\rho(y) \sim \langle \sigma^2 \rangle = \frac{1}{Z} \int [d\sigma]\, \sigma^2 \exp\left[-\frac{1}{T}\Gamma[\sigma]\right] \qquad (30)$$

where Z is the partition function

$$Z = \int [d\sigma] \; \exp\left[-\frac{1}{T}\Gamma(\sigma)\right] \tag{31'}$$

In order to reveal the critical character of the system it is sufficient to consider only the classical solutions $\sigma_c(y)$ with appropriate boundary conditions. We have:

$$\left.\frac{\delta\Gamma[\sigma]}{\delta\sigma}\right|_{\sigma=\sigma_c} = 0 \quad , \quad \sigma_c\left(-\frac{\Delta}{2}\right) = \sigma_c\left(\frac{\Delta}{2}\right) = 0 \tag{32}$$

From eqs. (32) we obtain a discrete set of solutions:

$$\sigma_c(y;m) \sim sn\left[\sqrt{\frac{-\beta/\alpha}{(1+m)}}\left(y+\frac{\Delta}{2}\right); m\right] \tag{33}$$

where $sn(u;m)$ is the Jacobi sinus elliptic function and the index m belongs to a finite set $\{m\}$ in the domain $0<m<1$.

Summing now the contributions of the solutions (33) we obtain for the inclusive density (Antoniou et al. 1989a).

$$\rho_c(y) = \langle\sigma^2(y)\rangle = \frac{\sum_{\{m\}} \sigma_c^2(y;m)\exp\left[-\frac{1}{T}\Gamma(m)\right]}{\sum_{\{m\}} \exp\left[-\frac{1}{T}\Gamma(m)\right]} \tag{34}$$

where the spectrum $\Gamma(m)=\Gamma[\sigma_c(y;m)]$ is given by eq. (29).

It is interesting to notice that the irregularities in the rapidity space generated by the pattern (33) of the classical solutions give rise to dynamical fluctuations in the density of particles $N^{(m)}(y)$ for any one event of the multihadron production process, as can be easily verified from the equation $N^{(m)}(y)=\sigma_c^2(y;m)$ near the critical point $(T\text{-}T_c, \Delta\text{-}o)$. These inhomogeneities are expected to survive in the inclusive distributions although the averaging procedure (34) has the tendency to smooth-out the inclusive density $\rho_c(y)$.

We see that the appearence of density fluctuations in the rapidity distributions is a natural consequence of the critical behaviour of the Feynman-Wilson fluid and therefore one expects to observe such non-statistical fluctuations in the experiments with relativistic ions if the quark-hadron phase transition is of second-order.

Moreover, in order to be able to compare the Ginzburg-Landau model with the experiments, one has to fix the temperature dependent coefficients α,β,γ. A natural choice is to adapt these coefficients to the Gross-Neven model which offers a suitable microscopic theory for the quark system in the two-dimensional space-time domain (fig. 4) of its evolution. With this extra assumption, the confinement-deconfinement critical temperature T_c and the chiral breaking critical temperature T_{ch}, satisfy the following equation (Antoniou et al. 1989a):

$$1 + \frac{(\tau_c T_c)^2}{\xi}\ln\left(\frac{T_c}{T_{ch}}\right) = 0 \tag{35}$$

In eq. (35) τ_c=1fm and $\xi=7\zeta(3)/8\pi^2$ where $\zeta(x)$ is the Riemann zeta function. For a typical value T_{ch}=200MeV we find T_c=180MeV and this is an encouraging result since it shows that T_c is smaller but very close to the chiral breaking critical value T_{ch} as expected in any dynamical study of the phase transitions in the quark matter.

The second, complementary but somewhat more general approach (Antoniou et al. 1984) to the problem of density fluctuations in the critical Feynman-Wilson fluid is to impose Kadanoff scaling on the grand-canonical partition function $Q(z,\Delta)$ near the critical point ($T=T_c, z\to 1$):

$$\ln Q(z,\Delta) = h\left[(z-1)^{\frac{\kappa}{\kappa-1}} \Delta \right] \quad (\Delta \to \infty, z \to 1) \qquad (36)$$

The index $\kappa>1$ is a critical exponent which specifies the behaviour of the order parameter (density) ρ as a function of the ordering field (pressure) p, near the critical point:

$$\rho - \rho_c = (p - p_c)^{1/\kappa} \qquad (37)$$

It is remarkable that the scaling property (36) leads to a simple, factorizable solution for the rapidity distributions given by the iteration of the exclusive kernel $K_c(y)$ which has a Laplace transform with the following structure:

$$k_c(\xi) = \exp\left(-b\xi^{1-\eta}\right) \quad \left(b>0, \eta = \frac{1}{\kappa}\right) \qquad (38)$$

The physical consequences of the self-similarity property (36) on the hadronic phase near the critical point are (a) the validity of KNO scaling suggesting long range rapidity correlations and (b) the intermittent behaviour of the multiplicity distribution suggesting the occurence of strong fluctuations in very small domains of the longitudinal phase space. In fact with the kernel (38) one obtains:

$$\langle N \rangle \frac{Z_N(\Delta)}{Q(\Delta)} = \beta x^{\kappa/2} \exp\left(-\alpha x^{\kappa}\right) \qquad (39)$$

where α,β are constants, $Z_N(\Delta)$ is the canonical partition function, $x=N/\langle n\rangle$ and $Q(\Delta)=Q(z=1,\Delta)$.

With eq. (39) KNO scaling is established and the corresponding scaling function is determined. On the other hand, in this theory, the Bialas-Peshanski moment (Bialas et al. 1986) F_p have an intermittent behaviour expressed by a power law in decreasing rapidity intervals δ (Antoniou et al. 1989b)

$$F_p(\delta;\Delta) = p! \frac{\left[\Gamma(1-\eta)\right]^{p-1}}{\Gamma\left[(p+1)(1-\eta)+2\eta\right]} \left(\frac{4\delta}{\Delta}\right)^{-\eta(p-1)} \qquad (40)$$

where $\delta/\Delta \ll 1$.

The behaviour (40) corresponds to a single fractal dimension $d_F=1-\eta$ and this pattern of the normalized factorial moments is expected to emerge from the multiplicity distribution in relativistic ion collisions, if a second-order quark-hadron phase transition occurs in these processes.

References

Alcock C. et al. Proc. of the 7th Int. Conference of Ultra-Relativistic nucleus-nucleus collisions, Nuclear Physics 498A, 301 (1989)

Alexopoulos T. et al. Phys. Rev. Letters 60, 1622 (1988)

Antoniou N.G., Karanikas A.I., Vlassopoulos S.D.P. Phys. Rev. 29D, 1470 (1984)

Antoniou N.G., Karanikas A.I., Papadopoulos C.G.: "The Gross-Neveu model and quark-hadron phase transition" preprint, University of Athens (1989a)

Antonioy N.G., Argyres E.N., Papadopoulos C.G., Vlassopoulos S.D.P. private communication (1989b)

Bacilieri P. et al. Physical Review Letters 61, 1545 (1988)

Bialas A. and Peshanski R. Nucl. Physics B273, 703 (1986)

Bjorken J.D. Physical Review D27, 140 (1983)

Gavai R.V. et al. CERN preprint TH. 5530/89

Hwu R. and Kajentie K. Physical Review D32, 1109 (1985)

Kajentie K. et al. Physical Review D34, 811 (1986)

Kapusta K. et al. Physical Review D33, 1304 (1986)

Karch F., "The QCD plasma Physics" CERN preprint TH. 5279/1989

Kittel W., Proc. of 24th Int. Conf. On High Energy Physics, eds. R. Kotthaus and J.H. Kuhn, Springer-Verlag, p. 625 (1989)

Koch P. et al. Phys. Rep. 142, 167 (1986)

Lee T.D. Proc. of an International Symposium on Statistical Mechanics of quarks and hadrons, ed. H. Satz, North. Holland, p. 3 (1981)

Matsui T. and Satz H. Physics Letters 178B, 416 (1986)

Pratt S. Physical Review D33, 1314 (1986)

Ryskin M.G. Physics Letters B208, 303 (1988)

Satz H. Ann. Rev. Nucl. Science 35 (1985)

Scalapino D.J. and Sugar R.L. Physical Review D8, 2284 (1973)

Sinha B. Physics Letters 128B, 91 (1983)

Van Hove L. Physics Letters 118B, 138 (1982)

Wang X.N. and R.C. Hwa Phys. Rev. D39, 187 (1989)

Wilson K.G. Phys. Rev. D10, 2445 (1974)

Yamashita et. al. Proc. of the International Symposium On Cosmic rays and particle physics, p. 30, Tokyo 1984

Inst. Phys. Conf. Ser. No 105
Paper presented at Int. Conf. on Spectroscopy of Heavy Nuclei, Crete, Greece, 1989

Polarized ^{23}Na-induced reactions with ^{208}Pb at 8 MeV/nucleon

O Karban, C O Blyth, H D Choi, J B A England, S J Hall, S Roman and G Tungate
University of Birmingham, Birmingham B15 2TT, UK
K A Connell, M A Nagarajan and N Rowley
SERC Daresbury Laboratory, Daresbury, Warrington WA4 4AD, UK
N J Davis and A C Shotter
Department of Physics, University of Edinburgh, UK
S E Darden
Department of Physics, University of Notre Dame, USA

It has been established in numerous experimental studies that nuclear deformation plays an important role in heavy ion inreractions. One of the most direct methods to investigate these effects is an application of polarized beams where the spin and thereby the mass alignment can be chosen at the source. The present experiment was performed with beams of tensor and vector polarized ^{23}Na interacting with the ^{208}Pb target at 170 and 196 MeV to study the four analysing powers in quasi-elastic[1]) and reaction channels[2]).

The experimental data for the quasi-elastic process are shown in fig.1 where it can be seen (solid lines) that the three second-rank tensor analysing powers are not independent but follow the so-called "shape effect" relations

$$T_{2q}(\theta) = - 2(4/5\pi)^{1/2} Y_{2q}(\mathbf{r}_o)^{T} T_{20}(\theta), \quad q = 0,1,2$$

These relations, first observed in several experiments with aligned ^{7}Li beams, have been recently derived theoretically by Gomez-Camacho and Johnson[3]) who introduced the isocentrifugal approximation and assumed momentum independence of the coupling interaction (the Tidal Symmetry model). In this approach, the coupled equations for the elastic scattering can be decoupled and the experiemntal quantities calculated. The dashed lines in fig.1 show such predictions using phenomenological potential derived from fits to the cross section data and a deformation length $\beta R = 1.40$ fm for the projectile. The apparent disagreement with the $^{T}T_{20}$ and $^{T}{}_{21}$ data is a consequence of neglecting the inelastic scattering contribution (0.44 MeV, $J^{\pi} = 5/2^{+}$ state of ^{23}Na).

Applying the adiabatic model[4]) and assuming the tidal symmetry to be valid also for the inelastic processes[3]) the equations become fully decoupled in the tidal spin space, Predictions of the combined effect (dotted lines in fig.1) are now in good agreement with the experiment. Furthemore, in the tidal symmetry approach the vector analysing power iT_{11} vanishes and this feature is also supported by the present data.

An alternative method to determine the nucleus-nucleus interaction potential is to use the double-folding model for a deformed projectile and spherical target[5])

$$V(\mathbf{r},\mathbf{r}\cdot\mathbf{P}) = \int \rho_P(\mathbf{r}_1,\mathbf{r}_1.\mathbf{P})\ \rho_T(\mathbf{r}_2)\ v(\mathbf{r}+\mathbf{r}_2-\mathbf{r}_1)\ d\mathbf{r}_1 d\mathbf{r}_2$$

where ρ_P, ρ_T are the projectile and target matter distribution and v denotes the effective nucleon-nucleon interaction.The monopole and quadrupole terms in ρ_P are deduced from fits to electron scattering data[6]) and the finite range M3Y interaction was used to calculate the above integral with a complex normalisation factor N as a free parameter. A search on the latter resulted in Re(N) = 0.996 and Im(N) = 0.78 and the corresponding fits are shown in fig.2. It is evident that the quasi-elastic data are consistent with the known spectroscopic quadrupole moment of ^{23}Na and the interaction potential deduced from the M3Y interaction.

The experimental arrangement also allowed to obtain information on cross sections, $^{T}T_{20}$ and T_{20} analysing powers for the 2<Z<13 ejectiles for the ^{23}Na + ^{208}Pb system. The character of the kinetic energy spectra changes from those peaked around the beam velocity for Ne, Na and Mg to relatively featureless distributions for lighter ejectiles.

The energy-integrated differential cross sections are presented in fig.3, showing forward-peaked distributions for Z<7 while a side peak begins to develop for Z>8. The latter feature has been interpreted[6]) as evidence for quasi-elastic processes dominating the reaction mechanism close to the grazing angle for ejectiles with A near to the projectile mass.

The angular distributions for the two analysing powers were also aberaged over the ejectile energy and are shown in fig.3 (smooth lines were drawn through the data points). For Z<5 the $^{T}T_{20}$ and T_{20} values are almost zero (only Li data shown) but large negative effects were observed for 6<Z<9. The enhanced cross section for m = $|1/2|$ substates of ^{23}Na is consistent with a prolate deformation of the projectile where for the same trajectory, the projectile-target overlap increases by $\Delta R \sim Q(^{23}\mathrm{Na})$. However, to explain the systematic change of sign in T_{20} around 50^0 it would be necessary to introduce some quantitativedescription if the collision process.

References

1) O. Karban et al., J.Phys.G: Nucl.Phys. **14** (1988) L261

2) O. Karban et al., Daresbury Ann.Rep. 1989 (to be published)

3) J.Gomez-Camacho and R.C.Johnson, J.Phys.G:Nucl.Phys. **14** (1988) 609

4) H.D.Choi et al., J.Phys.G:Nucl.Phys. **15** (1989) 457

5) K.-H. Mobius, Z.Phys. **A310** (1983) 159

6) Ch.Egelhaaf et al., Nucl.Phys. **A405** (1983) 397

H.Takai et al., Phys.Rev. **C38** (1988) 1247

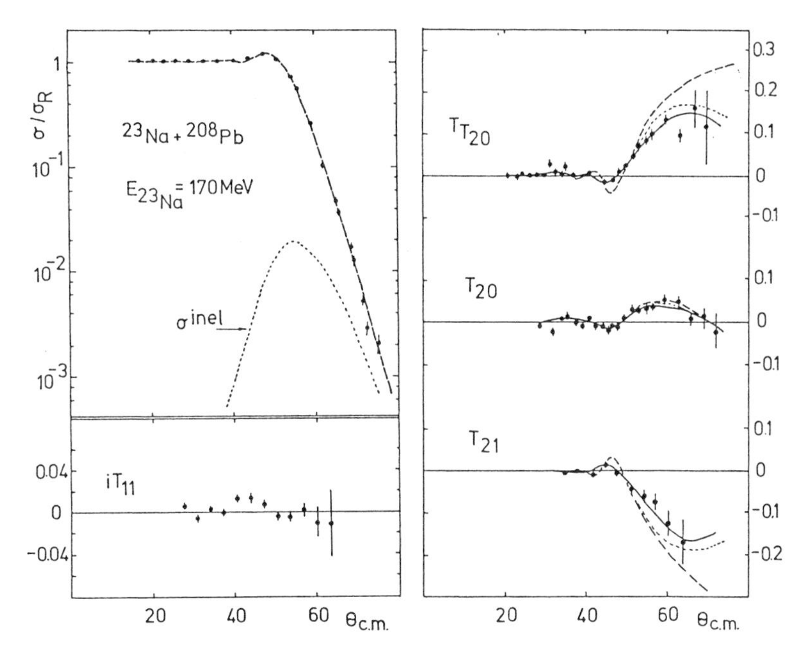

Fig.1. The $^{23}Na + ^{208}Pb$ scattering data at 170 MeV compared with the shape-effect relations(solid lines) and predictions of the tidal symmetry model for elastic scattering (dashed lines) and quasi-elastic scattering (dotted lines).

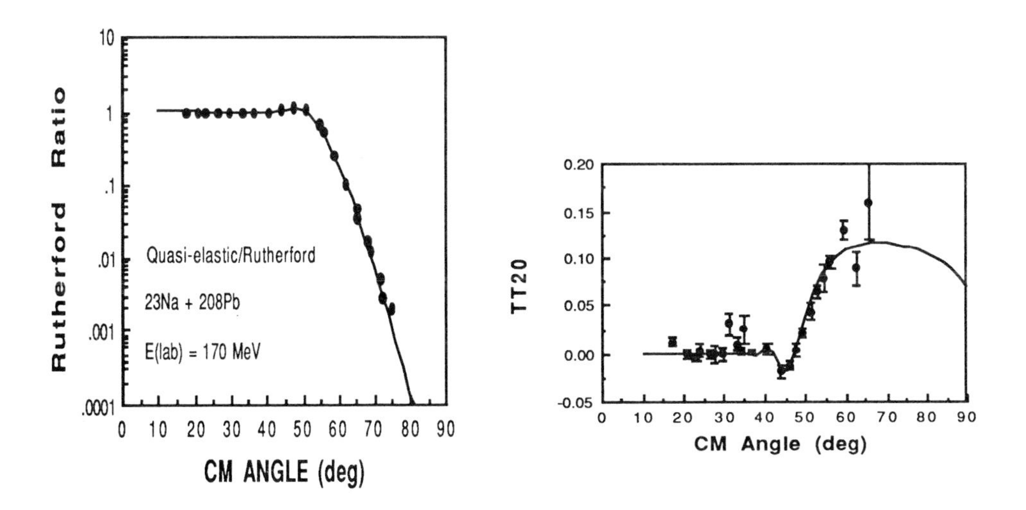

Fig.2. The calculated quasi-elastic cross section and tensor analysing power $^{T}T_{20}$ using the double folding model, compared with the experimental data.

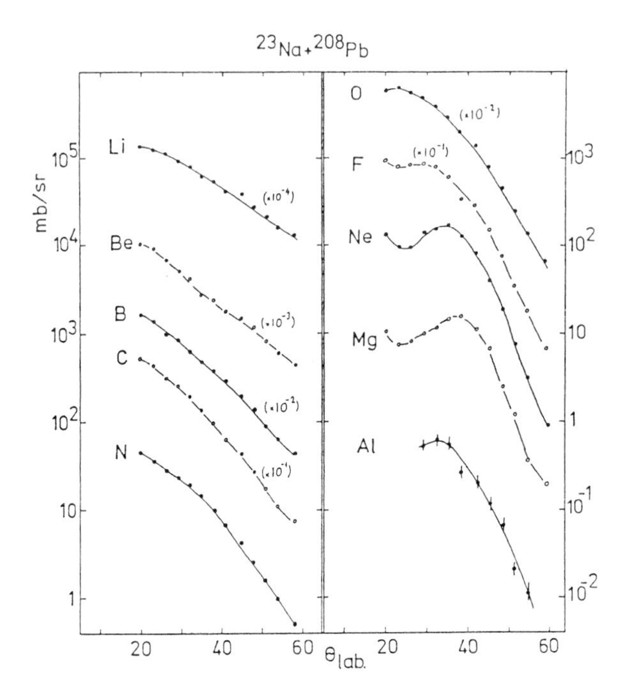

Fig.3. Energy integrated differential cross sections for Z = 3 to 13 ejectiles from the ^{23}Na + ^{208}Pb reaction. Smooth lines were drawn through the data points.

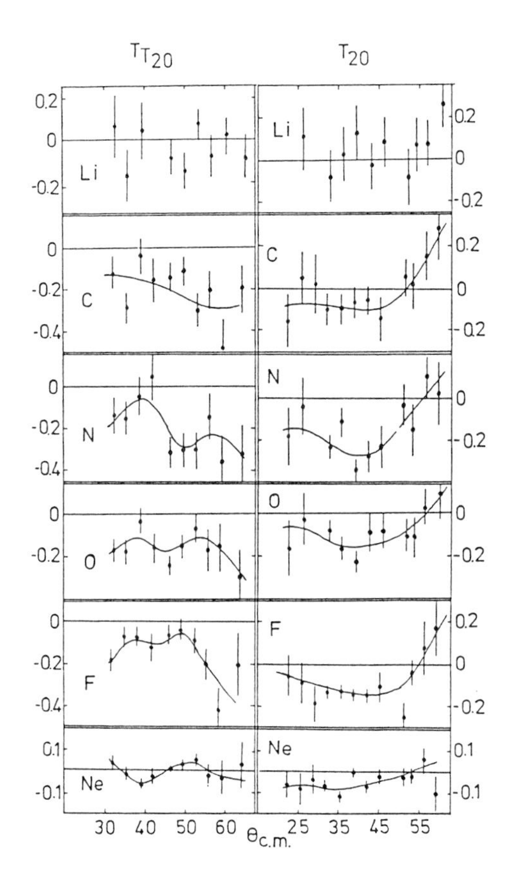

Fig.4. Energy-averaged angular distributions of the $^{T}T_{20}$ and T_{20} analysing powers for several ejectiles from the ^{23}Na + ^{208}Pb reaction. Smooth lines were drawn through the data points.

Inst. Phys. Conf. Ser. No 105
Paper presented at Int. Conf. on Spectroscopy of Heavy Nuclei, Crete, Greece, 1989

Systematic study of M1 strength in rare-earth nuclei: a microscopic approach

C De Coster[†] and K Heyde

Institute for Nuclear Physics, Proeftuinstraat 86, B-9000 Gent, Belgium
[†] Aspirant of the N.F.W.O.

ABSTRACT: A systematic study of magnetic dipole strength in the rare- earth region is carried out, with special attention to the series of $^{148\text{-}160}$Gd isotopes. Within a quasiparticle description for the Nilsson model, all possible two-quasiparticle $K^{\pi}=1^{+}$ states are constructed and calculating the corresponding B(M1) reduced transition probabilities, information on the total summed M1 strength within different energy regions and for the orbital and spin parts separately is obtained. Relations to quadrupole ground state deformation are pointed out.

1. INTRODUCTION

The study of low-lying 1^{+} excitations has been stimulated in recent years by the experimental discovery of a mainly orbital strong M1 excitation at approximately 3 MeV in ^{156}Gd by Richter and coworkers in Darmstadt (Bohle 1984). It was interpreted as a so-called "*scissors mode*", which had been predicted by Lo Iudice and Palumbo (1978) in their two-rotor-model (TRM). In this model, protons and neutrons constitute deformed but axially symmetric rigid bodies, with symmetry axes displaced over a certain angle and performing out-of-phase rotational oscillations. An analogous picture was obtained by Faessler and Nojarov (1986), starting from the generalized Bohr-Mottelson model.

Another, more successful macroscopic approach is given within the proton-neutron interacting boson model (IBM-2) (Iachello and Arima 1987), in which valence nucleons are treated in pairs as s(L=0) and d(L=2) bosons. Taking the charge degree of freedom into account, one can construct states which are non-symmetric under interchange of proton and neutron bosons, i.e. the so-called mixed- symmetry states. The corresponding geometrical picture is very analogous to the one obtained in the TRM, but here only valence nucleons take part in the motion. All macroscopic models essentially predict one low-lying 1^{+} excitation at approximately 3 MeV and seem to overestimate the strength.

As numerous new measurements have been done in (e,e'), (p,p') and (γ,γ') experiments, these low-lying 1^{+} excitations are known to occur in nuclei ranging from ^{46}Ti to ^{238}U (Heyde 1989). Out of comparison of (e,e') and (p,p') results information on the relative importance of orbital and spin contributions can be deduced. High resolution (γ,γ') NRF experiments showed that the 1^{+} mode is much more fragmented than originally assumed. To reproduce fragmentation, one has to turn to microscopic approaches, such as shell-model calculations for light near-to-closed-shell nuclei (Castel and Zamick 1987) and RPA and QRPA calculations for heavier nuclei (Faessler and Nojarov 1987). Most calculations done by now use schematic interactions based on the idea of symmetry restoration (Marshalek and Weneser 1969, Pyatov and Chernei 1973).

Since it is not possible within such a schematic approach to reproduce the details of the observed M1 patterns, we first study the unperturbed intrinsic 2qp picture in order to determine the total M1 strength present as well as obtaining a qualitative description of the underlying microscopic structure of the magnetic dipole excitations in different energy regions. Also, the specific character of the M1 strength (orbital- spin) as a function of the equilibrium quadrupole deformation in the ground state is mapped out.

2. TWO QUASI-PARTICLE CALCULATIONS

We start from a Hamiltonian

$$\hat{H} = \hat{H}_\pi + \hat{H}_\nu \quad , \tag{2.1}$$

where $\hat{H}_\pi(\hat{H}_\nu)$ describes the system of protons(neutrons) separately through a deformed mean field and a short range residual pairing interaction. One of the most successful approaches in describing the single-particle motion within a deformed nucleus is the Nilsson model (Nilsson 1955), where one starts from an axially symmetric harmonic oscillator potential. Using the parameterization of the single-particle Hamiltonian, as described by Nilsson et al (1969), the quadrupole deformed field is characterized by the parameter ε_2. Taking into account the short-range pairing correlations in the independent single-particle motion for the deformed field, the elementary modes become the one quasi- particle excitations. Here, a constant pairing force was used, following the prescription of Nilsson et al (1969).

Since our main interest centers around the magnetic dipole properties in the rare-earth nuclei, we construct $K^\pi = 1^+$ intrinsic excitations using 2qp product states (correctly antisymmetrized). So we set up the basis

$$\left\{ |\tilde{o}>_\pi \otimes |K^\pi{=}1^+(i)>_\nu \ , \ |K^\pi{=}1^+(j)>_\pi \otimes |\tilde{o}>_\nu \right\} \quad , \tag{2.2}$$

where

$$|K^\pi{=}1^+(i)>_\rho \equiv \alpha^+_{N_i\Omega_i,\rho}\ \alpha^+_{N_i'\Omega_i',\rho}\ |\tilde{o}> \quad , \quad \rho = \pi,\nu \ , \tag{2.3}$$

with the conditions

$$(-1)^{N_i + N_i'} = 1 \quad ,$$

$$\Omega_i + \Omega_i' = 1 \qquad \text{or} \qquad |\Omega_i - \Omega_i'| = 1 \quad . \tag{2.4}$$

Starting from the above basis of $K^\pi = 1^+$ 2qp configurations, we can calculate the magnetic dipole transition matrix elements using the unified rotational model (axial symmetry). The following reduced transition probabilities are obtained (with $|\overline{N\Omega}>$ the time-reversed of $|N\Omega>$)

$$B(M1\ ;\ 0^+ \rightarrow 1_i^+) = 3/4\pi\ (u_i v_i' - u_i' v_i)^2$$

$$|< N_i \Omega_i | g_\ell \hat{\ell}_+ + g_s \hat{s}_+ | \overline{N_i' \Omega_i'} >|^2 \quad . \qquad (2.5)$$

The gyromagnetic factors are taken as $g_\ell = g_\ell$(free), $g_s = 0.7\ g_s$(free), which seems to give an overall good agreement throughout the whole nuclear mass region (Paar 1973). We have corrected for the spurious components in the wave functions, corresponding with a rotation of the nucleus as a whole, by orthogonalizing the total $K^\pi = 1^+$ basis with respect to the normalized state $\mathcal{N}\ \hat{J}_+ |\tilde{0}>$. Finally, we stress that the calculation of the unperturbed picture of M1 strength is performed at the equilibrium quadrupole deformation ε_2, as listed by Möller and Nix (1986).

3. RESULTS

We have studied the rare-earth region from the Ce to the Pt nuclei, thereby covering nearly the whole $82 \leq N \leq 126$ neutron shell. A more detailed investigation was done on the series of $^{148-160}$Gd isotopes, obtaining thus a representative view of the underlying microscopic structure of the unperturbed $K^\pi = 1^+$ excitations. We here quote the most important results with the emphasis on the global features. A more extensive report on the results is given by De Coster and Heyde (1989).

The magnetic dipole strength is concentrated within a relatively small energy region. But still, the unperturbed picture is characterized by an appreciable fragmentation of

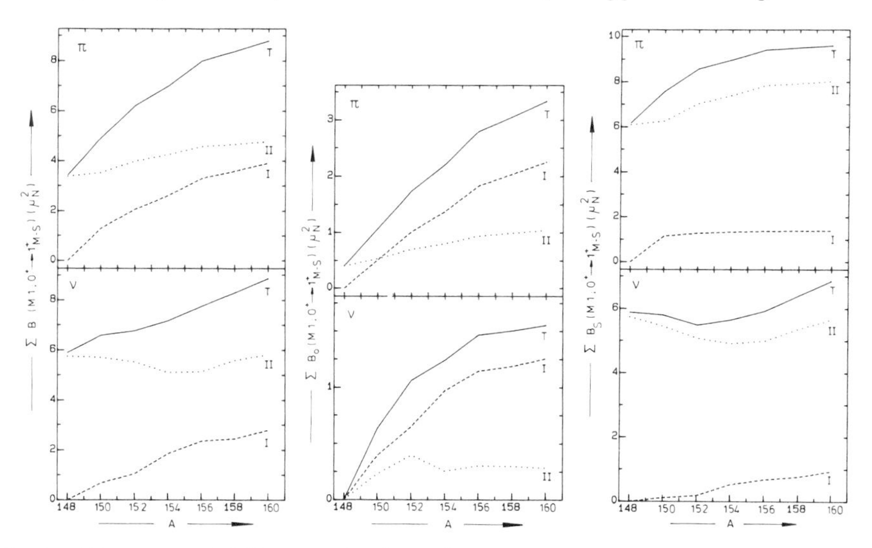

Fig. 1. The summed M1 strength (above for proton states, below for neutron states) is given for the series of Gd isotopes, over different energy regions: dashed line, region I ($E_X \leq 4$ MeV); dotted line, region II (4 MeV $< E_X \leq 9$ MeV); full line, total amount of strength. On the left, the total B(M1) values, in the middle the orbital and on the right the spin parts are shown.

strength for neutron as well as for proton states, due to the relatively high density of 2qp states. More than 95% of the total strength appears to be due to transitions into $K^{\pi} = 1^{+}$ 2qp levels below 9 MeV. Moreover, one can essentially distinguish between two groups, namely region I with $E_x \leq 4$ MeV and region II with 4 MeV $< E_x \leq 9$ MeV. In region I the transitions are mainly of the type $(\ell, j \rightarrow \ell, j)$,giving rise to a dominantly orbital transition, while in region II mainly spin-flip type transitions $(\ell, j = \ell + 1/2 \rightarrow \ell, j = \ell - 1/2)$ result. The concentration of orbital (spin) strength in region I (II) is very clear when inspecting the results for the summed total, orbital and spin strength within these regions for the series of Gd nuclei (see Fig.1).

One also observes in Fig.1 that the (mainly orbital) strength in region I increases with increasing mass number of the Gd isotopes, or increasing deformation. The link between equilibrium quadrupole deformation ε_2 and growth of orbital strength is also clear from the similarity between the corresponding landscape patterns over the rare-earth region (see Fig.2).

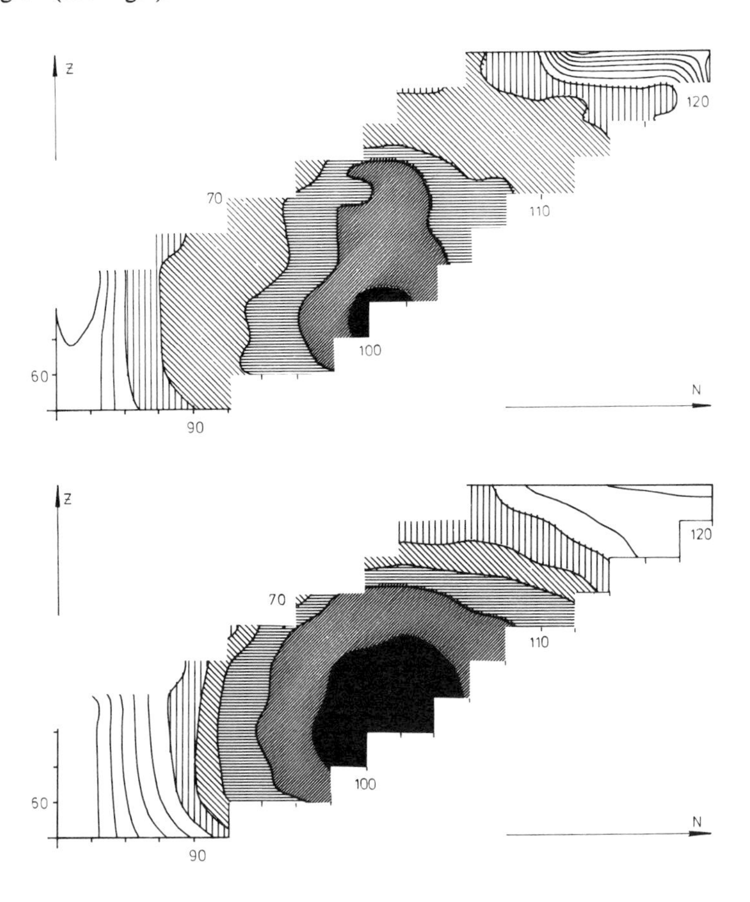

Fig.2. Landscape patterns giving the equilibrium deformation ε_2 (upper) and the orbital summed strength (lower) for nuclei in the rare- earth region. The darkest region corresponds to the largest value; resp. from ε_2 = 0.27, 0.26, 0.25 down by steps of 0.05 and from $\Sigma B_o(M1) = 5\,\mu_N^2$ down by steps of $0.5\,\mu_N^2$.

To understand this link, one has to realize that orbital strength mainly originates from transitions of type (ℓ,j → ℓ,j). For small deformations, a group of Nilsson orbits having the same spherical limit (ℓ,j) is almost degenerate in energy. Therefore, the occupation probabilities v^2 are very similar for this group of states. This then gives rise to a very small "pairing factor" $(u_1v_2-u_2v_1)^2$ in the transition probability (see eq.(2.5)). As deformation increases, the Nilsson orbits separate out in energy, thus having more different occupation probabilities so that the transition strength is enhanced, especially for transitions involving 1qp states near to the Fermi level. This argument is shown schematically in Fig.3 for ^{156}Gd. We indeed find for this nucleus with an equilibrium deformation $\varepsilon_2=0.245$ that the transition $(1h_{11/2})5/2^-$ - $(1h_{11/2})7/2^-$, corresponding to states below and above the Fermi level respectively, has a large B(M1) value.

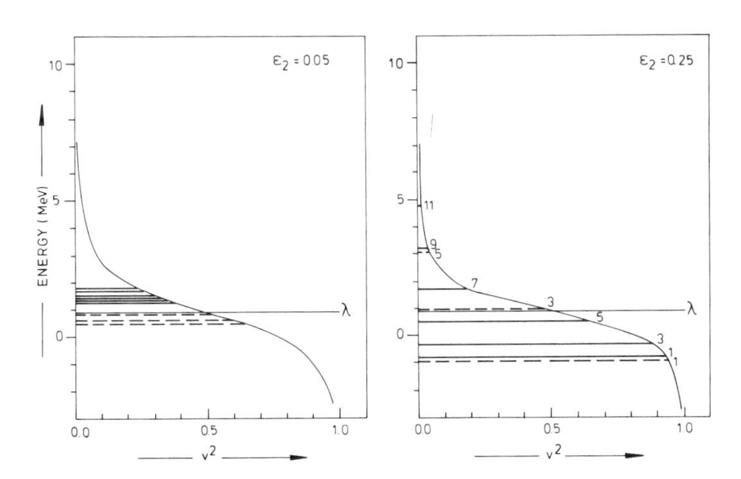

Fig. 3. Schematic picture of the influence of deformation on the (ℓ,j → ℓ,j) transition strength. Two groups of (ℓ,j) Nilsson orbits are given: $1h_{11/2}$ (full lines) and $2d_{5/2}$ (dashed lines). The length of the lines corresponding with the different levels indicates the occupation probability. The Fermi-level is indicated as λ. At the right hand side, the integer values give the double physical value of the Nilsson projection quantum number Ω of the orbit considered.

Due to the fact that the orbits in the neighbourhood of the Fermi-level give major contributions to the total summed M1 strength, one can interpret this description as a kind of "*valence model*", i.e. a microscopic counterpart of the IBM-2. Indeed, orbital summed strength is very similar to what is obtained in the latter model (De Coster and Heyde 1989).

As far as spin-flip transitions (ℓ,j = ℓ + 1/2 → ℓ,j = ℓ-1/2) are concerned, it is clear that - since the energy separation of the Nilsson orbits giving rise to M1 transitions remains fairly constant with deformation - the strength is much less sensitive to deformation effects (see also Fig.1).

4. CONCLUSION

We stress that, since our calculation only provides the underlying "unperturbed" picture, it makes no sense to compare the detailed strength distribution with the observed spectra. However, there is a general correspondence between regions where appreciable amounts of B(M1) strength are located. This analysis in terms of 2qp $K^{\pi}=1^{+}$ intrinsic excitations already points towards a region of mainly orbital transitions below 4 MeV, and a spin-flip resonance around 6 MeV.

We expect that turning on the residual interaction will support this picture. Due to the repulsive $\sigma.\tau$ interaction, spin strength will become shifted to higher energies, thus enhancing the orbital-to-spin ratio in the low-energy region below 4 MeV, where in our calculation a certain amount of spin strength is still present. Possible experimental evidence for such a spin resonance around 7-8 MeV is found in (p,p') measurements (Frekers et al 1989). Moreover, as far as the (ℓ,j → ℓ,j) group of transitions is concerned, diagonalization may give rise to, on one hand, rather pure 2qp excitations, as have been observed in transfer studies on ^{165}Ho by Freeman et al. (1989) around 2.5 to 3 MeV and on the other hand, a more "collective" isovector orbital transition, where collectivity is to be understood as a coherent contribution of proton 2qp states, out-of-phase with a coherent contribution of neutron 2qp states. To our opinion, even recent microscopic calculations (Hilton et al 1986, Hamamoto and Rönström 1987, Sugawara-Tanabe and Arima 1988, Speth and Zawischa 1988, Lo Iudice and Richter 1989) cannot give a firm conclusion about the possible existence of such a state, so that more detailed and systematic studies concerning the effects of the residual interaction are still worthwhile. Such work is in progress.

The authors are most grateful to A.Richter and U.Kneissl for many stimulating discussions. They wish to thank the N.F.W.O. and I.I.KW. for financial support which made this research possible.

REFERENCES

Bohle D. et al. 1984 *Phys.Lett.* **B137** 27.
Castel B. and Zamick L. 1987 *Phys.Rep.* **148** 217.
De Coster C. and Heyde K. 1989 *to be published.*
Faessler A. and Nojarov R. 1986 *Phys.Lett.* **B166** 367.
Faessler A. and Nojarov R. 1987 *Prog.Part.Nucl.Phys.* **19** 167.
Freeman S.J. et al 1989 *Phys.Lett.* **B222** 347.
Frekers D., Wörtche H. and Richter A. 1989 *private communication.*
Hamamoto I. and Rönström C. 1987 *Phys.Lett.* **B194** 6.
Heyde K. 1989 *Int.Journ.Mod.Phys.* **A4** 2063 and references therein.
Hilton R.R. et al 1986 *Proc.Int.Nucl.Phys. Seminar on Microscopic Approach to Nuclear Structure Calculations*
ed. Covello A. (Bologna : Soc.Ital.di Fisica) pp.357
Iachello F. and Arima A. 1987 *The Interacting Boson Model*, eds. Landshoff V.P., Mc.Crea W.H., Sciama D.W., and Weinberg S. (Cambrigde Monographs on Mathematical Physics) and references therein.
Lo Iudice N. and Palumbo F. 1978 *Phys.Rev.Lett.* **41** 1532.
Lo Iudice N. and Richter A. 1989 *to be published.*
Marshalek E.R. and Weneser J. 1969 *Ann.Phys.* **53** 569.
Möller P. and Nix J.R. 1986 *Preprint* **LA-UR-86-3983.**
Nilsson S.G. 1955 *Det.Kong.Danske Vid.Selsk.Mat.-fys.Med.* **29** 1.
Nilsson S.G. et al 1969 *Nucl.Phys.* **A131** 1.
Paar V. 1973 *Nucl.Phys.* **A211** 29.
Pyatov N.I. and Chernei M.I. 1973 *Sov.Journ.Nucl.Phys.* **16** 514.
Speth J. and Zawischa D. *Phys.Lett.* **B211** 247.
Sugawara-Tanabe K. and Arima A. 1988 *Phys.Lett.* **B206** 573.

Inst. Phys. Conf. Ser. No 105
Paper presented at Int. Conf. on Spectroscopy of Heavy Nuclei, Crete, Greece, 1989

Probing exotic shapes of nuclei with heavy ion transfer reactions

P A Butler, T H Hoare, R J Poynter*, N Clarkson, G D Jones and C A White

Oliver Lodge Laboratory, University of Liverpool, Liverpool L69 3BX

J R Hughes, N S Jarvis, S M Mullins+, P H Regan, R Wadsworth and D L Watson

Department of Physics, University of York, York YO1 5DD

R A Cunningham and J Simpson
Science and Engineering Research Council, Daresbury, Warrington WA4 4AD

D Cline
Nuclear Structure Research Lab, University of Rochester, Rochester NY 14627

L Goettig
Institute of Experimental Physics, Warsaw University , 00-681 Warsaw

S Juutinen
Department of Physics, University of Jyväskylä, 40100 Jyväskylä 10

* now at University of York
\+ now at University of Liverpool

ABSTRACT: The application of heavy ion transfer reactions to the study of the octupole deformed nucleus ^{224}Ra is presented, together with the results of an attempt to populate the fission isomeric state in ^{236}U.

1. INTRODUCTION

In this talk, I will discuss the application of heavy ion induced transfer reactions to the spectroscopic study of nuclei. The reactions employed are

$$^{58}\mathrm{Ni} + {}^{A}\mathrm{X} \rightarrow {}^{60}\mathrm{Ni} + {}^{A-2}\mathrm{X}$$

carried out at bombarding energies close to the interaction barrier for the particular reaction. Detection of backscattered Ni-like fragments near 180° scattering angle ensures that the relative velocity at the moment of transfer is close to zero, so that the induced angular momentum of the target-like fragment arises only from Coulomb excitation and from the angular momentum of the transferred particles. Studies of similar reactions carried out using the spin spectrometer at the Holifield Facility, Oak Ridge have shown that high angular momentum (15-25ħ) states in the residual system are populated at low excitation energies, see fig.

1. These conditions are ideal for the study of the low lying structure of heavy nuclei. I will discuss their application to the study of two nuclei which can posses exotic shapes: i) ^{224}Ra which is predicted to be one of the best examples of octupole deformation near its ground state and ii) ^{236}U which possesses a superdeformed 2nd minimum in its potential energy surface close to the ground state.

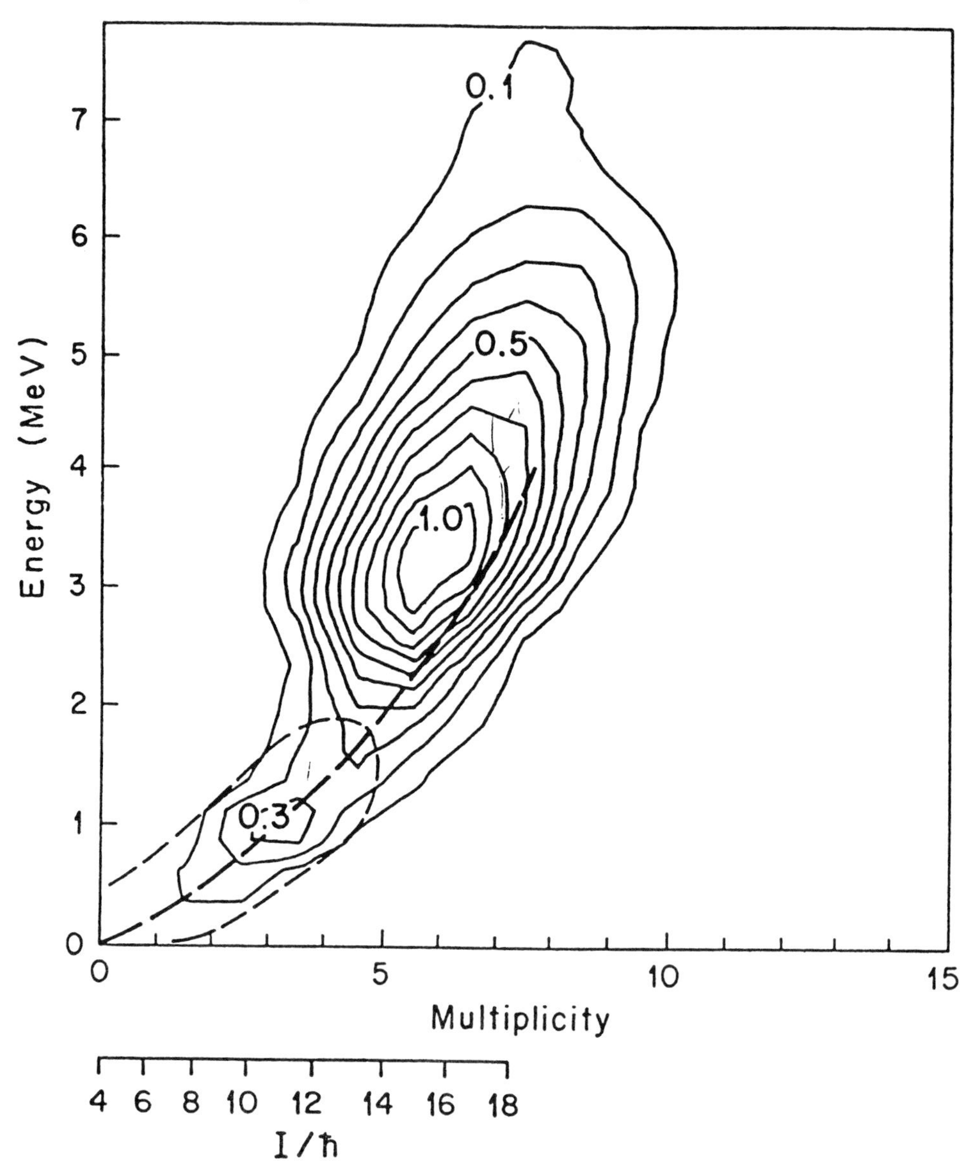

Figure 1. The total energy-multiplicity distribution for the transfer reaction ^{161}Dy(^{58}Ni,^{59}Ni)^{160}Dy (solid lines) and the inelastic reaction ^{161}Dy(^{58}Ni,^{58}Ni)^{161}Dy (dashed line, 0.1 contour only), taken from Guidry et al (1985)

2. STUDY OF ^{224}Ra

In the Ra-Th region, there is now a large body of experimental data which reveals the existence of strong octupole correlations around N~134, e.g. [Schüler et al (1986), Dahlinger et al (1988)]. These data are usually explained in terms of the geometric quadrupole-octupole model [Leander et al (1986), Nazarewicz et al (1987)], which is particularly successful in describing the behaviour of the Th isotopes with N = 130-138, which are experimentally well documented in terms of both energy levels and electromagnetic transition probabilities [Schüler (1986), Dahlinger (1988)]. For the Ra isotopes the lack of suitable targets has meant that extensive data only exist for ^{218,220}Ra and ^{226}Ra; very little data have been reported for the isotopes ^{222,224}Ra which should demonstrate the strongest octupole effects [Nazarewicz et al (1987)]. I will discuss the study of one of these nuclei, ^{224}Ra, in which γ-ray spectroscopy following the reaction 315 MeV ^{226}Ra(^{58}Ni,^{60}Ni)^{224}Ra (Q = +9.1 MeV) has been applied. The ^{58}Ni beam was provided by the N.S.F., Daresbury. In this experiment γ-rays were observed in POLYTESSA containing 22 escape suppressed Ge detectors and coincident backscattered projectile-like fragments were detected using a large solid angle position sensitive avalanche detector. The fragment detector consisted of 44 electrically isolated sections distributed over the four sides of a truncated pyramid. The method of determining which section was involved in an event, and hence the (θ, ϕ) scattering angles of the fragment, is similar to that described by Butler et al (1985). This information was used to make Doppler shift corrections for the coincident γ-ray signal for each event, thus reducing the Doppler broadening to values limited by detector solid angle effects. The target used in this experiment has been described by Poynter et al (1989). A total of 9 x 10^4 particle-γ-γ events were recorded. Various γ spectra are shown in fig. 2a) → 2d). Difficulty in identifying ^{224}Ra transitions was experienced because of the similarity between the energies of ^{224}Ra transitions and those in ^{226}Ra [Wollersheim et al (1987)], which were populated by inelastic scattering, see fig. 2a). At this bombarding energy the (2n) pick up cross section is ~15% of that for inelastic scattering. Unambiguous identification was achieved by requiring that at least one γ-ray in a coincident event was the 1332 keV (2^+ - 0^+) transition in ^{60}Ni, Doppler-shift corrected assuming it was emitted from the ^{60}Ni fragment. To obtain the coincident ^{224}Ra spectrum (fig. 2b) the other γ-ray was corrected assuming it was emitted from the ^{224}Ra fragment. Typical Ra-corrected spectra, in coincidence with specific transitions in ^{224}Ra, are also shown: fig. 2c) in coincidence with the 228 keV (6^+ → 4^+) transition, and fig. 2d) with the 265 keV transition (assigned 9^- → 7^-). The deduced decay scheme is shown in fig. 3. The spin and parity assignments are based on α-decay measurements, on the level decay properties and on the known rotational behaviour of nuclei in this mass region. Fig. 2d) also shows that the branching of the 7^- level to the 6^+ level (E_γ=162 keV) must be weak, which suggests that the B(E1)/B(E2) ratios are depressed in this nucleus, relative to other Ra isotopes. No evidence is seen for any other E1 transitions connecting the higher lying members of the negative and positive parity band.

Our data allow average values for the intrinsic electric dipole moment Q_1 in the light Ra nuclei to be compared with predictions of the geometric model [Leander et al (1986)]. This moment is determined from the B(E1) transition probability:

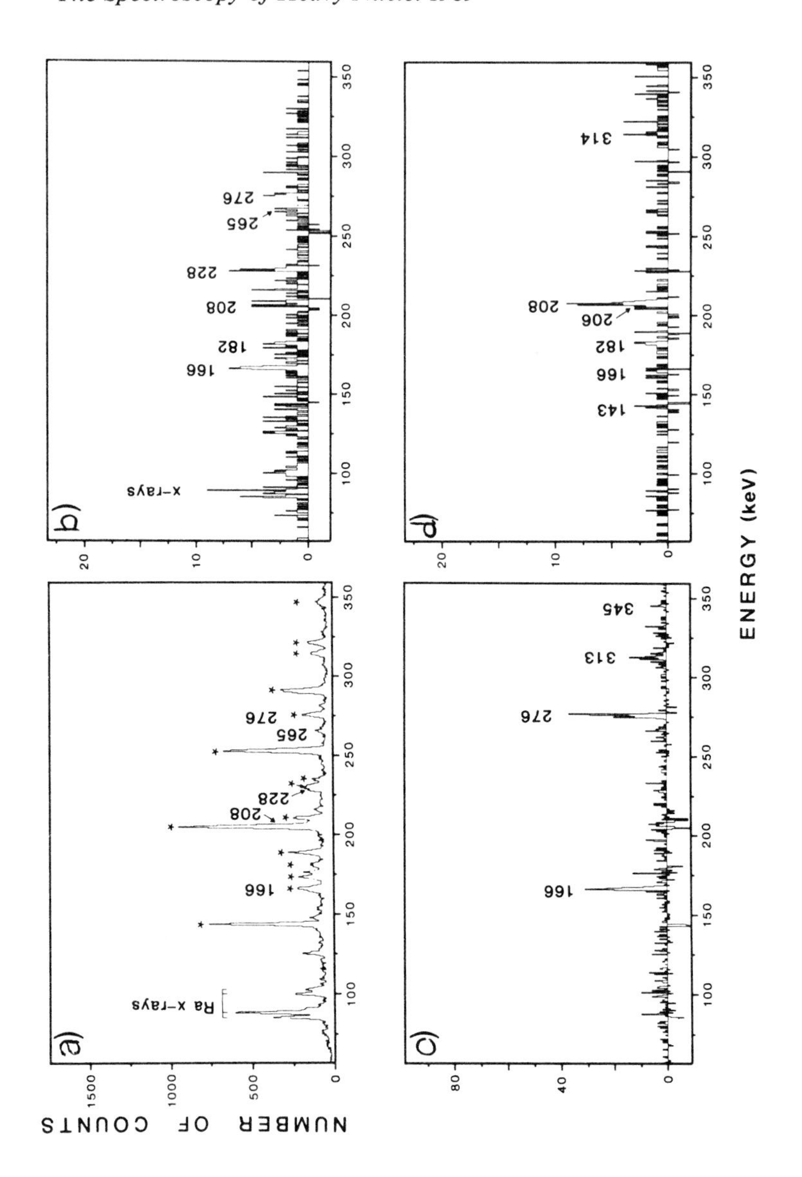

Figure 2

a) Spectrum of γ-rays following the ^{226}Ra + ^{58}Ni (E_{Ni} = 315 MeV) reaction, following the detection of at least two γ-rays and the detection of the backscattered projectile-like fragment. Peaks denoted by a star arise from transitions in ^{226}Ra, peaks labelled with transition energy (in keV) are assigned to ^{224}Ra.

b) As a), spectrum is gated on the ^{60}Ni 1332 keV ($2^+ \rightarrow 0^+$) transition.

c) As a), spectrum is gated on the ^{224}Ra 228 keV ($6^+ \rightarrow 4^+$) transition.

d) As a), spectrum is gated on the ^{224}Ra 265 keV (assigned $9^- \rightarrow 7^-$) transition.

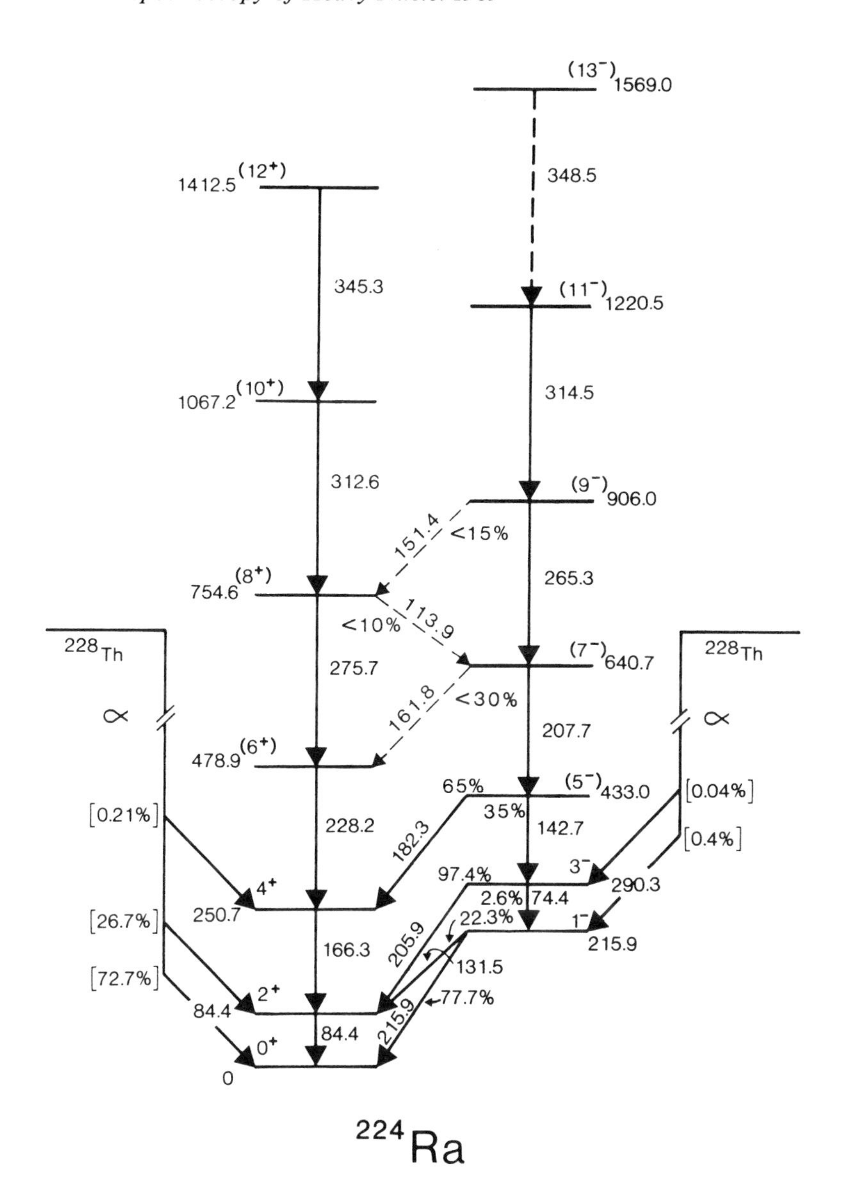

Figure 3. Level scheme of ^{224}Ra, deduced from α-decay and reaction data. The transition energies and level energies (given in keV) have errors ±.1 keV for those occurring below 500 keV in excitation energy, ±.2 keV for those above. Gamma-ray branching ratios are given for the 1^-, 3^- and 5^- levels, the absolute errors are ±0.3%, ±0.6% and ±12% respectively. Limits (two standard deviations) are given for the 7^-, 8^+ and 9^- levels. The branching ratios for the α-decay of ^{228}Th are also given (in square brackets), taken from Lederer and Shirley (1978).

$$B(E1,\ I_i \rightarrow I_f) = \frac{3}{4\pi} < I_i010|I_f0 >^2\ Q_1^2$$

The experimental values of Q_1 for ^{224}Ra are deduced from the B(E1)/B(E2) ratios (or limits) measured in this work. The calculated values of Leander (1986) for Q_1 at about a spin of 8ħ are shown in fig. 4 together with the known experimental values for the Ra isotopes. These

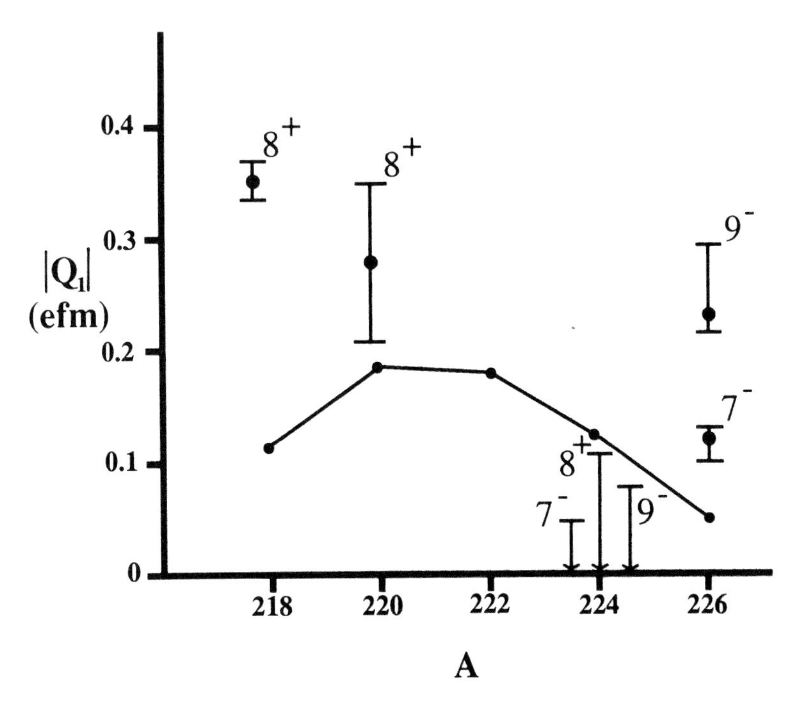

Figure 4
Experimental and calculated values of the electric dipole moment Q_1 for Ra isotopes, at $I_i \sim 8\hbar$. Straight lines join the calculated [Leander et al (1986)] points to guide the eye.

calculations assume that the bulk properties of the nucleus are described by the liquid drop model with shape parameterised by β_i ($i=1\rightarrow6$), but take into account quantal effects arising from single particle motion in a Woods-Saxon potential and arising from residual interactions. The calculations indicate that the shell corrections are large and can cancel out the contribution from the liquid-drop term at certain values of Z, N and I. In the Ra isotopes, we observe experimentally that Q_1 is very much smaller for ^{224}Ra than either ^{226}Ra or ^{220}Ra, even at moderate spin values. This behaviour is not reproduced by the calculations of Leander et al (1986) (see fig. 4) although it lends support to the assumption of a dominant role of shell effects in the description of the E1 moments for the light actinides. It is not clear whether the small value of Q_1 in ^{224}Ra arises from the cancellation of the liquid drop and shell terms, predicted to occur in ^{226}Ra [Leander et al (1986)], or from large fluctuations in the shell contribution itself, the liquid drop contribution being insignificant [Dorso et al (1986)]. A recent application of Hartree-Fock methods to this mass region has predicted that the B(E1, $1^- \rightarrow 0^+$) values for the Ra isotopes should exhibit a sharp minimum at A = 224 [Egido and Robledo (1989)], but these calculations have not yet been extended to higher spins.

3. STUDY OF ^{236}U

The existence of a 2nd Minimum in the nuclear potential energy surface with a 2:1 axis ratio has been experimentally well documented for nuclei with Z=92-97 and N=141-151 (e.g. Metag et al (1980)). In these nuclei the 2nd minimum arises because the liquid drop contribution to the total nuclear potential is weakened by the presence of a large Coulomb component and is easily modulated by the strong shell effects. In this case the superdeformed state which lies lowest in energy (typically 2-3 MeV above the ground state) has no collective angular momentum and is isomeric. This is in contrast to the behaviour of the observed superdeformed bands in nuclei with Z=64-66, N=84-86 and Z=58-60, N=73-77 (Nolan and Twin (1988)) where the corresponding liquid drop term is weakened by centrifugal effects, and the highly deformed structure is only seen at high angular momenta. In principle, the properties of the transuranic 'fission isomers', especially their stability at low spin, make them the ideal laboratory for studying the effect of the superdeformed mean field on single particle structure and on pairing. These studies have been impeded, however, by the low cross sections for the population of the 2nd minimum states, typically 1μb for light ion induced reactions such as (α,2n) or (d,p).

We have attempted to improve the yield of the fission isomeric states by employing heavy-ion induced transfer reactions. These reactions directly excite states of moderate angular momentum (~25ħ) lying close to the yrast line, which should be the optimum situation for populating the superdeformed states which become yrast at J~30ħ, see fig 5. In the experiment described here we employed the reaction ^{238}U(^{58}Ni,^{60}Ni)^{236}U (Q=+9.1 MeV) at E_{Ni}=325 MeV. The experimental arrangement (see fig. 6) consisted of a position sensitive avalanche detector (described in the previous section) in the backward quadrant and a large solid angle avalanche counter, with an opening angle of 20° in the forward direction. The position sensitive avalanche counter in the backward hemisphere served two purposes: i) to define the inelastic scattering + transfer channel, ii) To enable Doppler corrections to be applied to coincident γ-spectra, so that the transfer channel leading to the fission isomer of interest (^{236}U in this case) could be identified and its total yield measured. A time spectrum for detection of delayed events was generated following: i) a prompt coincidence between the backward and forward detectors (i.e. Ni-U); ii) a delayed coincidence between the backward and forward detectors (ie. Ni-isomeric fission). In a test experiment we observed a delayed component which could be fitted by an exponential decay with a lifetime consistent with that of the fission isomer in ^{236}U (τ=165ns).

Note In an experiment carried out recently, we were able to ascertain that the observed events were NOT from the decay of the fission isomers in ^{236}U. The analysis of the data is in progress to determine the upper limit for the population of this state.

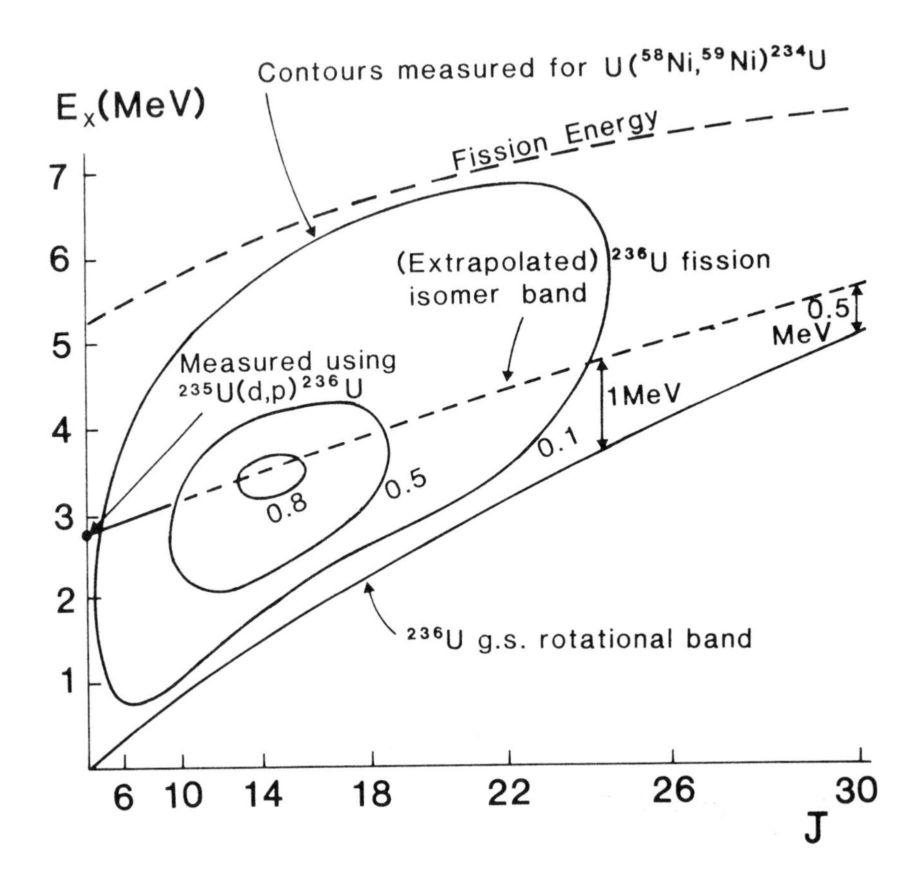

Figure 5
Schematic population of ^{236}U following the $^{238}U(^{58}Ni,^{60}Ni)^{236}U$ reaction, in relation to the ground state rotational band and the expected behaviour of the rotational band built on the fission isomer. The population contours have been measured following the $^{235}U(^{58}Ni,^{59}Ni)^{234}U$ reaction (Wu et al 1987). The excitation energies of ground state rotational band in ^{236}U are taken from the Owen et al (1982), and the energy of fission isomer in ^{236}U is taken from Schirmer et al (1988).

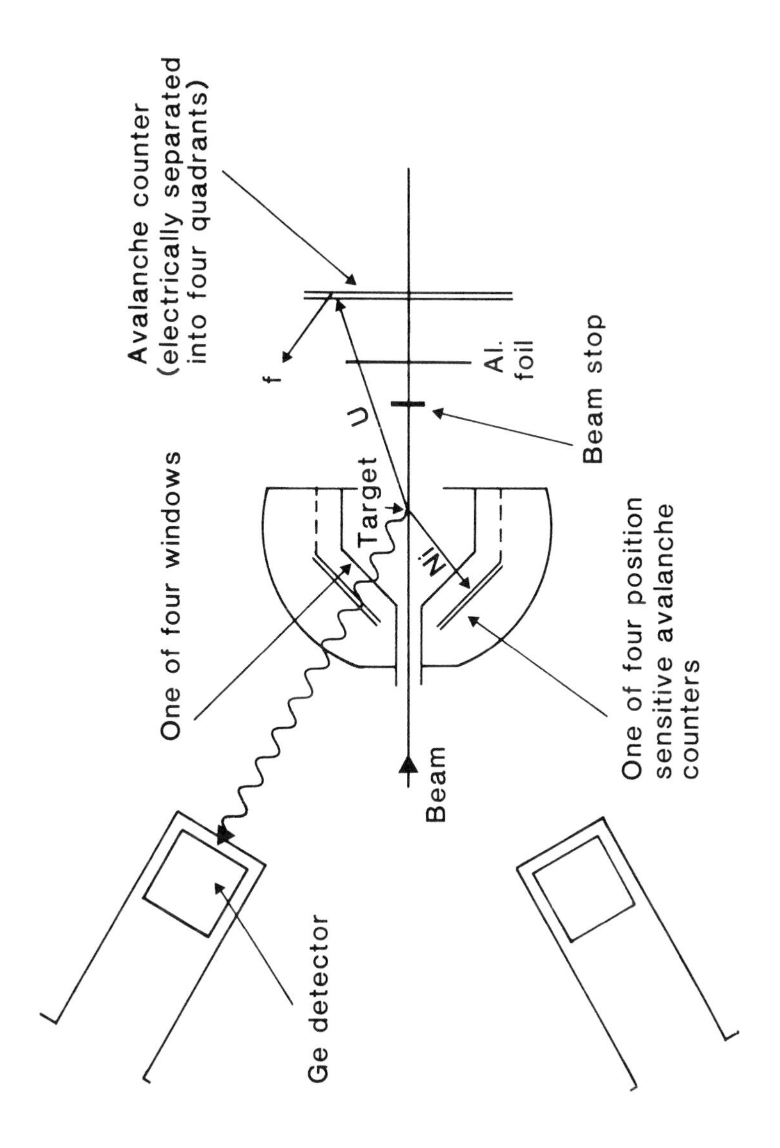

Figure 6
Experimental arrangement used to attempt to populate the fission isomer of ^{236}U using the $^{238}U(^{58}Ni,^{60}Ni)^{236}U$ reaction.

References

P.A. Butler et al; 1985 Nucl. Instrum. & Meth. **A239** 221
M. Dahlinger et al; 1988 Nucl. Phys. **A484** 337
C.O. Dorso et al; 1986 Nucl. Phys. **A51** 189
J.L. Egido and L.M. Robledo; 1989 Nucl. Phys. **A494** 85
M.W. Guidry et al; 1985 Phys. Lett **163B** 79
G.A. Leander et al; 1986 Nucl. Phys. **A453** 58
C.M. Lederer and V.S. Shirley, 1978 Table of Isotopes 7th edition, Wiley (New York)
V. Metag et al; 1980 Phys. Reports **65** 2
W. Nazarewicz et al; 1987 Nucl. Phys. **A467** 437
P.J. Nolan and P.J. Twin; 1988 Ann. Rev. Nuc. Part. Phys. **38**
H. Owen et al; 1982 Nucl. Phys. **A388** 421
R.J. Poynter et al; 1987-88 Daresbury Annual Report p.78, (ibid) p.83
J. Schirmer et al; 1988 Proc. Conf. High Spin and Shapes, Argonne, p.78
P. Schüler et al; 1986 Phys. Lett. **B174** 241
H.J. Wollersheim et al; 1987 Int. Conf. Nucl. Struct. Stat. Dyn. Mom. II, Melbourne, ed. H.H. Bolotin p.88
C.Y. Wu et al; 1987 Phys. Lett. **B188** 25

Inst. Phys. Conf. Ser. No 105
Paper presented at Int. Conf. on Spectroscopy of Heavy Nuclei, Crete, Greece, 1989

Cosmic ray muon background of the ^{205}Tl solar neutrino experiment—studied with high energy muons at CERN

S. Neumaier, E. Nolte, H. Morinaga
Faculty of Physics, Technical University of Munich, 8046 Garching, FRG

Abstract. The feasibility of the ^{205}Tl solar neutrino experiment depends on the ratio of neutrino induced reactions to background reactions. To study the production of ^{205}Pb induced by cosmic ray muons deep under ground, we exposed various samples of natural thallium, bismuth, lead and lorandite to the high energy muon beam (M2) of CERN. Targets were irradiated for several days with 280 and 200 GeV positive muons, respectively. Yields for the production of about 30 radioisotopes were measured. Results and conclusions are presented.

1. Introduction

Since first results of the ^{37}Cl solar neutrino experiment became available, the discrepancy between the predicted capture rate by the standard solar model and the measured ^{37}Ar production rate is known as the "solar neutrino problem". To solve this problem which is of fundamental interest in modern physics, neutrino experiments with low threshold energies and high capture rates are necessary. In 1976, the $^{205}\mathrm{Tl}(\nu, e^-)^{205}\mathrm{Pb}$ solar neutrino experiment with a high capture rate and an energy threshold of only 52 keV was proposed by Freedman and coworkers (Freedman *et al* (1976)).

The main problems in using the Allchar thallium deposit as a geochemical solar neutrino detector were discussed in detail in two workshops, (ed. Nolte (1986) and eds. Nolte and Pavićević (1988)).

Some of these problems - e.g. ore processing, chemical separation of lead from the lorandite mineral and the detection of ^{205}Pb by accelerator mass spectrometry (AMS) - should be solvable by technical improvements.

The ^{205}Tl solar neutrino experiment on the other hand, depends on the signal to background ratio which is determined by the chemical composition and the effective shielding against cosmic radiation since the formation of the mineral; conditions that can not be altered. Therefore the background problem is of main importance for the feasibility of the ^{205}Tl solar neutrino experiment.

At the present depth of the Allchar thallium deposit (about 330 mwe), reactions due to cosmic ray muons are considered to be the dominant source of background.

To determine the production rate of ^{205}Pb by cosmic ray muons we exposed massive targets to the (M2) muon beam of CERN. After several days of irradiation, the produced activities were measured "off-beam" with a high purity germanium detector and a passive shielding of 12 cm of lead. Via the measured production rate of ^{203}Pb from natural thallium, a crude estimation of the ^{205}Pb production rate has been deduced. Preliminary results are in rough agreement with former predictions on the fast muon background of the thallium solar neutrino experiment.

2. Background reactions of the geochemical ^{205}Tl solar neutrino experiment

Thallium in the Allchar mine (Yugoslavia) is mainly found as small grains of the rare lorandite ($TlAsS_2$) mineral. The age of the orebody is estimated by geologists to be 4.8 million years. The expected neutrino signal (^{205}Pb concentration in lorandite) therefore is determined by the time average of the solar neutrino flux over this period of time. In every geochemical solar neutrino experiment, background reactions due to natural radioactivity and cosmic ray muons have to be considered carefully. In the case of the Allchar thallium deposit natural radioactivity and stopped muons seem to play an subordinate role, while fast muons contribute dominantly to background.

Sources of background reactions:

- natural radioactivity:
  The decay of Uranium and Thorium produces α-particles and fission neutrons. Subsequent (α,n), (α,p) and (n,p) reactions lead to reaction chains as

$$\begin{array}{ccc} (\alpha,n) & \rightarrow & ^{204}Pb(n,\gamma)\,^{205}Pb \\ & \rightarrow & ^{206}Pb(n,2n)^{205}Pb \\ ^{32}S(\alpha,p)^{35}Cl & \rightarrow & ^{205}Tl(p,n)^{205}Pb \end{array}$$

- stopped muons:
  The number of stopped muons decreases more rapidly with depth than the fast muon component (Charalambus 1971). Therefore stopped muons contribute to the background only at shallow depths via the production of secondary particles and the $^{209}Bi(\mu_{stopped}, 4n)^{205}Pb$ reaction.

- fast muons:
  Even at great depth, fast muons play an important role in background production. In interactions with the surrounding rock, nuclear and electromagnetic cascades are induced, producing nucleons, pions and photons. Subsequent nuclear reactions like

$$^{205}Tl(p,n)^{205}Pb, \quad ^{205}Tl(\pi^+,\pi^0)^{205}Pb \text{ and } ^{205}Tl(\gamma,\pi^-)^{205}Pb$$

  transform target nuclei into ^{205}Pb.

A direct calculation of the ^{205}Pb production by cosmic ray muons is complicated and uncertain because of the lack of complete differential spectra and cross sections of all secondary particles.

Therefore an attempt was made, to simulate the complex reaction chains by an accelerator experiment.

3. The high energy muon channel at CERN

The differential muon spectra under ground were measured and calculated several times. According to Zatsepin *et al* (1981) it can be described by a quasi power formula. The mean muon energy increases with depth.

At 330 mwe (mwe = meters water equivalent) the mean energy of the cosmic ray muon spectrum is about 60 GeV. Assuming a constant erosion rate of 100 m/million years the depth of the Allchar ore at the end of the mineralisation process ($4.8 \cdot 10^6$ years ago) was 1600 mwe. The mean energy at this depth is roughly 200 GeV.

Muons of comparably high energy are only available at a few high energy accelerators, e.g CERN and FermiLab.

The CERN (M2) muon beam provides muon energies of 90 GeV up to 280 GeV. The available (time averaged) muon flux is of the order of $2 \cdot 10^4 cm^{-2} sec^{-1}$. At the present depth of the Allchar thallium deposit, the muon intensity is only $5 \cdot 10^{-5} cm^{-2} sec^{-1}$. During the whole age of the mineral, less than $7.6 \cdot 10^9$ muons have passed through an $1\,cm^2$ extended sample of lorandite. The same amount of muons can be achieved by an irradiation of about 5 days at CERN.

4. Measurements and Results

A 5m long block of concrete was put into the beam to certify the production of secondary particles by nuclear and electromagnetic cascades.

Targets of natural thallium, bismuth, lead and lorandite were irradiated for several days, respectively. The cross sections of the relevant muon induced reactions are only some μb. To get detectable amounts of radioactivity, the target masses were of the order of hundreds of grams. After several days of irradiation, the targets were measured "off-beam" using a high purity germanium detector with 44% relative efficiency (resolution: 1.95 keV at 1.33 MeV) and a passive shielding against natural radioactivity of 12 cm of lead. For each target, γ-spectra were taken every 20 hours for at least one week. More than 30 radioisotopes have been identified.

The long half-life of ^{205}Pb ($T_{1/2} = 1.5 \cdot 10^7$ a) prevents the determination of this isotope by conventional methods. Therefore similar transitions in the vicinity of ^{205}Tl have been studied. As a first approximation the production of ^{205}Pb from natural thallium is regarded to be equal to the production of ^{203}Pb from ^{203}Tl.

The main contribution to the ^{203}Pb production from natural thallium is due to $^{205}Tl(p,3n)^{205}Pb$ rather than to $^{203}Tl(p,n)^{203}Pb$ reactions. Assuming an $E^{-1.7}$

energy dependence of the secondary proton spectrum, a preliminary result for the production cross section of ^{205}Pb by cosmic ray muons (at 330 mwe) of about 2 μb has been derived.

This result is in rough agreement with former predictions and shows, that at the present depth of the Allchar thallium deposit the shielding against cosmic ray muons would be insufficient.

In order to reduce the undesired contribution of ^{205}Tl to the ^{203}Pb production rate and to get rid of assumptions on the secondary particle spectra, it is necessary to use highly enriched targets of ^{203}Tl.

5. Conclusions

It has been demonstrated, that high energy accelerators are suitable tools to study complex reactions of cosmic ray muons with matter. This is of special interest for the investigation of background reactions concerning solar neutrino experiments. First results from an irradiation with 200 GeV muons at CERN are in rough agreement with former estimations if assumptions on the composition of secondary particles, the shape of secondary proton spectra, and (p,n)- cross sections are correct. To get rid of these assumptions, it is necessary to use isotopically enriched targets.

Acknowledgments

We would like to thank D. von Harrach, F. Klein (The NMC-Collaboration) and D. Vignaud (The Gallex-Collaboration) for their support and hospitality at CERN.

We are grateful to the students M. Meier and C. Seidl who helped us several times to perform the experiments and analyse the data.

References

Charalambus S 1971, Nucl. Phys. **A166** 145

Freedman M S, Stevens C M, Horwitz E P, Fuchs L H, Lerner J S, Goodman L S, Childs W J and Hessler J 1976, Science **193** 1117

Nolte E 1986, Proceedings of the Workshop on the feasibility of the Solar Neutrino Detection with ^{205}Pb by geochemical and accelerator mass spectroscopical measurements, Report GSI-86-9

Nolte E and Pavićević M K 1988, Proceedings of the International Conference on Solar Neutrino Detection with ^{205}Tl, and Related Topics NIM **A271(2)** 237

Zatsepin G T, Kopylov A V and Shirokova E K 1981, Sov. J. Nucl. Phys. **33(2)** 200

Inst. Phys. Conf. Ser. No 105
Paper presented at Int. Conf. on Spectroscopy of Heavy Nuclei, Crete, Greece, 1989

Test of Pauli exclusion principle

E. Nolte, T. Faestermann, H. Gail, A. Gillitzer, G. Korschinek, D. Müller, R. Scheuer, V. M. Novikov, A. A. Pomansky
Faculty of Physics, Technical University of Munich, 8046 Garching, FRG
Institute for Nuclear Research of the USSR, Academy of Sciences, Moscow, USSR

Abstract. Accelerator mass spectrometry was used to test the validity of the Pauli exclusion principle for atomic electrons. Limits for the concentration of anomalous atoms where three electrons are occupying the K shell thus violating the Pauli exclusion principle were obtained. In the case of anomalous $^{20}\widetilde{Ne}$ atoms, a limit of $\frac{^{20}\widetilde{Ne}}{^{20}Ne} < 2{\cdot}10^{-21}$ was measured with a time-of-flight technique. In the case of anomalous $^{36}\widetilde{Ar}$ atoms, a limit of $\frac{^{36}\widetilde{Ar}}{^{36}Ar} < 4{\cdot}10^{-17}$ was obtained for hydrogenlike ions. The limits can be improved by two to three orders of magnitude.

1. Introduction

In quantum theories of gravity, the Pauli exclusion principle (PEP) can be violated (Kuzmin (1984)). If the PEP is violated the wavefunction which describes the atomic electrons or the wave function of the nucleons in the nucleus is not completely antisymmetric under the exchange of identical particles. This means that e.g. three electrons can be in the K shell or three protons or three neutrons in the $1s_{1/2}$ shell. Several theoreticians have considered a possible violation of the Pauli principle (Amado and Primakoff (1980), Gavrin et al. (1988), Govorkov (1989), Ignatiev and Kuzmin (1987), Okun (1988)) and related problems of parastatistics, i.e. non-Fermi or non-Bose statistics (Govorkov (1983), Green (1953), Greenberg and Mohapatra (1987 and 1989), Messiah and Greenberg (1964), Rubakov and Spiridonov (1988); see also references cited there).

Only few experiments were performed to test the PEP. Most of these experiments concerned atomic electrons. Reines and Sobel (1974) used the data of an experiment in which a limit on the lifetime of the electron was set (Moe and Reines (1965)) to deduce a lifetime of $> 6 \cdot 10^{19}$ a for Pauli violating transitions. In an analogous measurement, a lower lifetime limit of $3 \cdot 10^{22}$ a was obtained (Kovalchuk et al. (1979)). In a conduction electron experiment, Ramberg and Snow (1989) interpreted the non-observation of X-rays in terms of a parameter β^2 describing violation of the PEP, $\beta^2 \leq 1.6 \cdot 10^{-26}$.

In two papers known to us, the PEP for nucleons was investigated. In a ^{12}C sample, Logan and Ljubicic (1979) looked for γ-rays from a transition of nucleons from the 2p to the occupied 1s shell. They obtained a lower lifetime limit of $2 \cdot 10^{20}$ a. Plaga (1989) studied the contribution of a bound proton-proton system where the two protons are in a symmetric state thus violating the PEP to the hydrogen burning rate in the sun. He deduced a PEP violating parameter $\epsilon^2/2 < 1.6 \cdot 10^{-15}$.

The interpretation of experimental non-observations in the case of the PEP is not unique. Amado and Primakoff (1980) pointed out that transitions from a state with Fermi statistics to another state with anomalous statistics, non-Fermi statistics, is forbidden. This means that experiments looking for γ- or X-rays from these transitions do not test the PEP and that, in this sense, lifetimes for PEP violating transitions cannot be deduced. On the basis of general assumptions, Govorkov (1989) showed that a "small" violation of the PEP is not admissible.

In the present paper we describe very sensitive experiments with which we measure limits of concentrations of anomalous atoms in which the PEP for atomic electrons is violated. The detection method is accelerator mass spectrometry (AMS).

2. Principle of the Experiment

If the PEP is violated for atomic electrons we can find e.g. three electrons in the K shell. This situation is shown for the case of neon in Fig. 1. This anomalous neon, $\widetilde{Ne}$, has then only 7 electrons in the L shell, as fluorine. Since the chemical behaviour of an atom is determined by the outer valence electrons, it is expected that the anomalous neon, $\widetilde{Ne}$, is chemically as fluorine. More generally, anomalous atoms of the element Z in which the PEP is violated in a way that three electrons occupy the K shell behave chemically like the atoms of the element Z-1. The principle of our method is to look for the concentration of anomalous atoms of atomic number Z within a natural sample of the element with atomic number Z-1.

Because of its high sensitivity, we have chosen AMS as the detection method. Two AMS set-ups were used. The main components are negative ion source, Tandem accelerator and time-of-flight set-up or negative ion source, Tandem accelerator, radiofrequency postaccelerator and detector. With these AMS set-ups, a given mass of an isotope is selected.

In order to deduce small limits for the PEP, pairs of elements are considered where the ratio of the cosmic abundances P(Z)/P(Z-1) is high. The cosmic abundances of elements up to nickel are shown in Fig. 2. The data are taken from Cameron (1970). From the measured ratio of concentrations $[\widetilde{^AZ}]/[Z\text{-}1]$, we deduce the concentration of anomalous atoms, $[\widetilde{^AZ}]$, with respect to the concentration of normal atoms, $[^AZ]$, to be

$$\frac{[\widetilde{^AZ}]}{[^AZ]} = \frac{[\widetilde{^AZ}]}{[Z-1]} \cdot \frac{P(Z-1)}{P(Z) \cdot P(A,Z)},$$

where P(A, Z) is the cosmic isotopic abundance of the isotope with mass A of the element Z.

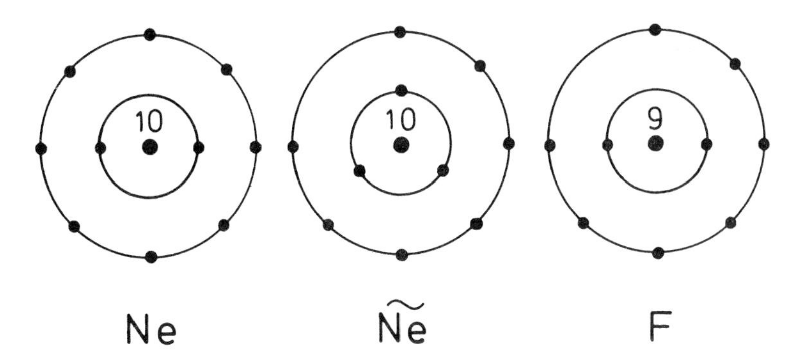

Figure 1: Configurations of atomic electrons for neon, fluorine and anomalous neon, $\widetilde{Ne}$, where three electrons are in the K shell.

If negative ions are extracted from the ion source as in the present work, no background events from possible impurities of normal atoms, ${}^{A}Z$, are expected when the element with atomic number Z is a noble gas. No negative ions of noble gases are known exept for He. The negative search for Ar^- ions is described in Kubik et al. (1984a). With these considerations, low limits for PEP violation are expected from the search for anomalous ${}^{20}\widetilde{Ne}$ in fluorine or for anomalous ${}^{36}\widetilde{Ar}$ in chlorine samples.

Two assumptions are made:

1. the chemical behaviour of anomalous atoms of the element Z is the same as of normal atoms of the element Z-1;

2. negative ions of anomalous noble gas atoms exist.

3. Method of Detection

For the detection of the anomalous ${}^{20}\widetilde{Ne}$ ions AMS with the time-of-flight set-up (Müller et al. (1989)) at the Munich accelerator laboratory was used (Fig. 3). The start signal is obtained from a channel plate detector. The ions are identified by measurements of energy loss with an ionization chamber and of residual energy with a surface barrier detector from which also the stop signal is derived. The distance from the start to the stop detector is 2.7 m.

The anomalous ${}^{36}\widetilde{Ar}$ ions were detected using AMS with postaccelerated heavy ion beams (see e.g.: Kubik et al. (1984a and 1984b)). This AMS set-up is frequently used at the Munich accelerator laboratory to identify the radioisotopes ^{26}Al, ^{36}Cl,

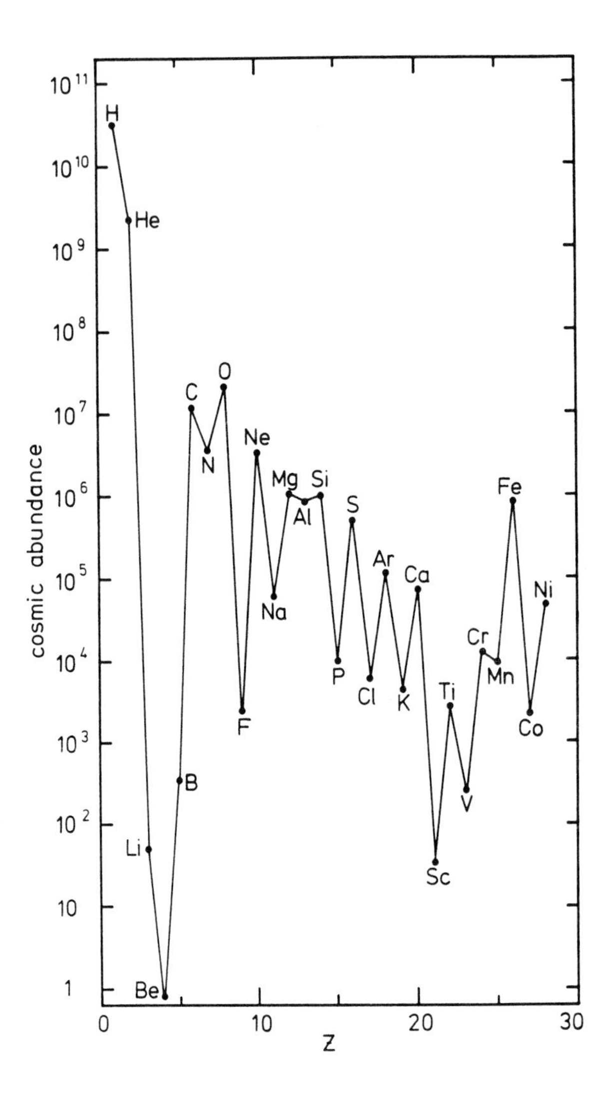

Figure 2: Cosmic abundances (Cameron (1970)) of elements up to nickel.

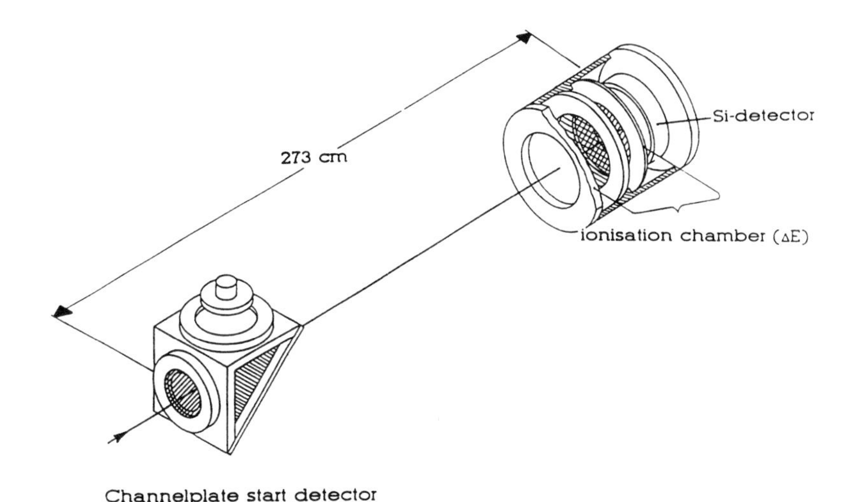

Figure 3: AMS measurement with the time-of-flight set-up.

^{41}Ca, ^{53}Mn and ^{59}Ni by the method of complete stripping. In the case of $^{36}\widetilde{Ar}$, hydrogenlike ions in the charge state 17^{+} were looked for.

4. Measurements and Results

For the search for anomalous Ne atoms, a natural CaF_2 sample was used in the ion source. Negative fluorine ions were extracted from the ion souce. All settings were for mass 20. After a magnetic 90° and electric 12° deflection, the ions were injected into the Tandem, stripped to charge state 5^{+} in the Terminal and accelerated to 46 MeV. In this run, typical F^{-} currents were about 10μA and the transmission from the ion source to the detector was 8%. A two dimensional spectrum E vs. v^2 (E energy, v velocity) is shown in Fig. 4. In this representation, the slope of the linear functions is proportional to the mass. Fig. 5 shows the mass spectrum obtained as $\frac{2 \cdot E}{v^2}$. No mass 20 event of neon was observed. This result yields an upper limit for the $^{20}\widetilde{Ne}$ concentration

$$\frac{[^{20}\widetilde{Ne}]}{[F]} < 6 \cdot 10^{-18}.$$

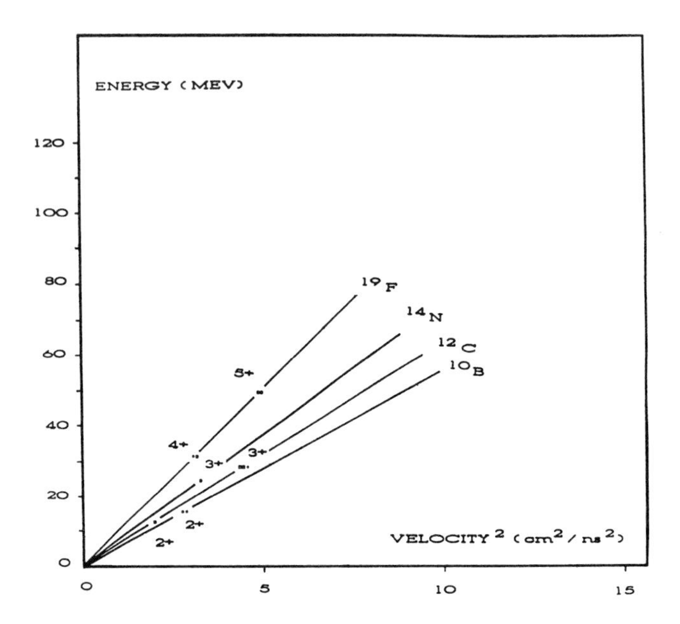

Figure 4: Two-dimensional spectrum: energy vs. (velocity)^2 measured for the search for anomalous $^{20}\widetilde{Ne}$ ions.

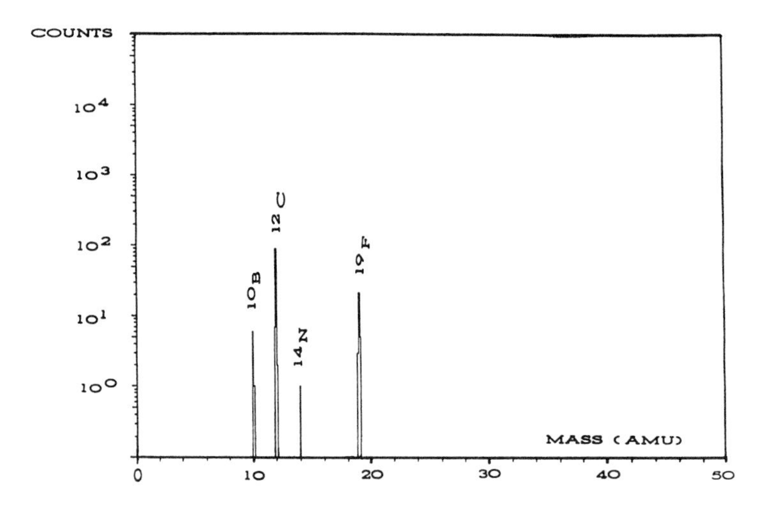

Figure 5: Mass spectrum obtained as $\frac{2 \cdot E}{v^2}$ (E energy, v velocity)

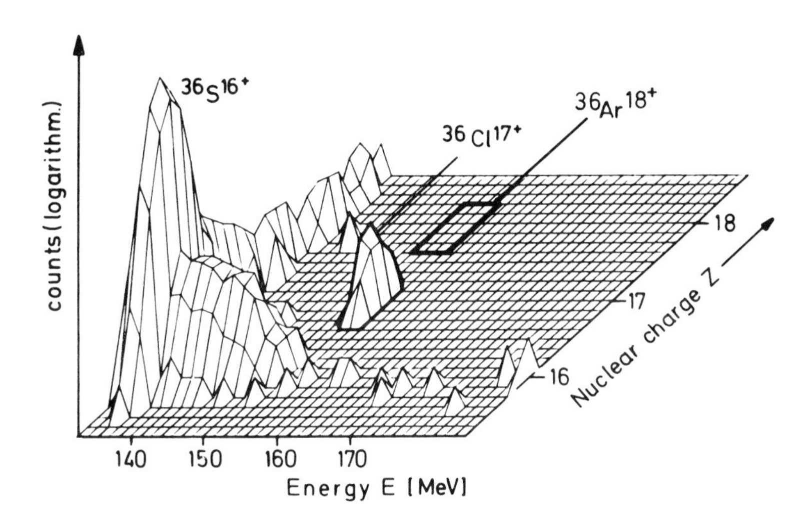

Figure 6: Two-dimensional spectrum: Z vs. E measured for the search for anomalous $^{36}\widetilde{Ar}$ ions.

With the cosmic abundances (Cameron (1970))

$$\frac{P(Ne)}{P(F)} = 3300 \qquad and \qquad P(^{20}Ne) = 90\%,$$

we obtain

$$\frac{[^{20}\widetilde{Ne}]}{[^{20}Ne]} < \frac{6 \cdot 10^{-18}}{3300 \cdot 0.9} = 2 \cdot 10^{-21}.$$

In the past, many AMS measurements were made with completely stripped ^{36}Cl ions. The sample in the ion source was AgCl. With the Tandem and the rf booster, the ^{36}Cl ions were accelerated to 4.3 MeV/ nucleon, analysed by a 90° magnet, stripped completely, analysed again by a 90° magnet and detected by a Bragg curve ionization chamber. The result of one experiment (Kubik et al. (1984a)) is shown in a two dimensional spectrum, Z vs. E (Fig. 6). No event of mass 36 ions in the charge state 17^+ with the energy loss ($\hat{=}$ Z) of argon ions was detected. This single run gives the upper limit for the concentration of anomalous $^{36}\widetilde{Ar}$ ions:

$$\frac{[^{36}\widetilde{Ar}]}{[Cl]} < 8 \cdot 10^{-16}.$$

With the cosmic abundances (Cameron (1970))

$$\frac{P(Ar)}{P(Cl)} = 22 \qquad and \qquad P(^{36}Ar) = 84\%,$$

we obtain

$$\frac{[^{36}\widetilde{Ar}]}{[^{36}Ar]} < \frac{8 \cdot 10^{-16}}{22 \cdot 0.84} = 4 \cdot 10^{-17}.$$

By analysing all ^{36}Cl data, this limit can be improved by three orders of magnitude. In the case of $^{20}\widetilde{Ne}$, by running the experiment with a higher current of more than 100 μA for several days and by improving the transmission to 30%, the limit given for $^{20}\widetilde{Ne}$ also can be improved by two to three orders of magnitude.

References

Amado R D and Primakoff H 1980, Phys. Rev. **C22** 1338

Cameron A G W 1970, Space Science Rev. **15** 121

Gavrin V N, Ignatiev A Yu and Kuzmin V Z 1988, Phys. Lett. **B206** 343

Govorkov A B 1983, Fiz. Elem. Chastits At Yadra **14** 1229

Govorkov A B 1989, Phys. Lett. **A137** 7

Green H S 1953, Phys. Rev. **90** 270

Greenberg O W and Mohapatra R N 1987, Phys. Rev. Lett. **59** 2507

Greenberg O W and Mohapatra R N 1989, Phys. Rev. Lett. **62** 712

Ignatiev A Yu and Kuzmin V A 1987, Yad. Fiz. **46** 786

Kovalchuk E L, Pomansky A A and Smolnikov A A 1979, Pis'ma Zh. Eksp. Teor. Fiz. **29** 145

Kubik P W, Korschinek G and Nolte E 1984a, Nucl. Instr. and Meth. **B1** 51

Kubik P W, Korschinek G, Nolte E, Ratzinger U, Ernst H, Teichmann S, Morinaga H, Wild E and Hille P 1984b Nucl. Instr. and Meth. **B5** 326

Kuzmin V Z 1984, Proc. 3. Sem. Quantum Gravity, eds. Markov M et al, (World Scientific, Singapore) p 270

Logan B A and Ljubicic A 1979, Phys. Rev. **C20** 1957

Messiah A M L and Greenberg O W 1964, Phys. Rev. **B136** 248

Moe M K and Reines F 1965 Phys. Rev. **B140** 992

Müller D, Faestermann T, Gillitzer A, Koerner H J, Korschinek G, and Scheuer R 1989 Nucl. Instr. and Meth., to be published

Okun L B 1988, Festschrift for Val Telegdi, ed. Winter K, (Elsevier Science Publ.) p 201

Plaga R 1989, Z. Phys. A, to be published

Ramberg E and Snow G A 1989, to be published

Reines F and Sobel H W 1974, Phys. Rev. Lett. **32** 954

Rubakov V A and Spiridonov V P 1988, Mod. Phys. Lett. **A3** 1337

Inst. Phys. Conf. Ser. No 105
Paper presented at Int. Conf. on Spectroscopy of Heavy Nuclei, Crete, Greece, 1989

Fission fragment spectroscopy

J.L. Durell

The Schuster Laboratory, University of Manchester, U.K.

ABSTRACT: New data on intermediate spin states in neutron-rich nuclei have been obtained by studies of prompt γ-rays emitted by fission fragments.

1. Introduction

The fission process has been long established as a mechanism for the production of neutron-rich nuclei. The predominant focus of study has been the β-decay of the primary fission products i.e. those nuclei produced after prompt neutron emission. Detailed studies have been made of β-decay strength functions, β-delayed neutron emission, and, most extensively, the spectroscopy of secondary fission products. These β-decay studies have not only produced wide-ranging and interesting nuclear structure information, but have also played a vital part in the establishing of fragment mass and charge distributions following fission. In contrast to this large body of data, the study of **prompt** γ-ray emission from primary fission products has been hampered by the technical difficulties of resolving individual γ-rays from the multitude produced by the primary products. In the 1970's pioneering work was carried out by Cheifetz et al.[1], at Berkeley, who studied prompt γ-rays following spontaneous fission. Many ground-state quasi-rotational states up to $J^\pi = 4^+$ or 6^+ were established for the first time. There followed a hiatus of many years before the technology of γ-ray spectroscopy improved sufficiently to allow more extensive studies.

With modern high-efficiency, multi-detector Ge arrays[2] it is now possible to investigate in some detail the prompt γ-radiation from primary products from spontaneous fission and heavy-ion induced fission. In the present work, using a variety of fissioning systems, yrast and near-yrast levels up to angular momenta in the range 12–15$\hbar$ have been established for even-even nuclei in the A=80–150 region. Prior work, particularly the β-decay studies of odd-odd nuclei, has provided the platform upon which our studies are based. This base has been supplied by the establishment of, at least, the energy of the first 2^+ state in many even-even neutron-rich nuclei. Given the power of $\gamma - \gamma$ coincidence techniques with modern Ge arrays, it is sufficient to know this one transition in a nucleus in order to build up a decay scheme.

The study of yrast and near-yrast structures in neutron-rich even-even nuclei is clearly of interest for the nuclear structure information that can be obtained. We have the opportunity to probe new regions in the N-Z plane which have unusually large numbers of neutrons. The extra neutrons will fill single-particle states different in character to those available to nuclei of the same atomic number lying near to the stability line: new intruder states can be occupied which may drive the nucleus toward large deformation; different combinations of single-particle orbitals become available, the residual interactions between which can lead to non-quadrupole deformations being important. Examples of these two situations will be presented in this report.

A further motivation for the study of the spectroscopy of primary fission products is that important information on the fission process itself can be obtained. One of the interesting problems in fission is the question of how the primary fragments share angular momentum. In spontaneous fission of an even-even nucleus the angular momentum arises from dynamics along the path to fission, together with the way in which this collective energy is equilibrated into internal excitation energy of the fragments. In heavy-ion induced fusion-fission a further contribution to angular momentum arises from the orbital motion of the projectile relative to the target nucleus. The major proportion of any angular momentum residing in a primary fission fragment will be retained by the primary product after prompt, post-fission neutron emission. Therefore the population pattern of levels in the primary products determined from prompt γ-ray intensities can provide reliable estimates[3] of primary fragment intrinsic angular momentum.

2. Primary Product Yields

A vital question to be addressed if one wishes to study a wide range of neutron-rich nuclei is precisely what nuclei will be populated in any given fission process. This will depend upon the following factors :-

The N/Z Ratio of the Fissioning System

In spontaneous fission this is clearly defined, but in heavy-ion induced fusion-fission this will be determined by the choice of projectile and target, together with the number of neutrons emitted from the compound nucleus prior to fission.

The Primary Fragment Mass Distribution

In most spontaneous or low energy (e.g. thermal neutron induced) fission reactions the primary mass distribution is asymmetric, with the heavy mass peak at $A\approx 144$. The light fragment peak is therefore at the mass of the fissioning system minus 144. There is little or no yield at the symmetric mass split. In heavy-ion induced fission, where shell effects are washed out by the high excitation energy of the fissioning system, the mass distribution is symmetric, peaking at half the fissioning nucleus mass. The distribution has a fwhm of typically $\sim$30 amu. For any given mass the distribution in atomic number is very narrow. Even for primary products, formed after the broadening effect of neutron emission, there are only 3 or 4 elements produced with any intensity at a given mass.

Post-Fission Neutron Emission

In spontaneous fission this is strongly influenced by shell effects in the fission fragments, as evidenced by the sawtooth distribution of the average number of emitted neutrons as a function of mass, observed in the fission of ^{252}Cf. Near the yield maxima approximately two neutrons per fragment are emitted. Heavy–ion induced fusion–fission leads to a smooth dependence of the post–fission neutron emission as a function of primary fragment mass. The average number of neutrons emitted is around 2–3 per fragment, principally determined by the temperature of the primary fragments.

By properly taking into account the above three factors, one can reliably estimate the primary products most likely to be populated in a given reaction. To a certain extent the experimenter is able to control, by judicious choice of beam, target and beam energy, the range of nuclei that can be studied. Fig. 1 shows the loci of the most probable products for three fissioning systems studied in the present work.

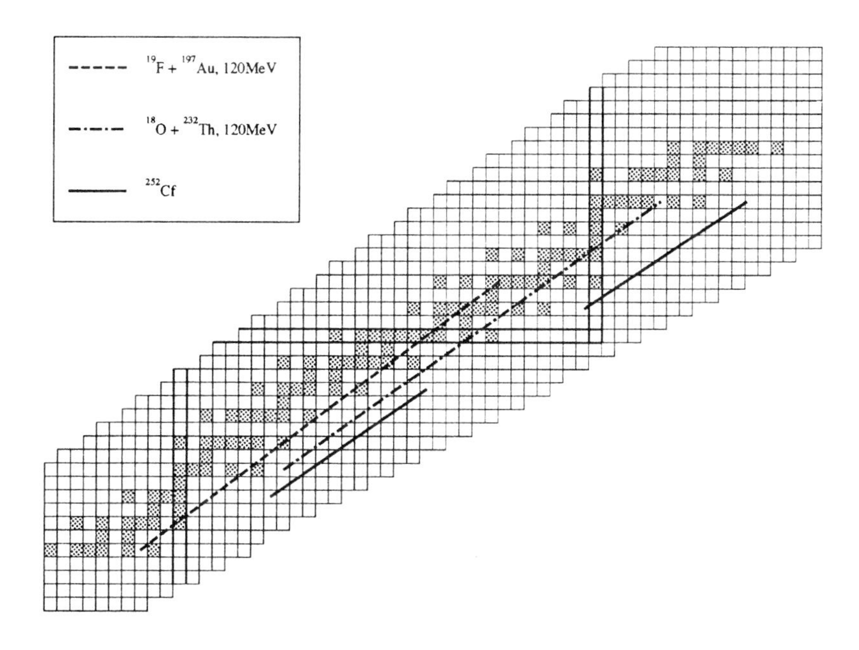

Figure 1: The loci of the most probable products for three fissioning systems, compared to the stable nuclei from Z=34 to 64

Table 1

Average Fissioning System	Production	Most Probable even–even Fragments	N/Z Ratio of Products
^{252}Cf	source	Mo + Ba	1.50(L); 1.56(H)
^{213}Ra	^{19}F + ^{197}Au	Ru + Ru	1.38
^{246}Cf	^{18}O + ^{232}Th	Cd + Sn	1.45
^{242}Am	^{7}Li + ^{238}U	Zr, Ba	1.50

3. Experimental Data

Particulars concerning the fissioning systems studied are presented in table 1. Data have been taken using the Argonne–Notre Dame Gamma–Ray Facility at the Argonne National Laboratory (ANL), and using the TESSA family of arrays at Daresbury Laboratory (DL). The analysis has proceeded through the use of the standard $\gamma - \gamma$ coincidence technique, in order to establish partial decay schemes, and to determine the intensities of γ–rays.

The first complication compared to (HI, xn) reactions arises from the fact that γ–ray cascades from 60 or more nuclei are produced at the same time. This clearly increases the problems of data analysis. A further difficulty arises from the existence of complementary fragments. When gating on the $2^+ - 0^+$ transition in a given primary product, not only will the prompt γ–rays within that nucleus be observed, but also γ–rays in two or three or more complementary fragments, since these γ–rays will also be in prompt coincidence. The construction of a decay scheme for one even–even nucleus therefore requires that the intense γ–rays from associated fragments are known.

The limitations in observing weak γ–ray peaks and placing the transitions in a decay scheme, arise from impurities in coincidence gates. These impurities arise from γ–rays in other (sometimes unknown) nuclei, which have the same energy as the known transition upon which the gate was set. These problems can be overcome using triple and higher–fold coincidences. The newly proposed γ–ray arrays will supply sufficient statistics in high–fold coincidence data to help solve this problem.

4. Neutron–Rich Nuclei in A=100–120 Region

One of the fruitful and interesting areas of investigation in the present work, has been the study of the neutron–rich isotopes of the elements from Zr to Cd. Using data from all the fissioning systems we have been able to build–up systematic information on the following nuclei: $^{96-102}$Zr, $^{98-108}$Mo, $^{102-116}$Pd and $^{112-122}$Cd.

Abrupt changes in nuclear structure occur in neutron–rich nuclei with A∼100. This is evident most clearly from the sharp changes in E_2, the energy of the first excited 2^+ state, as the neutron number increases. The isotopes of Zr show the most dramatic changes (see fig. 2), the energy of the first 2^+ state changing from 1223 keV in ^{98}Zr to 212 keV in ^{100}Zr. The downward trend is continued for ^{102}Zr in which E_2 = 152 keV. The measured lifetime[4)] of the 2^+ state of ^{100}Zr leads to a deduced $\beta_2 \approx 0.4$. There

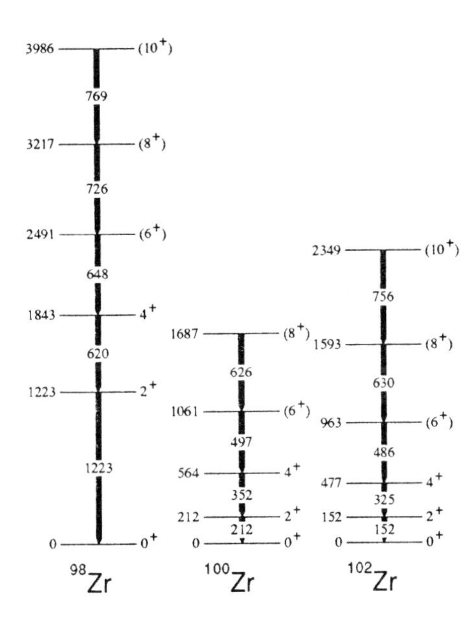

Figure 2: The yrast levels of the neutron-rich Zr isotopes

appears therefore to be a strong drive toward large deformation as neutron number increases. There was evidence from the energies of the 2^+ states in the known neutron-rich isotopes of Mo and Ru that the deformation also increases for these nuclei. The present work has established the yrast levels of the Zr, Mo and Ru isotopes up to $J = 10 - 14\hbar$. We therefore have the opportunity of looking in more detail at the structure changes. Figs. 3 and 4 show the systematics of the ground-state "bands" of the neutron-rich Mo and Ru isotopes. The higher-lying members of these bands have been established for the first time in the current work. The assignment of spins to these levels is based upon systematics. Many of the less neutron-rich isotopes populated in the fission reactions have been previously studied by α-particle or light-ion induced fusion-evaporation reactions, in which spin assignments have been made upon the basis of angular distributions. In these cases we can confirm that the predominant population of these nuclei as fission fragments is through the ground-state yrast cascade. This has been assumed to be true for all other even-even nuclei.

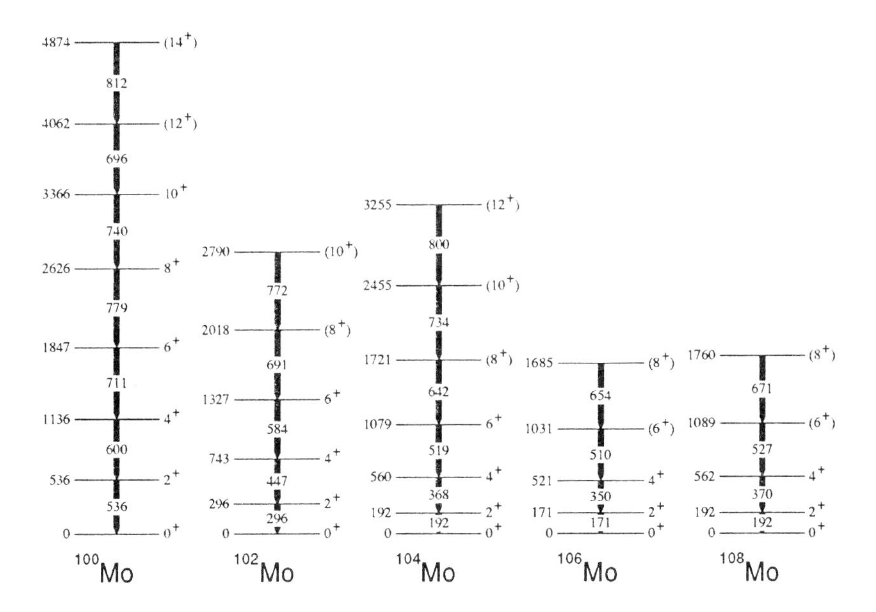

Figure 3: Systematics of the g.s. bands of neutron-rich Mo isotopes

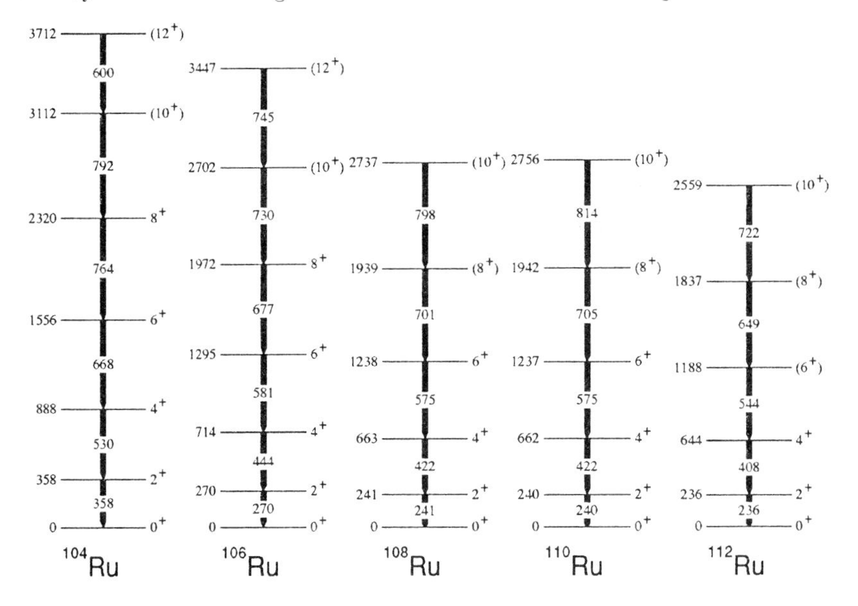

Figure 4: Systematics of the g.s. bands of neutron-rich Ru isotopes

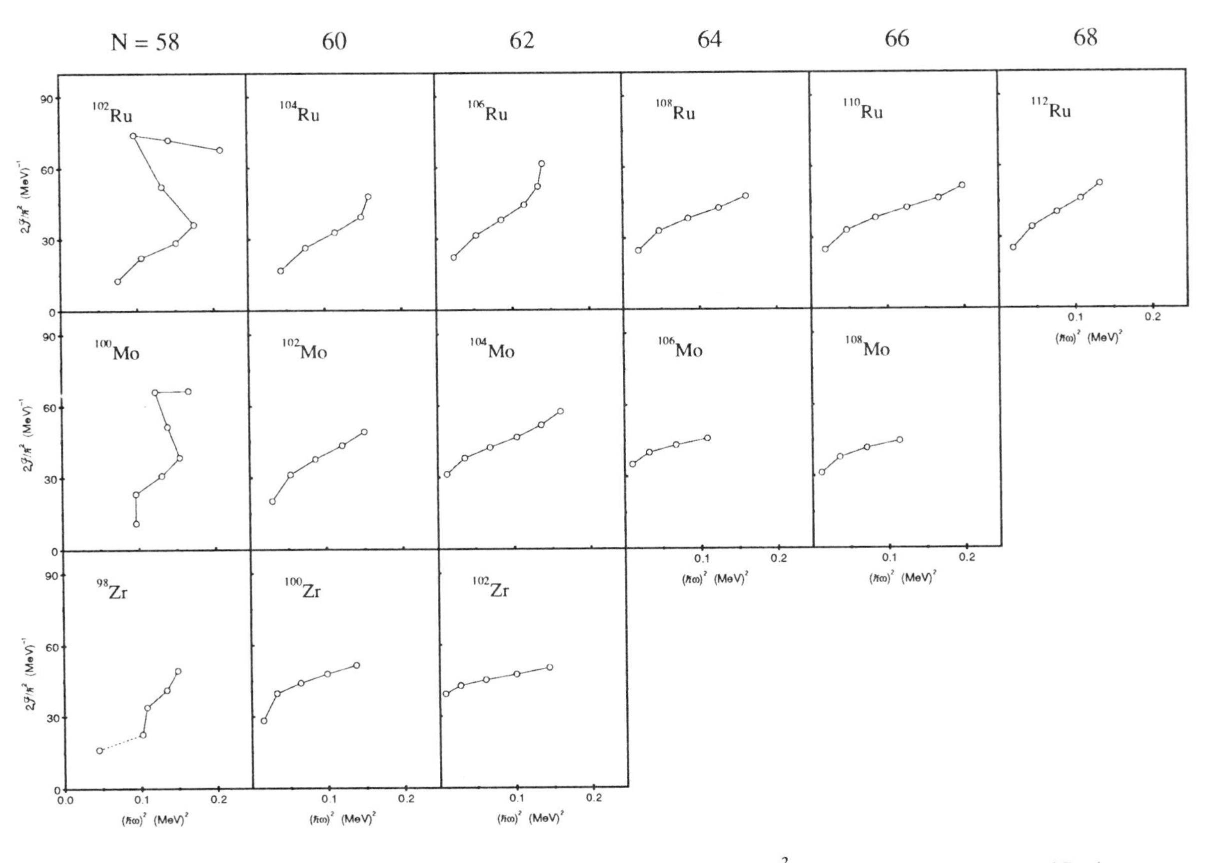

Figure 5: Moments of inertia plotted versus (angular frequency)^2 for neutron–rich Zr, Mo and Ru isotopes.

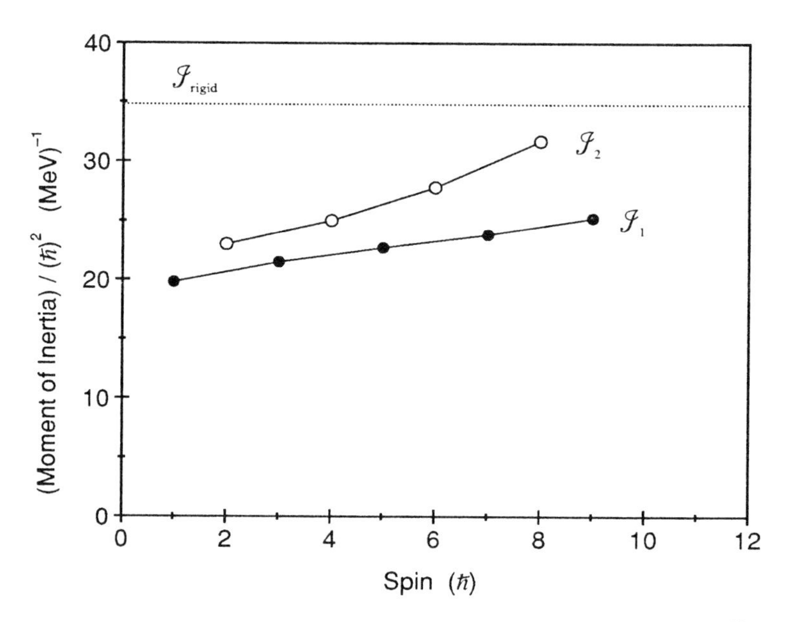

Figure 6: Moments of inertia $\mathcal{J}_1$ and $\mathcal{J}_2$ plotted versus spin for ^{102}Zr.

Fig. 5 summarises the information obtained on the yrast levels of Zr, Mo and Ru nuclei with neutron numbers ranging from N=58 to 68. It shows the kinematic moments of inertia versus the square of the rotational frequency. The curves illustrate the general trend towards good rotational patterns as the neutron number increases. The filling of the neutron major shell leads to significant population of the $h_{11/2}$ intruder state. This is clearly the candidate for the cause of the trend towards deformation in the ground state. The strange shapes of the curves at N=58 are suggestive of competition between spherical states based upon the Z=40 sub-shell closure, and deformed states associated with Z=38. It is now well-established in very neutron-deficient nuclei with A=76 and 80 that Z=38 is a magic number at large deformation. As more neutrons are added the deformed states become yrast and the degree of deformation clearly depends upon the details of the residual interactions between the neutrons and protons. It can be seen that as Z increases so does the number of neutrons required to minimise the excitation energy of the first 2^+ state. A most important open question remains: is ^{104}Zr more deformed that ^{102}Zr? It has not yet been possible to identify ^{104}Zr in our data, but we are searching to answer this question.

Even at the lowest spins, ^{102}Zr shows an yrast spectrum corresponding closely to that of a perfect rotor. Fig. 6 shows the kinematic and dynamic moments of inertia plotted against spin. These are seen to approach the rigid body value for a prolate spheroid with β_2 = 0.43. Their near constant and equal values suggest that there is very little aligned angular momentum. In terms of the standard formula

$$E_{\rm rot} = A\ I(I+1) + B\ I^2(I+1)^2$$

the B/A ratio, regarded as a measure of goodness of a rotor, is $\sim 1.0\ 10^{-5}$. Thus ^{102}Zr is the best example known of a good rotor at near zero spin and excitation energy.

What happens as more protons are added? How do the neutron-rich isotopes of Pd and Cd behave? It may well be expected that the approach to the Z=50 spherical magic number will lead to a stabilisation against the drive to deformation. In general terms this appears to be true, as shown in figs. 7 and 8, which show the systematics of the yrast sequences in Pd and Cd. There is very little change in the excitation energy of the first 2^+ states.

The Cd isotopes have long been considered to be amongst the best examples of vibrational nuclei. The vibrational characteristics of these nuclei have been established principally on the basis of the observation of the two-phonon triplet of states with $J = 0, 2, 4$. Recently, interest in these nuclei has been stimulated by the proposal of Aprahamian et al.[5], that members of the three-phonon quintuplet ($J = 0, 1, 2, 3, 4, 6$) in ^{118}Cd are populated in the β-decay of ^{118}Ag, produced following neutron-induced fission of ^{235}U. We have been able to extend the level scheme of ^{118}Cd, and also, those of ^{120,122}Cd, by the observation of prompt γ-rays following fission. These data were obtained from the ^{7}Li + ^{238}U experiment at Daresbury Laboratory.

Fig. 9 shows the partial decay scheme of ^{118}Cd established in the present work. Up to the level at 2223 keV the decay scheme agrees with that published by Aprahamian et al. The higher-lying levels have been observed for the first time in the present work. On the right-hand side of the figure are shown the levels populated in the β-decay of ^{118}Ag, but not observed in the prompt γ-ray study. This is consistent with these levels being non-yrast, low-spin states. The open arrows in the decay scheme indicate γ-rays not observed by us, but which must be there given the measured[5] branching ratios. The intensities of these γ-rays lie below the level of sensitivity of the present experiment. The non-observation of transitions from the 1929 keV level indicates that this level is not directly populated with significant intensity in the fission product.

Aprahamian et al. suggest that the 1916, 1929, 1936, 2074 and 2092 keV levels are the states of the three-phonon quintuplet. They determined the spins and parities of the 1916, 2074 and 2092 keV states as $J^\pi = 2^+$, 0^+ and 3^+ respectively. They propose that the 1929 and 1936 keV states are the $J = 4$ and 6 members of the multiplet. The present work demonstrates that the 1936 keV state is a member of the yrast sequence, and hence most probably does have $J^\pi = 6^+$. The non-population of the 1929 keV level in the fission process is consistent with it being a non-yrast 4^+state. In ref.[5] it is also remarked that the β-decay populates high-lying states with relatively enhanced γ-decays to the proposed quintuplet. In the present work we see two new states with this characteristic, which presumably have higher angular momentum. These states are the proposed 8^+ member of the yrast sequence, and the level at 2640 keV, both of which decay only to the 6^+ state. It is possible therefore that there do exist states in ^{118}Cd that contain components of a four-phonon structure in their wavefunctions.

Partial decay schemes for ^{120,122}Cd have also been established. The level scheme for ^{120}Cd (see fig. 10) shows similar characteristics to that for ^{118}Cd. Unfortunately no extra levels, of lower spin, over and above those shown in fig. 10 have been observed

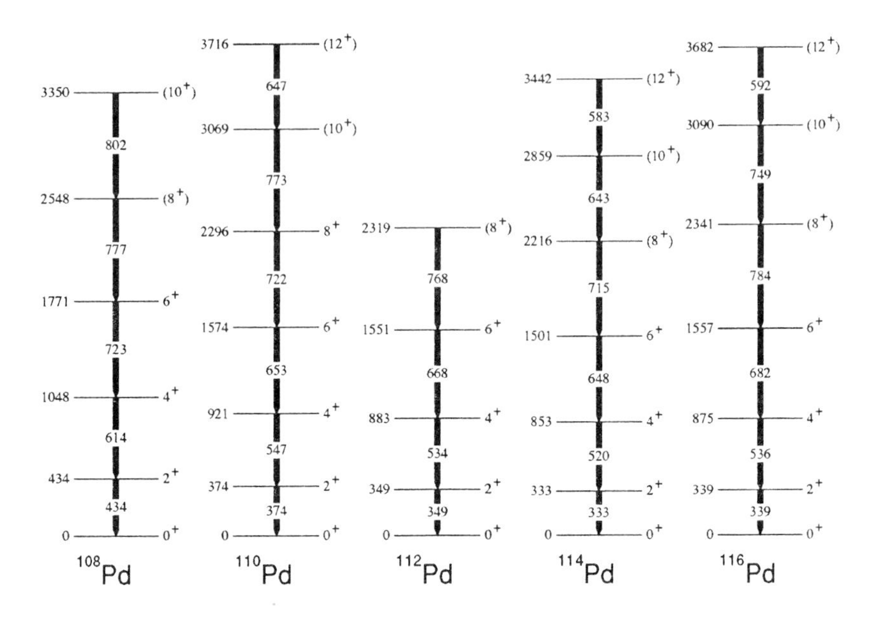

Figure 7: Systematics or the yrast states of neutron-rich Pd isotopes

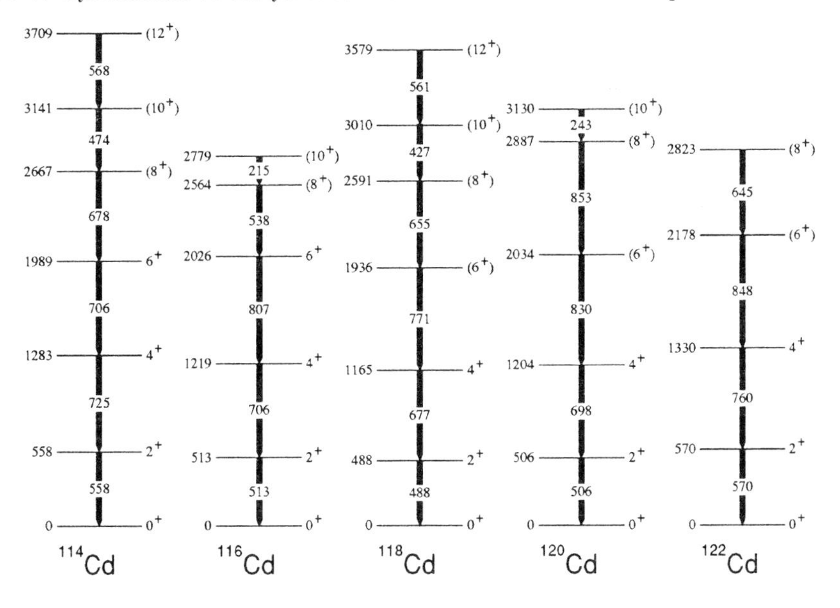

Figure 8: Systematics of the yrast states of neutron-rich Cd isotopes

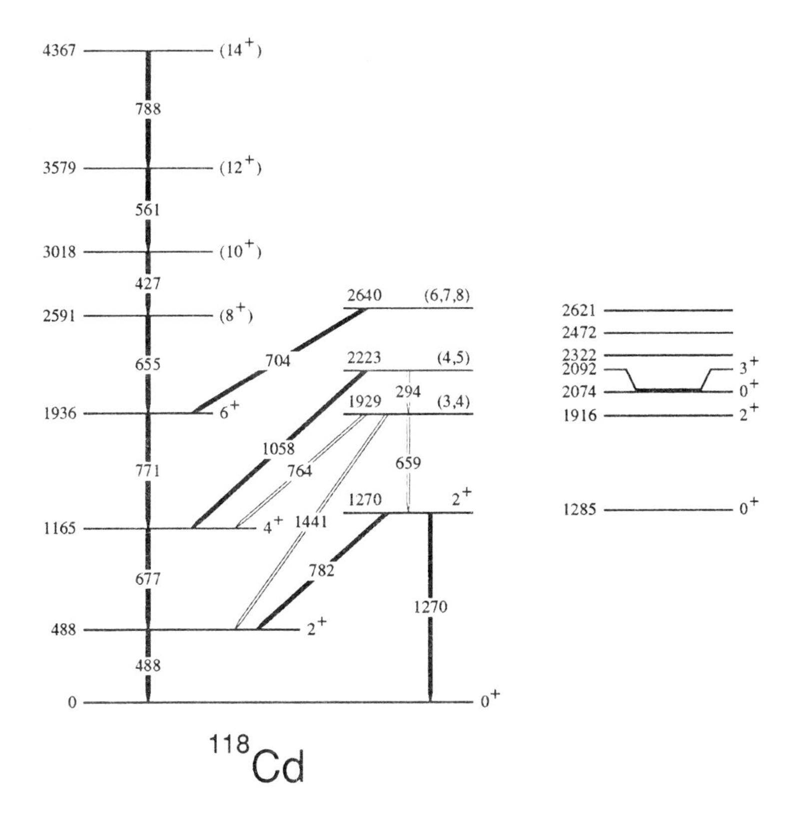

Figure 9: Partial decay scheme of ^{118}Cd determined from prompt γ-ray coincidence data. The levels on the right of the figure are those populated in β-decay

in β-decay. Prior to the present work only two γ-rays had been associated[6] with ^{122}Cd. Fig. 11 shows the level scheme determined from the fission study. This very neutron-rich isotope of Cd is only weakly produced, and hence the level scheme is not as extensive as those for the lighter isotopes.

It appears that our observation of the intermediate spin states in the neutron-rich isotopes of cadmium has provided some support to the harmonic vibrator description of ^{118}Cd proposed by Aprahamian et al. Whether the higher spin states can be reconciled with this description requires detailed calculation. As can be seen in fig. 8, there appears to be a change in the yrast sequence of the neutron-rich Cd isotopes at the proposed 10^+ level. This suggests a change in configuration, which may imply that deformed structures of the kind discussed above in the Zr, Mo and Ru nuclei may be coming into play. It should be noted, however, that such an abrupt change in the yrast sequence is not observed in the Pd nuclei (see fig. 7).

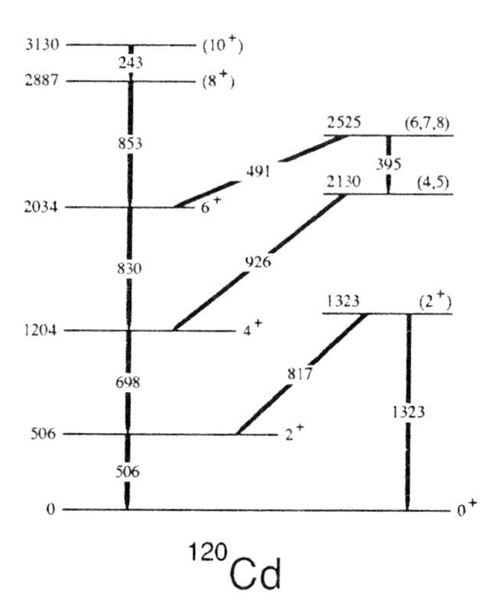

Figure 10: Partial decay scheme of ^{120}Cd

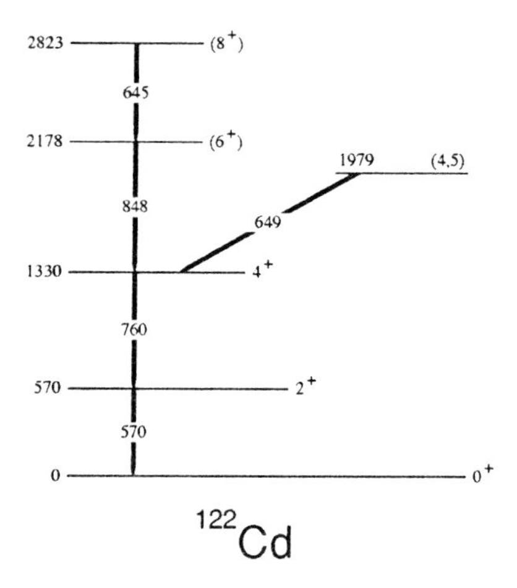

Figure 11: Partial decay scheme of ^{122}Cd. Only the 570 keV and 760 keV γ-rays have been reported previously.

For all of the nuclei discussed here it is clear that measurements of γ-ray transition rates would greatly advance our understanding. For most of these nuclei it is difficult to envisage how such measurements may be forthcoming.

5. Octupole Effects in Z~56, N~88 Nuclei

There has been considerable recent interest in nuclei which exhibit level sequences suggestive of rotational bands arising from reflection asymmetric systems. Such states were first observed[7] in the light actinides, and subsequently[8] in neutron–rich, deformed Ba and Ce isotopes. The latter study was carried out using a ^{252}Cf spontaneous fission source by observing the prompt γ–radiation. In all these cases, in the even–even nuclei, it appears that the reflection asymmetric structure is stabilised by nuclear rotation, since it is only above $J = 6$ that the classic band structure is approached. In nuclei near $Z = 56$, $N = 88$ the valence protons occupy levels based upon the spherical $h_{11/2}$ and $d_{5/2}$ orbits, while the valence neutrons are occupying $i_{13/2}$ and $f_{7/2}$ states. The residual interactions between the $\Delta j = \Delta \ell = 3$ single–particle states lead to strong octupole correlations which play an important role in the understanding of the rotational structures.

The spontaneous fission process is ideal for the study of ^{144,146}Ba and ^{146,148,150}Ce, since this process strongly produces such neutron–rich nuclei. However, heavy–ion induced fission provides a means of investigating less neutron–rich nuclei. We may therefore study the $N = 88$ nucleus ^{148}Nd for which octupole correlations should be important. We can also obtain data on lighter isotopes of Ce and Ba, which are more transitional in nature, whose yrast structures will be more complex.

The data to be presented here have been obtained from the ^{18}O + ^{232}Th and ^{7}Li + ^{238}U reactions carried out at Daresbury Laboratory. Fig. 12 shows the deduced level scheme for ^{148}Nd. It can be seen that between spins of 6 and 12 the expected sequence characteristic of reflection asymmetry is observed. The nucleus ^{148}Nd has $N = 88$, and has a level sequence similar to the Ce and Ba $N = 88$ isotones, confirming that at this neutron number such structures are most clearly developed.

Fig. 13 shows the partial decay schemes of the $N = 86$ nuclei ^{142}Ba and ^{144}Ce. In the case of ^{144}Ce we do not see any members of the positive parity yrast sequence above the proposed 6^+ state. This is in sharp contrast to all other nuclei in this mass region that have been populated as fission products. On the other hand, the proposed level scheme for ^{142}Ba shows similar behaviour[9] to other $N = 86$ nuclei, ^{146}Nd and ^{148}Sm, where interconnected positive and negative parity levels occur only over a limited range.

Further information may be obtained by making estimates of the transition dipole moments. This can be done by taking the appropriate transitions to be between members of a single rotational band with constant intrinsic quadrupole moment given by the measured B(E2: $2^+ - 0^+$) values. The intrinsic transition electric dipole moments so obtained are shown in table 2, where they are compared with theoretical predictions[10]. It should be emphasised that the experimental values are average values obtained from the branching ratios of states with $J \geq 7$. There is generally good agreement between the theoretical calculations and experimental estimates.

6. Conclusions

The study of fission fragment spectroscopy through prompt γ–radiation has provided a large body of new data on intermediate spin states in nuclei inaccessible by any

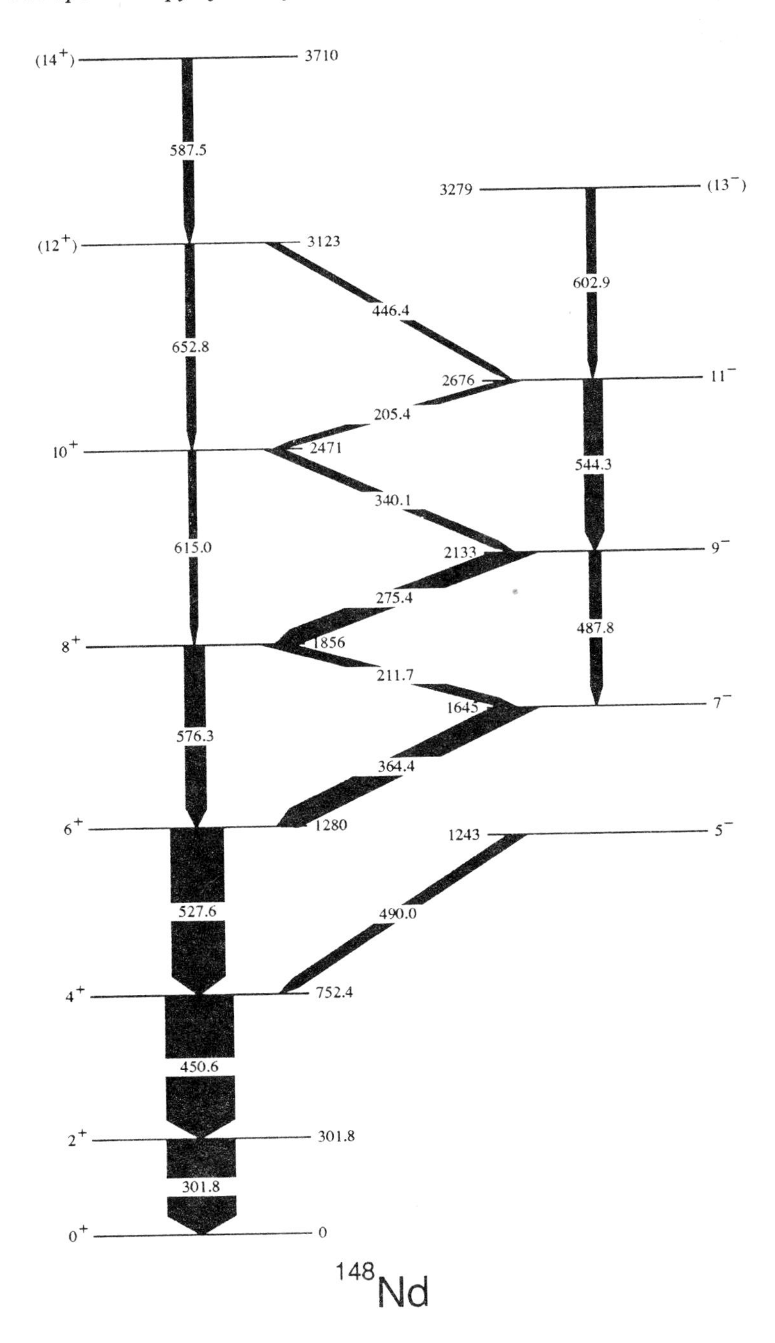

Figure 12: Decay Scheme of ^{148}Nd

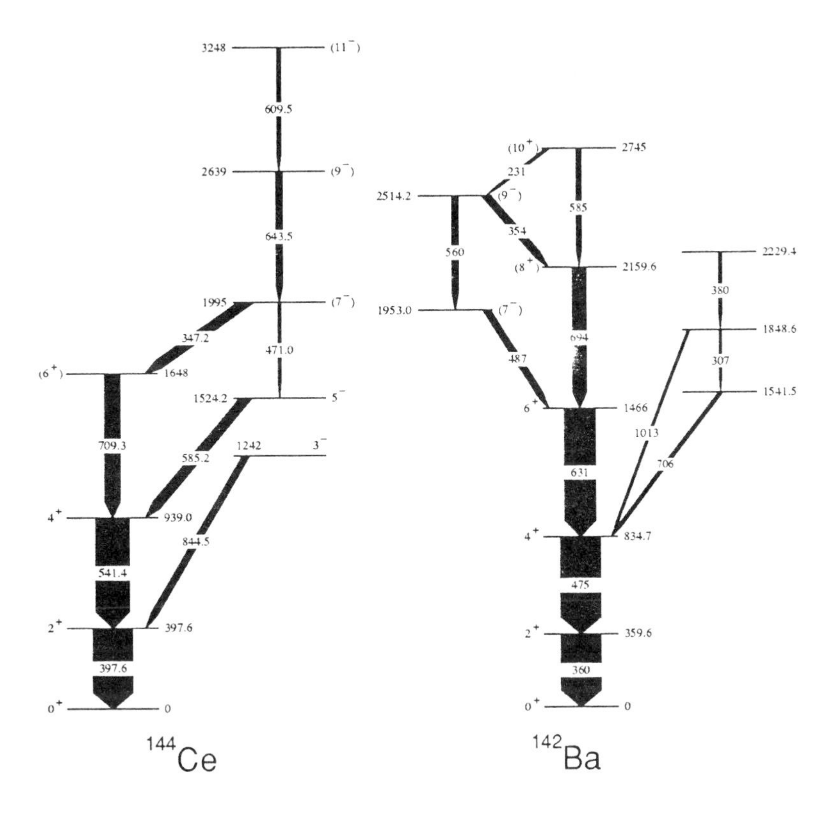

Figure 13: Decay schemes for ^{144}Ce and ^{142}Ba

Table 2

Nucleus	Do (exp) e fm	Do (theory) e fm
^{142}Ba	0.13 ± 0.02	—
^{144}Ba	0.13 ± 0.01	0.09
^{146}Ba	0.04 ± 0.01	0.03
^{144}Ce	0.17 ± 0.05	0.16
^{146}Ce	0.16 ± 0.04	0.18
^{146}Nd	0.18 ± 0.01	0.20
^{148}Nd	0.23 ± 0.03	0.22
^{150}Sm	0.20 ± 0.01	0.25

other reaction process. The information obtained extends and is complementary to the data gathered from β–delayed spectroscopy.

Two specific regions of nuclei have been discussed as examples of the way in which these new results can elucidate interesting nuclear structure topics. The present work has concentrated upon even–even nuclei. This arises from the basic simplicity of the γ-decay of even–even nuclei — essentially all the strength passes through the first 2^+ state. It is evident that the interplay between collective and single particle degrees of freedom will be more clearly revealed by the study of odd–A and odd–odd nuclei. This remains the next challenge.

Acknowledgements

The work presented in this paper was carried out in collaboration with the following colleagues at Manchester: Y. Abdelrahman, J.B. Fitzgerald, M.A.C. Hotchkis, A.S. Mowbray and W.R. Phillips.

We are grateful to I. Ahmad, R.V.F. Janssens and T.-L. Khoo at the Argonne National Laboratory for their part in the experiments carried out there.

References

1. E. Cheifetz et al., Phys. Rev. Lett. **25** (1970) 38
2. J. Simpson, Proceedings of this conference
3. Y. Abdelrahman et al., Phys. Lett. **199B** (1987) 504
4. E. Cheifetz et al., Nuclear Spectroscopy of Fission Products, IOP Conference Series 51 (1980) 193
5. A. Aprahamian et al., Phys. Rev. Lett. **59** (1987) 535
6. K. Fransson et al., Nucl. Instr. and Methods **113** (1973) 157
7. D. Ward et al., Nucl. Phys. **A406** (1983) 591,
   P. Schuler et al., Phys. Lett. **174B** (1986) 241
8. W.R. Phillips et al., Phys. Lett. **212B** (1988) 402; Phys. Rev. Lett. **57** (1986) 3257
9. W. Urban et al., Phys. Lett. **185B** (1987) 331; Phys. Lett. **200B** (1988) 424
10. W. Nazarewicz, Proc. Int. Conf. on Nuclear Structure through Static and Dynamic Moments (Melbourne, 1987) p. 180

Delayed-coincidence measurement of subnanosecond lifetimes in fission fragments

H. Ohm, M. Liang, G. Molnar and K. Sistemich

Institut für Kernphysik, Kernforschungsanlage Jülich, Postfach 1913,
D–5170 Jülich, FRG

ABSTRACT: Level lifetimes in short–lived fission fragments have been measured through delayed ßγ coincidences with a fast plastic scintillator and a BaF_2 crystal. A timing resolution of 160 ps FWHM (E_γ> 500 keV) facilitated the determination of lifetimes down to few picoseconds. Several short–lived fission products around mass number A=100 have been explored. An unusually large octupole strength of B(E3) = 66(+33,–16) spu has been derived from the 3_1^- lifetime measured in ^{96}Zr, which underlines the doubly magic character of this nucleus. The lifetime of the first excited level in the ^{101}Nb ground–state rotational band has been measured to 84(4) ps yielding a deformation of ß = 0.39(5).

1. INTRODUCTION

Level lifetimes in fission products far away from the line of ß stability can only be measured through delayed coincidences. The most common technique has been that with Ge detectors. Despite of the poor timing resolution of these detectors half–lives around 100 ps and even below have been determined [1,2] in special cases, albeit with large uncertainties. Higher precision in this range and access to even shorter half–lives is now possible with the new technique employing a fast plastic scintillator and a BaF_2 crystal in a delayed ßγ coincidence arrangement [3]. This technique has now been applied at the fission–product separator JOSEF of the KFA Jülich, FRG. Special attention has been paid to the A=100 region where lifetime data are still scarce although they are crucial for the understanding of the unusually sudden transition from spherical nuclear shapes to deformed ones.

2. EXPERIMENTS

The time information is obtained from the delay in the coincidences between the ß$^-$ rays feeding into the nucleus of interest and its γ radiation. The decay of fission products has been studied at the separator JOSEF of the KFA Jülich [4] which delivers fission products regardless of their chemical properties, thus giving access to nuclei like the refractory elements which are not easily available at ion–source separators. The separator delivers a range of several mass numbers simultaneously rather than a single nuclide. This allows determining the prompt time centroid vs. γ–ray energy curve using known lifetimes in neighbouring nuclei, when the centroid–shift method is applied. In this way, systematic errors are avoided which could occur in a prompt curve determination with external calibration sources. Separated fission products are collected in a gas filled thermalization chamber and deposited on a moving tape system by a gas jet. The activity is then transported to the detector arrangement in a start–stop mode with a properly chosen cycle time.

The ß detector consists of a 4 mm thick disk of NE111A with a diameter of 30 mm coupled to a XP2020 photomultiplier. High–energy ß particles deposit in this detector an energy of about 800 keV which leads to a broad peak in the pulse–height spectrum the position of which proved constant for all fission products measured in this experiment. The distance between the ß detector and the source was kept as short as 5 mm in order to avoid broadening of the timing distributions due to time–of–flight effects. Gamma radiation is measured with a BaF_2 crystal coupled to a XP2020Q photomultiplier tube. The shape of the crystal is that of a truncated cone with top and base diameters of 20 mm and 30 mm, respectively. The height of the crystal is 15 mm for γ–ray energies below 1 MeV and 30 mm for higher energies. The energy resolution at 662 keV is about 10% for both crystals. With the timing signals for both detectors taken from the 9th dynode the timing resolution in ßγ coincidence is 160 ps FWHM at γ–ray energies of about 550 keV.

Because of the limited energy resolution of the BaF_2 detector a Ge detector has been included in a ß$\gamma\gamma$ triple coincidence arrangement. The Ge detector selects the nuclide of interest which is necessary because of the complexity of the mixture of nuclei present in the radioactive source. Furthermore, together with the pulse–height information from the BaF_2 detector it defines the level the lifetime of which is actually measured. A narrow time gate of 200 ns was set on the Ge detector to keep the contribution of chance coincidences low. Time calibration of the fast–timing system was made with an Ortec 462 Time Calibrator. The pulse–height information for the ß particles and the γ rays detected in both γ–ray detectors together with the start–stop time difference between the two fast detectors was stored an magnetic tape event by event.

3. DATA ANALYSIS

The listmode data were sorted off–line with one gate set on the ΔE peak in the ß spectrum and several peak gates with corresponding background gates on either side in the γ–ray spectra both of the BaF_2 and the Ge detector. Time distributions were created by subtracting background distributions in both γ–ray energy directions.

Long lifetimes were determined with a computer program by fitting an exponential decay curve to the time distribution. For shorter lifetimes which are comparable to the timing resolution, a fit of an exponential decay folded with the shape of the prompt time distribution was performed. The prompt time distribution was well approximated by a gaussian distribution. The computer program allows fitting of two components to take care of possible contaminations due to complex peaks in the γ–ray spectra. The first component can be prompt or delayed. The second component is delayed with the option of an initial growth of the decay curve due to feeding of the long–lived level by a delayed γ–ray.

For level lifetimes shorter than the timing resolution the centroid–shift analysis was applied the prerequisite of which is the knowledge of the time–centroid position for prompt transitions. As this quantity usually shows some energy dependence a greater number of γ–rays emitted from levels with known lifetime is needed were the energies cover the whole range of interest. With this lifetime information the experimental time–centroid can be shifted back to the prompt position. In the present experiments γ transitions from fission products were used as internal calibration. In some cases where the lifetime was not known before the result from the slope analysis of the corresponding time distribution could be used for the correction, see Fig. 1.

4. RESULTS

The following data except those for ^{96}Zr have been measured during an experiment dedicated to ^{101}Nb and its close neighbours. Here the smaller BaF_2 crystal was used since most of the γ transitions to be measured are of low energy. The time walk between E_γ= 120 keV and more than 700 keV is less than 20 ps, see fig. 1 enabling a reliable centroid–shift analysis down to few picoseconds.

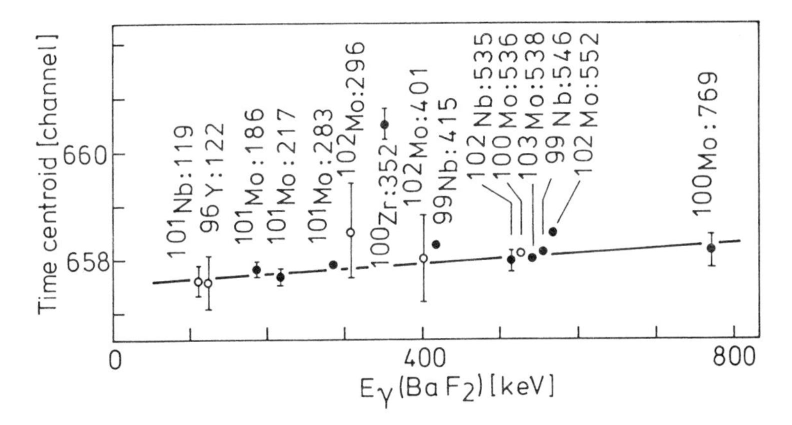

Fig. 1. Time–centroid positions for γ rays with a short delay time. Data points which are corrected for their known delay time are marked by open symbols. The distance of time centroids from the prompt curve gives the mean lifetime. Time calibration is 20.9 ps/channel.

^{100}Zr: With gates $E(BaF_2)/E(Ge)$ = 213 keV/352 keV and 213 keV/666 keV, see fig. 2, we have measured $t_{1/2}(2^+)$ = 544(37), in agreement with the recent result of [5]. This value corresponds to B(E2: $2^+ \rightarrow 0^+$) = 80(6) spu. From the centroid shift of the timing distribution with $E(BaF_2)/E(Ge)$ = 352 keV/213 keV we have derived the half–life of the next higher member of the g.s. band in ^{100}Zr for the first time: $t_{1/2}(4^+)$ = 37(4) ps. This translates into B(E2: $4^+ \rightarrow 2^+$) = 101(11) spu which confirms the strong deformation of ^{100}Zr and is basically consistent with the assumption of constant collectivity within the band. Here it must be pointed out that the B(E2: $2^+ \rightarrow 0^+$) value is supposed to be influenced by mixing of different shapes [5].

^{101}Nb: We have measured the half–life of the first excited rotational state in ^{101}Nb at 119 keV which permits determining the g.s. deformation of a neutron rich Nb isotope for the first time. With $E(BaF_2)$ = 119 keV and Ge gates at 554 keV, 760 keV, 1809 keV and 1838 keV, see fig. 3 we derived $t_{1/2}$ = 84(8) ps from the centroid shift and 84(4) ps from the slope analysis. With the mixing ratio $|\delta\,(E2/M1)|$ = 0.16(2) which was derived from a separate γ–x coincidence measurement [6] this converts into B(E2 : 119 keV) = 230(66) spu corresponding to a g.s. deformation $|\beta|$ = 0.39(5). Thus also in the Nb isotopic chain the isotope right at the onset of deformation (N=60) already reaches the magnitude of deformation close to $\beta \approx 0.4$ which is typical of all deformed neutron–rich nuclei in the A=100 region.

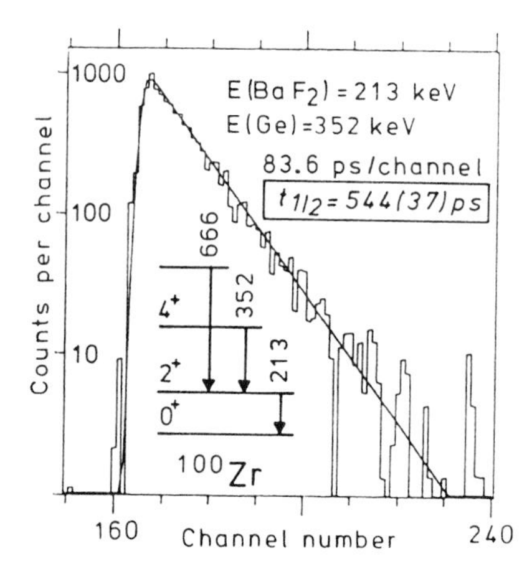

Fig. 2. Time distributions and fit for the $2^+ \to 0^+$ transition in ^{100}Zr.

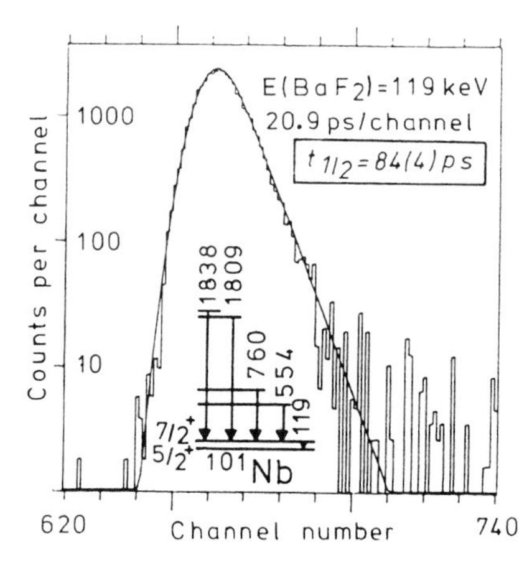

Fig. 3. Lifetime determination for the first excited state in the g.s. rotational band of ^{101}Nb. The time distribution was obtained with E(BaF_2) = 119 keV and a summation over the Ge gates with 554 keV, 760 keV, 1809 keV and 1838 keV.

<u>^{100}Mo</u>: In a recent investigation of the electric monopole strengths in the A=100 region, $t_{1/2}(0_2^+)$ in ^{100}Mo has been measured to be 1.58(4) ns [7] which is at variance with a recently published value of 3.1(1) ns [8]. We have determined $t_{1/2}(0_2^+)$ from the slope of the time distributions with gates E(BaF_2)/E(Ge) = 159 keV/769 keV (see fig. 4) and E(BaF_2)/E(Ge) = 535 keV/159 keV resulting in $t_{1/2}(0_2^+)$ = 1.65(4) ns.

<u>^{103}Mo</u>: The magnitude of the ground-state deformation of this nucleus was derived from the lifetime of the first excited state in the g.s. rotational band. The result of $t_{1/2}$ = 0.45(16) ns measured by [9] is in disagreement with a published value of $t_{1/2}$ = 1.7(3) ns [10]. From our data we derive $t_{1/2}$ = 0.44(2) ns, see fig. 5, reproducing the value from [9] with much higher precision, confirming the existence of a strong g.s. deformation in that nucleus. In fact, a value of $|\beta|$ = 0.34(3) is obtained for the N=60 isotone of ^{101}Nb.

<u>^{96}Zr:</u> The nucleus ^{96}Zr exhibits doubly magic features although it is very close to the region of strongly deformed nuclei. In a previous paper [2] we have presented new results on the lifetime of the 3^- state at 1898 keV. Though the accuracy was limited due to the poor timing resolution of the Ge detectors used at that time the result of $t_{1/2}(3^-)$ = 84(44) ps permitted the conclusion that ^{96}Zr exhibits a relatively high octupole collectivity.

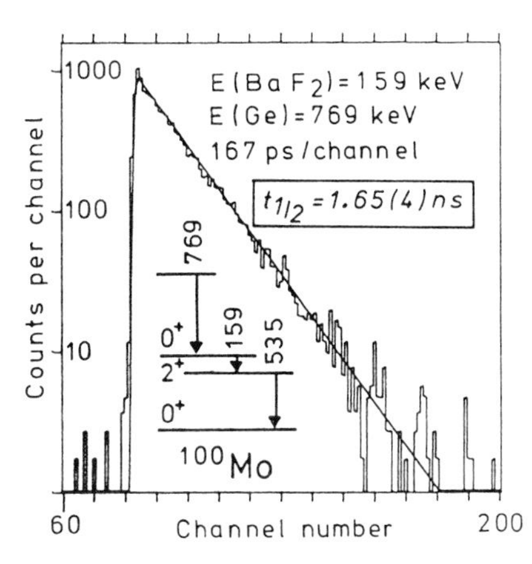

Fig. 4. Time distributions for the 0_2^+ state in ^{100}Mo.

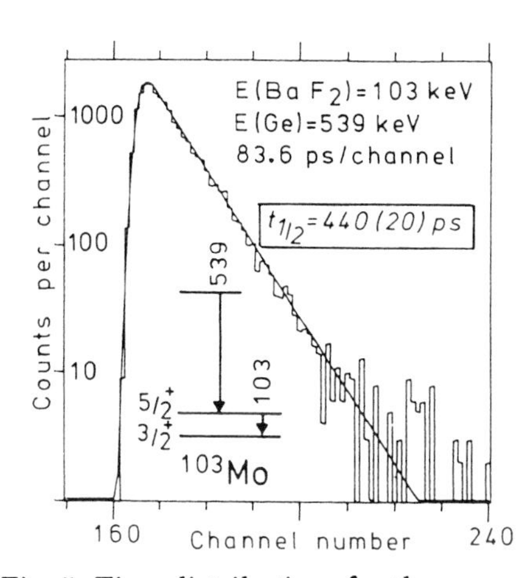

Fig. 5. Time distributions for the first excited rotational state in ^{103}Mo.

In order to check this result with higher precision the new timing technique was applied to study the β^- decay of the 8^+ isomer of ^{96}Y which predominantly feeds the 8^+ state in ^{96}Zr, see fig. 6. The bigger BaF_2 crystal was used to account for the relatively high energies of most of the ^{96}Zr γ transitions.

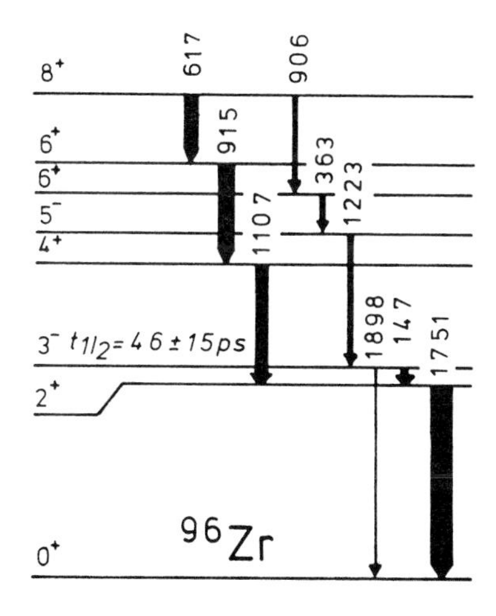

Fig. 6. Levels and γ transitions in ^{96}Zr used in the present analysis.

The 3^- lifetime was derived from the centroid shift of the timing distributions with $E_\gamma(BaF_2)$ = 1751 keV. Ge gates were set on all the strong transitions which directly or indirectly feed the 3^- level. As a prompt reference we took the distribution with $E_\gamma(BaF_2)$ = 1751 keV again but with Ge gates only on those γ transitions which directly or indirectly feed the 1751 keV level without populating the 3^- state. This technique avoids systematic errors due to walk corrections since the γ–ray energy in the BaF_2 detector is the same in both cases. It has been verified that there is no lifetime above ≈ 10 ps of one of the levels between the 8^+ and 3^- state which would systematically affect the lifetime result.

The new measurement gives $t_{1/2}(3^-)$ = 46(15) ps which still matches the previous value within the error bars, but is considerably shorter and more precise. The corresponding octupole collectivity is B(E3) = 66(+33,–16) spu. Such a high value is very unusual and it clearly exceeds all known B(E3) values for doubly–closed shell nuclei which are known for their high octupole collectivity.

In the time distributions of all those γ rays populated from the 8^+ decay a slope with the same decay constant was found which on the average corresponds to $t_{1/2}$ = 127(10) ps. This half–life has to be attributed to the 8^+ state at 4390 keV. It corresponds to B(E2 : $8^+ \rightarrow 6^+$, 617 keV) = 1.4(1) spu supporting the interpretation of the corresponding levels to be members of a band of essentially single–particle nature $g_{9/2}^{2}$ [11].

5. CONCLUSION

The new technique has opened up a new region of level lifetimes which had not been accessible before in the studies at JOSEF. New data around A=100 emphasize the most striking feature of this region where nuclei dominated by collective structures are in immediate neighbourhood to closed–shell nuclei. Further measurements in this mass region are in progress.

REFERENCES

[1] H. Ohm, G. Lhersonneau, K. Sistemich, B. Pfeiffer and K.–L. Kratz, Z. Phys. A327, 483 (1987)

[2] G. Molnar, H. Ohm, G. Lhersonneau and K. Sistemich, Z. Phys. A331, 97 (1988)

[3] M. Moszynski and H. Mach, Nucl. Instr. and Meth. A277, 407 (1989)

[4] H. Lawin, J. Eidens, J.W. Borgs, R. Fabbri, J.W. Grüter, G. Joswig, T.A. Khan, W.D. Lauppe, G. Sadler, H.A. Selic, M. Shaanan, K. Sistemich and P. Armbruster, Nucl. Instr. and Meth. 137, 103 (1976)

[5] H. Mach, M. Moszynski, R.L. Gill, F.K. Wohn, J.A. Winger, John C. Hill, G. Molnar and K. Sistemich, in print in Phys. Lett.

[6] U. Paffrath, Thesis, Universität zu Köln (1989)

[7] H. Mach, M. Moszynski, R.L. Gill, G. Molnar, F.K. Wohn, J.A. Winger and John C. Hill, to be published

[8] R.L. Estep, R.K. Sheline, D.J. Decman, E.A. Henry, L.G. Mann, R.A. Meyer, W. Stoeffl, L.E. Ussery and I. Kantele, Phys. Rev. C35, 1485 (1987)

[9] T. Seo, H. Lawin, G. Lhersonneau, R.A. Meyer, G. Menzen and K. Sistemich, Z. Phys. A320, 393 (1985)

[10] H.A. Selic, E. Cheifetz and J.B. Wilhelmy, Jül–Spez–99, 69 (1981), Annual Report 1980 of IKP, KFA Jülich, Germany

[11] M.L. Stolzenwald, G. Lhersonneau, S. Brant, G. Menzen and K. Sistemich, Z. Phys. A327, 359 (1987)

Inst. Phys. Conf. Ser. No 105
Paper presented at Int. Conf. on Spectroscopy of Heavy Nuclei, Crete, Greece, 1989

The mass and radius of exotic fragments

Yves Schutz
GANIL, BP 5027, 14021 CAEN, FRANCE

Abstract. Recent developments in the study of exotic nuclei are presented. A method to measure the ground-state mass is presented and the results are compared with standard models. Total reaction cross section measurements for exotic nuclei are also presented and interpreted in terms of matter distribution in the nucleus.

1. Introduction

Before I take you on this journey into the land of exoticism, I would like to give a few arguments as to why so much effort is expended to produce, identify and study nuclei on the borders of stability. Besides the pleasure and excitement of hunting for previously unknown species, there exists a real and fundamental need for knowledge of the existence and properties of nuclei far from the valley of stability. A broad systematic study of exotic nuclei should provide an understanding of the behavior of the nuclear force whe one exposes the isospin degree of freedom at its most extreme values. Of course, the purpose of the game is not to fill in systematically all the gaps—and there are many, only about one third to one half of the predicted stable nuclei have been discovered so far—but to choose those particular nuclei which will provide the strongest constraints on the various models which try to explain the properties of an assembly of neutrons and protons. For example, the demonstration of the existence of, let us say, ^{17}O will not change dramatically our knowledge of nuclear physics. If, however, we obtain the same information for ^{26}O, it is certainly not a trivial fact and will by itself put an additional constraint on the models. This points to nuclei in the vicinity of the neutron and proton drip lines as the most interesting region of the chart of nuclei. Because these limits are most easily reached at the beginning of the chart, the scope of my talk will be limited to light neutron and proton rich nuclei.

In the first part of the talk, I will briefly review the mechanism which is used to produce exotic nuclei at intermediate energies and the performances of the resultant exotic beams. I will then describe how we measure with a high accuracy the mass of the nuclei and indicate what we have learned from the systematic behavior of their binding energies. Finally, I will present a method which uses exotic beams to induce secondary reactions and describe how the results can be interpreted in terms of the matter distribution in the nucleus.

2. The production mechanism

Most known reaction mechanisms involving light or heavy projectiles are used to produced exotic nuclei. With the availability of high intensity beams at intermediate energy heavy-ion accelerators such as GANIL, a new experimental technique has appeared which greatly increases the number of nuclear species open to investigation. The mechanism of interest is a fragmentation-like process that yields a usable number of fragments, the rather low cross sections being counterbalanced at GANIL by intensities of about 10^{12} ions per second – intensities which have not been reached at accelerators of higher energies. This process is in many aspects similar to the fragmentation reactions observed with relativistic heavy-ions. The characteristics of the projectile-like fragments (PLF), as they have been observed at energies between 30 and 100 MeV/nucleon, are the following:

1. The PLF are emitted with velocities very close to the beam velocity.
2. The angular distribution is forward peaked, but broader than a pure fragmentation model would predict.
3. The isotopic distribution of the fragments reflects the N/Z ratio of not only the projectile but also of the target. The more neutron rich the projectile (target), the more neutron rich the fragments.

With these properties, the fragments are best collected and identified using fragment separators such as the LISE and SPEG spectrometer of GANIL. The unambiguous identification of the fragments is achieved by measuring the magnetic rigidity $B\rho$, the energy loss in a ΔE-E solid state detector and the time of flight (TOF) along a constant path. The combination of the following equations permits the determination of A and Z for each detected fragment,

$$B\rho = \frac{v \cdot A}{Q}$$

$$E \cdot \Delta E \propto A \cdot Z^2$$

$$TOF \propto \frac{A}{Q \cdot B\rho}$$

v is the velocity and Q the charge state (in most cases Q = Z). As an example, with the full momentum acceptance of SPEG (6%) and starting with a primary beam of 10^{12} ions per second, secondary beams of ^{39}Cl with 10^7 particles per second and ^{20}N with 10 particles per minutes can be produced. This produces enough exotic fragments to perform detailed studies of their ground-state properties, and for the most intense beams allows secondary reactions to be induced for which the cross section can be measured.

3. The masses of exotic nuclei

3.1. Mass and nuclear structure

Since 1935, the masses of almost all known stable isotopes have been measured with an accuracy of the order of 10^{-5} and von Weizsäcker elaborated his now famous semi-empirical mass formula. Systematic mass measurements have lead to improved accuracy (10^{-7} - 10^{-8}) and helped to refine the mass formulae including both macroscopic and microscopic aspects of the nucleus. As an example, one can mention the macroscopic-microscopic formula derived by Möller *et al* [1]. The macroscopic part of the nucleus is described in the framework of the finite-range droplet model, and has been corrected for microscopic effects using the Strutinsky method on single particle levels calculated in a folded-Yukawa single-particle potential. Figure 1. shows the results calculated for close to 1600 nuclei and makes a comparison with experimentally determined masses.

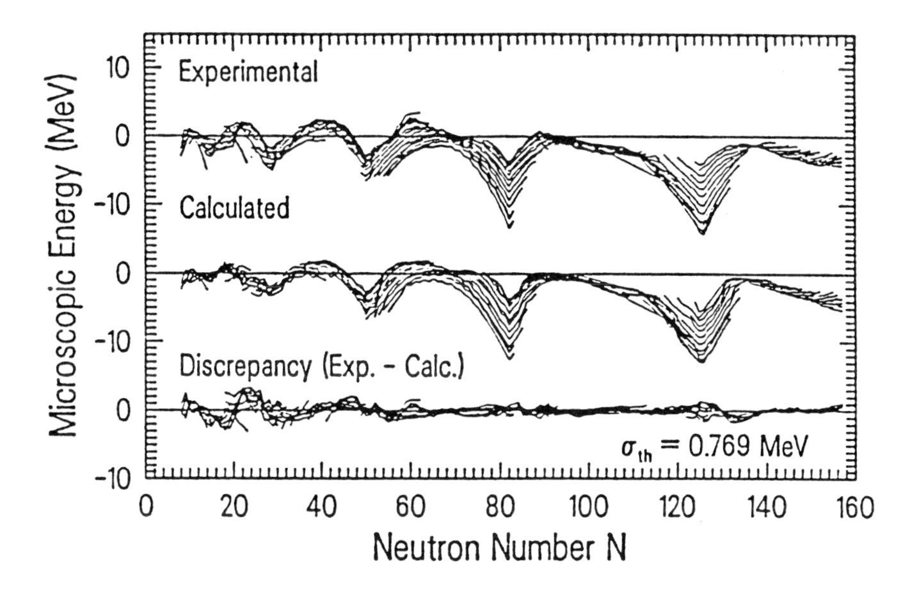

Figure 1. Comparison of experimental and calculated ground-state microscopic energies for 1593 nuclei.

Although there exists a global agreement between calculation and experiment, the average discrepancy is of the order of 800 keV, and reaches even a few MeV in some regions (N=20 for example). Note that most masses have been measured with an accuracy of better than 100 keV. The larger discrepancies are attributed to microscopic effects which are not present in the mass formula. A very serious test of such formulae is their ability to predict masses far from stability: will the parameters adjusted for stable nuclei or those close to stability still be valid? Does one need to introduce new concepts to interpret the properties of nuclear matter far from stability? We already have partial answers to these questions. We know that in general the predictive power of mass formula of any kind is poor and to reproduce new masses it is often necessary to introduce new parameters the origins of which are not always understood.

A different and more fundamental approach consists of solving the Schrödinger equation for an A-body system with a realistic nucleon-nucleon force. Because of the complexity of such a problem, one generally ends up, after several approximations, with an effective force and several free parameters. This effective interaction must contain the microscopic effects such as shell closure and deformation. In comparison mass formulae can account for these effects only when they have been found experimentally. The accuracy of the calculated masses within such a framework is generally worse than the ones obtained via mass formulae, but the predictive power is much better and one has a handle on the physics behind the different parameters.

From figure 1. one can already see that following the evolution of the mass along an isotopic chain reveals structural information concerning the nucleus such as shell closures. An other means to display such effects consists of following the two-neutron binding energy (S_{2n}) along an isotopic chain,

$$S_{2n} = M(A-2, Z) - M(A, Z) + 2 \cdot M_n c^2$$

S_{2n} is preferred to S_{1n} in order to avoid the odd-even structure due to the pairing energy. Two types of departure from a smooth decrease of S_{2n} with increasing neutron number (N) have been observed. First, a sharp drop is consistently associated with the crossing of a magic number. It corresponds to the onset of the filling of a new neutron shell of lower binding energy. Second, as observed in a few notable cases, the decrease of S_{2n} may level off and even turn upwards with increasing N. This reflects the fact that a new structure with a binding energy stronger than that extrapolated from less exotic nuclei has become available. This behavior is associated with the occurrence of deformation. A typical variation of S_{2n} is shown in figure 2. for rare earth nuclei: the shell closure at $N = 82$ and the region of deformation for $N \geq 90$ are clearly seen.

Whether one tries to compare the measured masses with theoretical predictions or one tries to reveal structural effects, it is necessary to achieve a measurement with the best possible accuracy. Most of the time a precision of the order of several hundred keV is required and sometimes less than 100 keV. This kind of precision remains experimentally possible and the next section will describe the method we have adopted.

3.2. The mass measurements [2,3]

The direct mass measurement experiments are performed using the beams delivered by the GANIL accelerators impinging on a thick tantalum target (150 - 350 mg/cm^2) located after the beam extraction of the last cyclotron. The resultant nuclei are then transported to the focal plane of SPEG, an energy-loss spectrometer (see figure 3.)

The beam transport parameters are adjusted in order to maximize the transmission of the line after the target. Moreover, the line from the target to the focal plane is doubly achromatic. To first order, this property implies a "pseudo-isochronism", i.e. the length of all trajectories will only (and weakly) depend on the momentum of the particles and not longer on the scattering angle or position on the target.

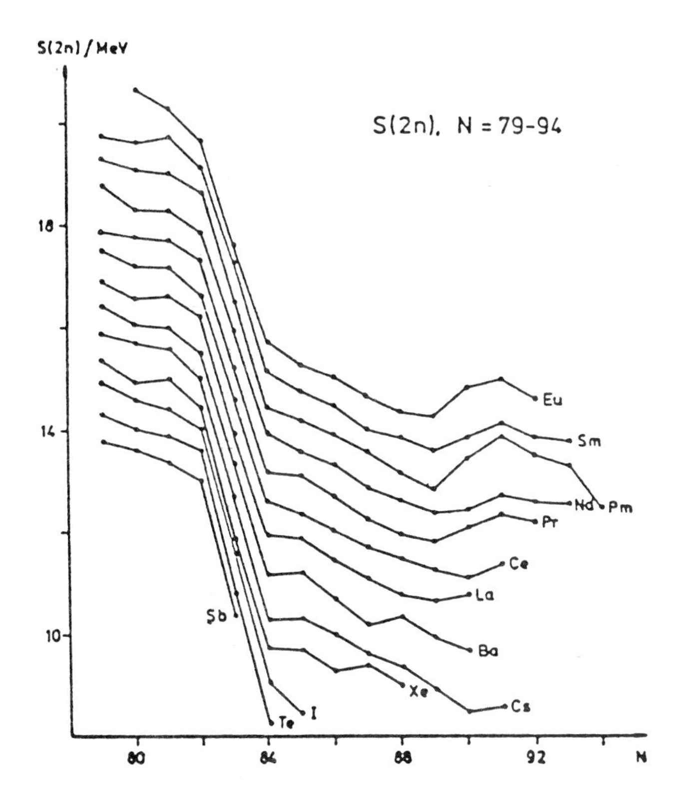

Figure 2. Two-neutron binding energies in the rare earth region

The mass values are then obtained from an accurate measurement for each particle of both the magnetic rigidity $B\rho$ and the time of flight t, which are related by,

$$B\rho = \frac{pc}{Q}$$

so that,

$$\frac{mc^2}{Q} \propto B\rho \cdot t$$

where p, m and Q are the momentum, mass and charge of the analyzed particles. The last relation is convenient to describe the principle of the method but must be corrected for relativistic effects. The magnetic rigidity of each nucleus is deduced from its dispersion in the horizontal plane at the standart target position, measured in a position sensitive parallel plate counter and corrected for finite image size as

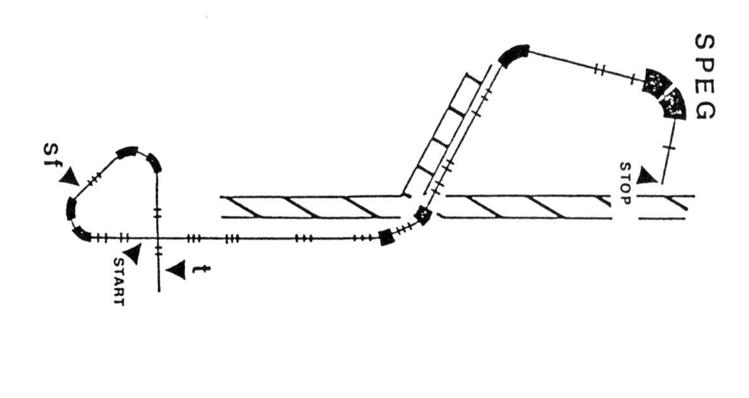

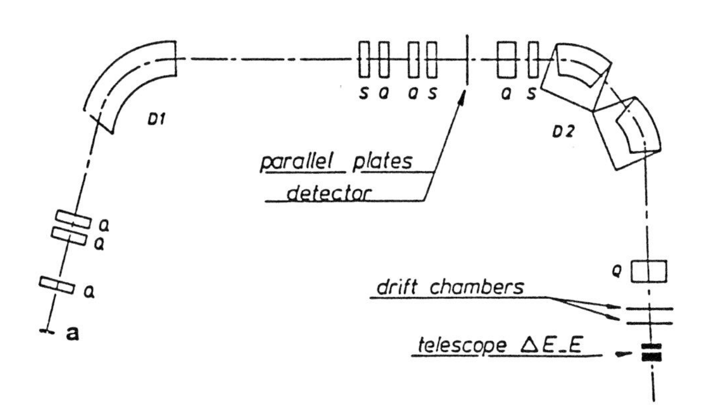

Figure 3. Experimental setup: the beam is shown at the exit of the second GANIL cyclotron; **t** represents the target position and **sf** the location of a stripping foil. At the focal plane of SPEG a ΔE-E telescope is set behind two drift chambers

determined from a double X,Y measurement obtained with a position sensitive μ-channel device and a Si-ΔE detector. For SPEG and its standard detection system a resolution of the order of 10^{-4} is achieved. Recent developments [4] on the detection system lead to an improved precision by a factor 2.

The fast signal of a micro-channel plates detector, set 80 meters away from the focal plane, is used as a direct START signal for the time of flight measurement. The STOP signal is obtained from the position sensitive μ-channel device in the focal plane. At velocities of the order of 30% of the speed of light, typical flight times are of the order of 1000 ns which can be measured with a resolution of 2.5·

10^{-4}. The final mass resolution which has been achieved with this set set up is equal to $5 \cdot 10^{-4}$.

Special care is taken in order to keep the systematic errors as low as possible. The dispersion obtained on mass values due to these errors is typically equal to $5 \cdot 10^{-6}$.

Many nuclei produced in the experiments have already been measured with an uncertainty of lower than 150 keV. These values extracted from the Audi-Wapstra tables are used as a reference from which we calculate in each case the uncertainty σ_{syst} introduced by the method. A mass is deduced from an interpolation between these references, with a total uncertainty σ,

$$\sigma = (\sigma^2_{syst} + \sigma^2_{stat})^{1/2}$$

where σ_{stat} stands for the statistical uncertainty. In the case of a mass deduced by extrapolation one needs to add an extrapolation error σ_e which may amount to up to $5 \cdot 10^{-6}$. In figure 4. this error is represented by the shaded area.

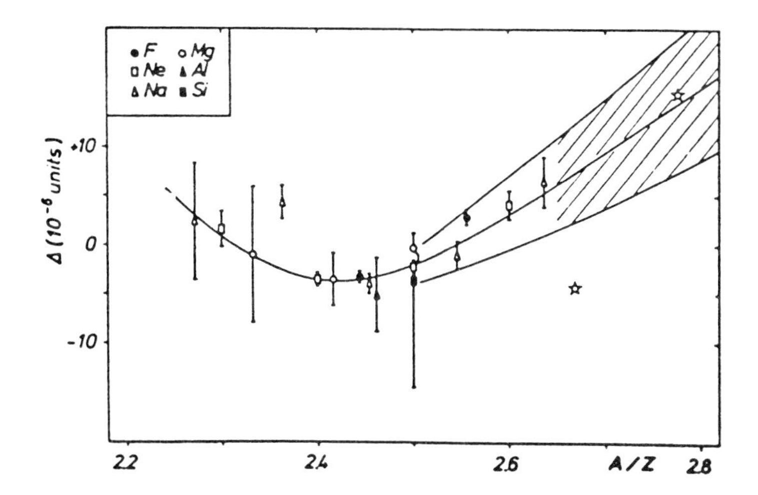

Figure 4. Deviation from one of the ratio between the measured mass and the tabulated mass as a function of A/Z. The error bars only stand for the statistical error.

Most of the new masses have been measured with uncertainties of between 500 keV and 140 keV. The most exotic nuclei, where the statistical error is the dominant one, could only be measured with an uncertainty of the order of 1 MeV.

3.3. The new masses measured at GANIL

The experiments have focused on the light neutron-rich nuclei. Various beams have been used in order to cover different regions in the chart of nuclei. Argon beams of 40 MeV/nucleon [2] and 60 MeV/nucleon [3] lead to data on the lighter nuclei whereas a Kr beam of 43 MeV/nucleon gave access to the heavier isotopes. With a given beam regions of the chart of the nuclides can be swept by selecting various $B\rho$ settings of the transport line and by changing the target thickness.

Figure 5. shows a typical identification matrix (A/Z,Z) obtained with the Kr beam. It is seen that in "one shot" it is possible to measure masses over a wide range of nuclei— from A/Z=2 to A/Z=3 for this example.

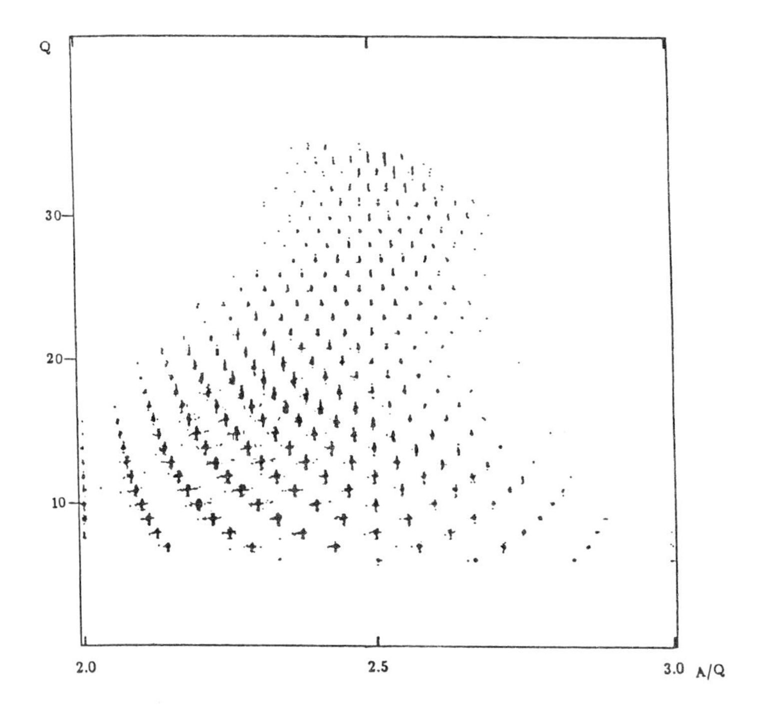

Figure 5. A/Z versus Z matrix measured with a ^{86}Kr beam at 43 MeV/nucleon

In this run we could measure the masses of isotopes with Z ranging from Z=18 (Ar) to Z=24 (Cr) and crossing the N=28 shell closure. In order to examine possible effects due to the microscopic structure for these nuclei, we have removed from the measured masses a macroscopic term calculated using the formalism proposed by Möller *et al* [1]. A zero deformation is assumed. The remaining part Δ, which should only contain information on the structure, is displayed in figure 6.

One clearly sees the shell closure effect as a the deep hole in the variation of Δ around the neutron number N=28. It is interesting to note that the absolute intensity of the minimum in Δ gradually decreases as the atomic number Z increases.

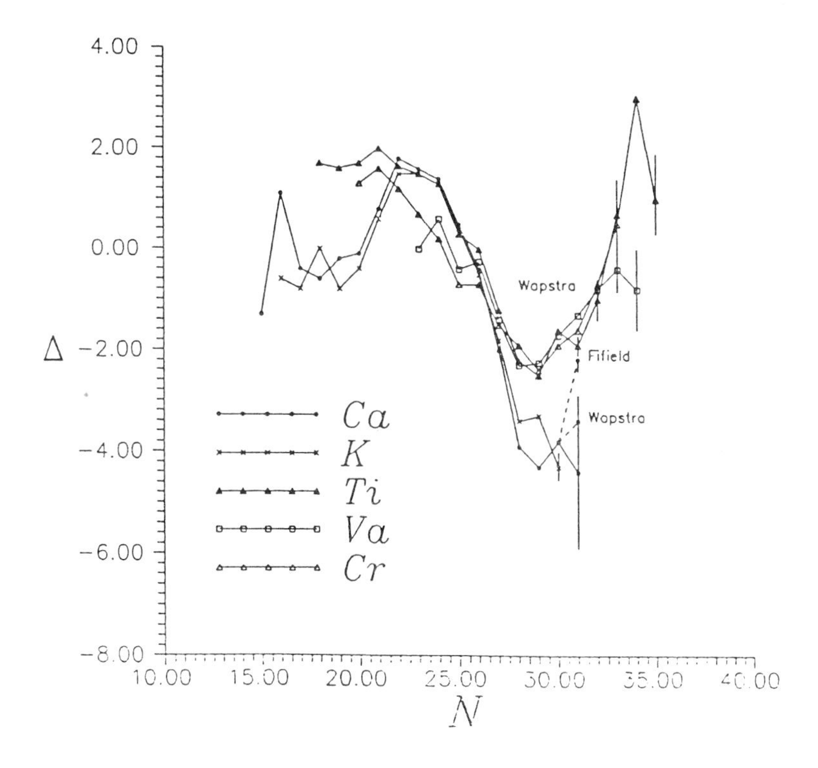

Figure 6. Difference between the measured mass and the macroscopic part of the mass formula [1] (calculated for a spherical shape) as a function of the neutron number.

This effect could indicate a modification of the neutron single-particle levels for increasing Z, or the onset of a deformation which would make the macroscopic term larger than the one calculated for a spherical shape.

The runs using the Argon beam enabled an extension of the systematic for the sd-shell nuclei toward the N=20 shell closure (see also in Vieira *et al* [5] for similar data). Figure 7. shows the results obtained for the oxygen isotopes when compared to predictions of several mass formulae. It is found that the heavier isotopes, $^{21-24}O$, are more bound than expected. Furthermore the S_{2n} value of ^{24}O is still quite large. It is thus an open question as to whether ^{26}O, not yet observed, is the last bound oxygen isotope as predicted, or if ^{28}O is also bound. Thus, even the identification, or not, of these nuclei will provide valuable information.

In the same figure the measured masses are compared to an exact calculation [7] using the shell model in the sd single-particle space (black squares). It is seen that the predictions for the most exotic isotopes are much better than those of any mass formula, which confirms the earlier assertion of the predictive powers of the

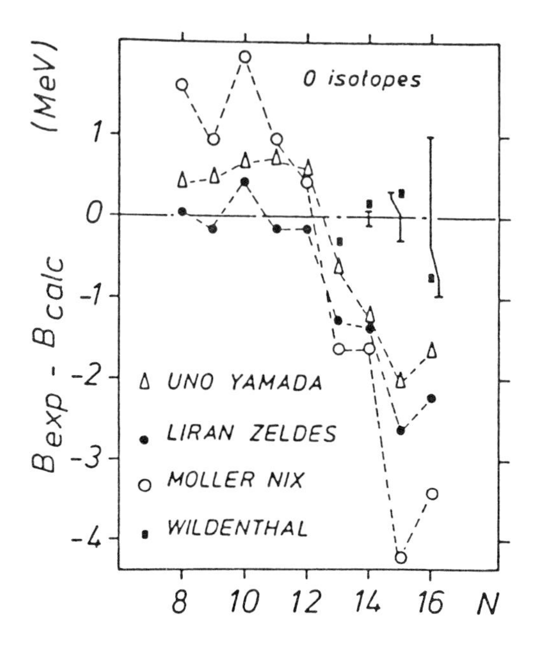

Figure 7. Difference between experimental mass excesses and predictions [6,7] for the oxygen isotopes.

two different approaches.

When reaching the N=20 neutron number for the heavier isotopes a rapid decrease in the variation of the S_{2n} curve is observed, as expected for a shell closure (figure 8). This trend is present down to the phosphorus (Z=15) and silicon (Z=14) isotopes. The situation changes for the sodium (Z=11) and the magnesium (Z=12) isotope where an strong increase of the two neutron binding energy has been observed [8] and interpreted as the onset of a new region of deformation.

With our results, each isotopic distribution is extended by one or more nuclei, some with good accuracy. The Fluorine (Z=9) and Neon (Z=10) isotopes are not exotic enough to test the shell closure. The magic nature of the N=20 neutron number, however, seems to be restored for the aluminum (Z=13) isotopes. ^{30}Na and ^{31}Na are found to be less bound than reported in reference [8], although the values agree with the errors. ^{32}Na was identified with a very low counting rate. As far as Magnesium isotopes are concerned, due to the rather large uncertainties, no final statement can be made about the evolution of the S_{2n} curve. A flattening cannot be excluded and would confirm the onset of deformation as expected from the low energy measured for the first 2^+ state of ^{32}Mg [9]. To clarify the situation and confirm the absence of the shell closure for the Magnesium and Sodium isotopes

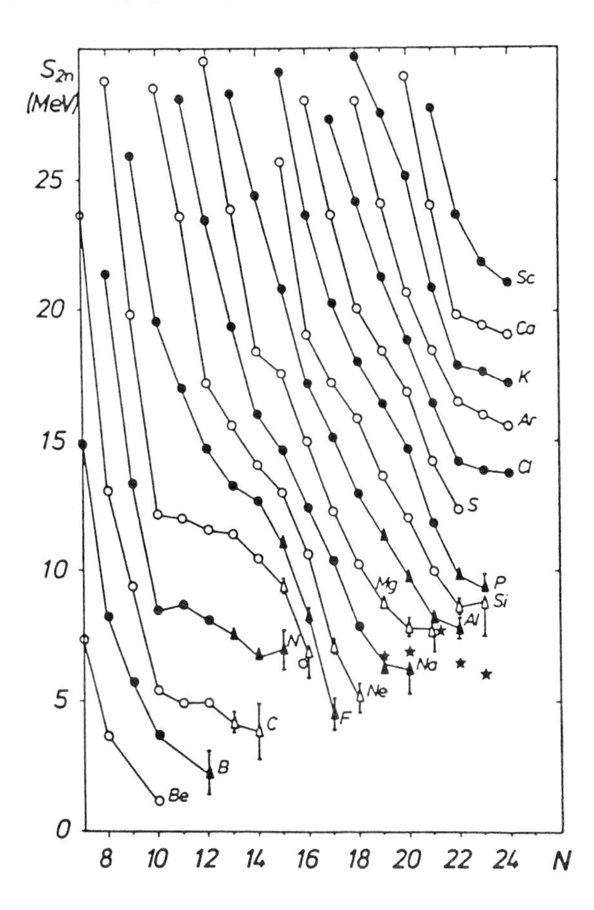

Figure 8. S_{2n} curves as a function of neutron number. The more neutron rich sodium isotopes ($N \geq 19$) measured at CERN [8] are represented by a star.

it is necessary to extend our measurement to larger N values for these isotopes with an improved precision.

4. The radius of exotic nuclei

4.1. The interaction cross section and nuclear radii

An other parameter of importance in testing various models is the radius of nuclei or more precisely their matter density distribution. If nuclear matter was incompressible, the radii of nuclei would follow a simple law in $A^{1/3}$. But we know that this is not the case and that the surface energy tends to shrink the nucleus whereas the coulomb energy and the loss of cohesion due to neutron excess cause the nucleus to dilate to a larger radius. Since the relative importance of these effects varies through the periodic table, some nuclei are smaller and some are larger than is predicted by the $A^{1/3}$ law.

An interesting question to answer is what the influence of a large proton or neutron excess is on the matter distribution and if it possible to observe a swelling of nuclei with a large isospin. A possible method capable of providing this kind of information involves the measurement of the interaction cross section which can then be related to the interaction radius,

$$\sigma_I = \pi \left[R_I^t + R_I^p\right]^2$$

where R_I^t and R_I^p are the target and projectile interaction radii respectively. It is less trivial to deduce from this a mass distribution. One way to do this is to calculate the interaction cross section using a Glauber model from which the effective root-mean-square radius is deduced [10]. While this approach is valid at high energy, there does not exist at intermediate energies a reasonable theory to connect the nucleon distribution to the interaction cross section or the reaction cross section. Instead, one can find an empirical formula which relates the cross section to the target and projectile mass. Such a formula has been derived by Kox *et al* [11] and parametrized for stable nuclei over a wide range of bombarding energies,

$$\sigma_r = \pi \cdot r_o^2 \left[A_p^{1/3} + A_t^{1/3} + \frac{a \cdot A_p^{1/3} \cdot A_t^{1/3}}{A_p^{1/3} + A_t^{1/3}} - C(E)\right]^2 \cdot \left[1 - \frac{B_c}{E_{cm}}\right]$$

where C(E) is an energy dependent term which reflects the increasing transparency of nuclei with increasing bombarding energy, and the last factor is the Coulomb term. Within this formulation r_0 is the reduced strong absorption radius and should be considered as a parameter independent of the bombarding energy and of the target and projectile mass. For stable nuclei, this parameter was found equal to 1.05 fm.

4.2. Measuring the reaction cross section

Our goal is to measure the reaction cross section using very low intensity radioactive beams. For the very exotic nuclei like ^9C or ^{20}N, the production rate is around 10 particles/minutes,which strongly limits the method and the quantities which can be measured. In order to increase the reaction probability, we have measured the energy-integrated reaction cross section. Since the beam is stopped in the target, the measured quantity is related to the energy-dependent cross section as follows,

$$\overline{\sigma_r} = \frac{1}{R} \cdot \int_{E_{max}}^{0} \sigma_r(E) \cdot \frac{dR}{dE}\, dE$$

where R is the range of the incident particle in the target.

The technique we have chosen to measure σ_r is the associated γ-ray method where it is assumed that each nuclear reaction (scattering processes excluded) is necessarily followed by the emission of at least one γ-ray — or a detectable energetic light particle. The probability of a reaction occuring is then deduced from the probability to detect a γ-ray as follow,

$$P_{reac} = \frac{P_\gamma \cdot \epsilon_\gamma}{N_{inc}}$$

where N_{inc} is the number of incident particles and ϵ_γ the γ-detection efficiency. The integrated cross section is obtained via the following relation,

$$\overline{\sigma_r} = \frac{-M_t \cdot \log(1 - P_{reac})}{N_A \cdot R}$$

where M_t is the target mass given in a.m.u. and N_A is the Avogadro number.

Combining the different relations, we can now easily extract the energy independent parameter r_0.

Secondary beams are produced as described earlier and transported to the focal plane of SPEG. A telescope consisting of two Si-surface barrier detectors of 50 μm (ΔE) was used for Z-identification. Time of flight together with Bρ and Z provide unique identification for all incident particles. A detector array consisting of 16 hexagonal NaI(Tl) crystals, covering nearly 4π, surrounded the target. The γ-detection efficiency for a single photon was 73%. This arrangement essentially detected the associated γ-rays and, with a lower probability, neutrons. To estimate the influence of the target, we have performed the measurement with two different targets, Si and Cu. Any possible influence of the bombarding energy has been checked by selecting several values for the Bρ of the transport line.

4.3. The results

All the available data on reaction cross sections available for isobars with A ranging from A=8 to A=15 have been analyzed as described in the previous section. This includes the data from Tanihata *et al* [10] which have been taken at 800 MeV/nucleon. We have also analyzed our data in two different ways. In order to deduce the reaction probability, the first approach takes into account the γ-rays as well the light particles detected in an almost 4π geometry, whereas the second approach only considers the γ-rays. The results of the latter should be compared with those of reference [12]. Recently we have developed a third method where we have direct access to the reaction cross section by measuring the Q-value of the particles after they have interacted in the target. The data from all three methods are displayed in figure 9. in terms of variation of r_0 for various isobars as a function of isospin.

The first conclusion one can draw from this figure is that the value of r_0 deduced from the different experiments and analysis agree rather nicely. Nevertheless one observes somewhat larger values when one also considers the charged particles as compared to the method which only take into account γ-rays. This effect becomes

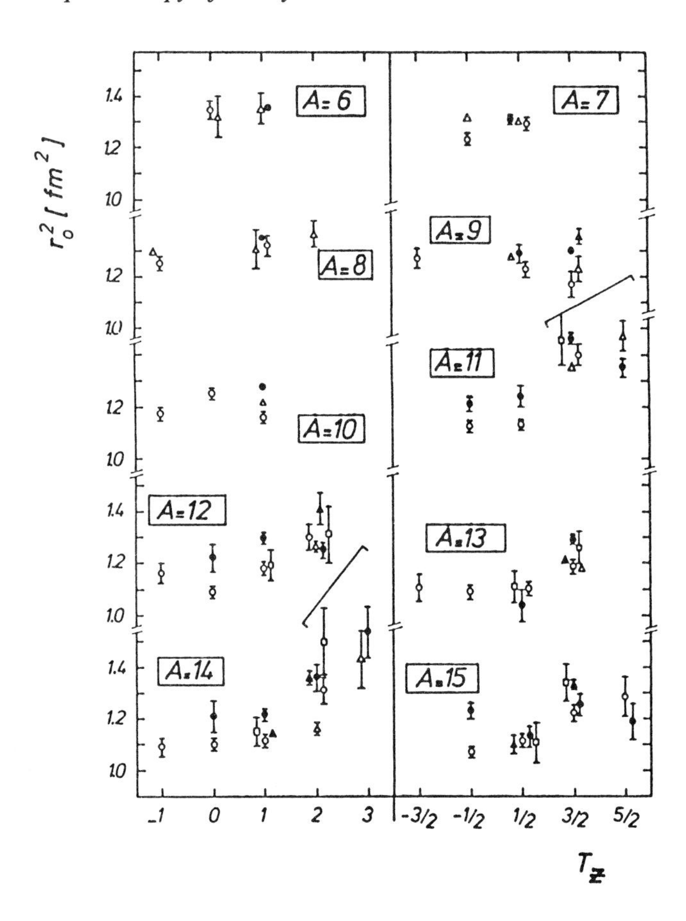

Figure 9. Variation of r_0 as a function of isospin: square=our work (1st method); open bullet our work (2nd method); closed bullet=reference [12]; open triangle=reference [10]; closed triangle=our work(direct method)

as compared to the method which only take into account γ-rays. This effect becomes stronger when one reaches the more neutron rich nuclei. Indeed for these nuclei, the breakup reaction leading to a nucleus in its ground-state or at low excitation energy plus one or several neutrons is expected to increase significantly. This is a severe limitation on the associated γ-ray technique, which in this case will underestimate the reaction cross section and thus r_0. It is best illustrated by the large discrepancy observed for ^{11}Li, where the value obtained via the γ-ray technique in ref. [12] is lower by three standard deviations than the r_0 deduced from the data of reference [10]. We have heard at this conference [13] that the group of Tanihata have confirmed their earlier result obtained for ^{11}Li.

All the data reveal a strong influence on the isospin for all the measured isobars (remember that the value of r_0 is expected to stay constant for any A and T_Z as deduced from the systematic over stable nuclei). One observes an increase of r_0 when one goes to the neutron or the proton rich side of any of the measured isobars. In order to have an idea of the mean behavior of all nuclei, independent of A, we have calculated an average deviation δ from the standard value r_0=1.05 fm given in reference [11]. We have added to the experimental data an artificial uncertainty of 5% in order to wash out shell effects for each element. The final deviation, δ, plotted as a function of the neutron excess is shown in figure 10. The shadowed area represents the results and the uncertainty in a linear fit to both of the neutron and proton-rich sides, assuming a $|N - Z|$ dependence of δ. We can see that the increase of r_0 is real and that the mean effect is of the order of 7% for N-Z=10.

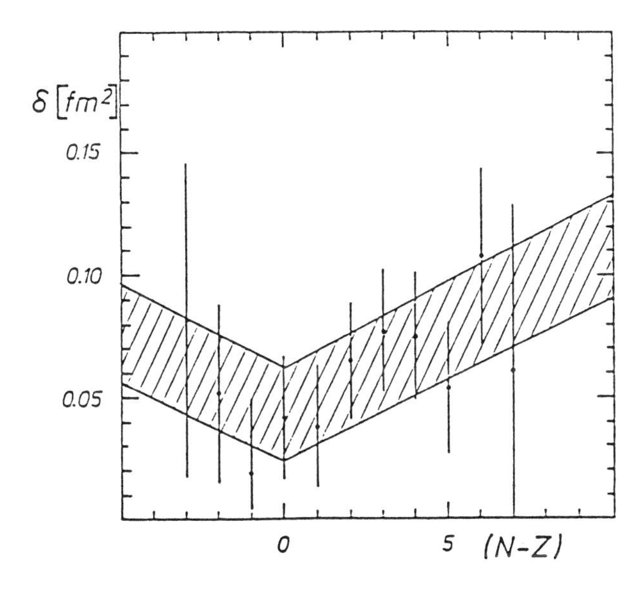

Figure 10. Average deviation δ as a function of (N-Z). The shadowed zone represents the result and the error of a linear fit to both sides, neutron and proton-rich, assuming a $|N - Z|$ dependence of r_0.

Recent self-consistent calculations [14] of light nuclei followed the evolution of the nuclear radii with neutron number. The results indicate that, whereas the root-mean-square (rms) radius of the matter distribution follows a $A^{1/3}$ law, the rms radius of the neutron distribution strongly increases with increasing neutron excess. This effect is due to the last barely bound neutrons which populates barely confined orbital of unusually large radius. Since, at the energies under consideration here, the total cross section is only sensitive to the tail of the matter distribution, the deduced increase of r_0 may be identified with the predicted increase of the rms radius of the neutron distribution.

It may also be possible that such exotic nuclei develop a strong deformation. Hartree-Fock calculations performed so far [10] are unable to reproduce the large values measured for the effective rms radius.

In his talk, P.G. Hansen [13] will develop the idea of the building-up of a giant neutron halo in nuclei close to the neutron drip line.

5. Conclusion

The use of the GANIL facility as an exotic nuclei "factory" has proven to be very successful in extending the domain of nuclei which can be experimentally accessed. Not only has it been possible to discover new exotic species close to the drip lines, but also, as a result of the excellent qualities of the exotic beams, it has proven possible to undertake detailed spectroscopic studies on these nuclei.

Combining a long flight-path with a very accurate measurement of magnetic rigidity has allowed the determination for the first time, and with a remarkable precision, the ground-state masses of a wide variety of nuclei at the limits of stability. A systematic study of the binding energies has proven very fruitful in modifying the various models which are used to predict this quantity. It has also revealed new structural effects such as the disappearance of the N=20 shell closure for the Sodium and possibly the Magnesium isotopes. The limiting factors in attaining a better precision for this kind of measurement are the rates of production of exotic nuclei, the efficiency with which they may be collected and the precision of the time of flight measurement. To solve the former problems we at GANIL have studied means to improve the intensity of the primary beam and to enhance the collection efficiency of the fragments. The solution should enable us to increase at least by two orders of magnitude the intensities of the exotic beams. To solve the problem of the flight time accuracy, we are currently studying a method to inject exotic beams into the second cyclotron. A typical flight path will then be of the order of 5,000 m and the expected corresponding precision in mass should be about $5 \cdot 10^{-6}$.

Finally, even with the low intensities, such as several counts/minutes, it has proven possible to measure total reaction cross sections. We have interpreted the results by reducing the cross section to an energy and mass independent parameter — the reduced strong absorption radius. We have shown that this parameter, expected to be constant, varies with isospin for all the isobars we have measured. This result can be tentatively interpreted as a swelling of nuclei with a large excess of neutrons or protons. Unfortunately, from our measurements it is not a straightforward task to make any statement about the real mass distribution in the nucleus. It is therefore necessary to develop microscopic theories which relate the nucleon distribution to the reaction cross section in a unified way from intermediate to high energy.

Acknowledgments

The work which has been presented here is the results of a collaboration of many people and institutes: W. Mittig, Y.S. (GANIL); L. Bianchi, B. Fernandez, J. Gastebois, A. Gillibert (Saclay); G. Audi, C. Stephan, L. Tassan-Got (Orsay); A. Cunsolo, A. Foti (Catania); M. Morjean, Y. Pranal (Bruyères-le-Chatel); A.C.C. Villari (São Paulo); Zhan Wen Long (Lanzhou); A. Artukh, Yu.E. Penionzkevich, S.M. Lukyanov (Dubna).

References

[1] P. Möller, W.D. Myers, W.J. Swiatecki and J. Treiner, Preprint LBL 22686
[2] A. Gillibert, L. B.anchi, A. Cunsolo, B. Fernandez, A. Foti, J. Gastebois, Ch. Gregoire, W. Mittig, A. Peghaire, Y. Schutz and C. Stephan, Phys; Lett. B, 176, (1984), 317
[3] A. Gillibert, W. Mittig, L. Bianchi, A. Cunsolo, B. Fernandez, A. Foti, J. Gastebois, Ch. Grégoire, Y. Schutz and C. Stephan, Phys. Lett. B, 192, (1987), 39
[4] A.C.C. Villari , W. Mittig, Y. Blumenfeld, A. Gillibert, P. Gangnan and L. Garreau NIM to be published
[5] D.J. Vieira, J.M. Wouters, K. Vaziri, R.H. Kraus, H. Wollnik, G.W. Butler, F.K. Wohn and A.H. Wapstra, Phys. Rev. Lett. 57, (1986), 3253
[6] S. Liran and N. Zeldes, At. Data Nucl. Tables, 17, (1976) 431;
P. Möller and J.R. Nix, At. Data Nucl. Tables, 26, (1981), 165;
M. Uno and Y. Yamada, INS Report NUMA, 40, (1982)
[7] B.H. Wildenthal, M.S. Curtinand B.A. Brown, Phys. Rev. C, 28, (1983), 1343
[8] C. Détraz, M. Langevin, M.C. Goffri-Kouassi, D. Guillemaud, M. Epherre, G. Audi, C. Thibault and F. Touchard, Nucl. Phys. A, 394, (1983), 378
[9] D. Guillemaud-Müller, C. Détraz, M. Langevin, F. Naulin, M. de Saint-Simon, C. Thibault, F. Touchard and M. Epherre, Nucl. Phys. A, 426, (1984), 37.
[10] I. Tanihata, Nucl. Phys. A, 488, (1988), 113c
[11] S. Kox, A. Camp, R. Cherkaouin, A.J. Cole, N. Longequeue, J. Menet, C. Perrin and B. Viano, Nucl. Phys. A, 420, (1984), 162
[12] M.G. Saint-Laurent, R. Anne, D. Bazin, D. Guillemaud-Müller, U. Jahnke, Jin Gen Ming, A.C. Müller, J.F. Bruandet, F. Glasser, S. Kox, E. Liatard, Tsan Ung Chan, G.J. Costa, C. Heitz, Y. El-Masri, F. Hanappe, R. Bimbot, E. Arnold and R. Neugart, Z. Phys. A, 332, (1989), 457
[13] P.G. Hansen, this conference
[14] R.J. Lombard, Preprint Orsay, IPNO/TH 89-26

Inst. Phys. Conf. Ser. No 105
Paper presented at Int. Conf. on Spectroscopy of Heavy Nuclei, Crete, Greece, 1989

Shell-model justification of the interacting Boson–Fermion model for odd-mass nuclei

P. Van Isacker

SERC Daresbury Laboratory, Warrington WA4 4AD, England

ABSTRACT: An isospin-invariant boson-fermion model is proposed for odd-mass nuclei and its connection with the shell model is investigated in the case of neutrons and protons in the $f_{7/2}$ shell.

1 INTRODUCTION

One of the attractive features of the interacting boson model (IBM) of Arima and Iachello (1976; 1978; 1979) is its connection with two well-established nuclear structure models. On the one side, the IBM can be related to the collective model of Bohr and Mottelson by taking the limit of large boson numbers (classical limit), and, on the other side, the bosons of the IBM can be interpreted as correlated pairs of valence nucleons (Cooper pairs) and hence a connection with the nuclear shell model can be made. Although this relation with the shell model exists in principle, it appears to be exceedingly difficult to establish the connection in practice. One of the difficulties is that in medium-mass and heavy nuclei where the IBM has been applied traditionally, no exact shell-model calculations can be performed. The work of Elliott and collaborators, however, extended the IBM to lighter nuclei (Elliott and White, 1980; Evans *et al.*, 1985) and thus opened the way for a better microscopic understanding of the model (Thompson *et al*, 1987). This new version of the IBM (commonly referred to as IBM-3, since it considers three different kinds of bosons) applies to nuclei where valence neutrons and protons occupy the same shell and is by no means a trivial extension of earlier versions of the IBM since it critically involves the consideration of isospin.

In this contribution, a shell-model interpretation of the interacting boson-fermion model (IBFM), proposed by Iachello and Scholten (1979), will be discussed. In the IBFM, collective excitations of an odd-mass nucleus are described by coupling a fermion to a system of bosons. In order to arrive at a shell-model interpretation of the IBFM, the same path can be followed as the one outlined above for even-even nuclei. First, an isospin-invariant version IBFM-3 is constructed (section 2) and subsequently related to the shell model (section 3). This can then be used to provide a test of the IBFM-3 and in this contribution an example case in the $f_{7/2}$ shell will be discussed (section 4).

2 AN ISOSPIN-INVARIANT BOSON-FERMION MODEL

A nucleus with $2N+1$ valence nucleons is described in the IBFM-3 as a system of N bosons interacting with the last, odd fermion. The bosons have angular momenta 0 and 2 (s- and d-bosons) and form a triplet in isospin space, i.e. they have three components $M_{T_B} = -1$ (ν), $M_{T_B} = 0$ (δ) and $M_{T_B} = +1$ (π). The corresponding creation and annihilation operators are denoted by

$$s_\nu^\dagger, d_{\nu,m}^\dagger, s_\delta^\dagger, d_{\delta,m}^\dagger, s_\pi^\dagger, d_{\pi,m}^\dagger, \quad s_\nu, d_{\nu,m}, s_\delta, d_{\delta,m}, s_\pi, d_{\pi,m}, \qquad (m = 0, \pm1, \pm2). \tag{1}$$

Since these operators form isospin triplets, it is more convenient to denote them explicitly by their isospin labels,

$$b_{1,M_{T_B},l,m}^\dagger, \quad b_{1,M_{T_B},l,m}, \qquad (M_{T_B} = 0, \pm1; l = 0, 2; m = 0, \ldots, \pm l). \tag{2}$$

Spherical tensors are built from the creation operators in (2) and annihilation operators of the type

$$\tilde{b}_{1,M_{T_B},l,m} = (-)^{1-M_{T_B}+l-m} b_{1,-M_{T_B},l,-m}. \tag{3}$$

The fermion has angular momentum j and isospin 1/2 and can be a neutron ($m_t = -1/2$) or a proton ($m_t = +1/2$). The corresponding creation and annihilation operators are

$$a_{\nu,j,m}^\dagger, a_{\pi,j,m}^\dagger, \quad a_{\nu,j,m}, a_{\pi,j,m}, \qquad (m = \pm\tfrac{1}{2}, \ldots, \pm j) \tag{4}$$

As for the bosons, we introduce operators in isospin notation,

$$a_{\frac{1}{2},m_t,j,m}^\dagger, \quad a_{\frac{1}{2},m_t,j,m}, \qquad (m_t = \pm\tfrac{1}{2}; m = \pm\tfrac{1}{2}, \ldots, \pm j), \tag{5}$$

while the annihilation operators

$$\tilde{a}_{\frac{1}{2},m_t,j,m} = (-)^{\frac{1}{2}-m_t+l-m} a_{\frac{1}{2},-m_t,l,-m} \tag{6}$$

have the transformation properties of a spherical tensor.

Since the above operators, which constitute the building blocks of the IBFM-3, form complete multiplets in three-dimensional, physical space as well as in isospin space, it is possible to impose rotational invariance in both spaces. We may thus construct Hamiltonians in the IBFM-3 that are scalar in angular momentum and isospin, i.e. whose eigenstates have good J and T.

It is instructive to illustrate the properties of such eigenstates with the simple example of a system consisting of two neutrons and one proton with $M_T = -1/2$. In the IBFM-3 this system is described by coupling a single boson to the fermion and this coupling must be carried out in angular momentum as well as in isospin. Thus, three-nucleon states have the form

$$(b_{1,l}^\dagger a_{\frac{1}{2},j}^\dagger)^{(T,J)}_{M_T,M_J}|o\rangle = \sum_{M_{T_B},m_t} (1 M_{T_B}\, \tfrac{1}{2} m_t | T M_T)(b_{1,M_{T_B},l}^\dagger a_{\frac{1}{2},m_t,j}^\dagger)^{(J)}_{M_J}|o\rangle. \tag{7}$$

The total isospin T can be 1/2 and 3/2 and, inserting the appropriate values for the Clebsch-Gordan coefficients, one finds for $M_T = -1/2$

$$\begin{aligned}
(b_{1,l}^\dagger a_{\frac{1}{2},j}^\dagger)^{(\frac{1}{2},J)}_{M_T,M_J}|o\rangle &= \left(-\sqrt{\frac{2}{3}}(b_{\nu,l}^\dagger a_{\pi,j}^\dagger)^{(J)}_{M_J} + \sqrt{\frac{1}{3}}(b_{\delta,l}^\dagger a_{\nu,j}^\dagger)^{(J)}_{M_J}\right)|o\rangle, \\
(b_{1,l}^\dagger a_{\frac{1}{2},j}^\dagger)^{(\frac{3}{2},J)}_{M_T,M_J}|o\rangle &= \left(\sqrt{\frac{1}{3}}(b_{\nu,l}^\dagger a_{\pi,j}^\dagger)^{(J)}_{M_J} + \sqrt{\frac{2}{3}}(b_{\delta,l}^\dagger a_{\nu,j}^\dagger)^{(J)}_{M_J}\right)|o\rangle.
\end{aligned} \tag{8}$$

Equation (8) shows that, for the three-nucleon state to have good isospin, the wave function must have two components, one in which the odd fermion is a neutron and another in which it is a proton. It should be stressed that this feature of the IBFM-3 is a consequence of isospin invariance and that a similar conclusion is reached for systems with more particles. Thus, in general, isospin symmetry requires an odd-mass nucleus (suppose with an odd number of neutrons) to spend part of its time as an even-even core coupled to a neutron and the other part as an odd-odd core coupled to a proton.

Besides states, also operators can be constructed which have a definite transformation character under rotations in physical space and in isospin space. Most importantly, we consider here the Hamiltonian operator, which, in first approximation, can be assumed scalar under both transformations and takes on the form

$$H = H_B + H_F + V_{BF}, \tag{9}$$

where H_B is the boson Hamiltonian of the IBM-3 (Evans *et al.*, 1985), H_F is the fermion Hamiltonian (reducing to a constant for a single-j shell) and V_{BF} is the boson-fermion interaction of the form

$$\begin{aligned} V_{BF} &= \sum_{l,l',T,J} \langle 1l, \tfrac{1}{2}j; TJ | V_{BF} | 1l', \tfrac{1}{2}j; TJ \rangle \\ &\times \sqrt{(2J+1)(2T+1)} \left((b^\dagger_{1,l} a^\dagger_{\frac{1}{2},j})^{(T,J)} (\tilde{b}_{1,l'} \tilde{a}_{\frac{1}{2},j})^{(T,J)} \right)^{(0,0)}_{0,0} . \end{aligned} \tag{10}$$

3 CONNECTION WITH THE SHELL MODEL

We begin by establishing an appropriate classification scheme in the shell model. Since one is dealing with a system of neutrons and protons in a single j-shell, the seniority classification of Flowers (1952) can be used. Shell-model states in this scheme are classified according to

$$|j^n \beta(v,t) \gamma T M_T J M_J\rangle, \tag{11}$$

where v is the seniority (the number of nucleons in pairs not coupled to zero), t is the reduced isospin (the isospin of these unpaired nucleons) and β and γ are additional labels to distinguish multiplicities of (v,t) for given n and T and of J for given (v,t).

Next, a corresponding classification in the IBFM-3 must be found and we propose to label its states as

$$|n_s t_s L_s = 0, (n_d t_d \alpha L_d, \tfrac{1}{2}j) tJ; TM_T JM_J\rangle, \tag{12}$$

where n_s (n_d) is the number of $s(d)$-bosons, t_s (t_d) is the isospin of the $s(d)$-bosons, L_d is the angular momentum of the d-bosons (while L_s necessarily must be zero) and α is an additional label to distinguish multiplicities of L_d for given n_d and t_d. The state (12) is coupled in angular momentum and isospin according to $\vec{J} = \vec{L}_d \times \vec{j}$ and $\vec{T} = \vec{t}_s \times \vec{t}$ with $\vec{t} = \vec{t}_d \times \frac{1}{2}$.

To connect the shell model and the IBFM-3, we establish relations between quantum numbers appearing in (11) and (12). Some of these relations are easy to recognise. For instance, we clearly associate n, the number of valence nucleons, with the number of s- and d-bosons, $n = 2(n_s + n_d) + 1$. Likewise, in accordance with the definition of seniority, we have $v = 2n_d + 1$ and, as our notation has anticipated, the reduced isospin t in (11) must be identified with t in (12), the isospin resulting from coupling the d-bosons to the fermion. Other relations are more subtle and require a more careful analysis of the shell model and the IBFM-3 (Elliott *et al.*, 1988). However, for small numbers of nucleons, the quantum numbers n, v, t, T and J suffice to characterise the state (11) uniquely and, hence, in those cases the connection between the shell model and IBFM-3 is easily established.

4 A TEST IN THE $f_{7/2}$ SHELL

Using the mapping between the shell model and the IBFM-3, discussed in the previous section, results of the two models can now be compared. Since the shell model is a well-established theory of nuclear structure, this can be viewed as a test of the IBFM-3. The procedure, when applied to nuclei in the $f_{7/2}$ shell, involves several steps which are described in subsequent subsections.

4.1 Extraction of the Boson-Fermion Interaction

In a first step, the shell-model interaction is extracted from binding energies and excitation spectra of Ca and Sc isotopes. This is a standard procedure in shell-model calculations and leads to the matrix elements as used by Evans *et al.* (1985). Using this interaction, the boson energies and boson-boson and boson-fermion interactions can now be extracted from shell-model calculations for nucleon numbers $n = 2, 3, 4$. This has been described in detail by Thompson *et al.* (1987) for $n = 2$ and $n = 4$ for the extraction of boson energies and boson-boson interaction matrix elements. Exactly the same method can be used for obtaining the boson-fermion matrix elements from the $n = 3$ system (Evans *et al.*, 1989). In this analysis it is essential that the low-lying states in the shell-model calculation have large components that can be mapped onto IBFM-3 states. If this is not the case, the boson-fermion model consitutes a poor approximation to the shell model for a three-nucleon system and very likely will also fail for systems with more particles. The situation in the $f_{7/2}$ shell is illustrated in table 1, where the $(v,t) = (1, \frac{1}{2}), (3, \frac{1}{2})$ and non-boson components of shell model wave functions are shown for $T = 1/2$ (^{43}Sc). It is seen that all yrast states have small non-boson components ($\leq 5\%$) except for the case $J = 5/2$, where the two shell-model eigenstates are almost degenerate.

4.2 Energy Spectra and Wave Functions

Having deduced the IBFM-3 Hamiltonian (9) from shell-model calculations for $n = 2, 3, 4$, we may compute the spectra and wave functions for more complex systems and compare with exact shell-model results. We consider only nuclei for which the $f_{7/2}$ shell is less than half-full in both neutrons and protons, since otherwise the problem is best formulated using holes. This limits the calculations to two bosons and one fermion with $T = 1/2$ (^{45}Ti) and $T = 3/2$ (^{45}Sc) or three bosons and one fermion with $T = 1/2$ (^{47}V). The resulting IBFM-3 spectra are shown in the centre of figures 1, 2 and 3, with the shell-model spectrum on the left. In general, there is good agreement between shell model and IBFM-3. Note also that *absolute* energies are calculated, i.e. the comparison represents a test of the calculation of binding energies in the IBFM-3. The IBFM-2 spectra on the right are discussed in subsection 4.3.

Table 1: Wave function analysis of shell-model $T = 1/2$ states for three nucleons in the $f_{7/2}$ shell

	E_x	Components		
		(v, t)		
J	(MeV)	$(1, \frac{1}{2})$	$(3, \frac{1}{2})$	Non-boson
7/2	0.00	0.91	−0.41	−0.09
9/2	1.69	—	0.98	−0.21
11/2	2.35	—	0.96	−0.28
7/2	2.80	0.41	0.91	0.01
3/2	2.90	—	1	0
5/2	3.46	—	0.73	−0.68
5/2	3.98	—	0.68	0.73
9/2	4.13	—	0.21	0.98
7/2	4.43	0.08	−0.04	0.99
11/2	4.45	—	0.28	0.96

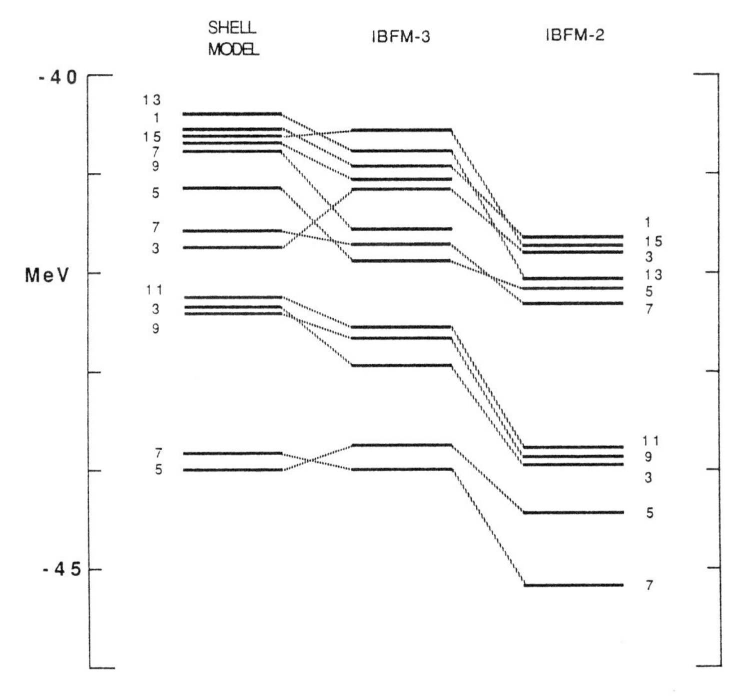

Figure 1: A comparison of IBFM and shell model for ^{45}Ti. Levels are labelled by $2J$.

Table 2: Wave function analysis of $T = 1/2$ shell-model and IBFM-3 states for five nucleons in the $f_{7/2}$ shell

J	Shell model					IBFM-3			
	s^2	sd		d^2	Non-	s^2	sd		d^2
		[1,1]	[2,0]		boson		[1,1]	[2,0]	
5/2	—	0.31	0.88	0.34	0.12	—	0.40	0.87	0.29
7/2	0.88	0.05		0.46	0.11	0.87	0.29		0.40
9/2	—	−0.26	0.81	0.47	0.24	—	−0.19	0.89	0.41
3/2	—	0.14	0.96	0.23	0.08	—	0.23	0.88	0.42
11/2	—	−0.13	0.95	0.17	0.23	—	0.00	0.93	0.37

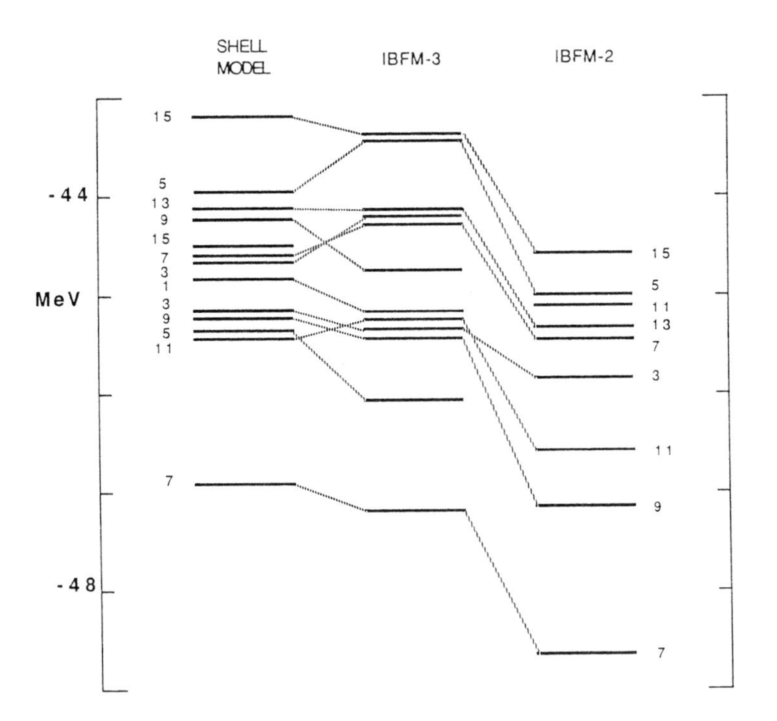

Figure 2: A comparison of IBFM and shell model for ^{45}Sc. Levels are labelled by $2J$.

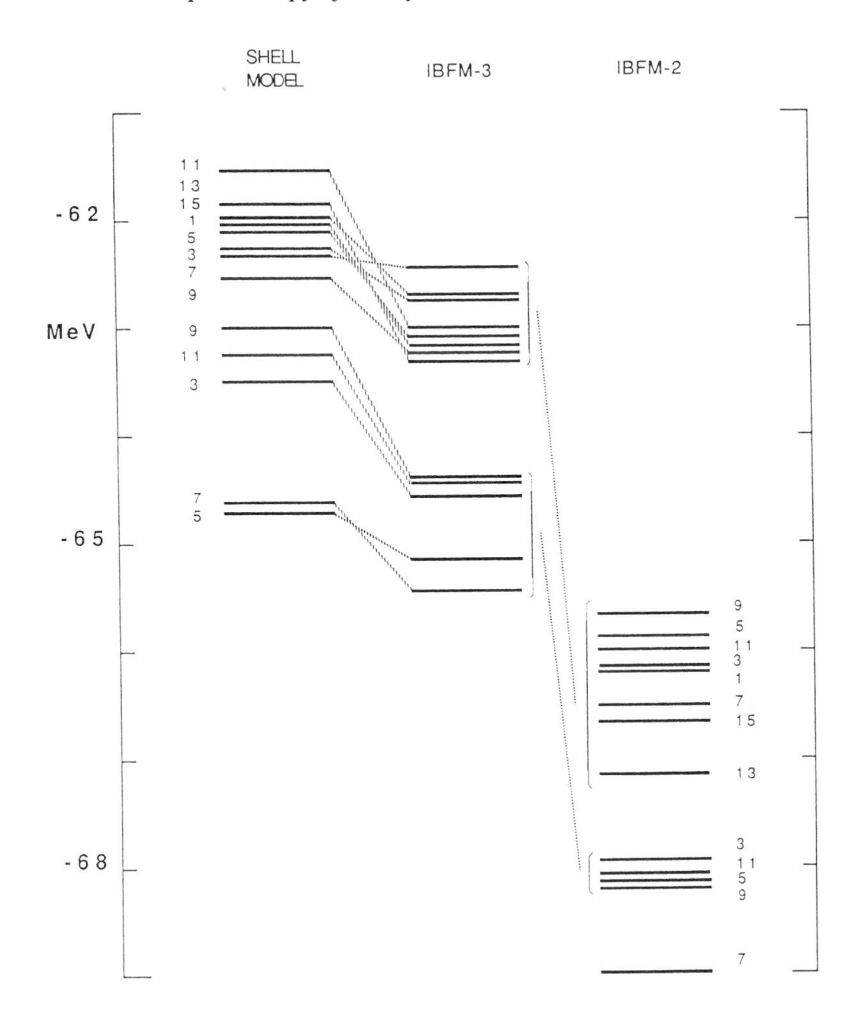

Figure 3: A comparison of IBFM and shell model for ^{47}V. Levels are labelled by $2J$.

Besides comparing the energy spectra, one should also check whether the wave functions in the two models are similar. This is illustrated in table 2 for some of the yrast states in ^{45}Ti. Two features should be noted from the table: (i) non-boson components in the shell-model calculation are small and (ii) the remaining components are in overall agreement with those found in the IBFM-3. Similar conclusions can be drawn for all low-energy states shown in figures 1-3.

4.3 Low-lying Mixed-Symmetry States in Odd-Mass Nuclei

Figures 1-3 also show results of an IBFM-2 calculation, that is, a calculation without δ-bosons. As argued earlier, the eigenstates resulting from the IBFM-2 do not have good isospin and it can also be seen from the figures that, in general, the absolute energies are not well reproduced.

Furthermore, one notes that in ^{45}Sc some low-lying shell-model and IBFM-3 states are absent from IBFM-2. To give a qualitative explanation of this feature, one may consider the example of a system with four neutrons and one proton. A basis in the IBFM-3 consists in this case of the states

$$\left((b^\dagger_{1,l}b^\dagger_{1,l'})^{(T_B,L_B)}a^\dagger_{\frac{1}{2},j}\right)^{(T,J)}_{M_T,M_J}|\mathrm{o}\rangle. \tag{13}$$

For $T = -M_T = 3/2$ these states are combinations of

$$\left((b^\dagger_{\nu,l}b^\dagger_{\nu,l'})^{(L_B)}a^\dagger_{\pi,j}\right)^{(J)}_{M_J}|\mathrm{o}\rangle, \tag{14}$$

and

$$\left((b^\dagger_{\nu,l}b^\dagger_{\delta,l'})^{(L_B)}a^\dagger_{\nu,j}\right)^{(J)}_{M_J}|\mathrm{o}\rangle. \tag{15}$$

Since $T = 3/2$, the isospin T_B of the bosons must be 1 or 2 and we expect states with $T_B = 1$ to be lowest in energy, since, in general, the energy of states tends to increase with isospin. However, $T_B = 1$ implies that the two-boson state must be antisymmetric under U(6) (Evans *et al.*, 1985) and this can only occur if the bosons are different . We are thus lead to the conclusion that the states (15), which only occur in IBFM-3 and not in IBFM-2 and which are antisymmetric (in general, mixed-symmetric) in U(6), are expected at lower energies than the states (14).

A careful analysis would reveal a more complicated situation, in which the occurrence of low-lying mixed-symmetry states also depends on the boson-fermion interaction and its dependence on T and J. We can summarise the result of such an analysis (Van Isacker *et al.*, 1988) by stating that low-lying mixed-symmetry states may arise in odd-mass nuclei with $\frac{1}{2}n - T$ odd, where n is the number of valence nucleons.

4.4 Boson-Fermion Exchange Interactions

Instead of the form (10), the boson-fermion interaction can be written in two alternative ways,

$$V_{BF} = \sum_{l,l',q,s} \alpha^{qs}_{ll'}\left((b^\dagger_{1,l}\tilde{b}_{1,l'})^{(q,s)}(a^\dagger_{\frac{1}{2},j}\tilde{a}_{\frac{1}{2},j})^{(q,s)}\right)^{(0,0)}_{0,0}, \tag{16}$$

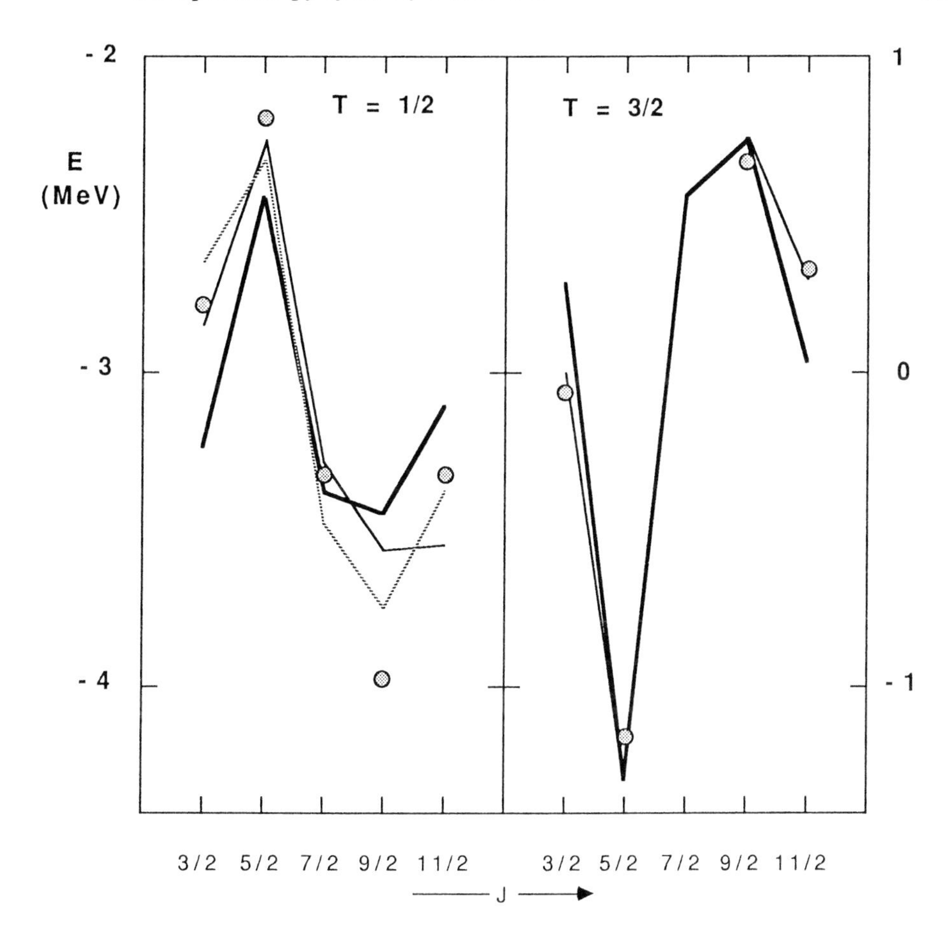

Figure 4: The boson-fermion interaction for $T = 1/2$ and $T = 3/2$. The matrix elements deduced from the $f_{7/2}$ shell model (circles) are compared with three fits with different interactions: (a) exchange + monopole (thick), (b) exchange + monopole + dipole (solid thin), (c) exchange + monopole + dipole + quadrupole (dotted).

and

$$V_{BF} = \sum_{l,l',p,r} \beta_{ll'}^{pr} \left((b_{1,l}^{\dagger} \tilde{a}_{\frac{1}{2},J})^{(p,r)} (a_{\frac{1}{2},J}^{\dagger} \tilde{b}_{1,l'})^{(p,r)} \right)_{0,0}^{(0,0)}, \tag{17}$$

referred to as the multipole and exchange expansions, respectively. Clearly, the three representations (10), (16) and (17) are equivalent: any set of matrix elements

$$\langle 1l, \tfrac{1}{2}j; TJ | V_{BF} | 1l', \tfrac{1}{2}j; TJ \rangle$$

can be converted into a set of $\alpha_{ll'}^{qs}$'s or $\beta_{ll'}^{pr}$'s and *vice versa*. Thus, if all the multipole coefficients $\alpha_{ll'}^{qs}$ (or all the exchange coefficients $\beta_{ll'}^{pr}$) are considered, the most general interaction can be parametrised with (16) or (17), but at the same time no simplicity is gained compared with the original parametrisation (10) in terms of interaction matrix elements. The strategy in phenomenological IBFM calculations (Scholten, 1985) is to select a few terms from the multipole and the exchange expansions to arrive at a simple parametrisation of the boson-fermion interaction. Since in the present work the boson-fermion interaction is derived from the shell model, it is thus possible to check whether this "realistic" interaction is consistent with phenomenological parametrisations.

In figure 4 are shown the matrix elements between dj-states as they are deduced from the $n = 3$ shell-model calculation (circles). The most conspicuous feature of these matrix elements is the difference between $T = 1/2$ and $T = 3/2$. The $T = 1/2$ matrix elements are centered around ≈ -3 MeV, whereas the $T = 3/2$ matrix elements (excluding $J = 7/2$) average to about zero. This difference can be obtained through an isospin-dependent monopole force. If one tries to explain deviations in the dj matrix elements from these average values, one finds that *all* terms in the multipole expansion (16) are needed and that none of them is dominant. This shows that a parametrisation in terms of a few multipole terms cannot reproduce the realistic interaction and that possibly some exchange term might be of importance.

In the exchange interaction used in previous IBFM calculations (see, e.g. Scholten, 1985) the boson and the fermion are coupled to angular momentum j (i.e. $r = j$ in (17)). The underlying physics is (Talmi, 1981; Scholten and Dieperink, 1981) that the operator $(b_l^{\dagger} \tilde{a}_j)^{(j)}$ has the same tensor character as the fermion and can be used to correct for Pauli exchange effects between the fermion and the bosons which have been ignored in the IBFM. The $r = j$ exchange operator can also be derived from the shell model using a seniority mapping procedure (Otsuka *et al.*, 1987). Similar arguments for the IBFM-3 lead to the conclusion that the important term in the exchange expansion has $(p, r) = (\frac{1}{2}, j)$ since the isospin of the fermion equals $1/2$.

Figure 4 compares the realistic interaction matrix elements with the result of a fit with parameters $\alpha_{22}^{00} = -17.3$ MeV, $\alpha_{22}^{10} = -20.4$ MeV and $\beta_{22}^{\frac{1}{2}J} = 9.85$ MeV. The average values of the $T = 1/2$ and $T = 3/2$ matrix elements are determined by the monopole parameters but deviations $\Delta V(T, J)$ from these averages are entirely due to the exchange parameter $\beta_{22}^{\frac{1}{2}J}$. The $(p, r) = (\frac{1}{2}, j)$ exchange interaction predicts deviations in the $T = 1/2$ and $T = 3/2$ cases to be correlated according to $\Delta V(T = 1/2, J) = -\frac{1}{2}\Delta V(T = 3/2, J)$. It is seen from figure 4 that this relation is approximately valid for the realistic boson-fermion matrix elements.

Figure 4 also shows fits to the realistic matrix elements if more terms of the multipole expansion are included. The solid, thin curve shows the effect of including dipole terms in the fitting process. This is equivalent to introducing a $J(J+1)$ term. The fit becomes near perfect for $T = 3/2$ and is significantly improved for $T = 1/2$. Finally, quadrupole terms are introduced to give the dotted line shown for $T = 1/2$. The dotted line is not drawn for $T = 3/2$ since it would be indistinguishable from the solid thin curve.

In summary, the realistic boson-fermion interaction in the $f_{7/2}$ shell has predominantly a monopole-plus-exchange character, in line with earlier findings (Otsuka *et al.*, 1987) which have been extended here to include isospin.

5 CONCLUSIONS

The work reported here has two obvious limitations: only the case of a single-j shell is discussed and, perhaps more significantly, only few-boson systems ($N = 2, 3$) are considered in the $f_{7/2}$ test because of the smallness of the shell. Nevertheless, within these limitations, it is shown that the results of the shell model and the IBFM-3 are in close agreement and one may hope that similar conclusions would apply to a more complex configuration.

The primary concern of this work was to give a shell-model justification of the interacting boson-fermion model. Because of that, our attention was focused on a mass region where shell-model calculations are readily available. However, one obvious advantage of the IB(F)M-3 over the shell model is that it can more easily extended to heavier nuclei. Of particular interest in this respect is the $28 \leq N, Z \leq 50$ mass region, where full shell-model calculations cannot be carried out in general. Based on the conclusions of the present work, it seems that one can now attempt a systematic study of nuclei in this region in the framework of the isospin-invariant interacting boson(-fermion) model.

ACKNOWLEDGEMENT

I wish to thank J. P. Elliott and J. A. Evans, my collaborators in the work reported here.

REFERENCES

Arima A and Iachello F 1976, *Ann. Phys.* (NY) **99** 253
Arima A and Iachello F 1978, *Ann. Phys.* (NY) **111** 201
Arima A and Iachello F 1979, *Ann. Phys.* (NY) **123** 468
Elliott J P, Evans J A and Van Isacker P 1988, *Nucl. Phys.* **A481** 245
Elliott J P and White A P 1980, *Phys. Lett.* **B97** 169
Evans J A, Elliott J P and Szpikowski S 1985, *Nucl. Phys.* **A435** 317
Evans J A, Van Isacker P and Evans J A 1989, *Nucl. Phys.* **A489** 269
Flowers B H 1952, *Proc. Roy. Soc.* **A212** 248
Otsuka T, Yoshida N, Van Isacker P, Arima A and Scholten O 1987, *Phys. Rev.* **C35** 328
Scholten O 1985, *Prog. Part. Nucl. Phys.* **14** 189

Scholten O and Dieperink A E L 1981 *Interacting Bose-Fermi Systems in Nuclei* ed F Iachello (New York: Plenum) p 343
Talmi I 1981 *Interacting Bose-Fermi Systems in Nuclei* ed F Iachello (New York: Plenum) p 329
Thompson M J, Elliott J P and Evans J A 1987, *Phys. Lett.* **B195** 511
Van Isacker P, Elliott J P and Evans J A 1988, *J. Phys.* **G14** L201

The Fermion dynamical symmetry model

Da Hsuan Feng
Department of Physics and Atmospheric Science, Drexel University
Philadelphia, Pennsylvania 19104-9984, U.S.A.

Abstract

A microscopic nuclear collective model, the Fermion Dynamical Symmetry Model (FDSM) is presented. The model is a symmetry dictated truncated spherical shell model. As such, it is tractable and can and has been applied to the "non-traditional territories": the heavy nuclei. Some recent developments, especially the mass calculations in the actinide region and a study of the BE(2) systematics in which one finds strong Pauli effects are discussed here.

1. Introduction

It is my great pleasure to present this talk on the *Fermion Dynamical Symmetry Model* (FDSM) (Wu et al., 1986; Wu et al.,1987a) in this conference held in this beautiful island of Crete. The FDSM is a microscopic model of nuclear collective excitations and was proposed three years ago by my colleagues ***Cheng-Li Wu, Mike W. Guidry, Jin-Quan Chen, Xuan-Gen Chen*** and myself. So in terms of nuclear models' time scale, it is still in its infancy. Many subsequent developements were in collaboration with *Hua Wu, Wei-Min Zhang, Zhen-Ping Li, Xiao-Ling Han, Joe Ginocchio and Rick Casten.*

The nucleus is a strongly interacting many-body system and has long posed a serious challenge for microscopical understanding. Of course, it is commonly recognized that the spherical shell model (Mayer, 1948; Haxel et al., 1949) is *the* fundamental model. Yet the shell model is successful primarily in what I would call the "traditional territories": light nuclei (e.g. s-d shell, see a discussion by Wildenthal [Wildenthal, 1985]) and low lying states of nuclei with a small number of valence nucleons. As soon as one moves away from the traditional territories, say in the medium-heavy to heavy weight nuclei, the encountered configuration becomes *too large, in fact astronomical,* even for today's computers. Therefore, numerous phenomenological models, geometric (Bohr and Mottelson, 1975) or algebraic (e.g. Arima and Iachello, 1975), have been introduced and they all in one way or another have deviated significantly from the original spherical shell model. Many people here have made excellent presentations about those models and I will not discuss such approaches here.

Let me deviate from tradition. Afterall, if indeed the spherical shell model is fundamental, then the accumulated wisdom of studying nuclear structure for the past 40 years should somehow be qualitatively, if not quantitatively, provide statements about properties of nuclei via the spherical shell model not in the traditional territories. For example, while it is fully recognized that Pauli principle must play a crucial role in the nuclear system (especially for the deformed region where the physics is a many particle system in a finite shell), its precise manifestation of this important quantum behavior remains elusive. The Fermion Dynamical Symmetry Model (FDSM) discussed in this talk is precisely *a prescription for the many-body solutions of the spherical shell model* in the heavy or medium-heavy mass regions. As a fully fermionic model, the FDSM has no ambiguity as to the role of Pauli principle and therefore some precise statements can indeed be made. I like to point out that in parallel to the development of the FDSM, there is another fermion model which concentrates primarily in the strongly deformed region. It is the pseudo *SU(3)* model, first proposed by Arima et al. and Hecht et al. some 20 years ago (see Hecht, 1985). Since the most active practitioner of the model, Gerry Draayer will give a talk later on (Draayer, 1990), therefore I will not comment much on it here.

2. What is the FDSM ?

In Fig. 1, I will outline the basic idea of the FDSM. It is well known that even for one major shell the "raw" shell model matrix H_{sh} is way beyond control. For collective excitations where the physics is strong configuration mixings, there are no obvious criterion to ignore the off-diagonal matrix elements. Hence, even if we were only interested in the nuclear structure in low energy (small box on the lower right hand corner of the matrix in Fig.1a), diagonalizing the entire matrix appears a must. This is why one traditionally employs either the geometrical model or the interacting boson model for such states. However, the FDSM transforms the entire shell model Hamiltonian into the so-called *k-i* basis in which the Hamiltonian matrix is generally smaller (see Fig.1b). Therefore, for the low energy structure, we need only to diagonalize H_{FDS} which is what we shall refer to as the FDSM Hamiltonian matrix H_{FDS}. The Hilbert space of the H_{FDS} is obviously severely truncated and thus the calculations can become tractable. It is in this sense that the FDSM is the shell model truncation in the *k-i* basis.

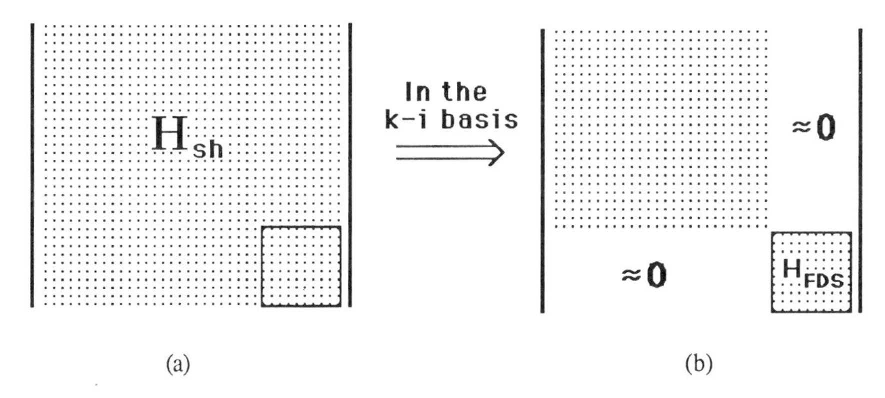

Fig.1 *The shell model Hamiltonian in the k-i basis*

Let me now briefly explain what the *k-i* basis is. *It is a single-particle basis with the angular momentum* ***j*** *decomposed into a "pseudo orbital angular momentum"* ***k*** *and a "pseudo-spin"* ***i*** *:*

$$\boldsymbol{j} = \boldsymbol{k}+\boldsymbol{i}\,, \qquad a^{\dagger}_{jm} = \sum_{m_k m_i} < k\, m_k\, i\, m_i \,/\, j\, m > b^{\dagger}_{k m_k i m_i} \qquad (1)$$

where $a^{\dagger}_{jm}$ and $b^{\dagger}_{km_kim_i}$ are the single-particle creation operators in the original spherical shell model and the *k-i* basis respectively. However, by itself, Eq.(1) is not addequate to uniquely define a basis since for a given j there are many possible k's and i's. So additional constraints are obviously needed if one wants to proceed. The k-i basis that is employed in the FDSM is a "constraint" k-i decomposition:

(1) The values of k and i should be such that all the j's in a major shell are reproduced, no more and no less.

(2) Either k must be 1 or i must be 3/2, unless the assignment is in contradiction with the first constraint, then k=0 should be chosen.

The first constraint is merely a statement of unitary transformation. With this condition the reverse transformation of Eq.(1) exists:

$$b^{\dagger}_{km_k im_i} = \sum_j \langle k\, m_k\, i\, m_i \,/\, j\, m \rangle\, a^{\dagger}_{jm} \tag{2}$$

Thus there is a one-to-one correspondence between the k-i basis and the original spherical shell model basis. The second constraint is motivated by the physical consideration that the nuclear low-lying collective excitations is dominated by the coherent S and D pairs. In the traditional pairing model, only the coherent S-pairs (condensate) are considered. This is a resulted from assuming that the nucleon-nucleon interaction favors the coupling of two identical nucleons' angular momenta to zero. Generally speaking, for the nuclear ground states, such a condensation is reasonably accurate. For excited states, however, it is far too simplistic and some form of generalization is called for. It appears to us that a natural way to generalize the pairing concept in order to include the D pairs is to assume that only part of the single-particle angular momentum, which we refer to as the "inert" part j_{inert}, is to couple to zero, leaving a small residual part of the single-particle angular momentum, which we refer to as the "active" part j_{active}, that allows two identical nucleons to form S and D pairs. Thus, we can choose the k-i decomposition such that one part is the active part and the other part is the inert part :

$$j = j_{active} + j_{inert} \tag{3}$$

In other words, the physical meaning of our *k-i* decomposition can be interpreted as the decomposition of the single particle angular momentum into active and inert parts instead of the conventional orbital and the spin parts. By definition, there are only two alternatives for the active angular momentum in such a k-i decomposition: for the active ***k***, *k must be 1*; for the active ***i***, *i must be 3/2*. This is why the second constraint is imposed. The corresponding coupling schemes are illustrated in Fig. 2.

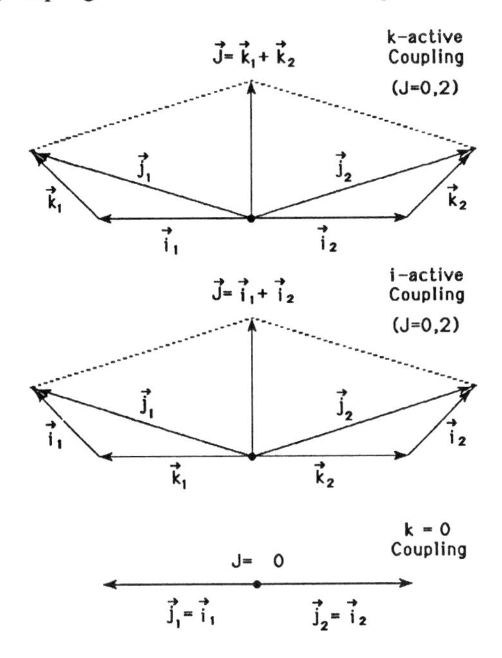

Fig. 2 *The coupling scheme in the FDSM.*

With these two constraints the *k-i* basis is then determined uniquely. The single-particle basis reclassified in terms of *k-i* quantum numbers is shown in Table 1 from which one sees that whether the normal parity levels should be *k*-($k=1$) or *i*-active ($i=3/2$) is *entirely determined by the shell itself,* while $k = 0$ always prevail for the abnormal parity level in each shell and thus *S*-pairs dominant. Any other assignment of *k* in the abnormal level will result in more then one abnormal states which violates the first constraint.

Table 1

Reclassification of Shell Model Single-Particle Basis

No.	1	2	3	4	5		6		7			8		
n	0	1	2	3	3	4	4	5	5	5	6	6	6	7
k	0	1	1	0	1	0	2	0	1	1	0	1	1	0
i	$\frac{1}{2}$	$\frac{1}{2}$	$\frac{3}{2}$	$\frac{7}{2}$	$\frac{3}{2}$	$\frac{9}{2}$	$\frac{3}{2}$	$\frac{11}{2}$	$\frac{1}{2}$	$\frac{7}{2}$	$\frac{13}{2}$	$\frac{3}{2}$	$\frac{9}{2}$	$\frac{15}{2}$
CONFIGURATION	$s_{1/2}$	$p_{1/2}$ $p_{3/2}$	$s_{1/2}$ $d_{3/2}$ $d_{5/2}$	$f_{7/2}$	$p_{1/2}$ $p_{3/2}$ $f_{5/2}$	$g_{9/2}$	$s_{1/2}$ $d_{3/2}$ $d_{5/2}$ $g_{7/2}$	$h_{11/2}$	$p_{1/2}$ $p_{3/2}$	$f_{5/2}$ $f_{7/2}$ $h_{9/2}$	$i_{13/2}$	$s_{1/2}$ $d_{3/2}$ $d_{5/2}$	$g_{7/2}$ $g_{9/2}$ $i_{11/2}$	$j_{15/2}$
SYM			$G_6G_8G_3$		$G_6G_8G_3$		G_8		G_6			G_6		
Ω_0	0	0	0	0	5		6		7			8		
Ω_1	1	3	6	4	6		10		15			21		
n	2	8	20	28	50		82		126			184		

No. labels the shell ordering, n, k, i label the principle, pseudo-orbit and pseudo-spin quantum numbers, and Ω_0 and Ω_1 are the pair degeneracies of the abnormal-parity and normal-parity levels for each shell. The number n means the maximum allowable nucleon number up to and including that particular major shell. The symbols $\mathbf{G_6}$, $\mathbf{G_8}$ and $\mathbf{G_3}$ are short-hand notation for the symmetries:

$G_6=(Sp_6^k \times SO_3^i) \times (SU_2 \times SO_3)$ (k-active)

$G_6=(SO_8^i \times SO_3^k) \times (SU_2 \times SO_3)$ (i-active)

$G_6=(SU_3^k \times SO_6^i) \times (SU_2 \times SO_3)$ (k/i-active)

where $SO_3{}^k$, $SO_3{}^i$, and SO_3, denote the rotational groups associated with angular momenta k, i, and j_0 (abnormal). For s-d shell there is no $SU_2 \times SO_3$ factor due to the absence of an abnormal-parity level.

The *k-i* basis has several many interesting properties. First of all it has a coherent *S-D* subspace which is believed to be the major portion of the space responsible for the low lying nuclear collective states. The space can be separated easily from the entire shell model space by simply "freezing" the inert part of the angular momenta (couple to zero). In the *k-i* basis the *S* and *D* pairs are defined as

$$S^{\dagger}=\sqrt{\Omega_{ki}/2}\,[b^{\dagger}_{ki}b^{\dagger}_{ki}]^{00}_{00},\quad D^{r}_{\mu}=\sqrt{\Omega_{k3/2}/2}\,[b^{\dagger}_{k3/2}b^{\dagger}_{k3/2}]^{0r}_{0\mu},\quad D^{r}_{\mu}=\sqrt{\Omega_{1i}/2}\,[b^{\dagger}_{1i}b^{\dagger}_{1i}]^{r0}_{\mu 0} \qquad (4)$$

(for any k and i) (for i-active) (for k-active)

where $\Omega_{ki}=(2k+1)(2i+1)/2$. For the shell with more then one i's (e.g. shells 7 & 8) Eq.(4) should sum over i's. These S and D pairs have a very simple structure in the k-i basis and yet they are highly coherent in the original shell model basis; in particularly, the S-pair is precisely the Cooper pair condensate. These pairs have the desired coherent property to represent nuclear collective motion.

Secondly, the space built up by such S and D pairs has many symmetries: all the abnormal levels in all shells have SU_2 (the quasi spin group) symmetry ; the normal-parity levels have either SO_8 or Sp_6 symmetries depending on whether the shell is i- ($i=3/2$) or k-active ($k=1$), and have multi-chain dynamical symmetries as shown in Fig.3 .

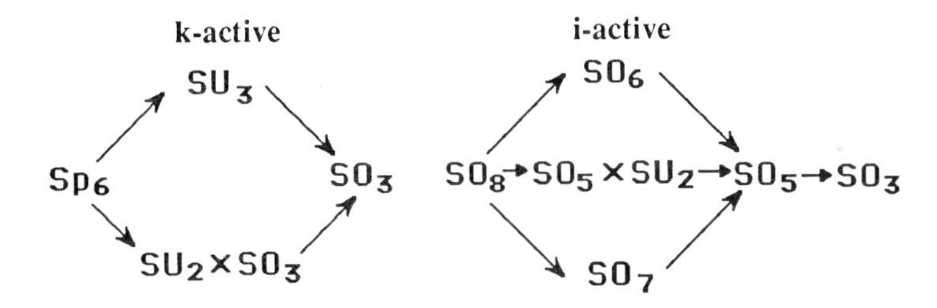

Fig. 3 *The group chains of the SO_6 and SO_8 symmetries*

In the S-D subspace, the most general FDSM Hamiltonian for identical particles can be written as follows:

$$H_{FDS} = e_0 n_0 + e_1 n_1 + H_a + H_n + H' \tag{5-1}$$

$$H_a = G_0 S^\dagger S + (B_0/4)\, n_0^2 \tag{5-2}$$

$$H_n = G_0 S^\dagger S + G_2 D^\dagger D + \sum_r B_r P^r \cdot P^r \tag{5-3}$$

$$H' = g_0 (S^\dagger S + S^\dagger S) + (b_0/2)\, n_0 n_1 \tag{5-4}$$

where the pair operator S and D are defined in Eq.(4) For the abnormal-parity level (k=0), the pair is denoted by $\mathcal{S}^{\dagger}$. The multipole operators P^r are defined as follows:

$$P^r_\mu = \sqrt{\Omega_{ki}/2}\,[b^\dagger_{ki}\tilde{b}_{ki}]^{r0}_{\mu 0}, \qquad P^r_\mu = \sqrt{\Omega_{k3/2}/2}\,[b^\dagger_{k3/2}\tilde{b}_{k3/2}]^{0r}_{0\mu} \qquad (6)$$

(for k =1 or 0) (for i-active)

The operators n_0 and n_1 which are the number operators for particles in abnormal-parity level and normal-parity levels respectively, equal to $2P^0_0$. Thus one can obtain analytical results for individual dynamical symmetry chain and see if the basic features of low-lying nuclear collective motion can be described by this *S-D* subspace. To apply the FDSM to realistic systems, where perfect symmetry is the exception rather then the rule, one must consider the all important neutron-proton interactions. For example, for the deformed region, the symmetry is viewed as the coupled SU_3 symmetries:

$$SU^{\pi}{}_3 x SU^{\nu}{}_3 \supset SU^{\pi+\nu}{}_3 \qquad (7)$$

The predictions of the spectra and electromagnetic transitions indicate that there is a *one to one* correspondence between the fermion dynamical symmetry and the well known nuclear collective modes: $SO_8 \supset SO_5 \times SU_2$ and $Sp_6 \supset SU_2 \times SO_3$ limits both correspond to vibrational modes deferring only in some details of transition rates; the $SO_8 \supset SO_6$ limit corresponds to a γ-soft rotor; and the $Sp_6 \supset SU_3$ limit corresponds to an axially-symmetric rotor. In addition, there is a $SO_8 \supset SO_7$ limit. For fixed particle number, its spectrum looks vibrational, but the energy spacing decreases as number of nucleons increases and the γ-transition rates look more like those for transitional nuclei. Such a vibrational mode has not been recognized before. However, following the FDSM prediction, the experimental evidence for such a mode was found (Casten et al., 1986): The nuclei Ru and Pd have precisely this kind of behavior, and they exist at just the right positions with valence neutrons in the SO_8 shell (shell 6, see Table 1).

It should be emphasized that all the fermion dynamical symmetries discussed above are direct consequences of the second constraint, which is where the physics begins. With different physical considerations we might impose different constraints on the *k-i* decomposition and will lead to different *k-i* assignments. If no truncation is employed then it does not matter since the different choices for the k-i assignment are only a matter of using a different basis (see Eq.(2)); the physics will not change. However our aim is to find a better truncation scheme by choosing the *k-i* decomposition and then freezing the inert part. Therefore, the nature of the second constraint is quite crucial: a different constraint will lead to a different truncated space and result in different dynamical symmetries. It also should be pointed out that the *k-i* basis with either $k = 1$ or $i = 3/2$, and the associated fermion dynamical symmetries shown in Fig. 3, were first discovered by Ginocchio (Ginocchio 78; Ginocchio 80) although it was not obvious then that such a basis could be a physical basis to describe realistic nuclei.

All these considerations suggest that the *S-D* subspace in the *k-i* basis is a good subspace. If the Hamiltonian of a nuclear system does have precisely SO_8 or Sp_6 symmetry, then the S-D subspace is exactly decoupled with the rest of the space; this means that all the coupling terms shown schematically in Fig.1b vanish. Also other coupling terms which will break the symmetries are generally nonzero. Still, the fact that even the symmetry limits in the *S-D* subspace already manifest the basic features of nuclear collective motion strongly suggests that the influence from the space outside of the *S-D* subspace should be either small, or at least rather smooth, so that the renormalization process for the effective interactions can take place. So in the final analysis, if we were to enlarge the model space slightly by including the configurations with a few unpaired nucleons plus *S-D* pairs, then there is the possibility of finding an effective interaction which can quantitatively describe the low lying nuclear structures in heavy nuclei, a goal which was already realized for light nuclei within the traditional shell model.

For this purpose, the model defines a new quantum number u,

u ≡ the number of nucleons which do not form coherent S and D pairs

which is called the "heritage" (the coherent *S* and *D* pairs are defined in Eq.(4)). The full fermion shell model space can therefore be "stratified" according to this number: $u = 0$ for the *S-D* subspace; $u = 1$ for *S* and *D* pairs plus an odd particle; $u = 2$ for *S* and *D* pairs plus a broken pair, and so on. For the low-lying low-spin states of heavy nuclei we may truncate the shell model space to the *S-D* subspace. For the high spin states, physics dictates that we consider the broken pair (or pairs), thereby enlarging the space somewhat. Such an enlargement renders the calculations still within reach of present computers if $u < 4$. Thus the FDSM provides a very convenient and physically reasonable truncation scheme to carry out detailed numerical computations within the philosophy of the spherical shell model in heavy-mass regions.

Since the FDSM is microscopic, it is testable. The theory may be tested at two levels. The first concerns the global predictions of nuclear systematics by the theory in the analytical symmetry limits: do the collective modes implied by the FDSM occur in a systematic fashion in the low energy nuclear structure? The second concerns microscopic calculations including numerical diagonalization of symmetry breaking terms: do the collective modes of the FDSM have the accurate microscopy to describe detail behaviors of individual nuclei ? We are now to the stage where we may conclude that the first test has been passed, and passed quite well (Han et al., 1987; Guidry et al., 1986a; Guidry et al., 1986b); the second has just started. In this section I shall concentrate on the first aspect. Unfortunately, time does not permit me to discuss some of the work recently carried out for the second. May I refer the interested readers to our papers (Vallieres and Wu, 1989).

II. The Dynamical Pauli Effect

For low lying states, once the valence shells are specified, the FDSM *uniquely* links the normal parity levels of neutrons or protons with the symmetry SO_8 or Sp_6. For the intruder level, the FDSM prescribes a quasi-spin symmetry (denoted in this talk as a script SU_2). The SU_3 depicts an axial rotation geometry (i.e. deformed structure). In this section, I will discuss the Pauli effect in the strongly deformed region which we have given the name Dynamical Pauli Effect (DPE).

The $Sp_6 \supset SU_3$ chain was the group chain where the DPE was first mathematically demonstrated. It's physical meaning can be understood very simply: Consider a system of spin $\frac{1}{2}$ fermions (say electrons); antisymmetrization demands that the Young diagram (the jargon of group theoretical methods) representing an irreducible representation (irrep) can at most have two columns, i.e. $s = \frac{1}{2}$ and $2s + 1 = 2$. Likewise, the Sp_6 symmetry is constructed from the pseudo-orbit $k = 1$. Since $(2i + 1)(2k + 1) = 2\Omega$ (where Ω is the degeneracy of the single particle levels), for Sp_6 we have $2i + 1 = \frac{2}{3}\Omega$. In other words, the maximum number of columns the Young diagram can have for this symmetry is $\frac{2}{3}\Omega$.. Therefore, when the number of valence nucleons $n \leq \frac{2}{3}\Omega$, the SU_3 irrep $(\lambda,\mu) = (n, 0)$ is allowed. This irrep gives the "standard" ground band of a deformed nucleus, i.e. a band of states with energies proportional to $J(J + 1)$ *where J is the angular momentum.* However, when n exceeds $\frac{2}{3}\Omega$, the "extra" boxes in the diagram, due to the Pauli principle, must be placed in the next row which means $\mu > 0$. When this happens, the representation $(n, 0)$ is *strictly forbidden* and therefore the entire band of states with this SU_3 irrep must vanish. Since each band has the same "intrinsic" SU_3 irrep, therefore when the Pauli principle "hits", it doesn't just eliminate one of the states in the band, but the *entire* band. I suspect that any "real-life" collective model which is built from fermions will have this feature. This is what I mean by the DPE. By the term *dynamic* I mean that this manifestation of the Pauli principle is strictly a consequence of the group symmetry Sp_6, which is through the restriction on the allowed irrep of a certain fermion dynamical symmetry chain (which in this case is $Sp_6 \supset SU_3$). It is worth emphasizing that deviation from this symmetry will cause the DPE to "wash out". Therefore experimental manifestation of it can serve as a sensitive test of the symmetry Sp_6 and the underlying dynamical assumption of the S and D fermion pairs.

The realization of the importance of the DPE came only after the proposal of the FDSM was made that the symmetry chain $Sp_6 \supset SU_3$ should not be applied to the entire physical shell, but only to the valence nucleons in normal parity levels. The reason is because if we were to interpret $\Omega = \sum(2j+1)/2$ as the "entire" physical shell degeneracy, then only nuclei with valence nucleon number n significantly less than the midshell value of Ω (i.e., $n \leq \frac{2}{3}\Omega$) can be deformed. The reason is that the "canonical" deformed behavior of nuclei has a ground band with an intrinsic structure represented by the SU_3 $(n,0)$ irrep, a pair of side (β- and γ-) bands with intrinsic structure $(n\text{-}4,2)$, and so on. Hence such an interpretation of Ω clearly contradicts the data since we know that well deformed nuclei occur near midshell (n certainly larger than $\frac{2}{3}\Omega$). However, as I have mentioned before, the FDSM separates nucleons into two camps: one for normal levels (n_1^σ) and one for the intruder level (n_0^σ) where the total number of valence nucleons $n^\sigma = n_1^\sigma + n_0^\sigma$ (There are occasions in this talk where I will talk about number of valence pairs N, i.e. $N_\sigma = \frac{1}{2} n^\sigma$, $N_1^\sigma = \frac{1}{2} n_1^\sigma$ *and* $N_0^\sigma = \frac{1}{2} n_0^\sigma$). With such a separation, even the most simplistic interpretation will show that even though $n^\sigma \leq \frac{2}{3}\Omega_1^\sigma$ (Ω_1^σ is just the degeneracy of normal parity levels and *not* the full degeneracy Ω^σ of the entire physical shell), the value of n^σ can still approach Ω^σ since some of the valence nucleons will be occupying the intruder level. Furthermore, for realistic n_1^σ and n_0^σ separation, there may be occasions where $n_1 > \frac{2}{3}\Omega_1^\sigma$. When that happens, a real "structural" rearrangement of the band structure should emerge. Therefore we see immediately that there are some definite consequences for the structure of nuclei in such a separation. It will be most interesting to see whether data are consistent with these consequences; this point will be discussed in the next section. A word about notation simplification: from now on I will drop the superscript σ on the various quantities unless its specification is necessary for clarity.

3. Nuclear mass - Applications of the FDSM-Strutinskii method

For the work reported here, we have received crucial help from **Xiao-Ling Han**. Accurate computation of nuclear masses is important not only for its own sake but also for the understanding of nuclear astrophysics. There are accurate calculations using the spherical shell model for light nuclei. In fact, there was at least one excellent talk in the conference. In other mass regions, the most extensive calculations of the masses utilized the well known Strutinskii's shell correction procedure. As an application of the FDSM, we have proposed another method to compute the nuclear masses which partly utilizes the Strutinskii's shell correction procedure. As we shall see, the physics of the problem emerges simply and naturally from this approach. I will call our approach here the FDSM-Strutinskii mass calculations. This part of the talk will concentrate on this discussion.

The first question of interest is why do we think mass calculations are interesting with the FDSM? As I have mentioned before, according to the FDSM, as long as the system is dominated by the $Sp_6 \supset SU_3$ dynamical symmetry, we should expect to find signals for the DPE lurking around. Therefore the more the system behaves according to the dynamical symmetry of $Sp_6 \supset SU_3$, the more obvious should the DPE be. Also, from experience one knows that the closest that one comes to a pure SU_3 state is the deformed ground state, therefore we thought that the best place to find a manifestation of the DPE is in the masses (i.e. ground state) of deformed nuclei.

Second question of interest is why we have chosen to study the actinide masses? At this point, it is useful to recall the relationship between the FDSM symmetries and the periodic table. In order to keep the notation as brief as possible, I will not mention the quasi-spin *SU(2)* of the intruder level clarity calls for it. The three main regions of heavy nuclei have the following symmetries: Actinides: $Sp^{\pi}_6 \times Sp^{\nu}_6$; Rare-earths: $Sp^{\pi}_6 \times SO^{\nu}_8$; Z, N=50-82 region: $SO^{\pi}_8 \times SO^{\nu}_8$. From this classification, we see that the most obvious region to seek the DPE is in the actinides where both valence neutrons and protons in the normal parity levels have Sp(6) symmetries.

As a first attempt, we assumed that the binding energy of actinide nuclei relative to ^{208}Pb is $B_g = \langle H_{FDS} \rangle$, where $\boldsymbol{H_{FDS}}$ is the most general FDSM Hamiltonian which include pairing, quadrupole - quadrupole and monopole - monopole interactions. A dynamical symmetry for normal parity neutrons and protons in the the deformed regions is assumed of the form

$$Sp^{\pi}{}_{6} \times Sp^{\nu}{}_{6} \supset SU^{\pi}{}_{3} \times SU^{\nu}{}_{3} \supset SU^{\pi+\nu}{}_{3} \tag{8}$$

For nucleons near the beginning or ends of shells (for protons or neutrons), the symmetry is assumed to be $SU^{\pi}{}_{2} \times SU^{\nu}{}_{2}$. Finally, the symmetry of the intruder parity level is taken to be the coupled (neutron-proton) quasi-spin symmetry. Our approach uses a simple first-order perturbation scheme: SU_2 perturbation on SU_3 and vice versa.

In order to carry out the mass calculations within the framework of the FDSM, it is crucial to first determine the distribution of n_1 and n_o (i.e. the valence nucleon numbers in the normal and the intruder levels). According to the model, the n_1 dependence on n can be derived by minimizing the ground state energies. With this minimization, a linear dependence $n_1 = a + bn$ is obtained (for details, see eqs. (5.15) and (5.20) in ref. 1). In principle the parameters a and b are fixed by the effective interactions of the FDSM model space. As it turns out, we do not find a sensitive dependence of the n_1 distribution to the physical shell, nor is it sensitive to the type of nucleons (protons or neutron): it is well described in all heavy nuclei by the universal formula $n_1 = 1.5 + 0.5n$ (which is equivalent to $N_1 = 0.75 + 0.5N$). As a check of its physical reasonableness, one can use the Nilsson model to obtain a similar universal formula (Figure 4).

With these assumptions the mass formula is remarkably simple (Han et al., 1987)

$$\begin{aligned} B_g(\alpha) = \quad & a(\alpha)N_\pi^2 + b(\alpha)N_\pi + c(\alpha)N_\nu^2 + d(\alpha)N_\nu \\ & + e(\alpha)N_\nu N_\pi + Q(\alpha) + k_\nu(\alpha)u_\nu + k_\pi(\alpha)u_\pi \end{aligned} \tag{9a}$$

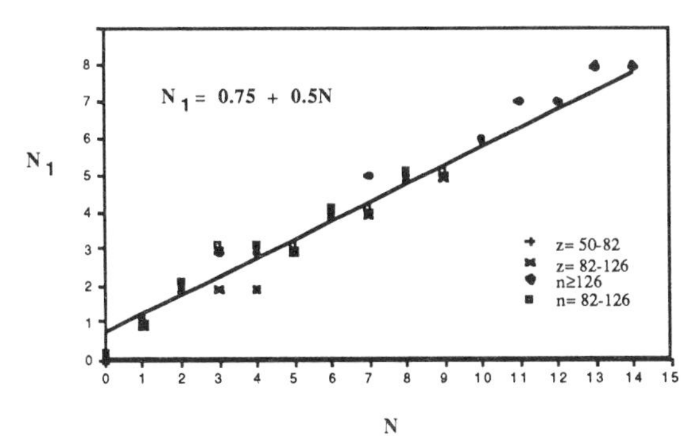

Fig. 4 *N_1 (pair number $=n_{1/2}$) obtained from Nilsson scheme as a function of the total valence pair number N. The straight line is for the equation given in the diagram. The points are obtained from the Nilsson scheme by assuming an appropriate deformation. The same N_1 dependence is assumed for both neutrons and protons.*

and

$$Q(\alpha) = \begin{cases} 0 & \alpha = SU(2) \\ \delta_\nu \Delta N_1^{\,n} + \delta_\pi \Delta N_1^{\,\pi} & \alpha = SU(3) \end{cases} \tag{9b}$$

$$\Delta N_1^{\,\nu} = \begin{cases} N_\nu - N_\nu^{\,crit} & N_\nu > N_\nu^{\,crit} \\ 0 & N_\nu \leq N_\nu^{\,crit} \end{cases} \tag{9c}$$

$$\Delta N_1^{\,\pi} = \begin{cases} N_\pi - N_\pi^{\,crit} & N_\pi > N_\pi^{\,crit} \\ 0 & N_\pi \leq N_\pi^{\,crit} \end{cases} \tag{9d}$$

In eqs. (9a) - (9d), the formula is related to pair numbers represented by N (and not the number n). Also, the symbol α stands for the symmetry group [SU_2 or SU_3]. For SU_2, the ground states are seniority zero, one or two, depending on whether we are referring to even-even, even-odd, or odd-odd nucleus respectively. Since the SU_3 irrep is $(\lambda^\sigma,\mu^\sigma) = (n_1^{\,s},0)$ or $(0, \bar{n}_1^{\,\sigma})$ for the ground state up to $n_1^\sigma (\bar{n}_1^\sigma) = \frac{2}{3}\Omega_1^\sigma$, where $\bar{n}_1^\sigma = 2\Omega_1^\sigma - n_1^\sigma$; Beyond this, DPE will force the ground state irrep to switch to

$$(\lambda^\sigma,\mu^\sigma) = (\tfrac{2}{3}\Omega_1^\sigma - 4\Delta N_1^\sigma \ , \ 2\Delta N_1^S) \tag{10a}$$

and

$$(\lambda,\mu) = (\lambda^\pi+\lambda^\nu, \mu^\pi+\mu^\nu) \tag{10b}$$

In Eq. (9) $Q(\alpha)$ is just the correction term due to the irrep change. The pair numbers N_1^σ is the total nucleon pair number (in normal and the abnormal parity levels) for the σ-type particles. The symbol N_σ^{crit} is the total number of σ–type particles for which the (σ–type) normal parity particles satisfy the condition $n_1{}^\sigma = \frac{2}{3}\,\Omega_1{}^\sigma$, i.e. particle number of the normal parity levels has reached its maximum for the $(n_1{}^\sigma,0)$ irrep to occur. Finally, the heritage quantum numbers for protons and neutrons are defined as u_π and u_ν respectively. They correspond to the number of nucleons which do not form coherent S and D pairs. For the ground states of even-even nuclei the heritage numbers are zero, while for the odd-neutron, or odd protons, u_σ ($\sigma = \pi$ or ν) = 1.

It should be emphasized that Eq. (9) is fully *microscopic* in that the coefficients a, b, c, d, e, δ_ν, δ_π, κ_ν, κ_π are explicitly given in terms of the effective interactions of the spherical shell model, for which the FDSM model is an approximate solution in a truncated space. The precise microscopic shell model effective interaction to use in the truncated space of the FDSM is still completely not known. However, the essential physics of the problem at hand can be studied by a semi-microscopic fit of the parameters: the mass formula is derived microscopically via the FDSM and the parameters are determined by fits to the ground-state masses. For SU(2) and SU(3), the parameters are (a = -0.795, b = 9.68, c = -0.185, d = 9.541, e = 1.247, k_ν = -0.902 and k_π = -1.096) and (a = -0.613, b = 8.10, c = -0.306, d = 12.26, e = 0.97, δ_ν = -0.622, δ_π = -2.26, κ_ν = -0.70 and κ_π = -0.6168) respectively (Wu et al., 1987b). All units are in MeV. The root mean square error for 332 known nuclei is 0.34 MeV (for comparison, the r.m.s. of Møller-Nix (MN)(Møller and Nix, 1981; Møller and Nix, 1989) mass calculation in actinide region is 0.79 MeV).

In Fig. 5, the discrepancies between experimental masses and the FDSM predictions ($\Delta B = E_{exp} - E_{FDSM}$) is plotted as a function of the total valence neutron pair number N_n (the SU_3 irrep correction term $Q(\alpha)$ has not yet been considered). Different lines correspond to different proton pair number N_p. It is clear from Fig. 5 that the overall agreement for $N_n \leq N_n^{crit} = 13$ and $N_p \leq N_p^{crit} = 8$ is excellent. This implies that for the ground state, the representations $(2N_1^n, 0)$ and $(2N_1^p, 0)$ are reasonable ones to describe the binding energy.

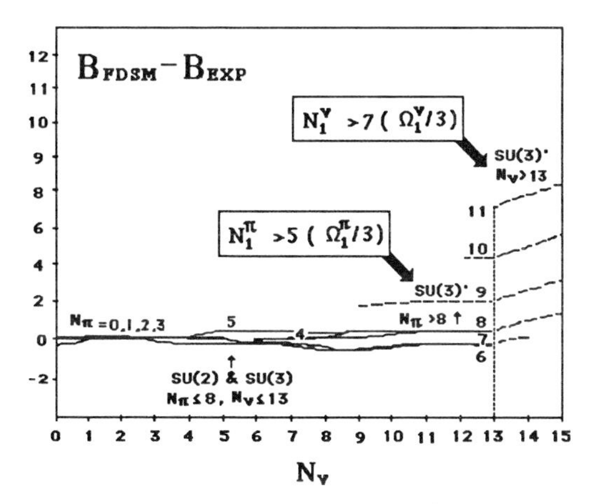

FIG. 5 *The difference in mass between the FDSM predictions and the data (ΔB) plotted as a function of neutron pair number(Ref. 2b). The various lines correspond to different proton pair number.*

However, beyond N_n^{crit} and N_p^{crit}, there is a clear and systematic increase in the discrepancy which is approximately linear in N_n and N_p. Such a linear dependence is expected from Eq.(9b). When the"correct" SU_3 irrep is used for $N_n > N_n^{crit} = 13$ and $N_p > N_p^{crit} = 8$ (i.e. including the $Q(a)$ term), the agreement is again excellent (in other words, ΔB is now of the same order of magnitude as those below 13 and 8 for neutron and proton pairs respectively). Hence, from the masses, the first signal of the DPE emerges.

It is known that the shell model Hamiltonian for the valence nucleons contains two parts: the single particle part for neutrons and protons and the two body interactions part. The single particle part of the Hamiltonian is

$$H_I = \sum^{\pi} \varepsilon_i n_i + \sum^{\nu} \varepsilon_i n_i \tag{11}$$

where ε_i and n_i denote the ith single particle (s.p.) level energy and occupancy respectively. The next part is the two body interaction term, which will be denoted as V. In principle, all residual nucleon-nucleon interactions should be included in this term. In the standard deformed mean-field approach, the energy of the ground state mass is replaced by $\sum \varepsilon_k$ (where k is the z-component of the single particle angular momentum in the intrinsic axis). For the strongly deformed actinides, there are of the order of 10^2 neutrons and protons. Therefore if each s.p. energy is roughly 90% accurate (the best we can hope for with an absolute magnitude of about 1 MeV or so), then the masses computed from (1) must also incur an uncertainty of the order of 10 MeV or so, which is an intolerable amount for any meaningful understanding of the masses. This prompted Strutinski to devise a method to minimize this uncertainty. To do so, he added and subtracted the "same" quantities from $\sum \varepsilon_k$ as follows:

$$\left\{ \sum \varepsilon_k - \sum < \varepsilon > \right\} + M^{def}_{Liq\ drop} \tag{12}$$

where the smooth part of the energy is supposed to be represented by a deformed liquid-drop formula $M^{def}_{Liq\ drop}$. The term $\left\{ \sum \varepsilon_k - \sum < \varepsilon > \right\}$ represents the fluctuation part of the energy and is called the energy shell correction. It can be computed following the standard Strutinski's procedure.

For the FDSM, however, there is no deformed mean field and one begins with the standard shell model Hamiltonian

$$H = H_I + V \tag{13}$$

where H_I is given by Eq.(11) (s.p. energies for valence protons and neutrons) and V the two body energy term. It must be remembered that, unlike Strutinski's deformed mean-field approach, the shell model Hamiltonian as defined here is in the spherical basis. The "deformed" mean field, which is produced by the various long-range two body interactions, is now treated in the FDSM directly without the use of a deformed mean field approximation. Nevertheless, the Strutinski's type approach is still required since the spherical shell model single-particle energies are also not accurately known enough to produce meaningful nuclear masses. The reason why we were able to obtain reasonable results in our first study using eq.(9) without separating the total mass into a liquid-drop part plus a shell correction part is because of the fact that all the parameters in Eq.(9) are phenomenologically determined. It can be seen that if the liquid drop part and the spherical shell correction part are smooth functions of Z and N, and note that $N_\pi/82$ and $N_\nu/126$ are small quantities for actinide nuclei, then making an expansion up to quadratic terms of $N_\pi/82$ and $N_\nu/126$ may not be a bad approximation. Thus, the liquid drop and the spherical shell correction parts have already been included effectively in the polynomial of the mass formula Eq.(9) through parameters renormalization. Of course, the liquid drop part is no doubt a smooth function of proton and neutron numbers. However, we do not a priori know whether the spherical shell correction is also smooth. The successes of Eq.(9) in fitting data suggest that it is, at least approximately so. Hence the main part of the energy fluctuation in the FDSM approach comes primarily from the symmetry dependence which is due to the competition between pairing and quadrupole-quadrupole interactions.

Recently, we have carried out a more refined study of the FDSM mass formula. This time the spherical single-partical shell correction are explicitly treated. To carry out the

Strutinski's program within the FDSM context, we must also "subtract" and "add" a smooth part of the single-particle energy to the Hamiltonian:

$$H = [\,(H_1 + V_p\,) - (\Sigma\langle\varepsilon\rangle_{sp} + V_p^{deg})\,] + M_{liq\ drop}^{sph} + V \tag{14}$$

In (14), the liquid drop part $M_{liq\ drop}^{sph}$ is taken from Møller-Nix paper. The term $\Sigma\langle\varepsilon\rangle_{sp}$ is the smooth part of the single-particle Hamiltonian which is calculated by the Fermi-gas model with the single-particle zero-point energies e_p and e_n determined from the maximun cancellation with H_1 , while H_1 is single-particle energy calculated from a Wood-Saxon potential. There are two pairing terms in equation (9): V_p represents the pairing energies for the Wood-Saxon (hence realistic) single-particle energy with the protons and neutrons pairing strengths chosen to reproduce the experimental pairing gaps; and V_p^{deg} are the pairing energies for the degenerate single-particle energy case with pairing strengths G_p and G_n determined from the maximum cancellation with V_p . Thus the term $[\,(H_1 + V_p\,) - (\Sigma\langle\varepsilon\rangle_{sph} + V_p^{deg})\,]$ in Eq.(14), which is the spherical single particle shell correction (with pairing correction) is represented from now on as $M_{sh}^{s.p}$ and can be computed without any free parameters. Futhermore, the expectation value of the FDSM effective interaction $\langle V\rangle$ can be expressed in terms of the more fundamental pairing and quadrupole-quadrupole strengths :

$$\begin{aligned}\langle V\rangle = a(\alpha) + b(\alpha)N_\pi + c(\alpha)N_\pi^2 + d(\alpha)N_\nu + e(\alpha)N_\nu^2 + f(\alpha)N_\nu N_\pi \\ + (G_0^\pi - G_2^\pi)\,V_p^\pi(\alpha) + (G_0^\nu - G_2^\nu)\,V_p^\nu(\alpha) \\ + (B_2^\pi - G_2^\pi)\,V_q^\pi(\alpha) + (B_2^\nu - G_2^\nu)\,V_q^\nu(\alpha) + B_2^{\pi\nu}\,V_q^{\pi\nu}(\alpha)\end{aligned} \tag{15}$$

where $V_p^\sigma(\alpha), V_q^\sigma(\alpha)$ $(\sigma = \pi, \nu)$, and $V_q^{\pi\nu}(\alpha)$ are the expectation values of pairing and quadruploe-quadrupole interactions; G_0^π and G_2^π are the monopole and quadropole pairing strengths; B_2^π, and $B_2^{\pi\nu}$ are the quadrupole-quadrupole strengths, while *a,b,c,d* and *e* are the linear combinations of the various monopole-monopole interaction strengths. Thus through the mass fitting a priliminary effective iteractions for the actinide region can be obtained. The results are really quite remarkable. The r. m. s. error from has reduced from the 0.34 MeV of of our preliminary try down to 0.2 MeV. The parameters are as following:

SU_2: a = 13.75 b = -5.043 c = 0.5057 d = -5.213 e = 0.3600 f = -0.32

SU_3: a = 5.5 b = -6.37 c = 0.427 d = -6.99 e = 0.388 f = -0.105

$(G_0^\pi - G_2^\pi) = -0.078$ $(G_0^\nu - G_2^\nu) = -0.044$ $(B_2^\pi - G_2^\pi) = 0.064$ $(B_2^\nu - G_2^\nu) = 0.044$ $B_2^{\pi\nu} = -0.0883$

At this point I would like to make two comparisons of the FDSM predicted masses with the well-known results of Møller-Nix. I will compare the so-called "shell corrections", and the single-neutron separation, as another way to display the DPE. Unfortunately, due to page limitations, I can only briefly describe them here and request interested readers to consult our forthcoming long paper on this subject.

First I will talk about the shell corrections. Let me say a few words about the approach of Møller-Nix (MN). MN employ what is known as the Strutinski procedure. This procedure separates the nuclear mass into a "smooth liquid drop" part and a fluctuating shell correction M_{sh}. In Fig. 6a, we show the FDSM-Strutinskii shell corrections calculated by subtracting the liquid-drop energy from the total binding energy. Obviously, These shell corrections are in excellent agreement with the data and the shell corrections of MN (see Fig.6b).

I would also like to present some results of single-neutron separation energies S_n. Again, we compare the FDSM results with those of MN (See Figs. 7a and 7b). We see that both approaches (FDSM and MN) are in agreement with the data. Certainly this is not surprising in view of the fact that the FDSM predictions of mass ΔB (Fig. 5) and M_{sh} (Figs. 6a and 6b) agree so well. However, what is truly surprising came when we "push" the our results beyond the realm of data (lines in Fig. 4 which are "beyond" the data). You

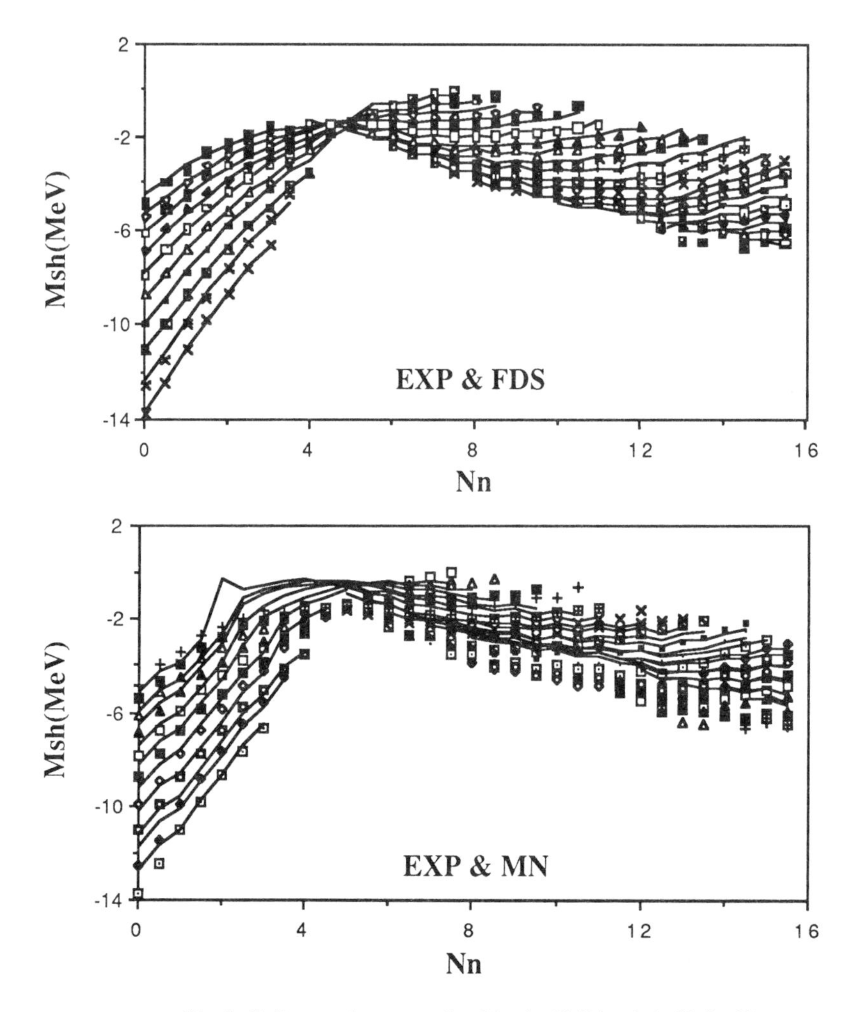

Fig. 6 *Shell corrections as predicted by the FDSM and the Møller-Nix.*

should notice that for the FDSM results (Fig. 4a), there are "kinks" in the S_n values around $N_n = \frac{1}{3}\Omega_1^n$ (N=152), $\frac{1}{2}\Omega_1^n$ (N=163), and $\frac{2}{3}\Omega_1^n$ (N=170). These kinks are natural consequence of the DPE. Yet, we noticed that for the MN results, there appear similar kinks at roughly the same positions in S_n! This is the reason why I conjectured in the beginning of this talk that the DPE may be a generic behavior of strongly deformed system.

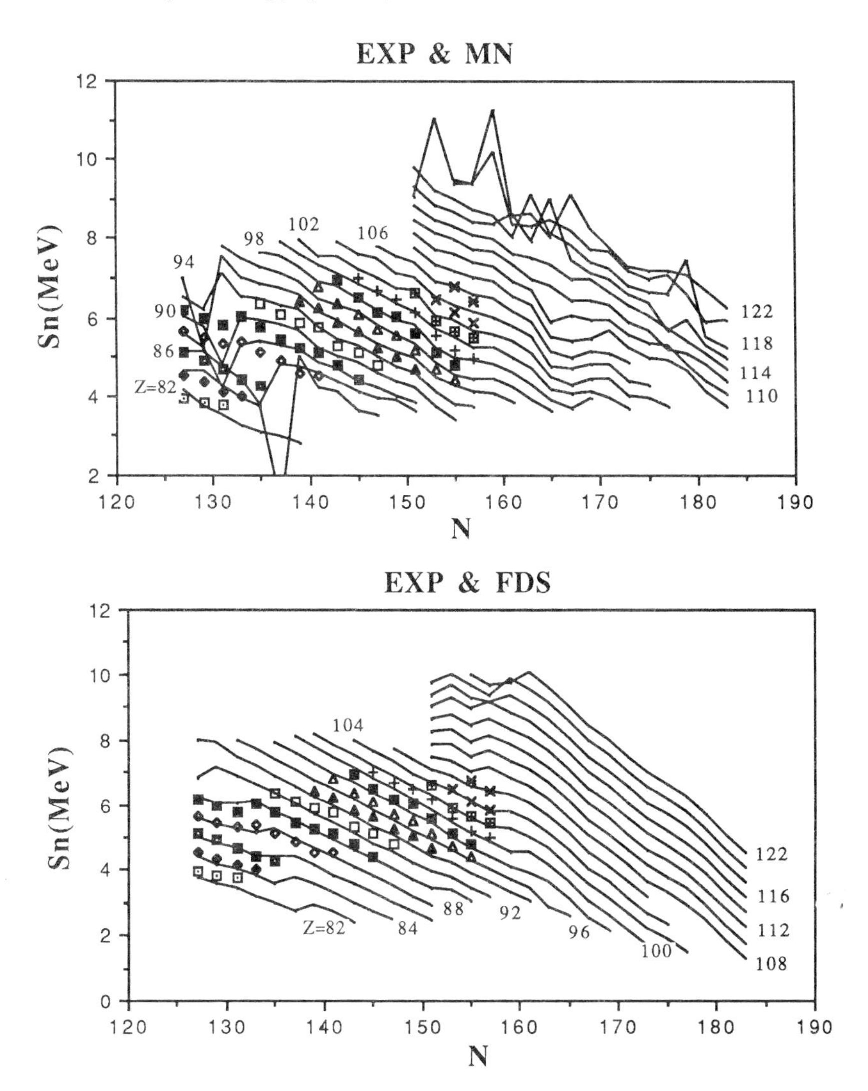

Fig. 7 *Comparisons of the single neutron separation energies vs. total neutron number. Plot on the top diagram is for the Moller-Nix calculations and for the bottom diagram is the FDSM calculations.*

4. Saturation of collectivity

In this section, I like to discuss a subject which has attracted much attention lately. This is the question of the saturation of the B(E2)'s. This work is also in collaboration with **Zhen-Ping Li**. The B(E2)'s are measurement (I should say here that this work was very much facilitated by the B(E2) systematic compilation of Raman et al.(Raman et al., 1987) of "nuclear collectivity" and therefore its saturation (as a function of valence neutrons or protons) must be nature's way of signalling a physical mechanism for this system. The observation was first made by Casten and coworkers in 1985 (Casten et al.,1985) in their phenomenological analysis of such data.

I would like to point out that the saturation is due to the Pauli principle. To show this most vividly, let me compare the FDSM, a model of *fermions*, with the IBM, a model of *bosons*. Even though the link between the IBM and the shell model is not transparent, it is commonly envisioned that the boson number should be the total number of valence nucleon pairs N (= $N_1 + N_o$). Therefore, according to the IBM, low-lying collective properties of nuclei should grow as a function of N. In its simplest implementation, it must predict no B(E2) saturation before midshell. The FDSM has three features distinct from the IBM: (1) Collective behavior is expected to depend more strongly on the number of normal-parity valence pairs N_1 than on the total number of valence pairs N; (2) there is a DPE; (3) the FDSM predicts a shell dependence for collectivity. It is of interest to establish whether these differences allow us to explain the saturation. In this discussion, we shall mainly concentrate on nuclei where non-spherical behavior is dominant.

The B(E2) formula for the FDSM or IBM are:

$$B(E2) = \frac{5e^2}{16\pi} Q_o{}^2, \qquad Q_o = Q^\pi + \frac{e_\nu}{e_\pi} Q^\nu \tag{16}$$

where Q_o is the intrinsic quadrupole moment and Q^π and Q^ν are the proton and neutron quadrupole matrix elements respectively. The parameter e_ν / e_π is the neutron and proton effective charge ratio. For simplicity we have set it equal to unity. In principle one can vary it, as Raman et al. (Raman et al., 1987) have done so recently. However the variation of this ratio does not alter the basic features presented here.

In the interest of simple calculations and for comparisons, we will employ minimal implementations (symmetry limits) for both IBM and FDSM calculations. In the IBM, there are two symmetry limits for deformed nuclei: SU_3 and SO_6. Unless we specify otherwise the term (for either bosons or fermions) "SU_3 limit" in this section means the coupled group chain $SU^{\pi}_3 \times SU^{\nu}_3 \supset SU^{\pi+\nu}_3$, with the decomposition $(\lambda^{\pi},\mu^{\pi}) \times (\lambda^{\nu},\mu^{\nu}) \supset (\lambda,\mu)$. Likewise, the designation "SO_6 limit" means the coupled group chain $SO^{\pi}_6 \times SO^{\nu}_6 \supset SO^{\pi+\nu}_6$,. For the IBM-2, the Q^{σ} ($\sigma = \pi$ or ν) of Eq. (16) are

$$Q^{\sigma}_{IBM} = q^B \sqrt{\frac{2N+3}{N}}\, N^{\sigma} \quad (SU_3 \text{ limit}) \tag{17-a}$$

$$Q^{\sigma}_{IBM} = q^B \sqrt{\frac{N+4}{N}}\, N^{\sigma} \quad (SO_6 \text{ limit}) \tag{17-b}$$

where $N = N^{\pi} + N^{\nu}$, N^{π} (N^{ν}) is the total valence proton (neutron) pair number, and q^B a constant. According to Eq. (17), the B(E2) formula in either limit is essentially the same; each is proportional to the square of the total number of bosons. Furthermore, since the building blocks of the IBM are bosons there is only one "universal" highest symmetry, $U^{\pi}_6 \times U^{\nu}_6$, for all physical shells. Therefore the IBM predicted B(E2)'s for deformed nuclei are shell independent except for the maximum boson number.

There are also two deformed FDSM symmetry limits (SU_3 and SO_6). However, the fermion pair number dependence for Q^{σ} differs for these limits:

$$Q^{\sigma}_{FDSM} = q^F U^{\sigma} \sqrt{C(\lambda^{\sigma},\mu^{\sigma})} \;\; \langle(\lambda,\mu)2(1,1)2||(\lambda,\mu)0\rangle \quad (SU_3 \text{ limit}) \tag{18-a}$$

$$Q^{\sigma}_{FDSM} = q^F \sqrt{\frac{N_1+4}{N_1}}\, N^{\sigma}_1 \quad (SO_6 \text{ limit}) \tag{18-b}$$

where $N_1 = N_1^\pi + N_1^\nu$, $N_1^\pi (N_1^\nu)$ represents the valence proton (neutron) pair number in the normal parity levels, and q^F a constant. In Eq. (18-a) $C(\lambda^\sigma,\mu^\sigma)$ is the eigenvalue of the SU(3) Casimir for the representation $(\lambda^\sigma,\mu^\sigma)$ and $<(\lambda,\mu)2(1,1)2||(\lambda,\mu)0>$ is the $SU^{\pi+\nu}{}_3 \supset SO^{\pi+\nu}{}_3$ isoscalar factor. The symbol U^σ is an SU_3 Racah coefficient. The appearance of these factors means that the system is *fermionic*! For the FDSM SO_6 limit, the Q^σ has the same IBM form except that N is now replaced by N_1 and N^σ by N_1^σ. For the SU_3 limit it can be shown that when $N_1^\sigma \leq \Omega_1^\sigma/3$ Eq.(18-a) can also be reduced to Eq.(17a), except for the replacement of N^σ by N_1^σ. However, for $N_1^\sigma > \Omega_1^\sigma/3$ the DPE comes into play, namely, the FDSM SU(3) representation $(\lambda^\sigma, \mu^\sigma)$ with $\lambda^\sigma+\mu^\sigma > \frac{2}{3}\Omega_1^\sigma$ is forbidden. Therefore, the N_1 dependence of Q^σ for higher SU_3 representations differs from that for the $(2N_1^\sigma, 0)$ representation. Let me say that, in principle, there could be DPE for other fermionic symmetries, but their onset need not occur at $\Omega_1/3$. This is what I meant earlier that the existence of this DPE is a crucial test of the FDSM. I hope to convince you that it is precisely the fermion behavior that allows us to understand the saturation effect.

According to the FDSM, the Q^π and Q^ν for the actinides should be determined from the SU_3 formula. For nuclei with $50 \leq z \leq 82$ and $50 \leq n \leq 82$ both neutrons and protons should have SO_8 symmetry. For rare earths, assuming that the wave functions for deformed nuclei are the completely symmetric functions of the direct product $SO^\pi{}_6 \times SU^\nu{}_3$, Q^π should be determined from the $SO^\pi{}_6$ formula, which is the same as Eq.(6-b) except that N_1 and N_1^σ are replaced by N_1^π, and $N_1^\pi/\sqrt{2}$, and Q^ν should come from the $SU^\nu{}_3$, formula, which is the same as Eq.(6-a) except that the Racah coefficient U^σ is replaced by $1/\sqrt{2}$ and (λ,μ) should be (λ^ν,μ^ν). Thus the B(E2) behavior predicted by the FDSM varies from shell to shell.

The B(E2) values for the Geometrical Model (GM) can also be calculated from Eq.(16). The method of determining Q_o is well documented in the literature and will not be described here. I will take the Møller-Nix calculations as representative. In Figs. 8(a-f),

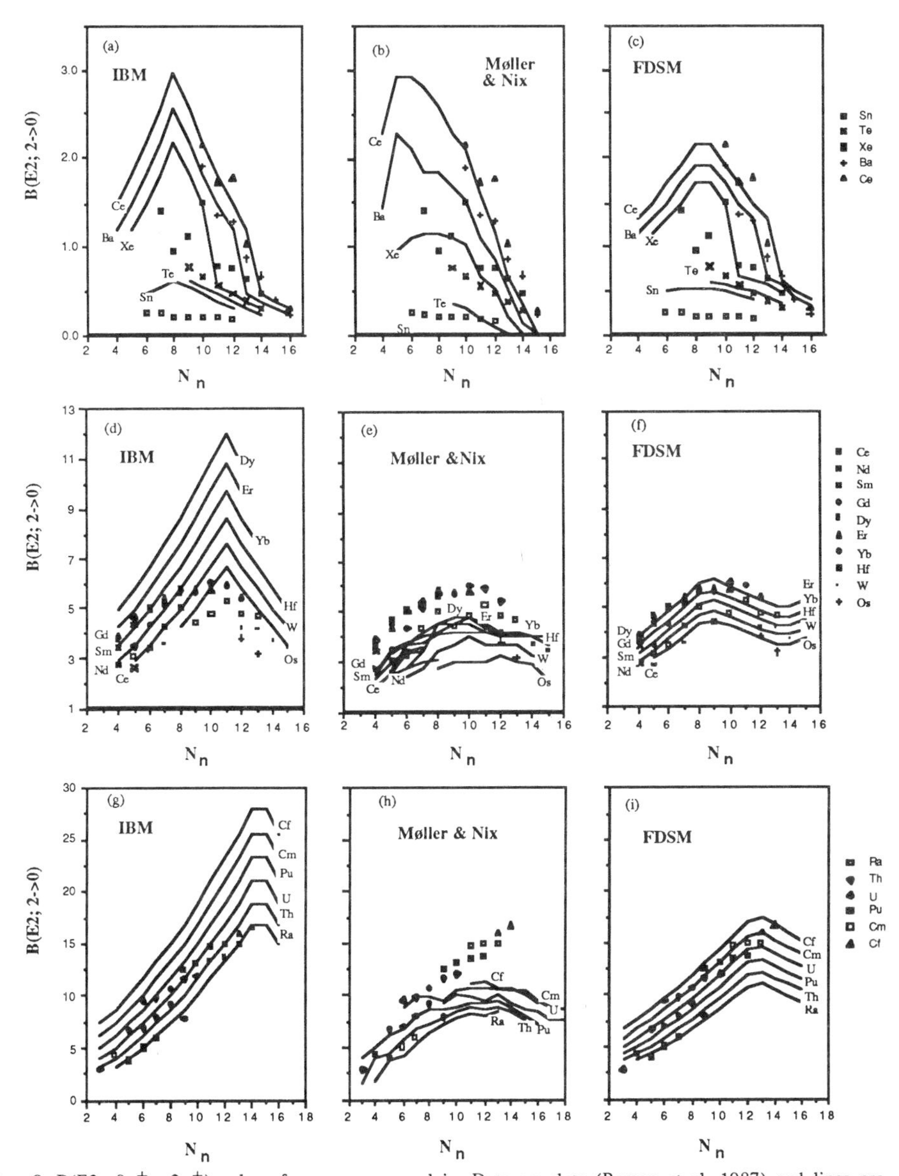

Fig. 8 B(E2; $0_1^+ \rightarrow 2_1^+$) values for even-even nuclei. Dots are data (Raman et al. 1987) and lines are calculations. For actinide and rare-earth mass regions only deformed nuclei are included. They are selected according to their E4/E2 ratio (E4/E2>2.9). For the $52 \leq (n,z) \leq 82$ mass region (Fig. a, b, and c) no selection is made since there are not as many data available. For the nuclei in this region, if E4/E2 < 2.4 symmetric SO(5) wave functions are used for the IBM and symmetric $SO^{\pi}(5) \times SO^{\nu}(5)$ wave functions are used for the FDSM. Otherwise, SO(6) symmetry is assumed for both the IBM and the FDSM. The parameter q is chosen to normalize at the beginning of the deformed regions as follows: for $52 \leq (n,z) \leq 82$, $q^F=0.501$, $q^B=0.395$; for rare-earths, $q^F=0.767$, $q^B=0.393$; and for actinides, $q^F=0.757$, $q^B=0.521$. The GM calculations of Figs. (b), (e), (h) are from Moller-Nix (Moller and Nix, 1981)

we compare the global predictions of the B(E2)'s by the FDSM, IBM, and GM with the data.

Although the GM predictions for this mass region are somewhat low, but they reproduce the overall features of the data; namely the position of the onset of saturation. I like to illustrate this effect of $\Omega_1^v/3$ by showing specifically the results of the Erbium and Curium isotopes (See Fig.9). One sees clearly that there is a clear DPE effect.

In the calculations we have chosen to determine the constants q^F and q^B for both the FDSM and the IBM at the beginning of the deformed region. We believe that this probably represents the most reasonable choice for the IBM. It is expected that the IBM boson approximation is more justified for nuclei with fewer particles in the shell, where the Pauli effects and the distinction between the normal and abnormal parity levels are less important; the choice of where to determine q^F for the FDSM is relatively immaterial. The selection of a different point to fix q^B could improve the appearance of the IBM results on the linear scale used for Fig. 8, but the essential feature of lacking in the Pauli effect for producing strong saturation would still be missed.

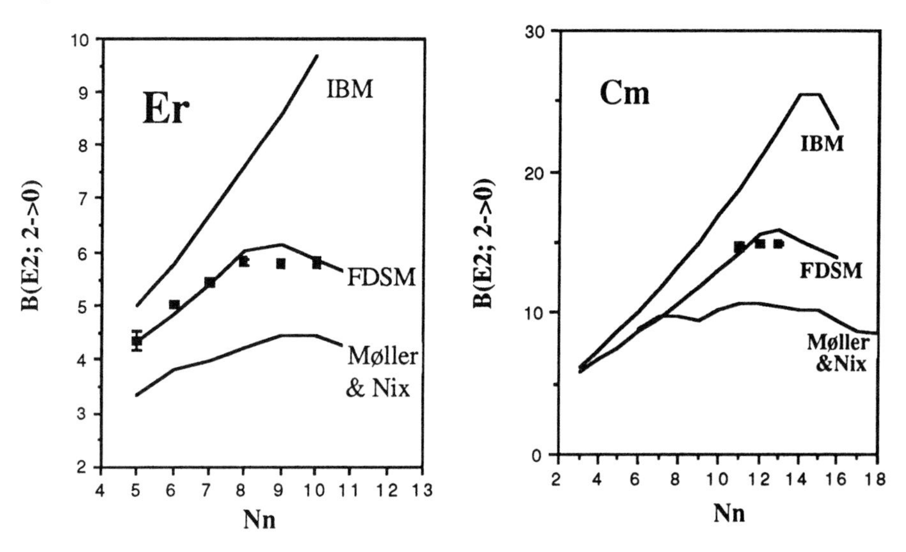

Fig. 9 *B(E2)'s plotted as a function of the neutron pair N_v for Er and Cm isotopes.*

In the B(E2)'s, we find that even in the lowest two states of even-even nuclei there are fermion effects which have a considerable influence on fundamental observables and which are a generic property of the system. It is an open question as to how much the B(E2) systematics can be improved by more detailed IBM calculations, but it seems likely that substantial numerical effort would be required to reproduce the quality of the *analytical* FDSM results presented here. Conversely, we have presented evidence in the B(E2) systematics that heavy nuclei (1) exhibit strong dynamical Pauli effects in the B(E2) values; (2) show a clear dependence of E2 collectivity on the number of normal-parity particles N_1; (3) exhibit a strong shell-dependence of the E2 collectivity. Each of these observations confirms a crucial assumption of the FDSM. Despite our emphasis on normal-parity orbitals in this discussion, it should be understood that the abnormal parity orbitals play a substantial indirect role in the quadrupole collectivity because the E2 saturation shown in Fig. 6 would occur much earlier in the shell if the abnormal orbitals did not accommodate some of the particles (see Ref.1). The presence of the abnormal-parity orbitals increases the effectiveness of the SU(3) collectivity in the normal-parity orbitals of the FDSM.

Let me briefly summarize my discussion about the saturation of B(E2)'s as follows: the building blocks of real nuclei are fermions, therefore if too many nucleons are added to the finite number of single particle levels in a shell, **the Pauli principle** forbids some configurations which would otherwise contribute to the collectivity. *Thus saturation of collectivity is expected to be a general feature of a finite fermion system.* The particle-hole symmetry in one major shell can be regarded as a natural saturation of collectivity due to the Pauli principle. The IBM exhibits a saturation of the B(E2) at half shell (although this is not the right place for the actinides and rare-earths) because of the assumption that a boson is actually a fermion pair with particle-hole symmetry in one major shell ; this Pauli effect is imposed on the IBM and is not an inherent property of a boson model.

5. Summary

The study of nuclear structure was traditionally carried out under the auspices of two models: the *shell model* and the *geometrical model.* The shell model is useful for determining the properties of nuclei near the doubly-closed shells where the dominant nuclear modes are single-particle like; the geometrical model is expected to work for nuclei far away from the doubly closed shells where collective modes dominate. It was later suggested that the collective motion in low energy can also be described by a U_6 boson model (IBM), without the necessity of introducing the concept of deformation. The FDSM now can *unify* all three previous models. I have emphasized from the very beginning that the FDSM can be regarded as the shell model in the truncated *k-i* basis. For the lower shells (e.g. the s-d shell) where the Hilbert space can be fully taken into consideration without truncation, by including all possible heritage quantum number u's, *the FDSM becomes identical to the traditional shell model.* For larger shells where the traditional shell model is not applicable, the FDSM can be employed as a truncation scheme for shell model calculations. Although I don't have time to discuss here, the *IBM is an approximation to the FDSM under the conditions that u=0 (no broken pairs), and that the Pauli effects are negligible* ($\Omega_1 \rightarrow \infty$). Also the Geometrical Model (GM) is another limit of the FDSM when $N_I \rightarrow \infty$ (classical limit). Recently we have also completed the connection to the HFB approach through the coherent state technique (Zhang et al., 1990). Some nuclear deformation calculations within the FDSM framework is along this line. I am sorry that I have no time to explain it but details can be found in reference (Wu et al., 1990). The relationships among these models are schematically shown in the "cartoon" of Fig.10. Although this "cartoon" is schematic, I dare say that it represents a concrete blueprint for unification and microscopic justification of the standard algebraic and geometrical models of low energy nuclear structure.

Acknowledgments: I would like to take this opportunity to thank the organizers for hosting this most informative conference. The work carried out in this talk is supported by the United States National Science Foundation.

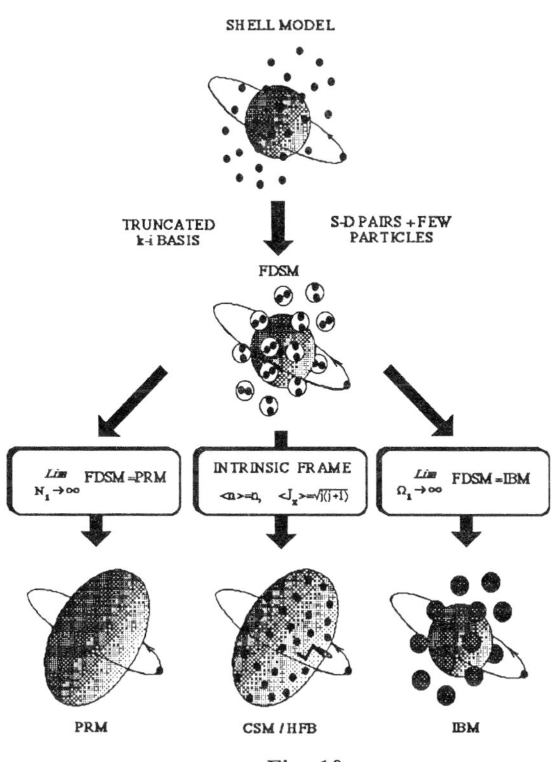

Fig. 10

References

Arima A. and F. Iachello, *Ann. Rev. of Nucl. Part. Sci. 31, 75(1981).*

Bohr A. and B. R. Mottelson, *Nuclear Structure Vol I and II,* (Benjamin, New York, (1975).

Casten R. F., W. Frank and P. von Brentano, *Nucl. Phys.* ***A444****, 133 (1985)*

Casten R. F., C.-L. Wu, D. H. Feng, J. N. Ginocchio and X.-L. Han, *Phys. Rev. Lett.* ***56****, 2578 (1986).*

Casten R. F., K. Heyde and A. Wolf, *Phys. Rev.* ***B208****, 33 (1988)*

Chen J.-Q., D. H. Feng and C.-L. Wu, *Phys. Rev.* ***C34****, 2269 (1986).*

Draayer, J. D., *in these proceedings (1990)*

Feng, D. H., C.-L. Wu, M. W. Guidry and Z.-P. Li, *Phys Lett.* ***B209****, 157 (1988)*

Ginocchio J. N., *Phys. Lett.* ***B79****, 173(1978)*

Ginocchio J. N., *Ann. of Phys.* ***126****, 234(1980).*

Guidry M. W., C.-L. Wu, D. H. Feng, J. N. Ginocchio, X.-G. Chen and J.-Q. Chen, *Phys. Lett.* ***176B****, 1 (1986a).*

Guidry M. W., C.-L. Wu, Z.-P. Li, D. H. Feng and J. N. Ginocchio, *Phys. Lett.* ***187B****, 210 (1986b).*

Han X.-L., M. W. Guidry, D. H. Feng, K.-X. Wang and C.-L. Wu, *Phys. Lett.* ***192B****, 253 (1987).*

Haxel O., J. H. D. Jensen and H. E. Suess, *Phys. Rev.* ***75****, 1766(1949)*;

Hecht K. T., *Notas de Fisica VIII (1985).*

Li Z.-P. et al. (to be published).

Mayer M., Phys. Rev. ***74****, 235(1948)*

Møller P. and J. R. Nix, *Atom. Nucl. Data Tables* ***26****,165(1981).*

Møller P. and J. R. Nix, *Atom. Nucl. Data Tables, in press (1989).*

Raman S., C. H. Marlarky, W. T. Milner, C. W. Nestor and P. H. Stelson, *Atom. Nucl. Data Tables* ***36****,1(1987).*

Sakai M., *Atom. Nucl. Data Tables* ***31****,399(1984).*

Wildenthal B. H. in *Proceedings of the Conference on Nuclear Shell Model*, M.Vallieres and B.H.Wildenthal (eds.) (World Scientific Publications, Singapore, (1985) and refs. therein.

Wu C.-L., D. H. Feng, X.-G. Chen, J.-Q. Chen and M. W. Guidry, *Phys. Lett.* ***168B****,313 (1986)*

Wu C.-L., D. H. Feng, X.-G. Chen, J.-Q. Chen and M. W. Guidry, *Phys. Rev.* ***C36****, 1157 (1987)*

Wu C.-L., X.-L. Han, Z.-P. Li, M. W. Guidry and D. H. Feng, *Phys. Lett.* ***194B****, 447 (1987)*

Wu C.-L., W. -M. Zhang, D. H. Feng, M. W. Guidry, and L. L. Riedinger (to be published) (1990)

Wu H. and M. Vallieres, *Phys. Rev* ***C39****, 1066 (1989)*

Wu C. L. et al.(to be published).

Zhang W.-M., D. H. Feng, and J. N. Ginocchio, *Phys. Rev, Lett.* ***59****, 2032(1987)*

Zhang W.-M., D. H. Feng, and J. N. Ginocchio, *Phys.Rev.* ***C37****,1281 (1988)*

Zhang W.-M., C.-L. Wu, D. H. Feng, M. W. Guidry,and J. N. Ginocchio, *Phys. Rev.* ***C38****, 1475 (1988)*;

Zhang W.-M., D. H. Feng, C.-L. Wu, H. Wu, and J. N. Ginocchio, *Nucl. Phys* (1989) (in press).

Zhang W.-M., D. H. Feng and R. Gilmore, *Rev. Mod. Phys (submitted) (1990)*

Some aspects of the phenomenological application of the interacting Boson model in the actinide region and its shell-model foundations

D Bonatsos [1], D P Menezes [2], J Rikovska [3] and N Yoshinaga [2]

[1] Institut für Theoretische Physik, Universität Tübingen, Auf der Morgenstelle 14, D–7400 Tübingen, F R Germany
[2] Department of Theoretical Physics, University of Oxford, 1 Keble Road, Oxford OX1 3NP, UK
[3] Clarendon Laboratory, Parks Road, Oxford OX1 3PU, UK

ABSTRACT: A wide range of deformed actinides is described in terms of an Interacting Boson Model (IBM) Hamiltonian with three smoothly varying parameters. The presence of subshell closures at N=142 and N=152 requires the introduction of effective boson numbers for the fitting of B(E2:$0_g^+ \rightarrow 2_g^+$) values as well as the use of s' and d' bosons for avoiding the discrepancy between the experimental and calculated β_1 and γ_1 bands. Several boson mappings are tested using a pairing plus quadrupole–quadrupole Hamiltonian in a single $j = \frac{23}{2}$ shell, with satisfactory results.

1. INTRODUCTION

The purpose of this paper is twofold. In the first part a wide range of deformed actinides is studied in the framework of the Interacting Boson Model–1 (IBM–1) (Iachello and Arima 1987, Bonatsos 1988a). In the second part several boson mappings (an essential tool in establishing a connection between the phenomenological IBM and the shell model) are tested by using a pairing plus quadrupole–quadrupole Hamiltonian in a single j=$\frac{23}{2}$ shell.

2. IBM PHENOMENOLOGY IN THE ACTINIDE REGION

The heavy actinides form an extensive region of large deformation. Therefore they are an ideal test–ground for the deformed (SU(3)) limit of the Interacting Boson Model (Iachello and Arima 1987, Bonatsos 1988a). In the simplest version of the model (IBM–1) only s-bosons (with J=0) and d-bosons (with J=2) are used.

A large number of medium and heavy even–even nuclei has been described in terms of the Hamiltonian (Cizewski and Dieperink 1985, Casten *et al.* 1985)

$$H = \epsilon_d n_d + kQ \cdot Q, \tag{1}$$

where n_d is the number of d-bosons and Q is the quadrupole operator given by

$$Q = (d^+ s + s^+ \tilde{d})^{(2)} + x(d^+ \tilde{d})^{(2)}. \tag{2}$$

ϵ_d, k and x are free parameters, fitted to the experimental data. When $x = -\frac{\sqrt{7}}{2}$, the quadrupole operator is a generator of SU(3), while in the case $x = 0$ it is a generator of O(6). Since the first term of Eq. (1) is of vibrational (U(5)) character, while the second one is of rotational (SU(3)) character, this Hamiltonian can fit a wide range of nuclei (Casten *et al.* 1985). In particular it has been successfully used in describing the transition from the spherical to the deformed region in the light actinides (Cizewski and Dieperink 1985). However, its ability to describe the deformed actinides has been questioned in the literature very early (Jänecke *et al.* 1981).

For deformed nuclei one can use the Hamiltonian (Zhang *et al.* 1985)

$$H = -kQ \cdot Q + k'L \cdot L, \tag{3}$$

where Q is the quadrupole operator of Eq. (2) and L is the angular momentum operator given by

$$L = \sqrt{10}(d^+ \tilde{d})^{(1)}. \tag{4}$$

This Hamiltonian has been used for fitting 5 deformed actinides (Zhang *et al.* 1985) with considerable success. However, when going from one nucleus to another, the values of the parameters change non-negligibly. In particular, one has to break the SU(3) symmetry (by allowing $x \neq -\frac{\sqrt{7}}{2}$) in order to account for the breaking of the degeneracy between the β_1 and the γ_1 bands shown by the experimental data (Sakai 1984). This breaking of the degeneracy does *not* approach zero as deformation increases. Therefore x does *not* approach its SU(3) value with increasing deformation. This is a feature we would like to avoid. If there is a sensible limiting symmetry, one should obtain smooth limiting behaviour when approaching it. If the SU(3) limit of IBM is a reasonable way to describe deformed nuclei, its free parameters should exhibit a smooth behaviour with increasing deformation. A nice example of smooth limiting behaviour is shown by the displacement of the odd levels of the γ_1 band relative to the even ones. This displacement approaches zero with increasing deformation (Bonatsos 1988b).

For studying the deformed actinides we have used the Hamiltonian (Rikovska and Bonatsos 1988)

$$H = kQ \cdot Q + k'L \cdot L + k''P^+ \cdot P, \tag{5}$$

where Q and L are given by Eqs. (2) and (4), while the pairing operator is given by

$$P = \frac{1}{2}(\tilde{d} \cdot \tilde{d}) - \frac{1}{2}(s \cdot s). \tag{6}$$

In Eq. (2) we fix x to its SU(3) value ($x = -\frac{\sqrt{7}}{2}$). Therefore the SU(3) symmetry in the Hamiltonian of Eq. (5) is broken only by the last term.

Using this Hamiltonian we fitted 21 deformed actinides, ranging from ^{224}Ra to ^{250}Cf. These actinides have been arranged in order of increasing collectivity, which is the same as increasing B(E2:$0_g^+ \to 2_g^+$) or increasing $R_4 = \frac{E(4)}{E(2)}$. The results of the fits of the four first members of the Yrast band are compared to the experimental data in Figure 1, while in Figure 2 the calculated β_1 and γ_1 bands are compared to the experimental ones. The parameters obtained from the fits are shown in Figure 3. We remark that two of the parameters (including the coefficient of the only SU(3) symmetry–breaking term) remain constant over the whole region, while the coefficient of the quadrupole–quadrupole term changes smoothly with increasing deformation, as expected. The same Hamiltonian has been used earlier for fitting 4 uranium isotopes (Maino *et al.* 1981), the results being consistent with the ones of the present calculation.

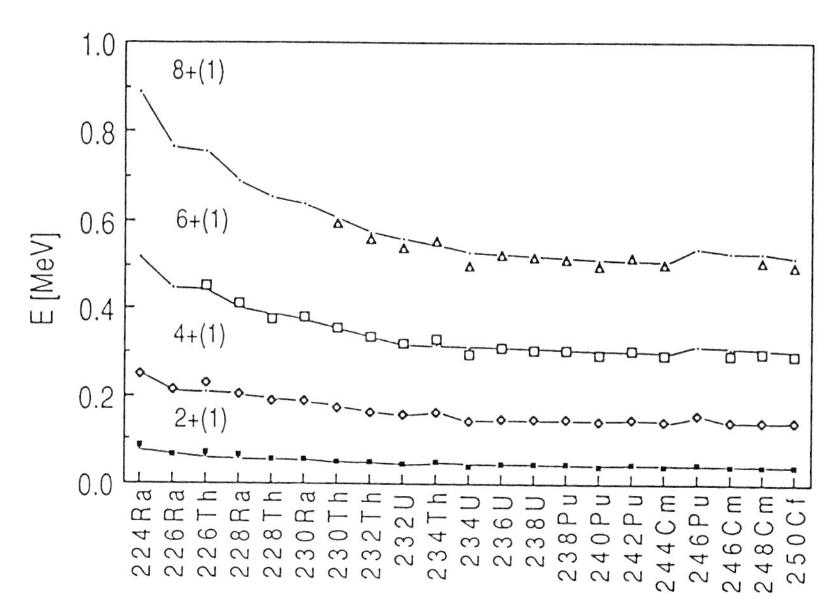

FIG. 1. Experimental and calculated ground state bands of deformed actinides. The calculations were performed with the Hamiltonian of Eq. (5). The parameters of the fits are shown in Figure 3, while the corresponding boson numbers are shown in Figure 4. Calculated values are represented by dots connected by solid lines, while experimental data are represented by symbols. Experimental data are taken from Sakai (1984).

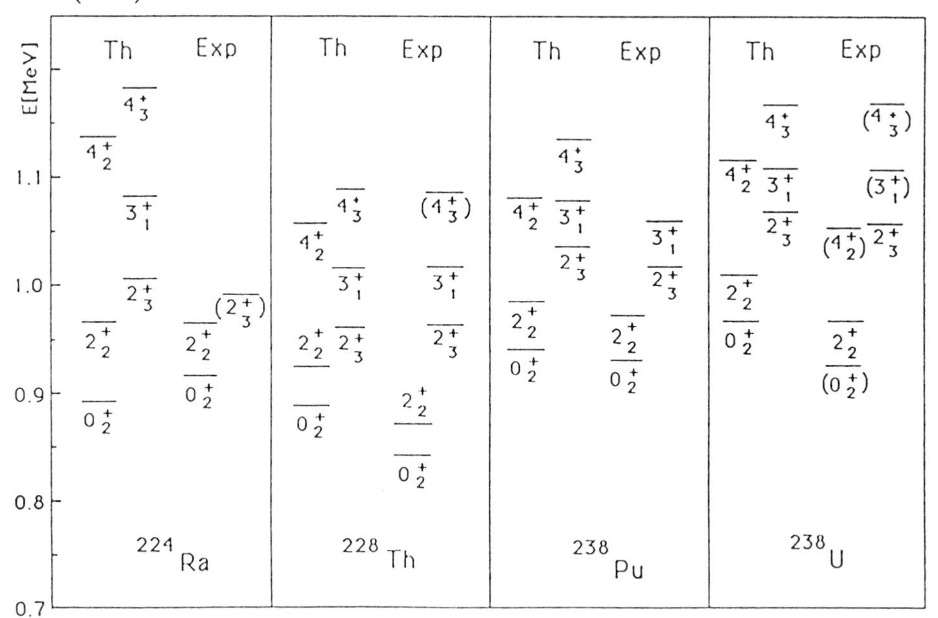

FIG. 2. As Figure 1, but for the β_1 and γ_1 bands of deformed actinides.

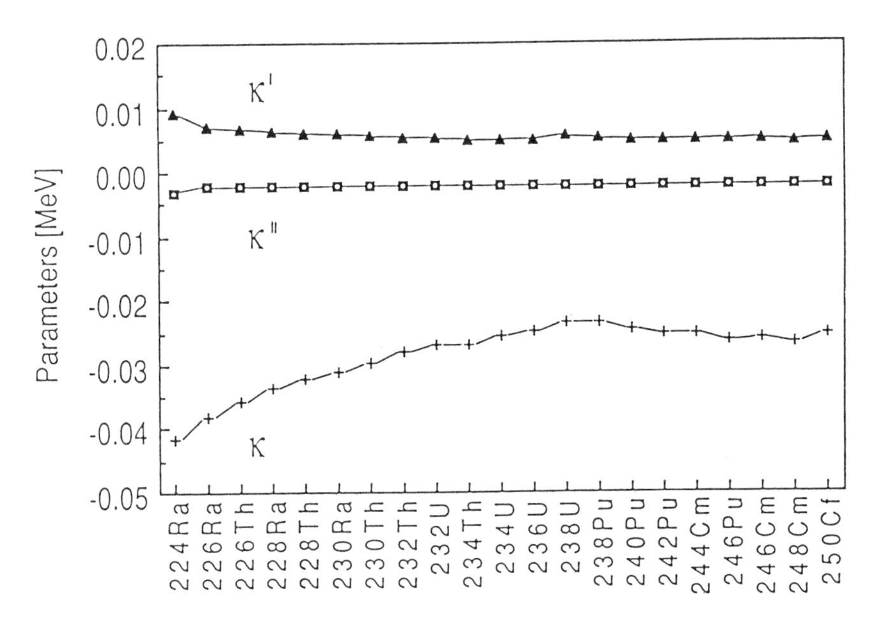

FIG. 3. Parameter values used in the Hamiltonian of Eq. (5) for obtaining the fits of Figures 1 and 2.

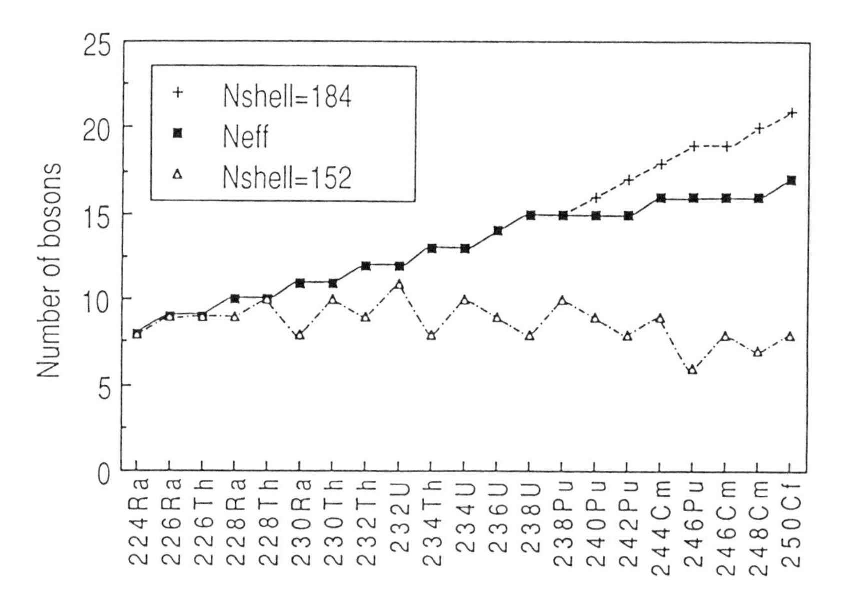

FIG. 4. Boson numbers used in obtaining the fits of Figures 1 and 2. The upper curve corresponds to boson numbers calculated by ignoring the N=152 subshell closure, while in the case of the lower curve the N=152 closure was treated as a major shell closure. The middle curve corresponds to the fitted effective boson numbers.

In addition to spectra, it is essential to fit B(E2) transition probabilities (Raman *et al.* 1987) as well. The corresponding quadrupole transition operator has the form

$$T(E2) = e[(d^+s + s^+\tilde{d})^{(2)} + x'(d^+\tilde{d})^{(2)}], \tag{7}$$

This operator is not necessarily the same as the quadrupole operator of Eq. (2); the two operators have similar mathematical structure but different physical origins. In the SU(3) limit one has $x' = -\frac{\sqrt{7}}{2}$. In the SU(3) limit, however, the transitions $\gamma_1 \to$ *ground* and $\beta_1 \to$ *ground* are forbidden, and only the transitions $\gamma_1 \to \beta_1$ are allowed. Experimentally it is known that the $\gamma_1 \to$ *ground* transitions are quite strong, while the $\beta_1 \to$ *ground* transitions are small and the $\gamma_1 \to \beta_1$ ones are difficult to measure (Warner and Casten 1983). One therefore needs to break the SU(3) symmetry of the $T(E2)$ operator in order to fit these transitions. Zhang *et al.* (1985), for example, used $x' = 0$.

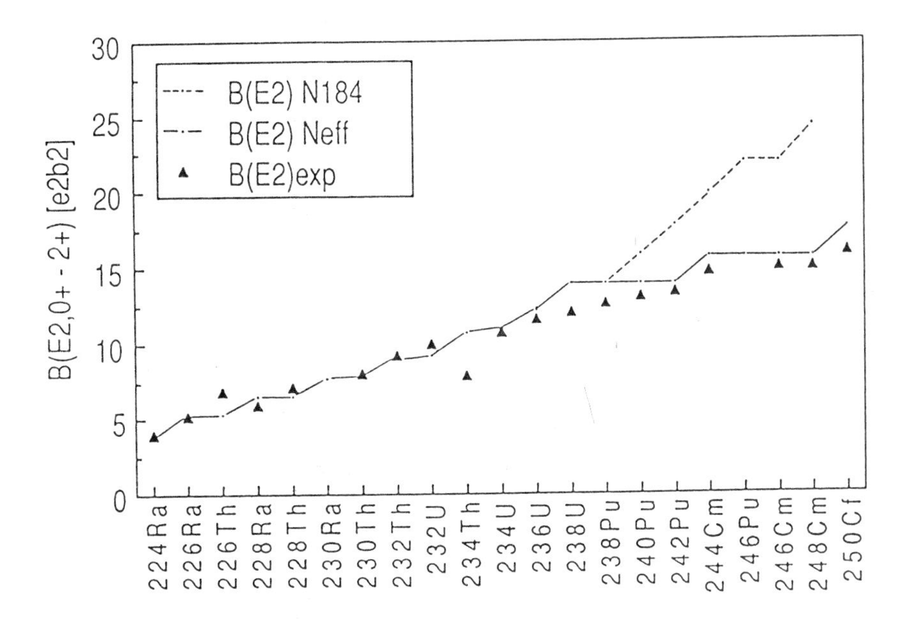

FIG. 5. Experimental and calculated B(E2:$0_g^+ \to 2_g^+$) values for deformed actinides. No subshell closure was used for the points forming the upper curve (dots connected by dashed lines), while for the points forming the lower curve (dots connected by solid lines) the presence of the N=152 subshell closure was taken into account by using the effective boson numbers shown in Figure 4. Since the results of the calculation using T(E2) with unbroken SU(3) symmetry ($x = -\frac{\sqrt{7}}{2}$) and the results of the calculation with broken SU(3) symmetry ($x' = -2.362$)) practically coincide, only one point is shown in each case. The normalization factor e in Eq. (7) has been fitted to ^{224}Ra and had the value $e = 0.162$ eb in the first case and the value $e = 0.130$ eb in the latter. Experimental data (represented by triangles) are taken from Raman *et al.*(1987).

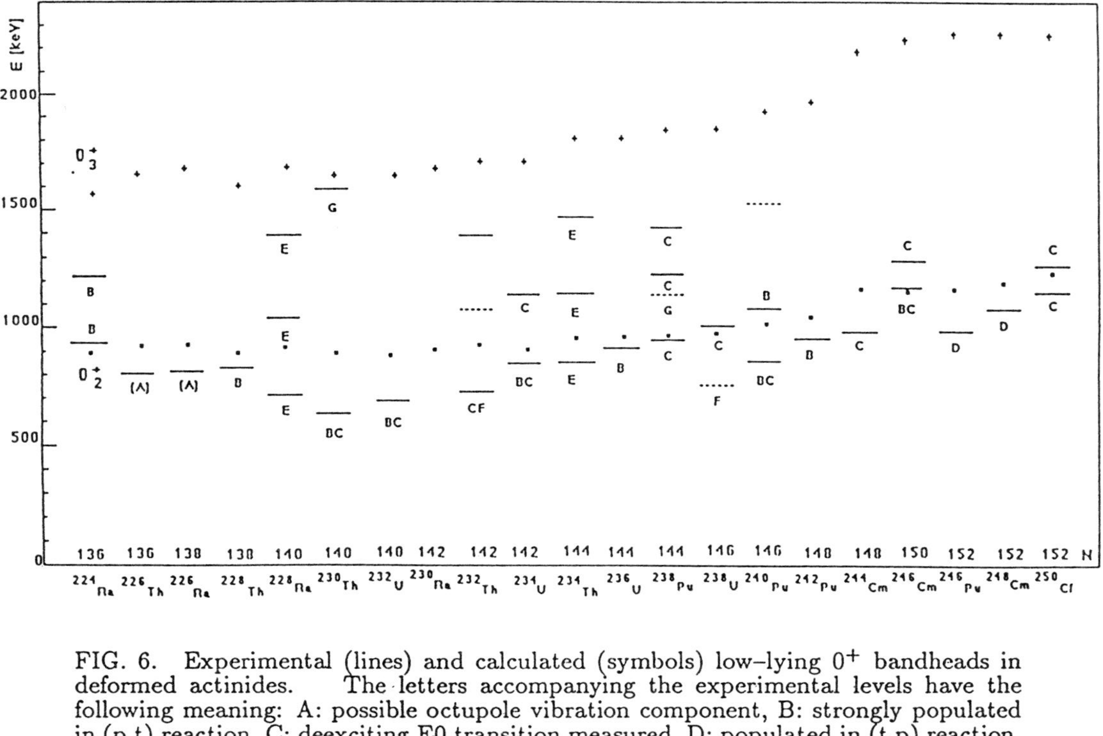

FIG. 6. Experimental (lines) and calculated (symbols) low–lying 0^+ bandheads in deformed actinides. The letters accompanying the experimental levels have the following meaning: A: possible octupole vibration component, B: strongly populated in (p,t) reaction, C: deexciting E0 transition measured, D: populated in (t,p) reaction, E: populated in (d,^{6}Li) reaction, F: seen in Coulomb excitation. Theoretical results come from the calculation described in the text. The calculated 0_2^+ and 0_3^+ states are shown.

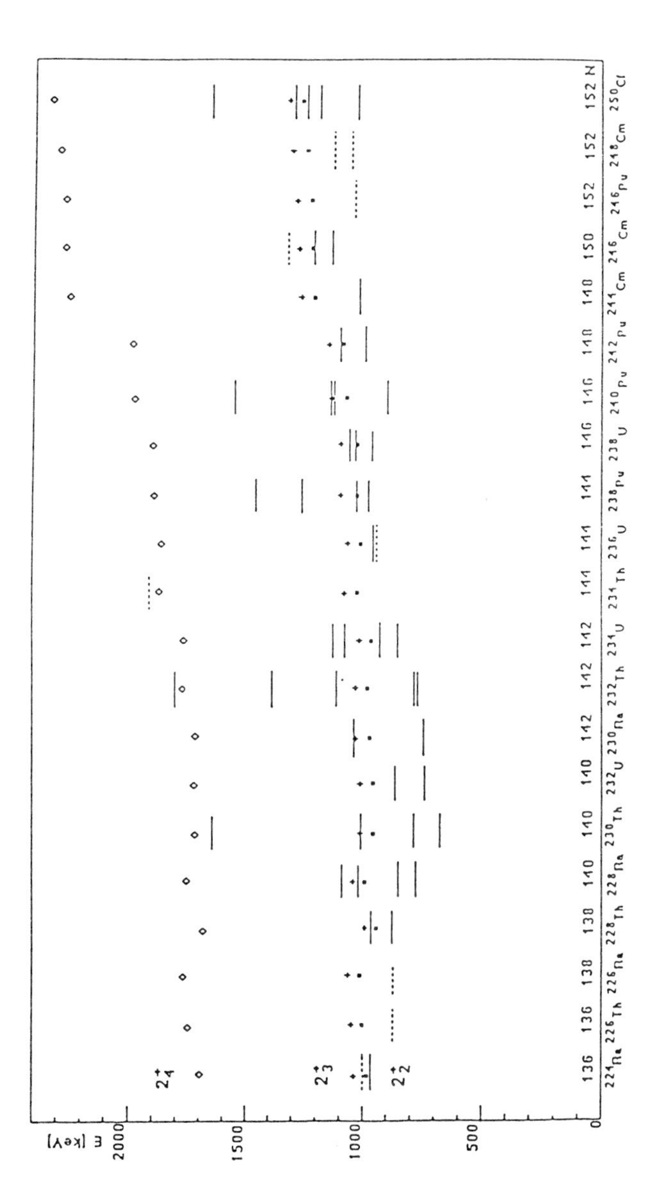

FIG. 7. Same as Figure 6, but for the low-lying 2^+ bandheads. The calculated 2_2^+, 2_3^+ and 2_4^+ states are shown.

We first try to fit the B(E2:$0_g^+ \rightarrow 2_g^+$) values. No need to break the SU(3) symmetry exists in this case. When counting the bosons in the usual way (taking into account the whole N=126–184 shell) we get the boson numbers shown in Figure 4 and the B(E2:$0_g^+ \rightarrow 2_g^+$) values shown in Figure 5. We remark that for the heaviest actinides the theoretical predictions overshoot the experimental values. On the other hand if one handles the well-known subshell closure at N=152 as a major shell closure, one gets the boson numbers shown in Figure 4 and a similar picture for the B(E2:$0_g^+ \rightarrow 2_g^+$) values. Thus in this case one underestimates the experimental data. For the heaviest actinides we used effective boson numbers (shown in Figure 4) which we fitted to the data (shown in Figure 5). The effective boson numbers are smaller than the ones obtained from the full N=126–184 shell, but larger than the ones obtained by handling the N=152 sushell closure as a major shell closure.

It is interesting to compare these findings for the effective boson numbers to the results obtained by Maino and Ventura (1984), who used the Nilsson–BCS approach. They find that in shells where subshell closures exist, the effective boson numbers get values intermediate between the ones corresponding to the full shell (ignoring the subshell closure) and the ones corresponding to handling the subshell closure as a major shell closure. The particular findings shown in Figure 4 have been reconfirmed by Maino (1989) using the Nilsson–BCS approach. It is very encouraging that two completely different methods give very similar answers to the question of effective boson numbers. It must be noted that the B(E2) transition probabilities depend very sensitively on the effective boson numbers, while there is no visible change occuring in the spectra. By inverting the argument, effective boson numbers can be determined very accurately by fitting B(E2) transition probabilities.

As we have already mentioned, it is necessary to break the SU(3) symmetry of the $T(E2)$ transition operator in order to obtain non-vanishing $\beta_1 \rightarrow ground$ and $\gamma_1 \rightarrow ground$ transitions. An example has been given by Rikovska and Bonatsos (1988). We only mention here that the breaking of the SU(3) symmetry in $T(E2)$ has little effect on B(E2:$0_g^+ \rightarrow 2_g^+$), while it changes B(E2:$0_g^+ \rightarrow 2_\beta^+$) and B(E2:$0_g^+ \rightarrow 2_\gamma^+$) by two orders of magnitude.

In some heavy actinides the identification of the experimental β_1 and γ_1 bands is a non-trivial task, because of the presence of several low-lying K=0 and K=2 bands. Therefore we tried to fit the parameters of the Hamiltonian of Eq. (5) to the ground state spectra and the B(E2:$0_g^+ \rightarrow 2_g^+$) and B(E2:$2_g^+ \rightarrow 4_g^+$) values. Again we kept x equal to its SU(3) value and used effective boson numbers beyond N=146. The lowest two K=0^+ bandheads obtained from this calculation are compared to the experimental K=0^+ bandheads in Figure 6. In Figure 7 the calculated and experimental K=2^+ bandheads are shown. In both cases we remark that the lowest experimental bandheads are much lower than the lowest calculated bandheads in the region 138<N<142, as well as around N=152, while for the rest of the nuclei under study the lowest experimental and calculated bandheads are in good agreement. These results indicate that in the regions where the systematic discrepancies occur the lowest experimental bandheads are outside the IBM-1 space. These systematic discrepancies are signs of intruder levels occuring near the subshell closures at N=142 and N=152. From Figures 6 and 7 it is also clear that Zhang *et al.* (1985) have used in their fits the intruder bands as the β_1 and γ_1 bands.

In order to describe intruder states in the vicinity of sushell closures one needs to introduce s' and d' bosons in addition to the usual s and d bosons. The Hamiltonian will then have, in addition to the usual sd part, an $s'd'$ part as well as a part mixing sd and $s'd'$ (Van Isacker *et al.* 1982). The g boson is also useful in these heavy nuclei, since i) it resolves the problem of the too low cut-off in the Yrast band, ii) allows for the existence of low-lying K=odd bands, iii) improves the results for the B(E2) transition probabilities (Wu 1982). We are now undertaking a calculation including s', d' and g bosons in addition to the usual s and d bosons in order to resolve the problem of intruder states.

In symmary, our main results are:

i) We have demonstrated that a wide range of deformed actinides can be described in terms of a simple IBM-1 Hamiltonian with constant or smoothly varying parameters. In particular, the coefficient of the only term breaking the SU(3) symmetry remains constant.

ii) We have shown that effective boson numbers, resulting from the existence of subshell closures, can be determined very accurately by fitting B(E2) transition probabilities. The resulting effective boson numbers have been independently confirmed through use of the Nilsson–BCS approach.

iii) We have demonstrated the need for s' and d' bosons (in addition to the s and d bosons) in the effort to fit the intruder levels appearing in the vicinity of the N=142 and N=152 subshell closures, where the identification of the experimental β_1 and γ_1 bands is not straightforward.

3. SHELL-MODEL FOUNDATIONS OF IBM

The phenomenological success of the Interacting Boson Model has caused interest in making a connection between it and the shell-model. One major step in this effort is the boson mapping, i.e. the transition from the fermion space (used in the shell-model) to the boson space (used in IBM). Many boson mapping methods exist. Here we will limit ourselves to methods which conserve the number of particles, since the number of bosons in IBM is fixed for each nucleus. Some of the existing boson mapping procedures are:

i) The method of Otsuka, Arima and Iachello (OAI) (Otsuka *et. al.* 1978). In this method the mapping is achieved by requiring that the matrix elements of the mapped operators in the boson space be equal to the matrix elements of the original fermion operators in the fermion space.

ii) The method of Bonatsos, Klein and Li (BKL) (Bonatsos *et al.* 1984). Here the mapping is achieved by requiring that the boson images of the fermion pair and multipole operators satisfy the same commutation relations as their fermionic archetypes.

iii) The method of Zirnbauer and Brink (Zirnbauer and Brink 1982, Zirnbauer 1984). In contrast to the previous two methods, which deal with hermitian boson mappings, this method makes direct use of the non-hermitian Dyson mapping.

Before using these methods for calculations related to realistic nuclei, it is essential to check their accuracy. Some of the relevant questions and problems are listed here:

i) The OAI method is known to violate some of the commutation relations (Matsuyanagi 1982). Is it a serious disadvantage of the method or is this fact a result of the (unavoidable) approximations used in the method, like the limitation to s and d bosons only? How do the results of this method compare to the ones provided by the BKL method, which is based on the preservation of the commutation relations?

ii) In the BKL method in lowest approximation only s and d bosons are included. In the next order of approximation, g bosons come in. How good is the lowest order approximation? How are the results changed by the inclusion of the g boson?

iii) Both the OAI and BKL methods are based on the concept of seniority. Therefore they are expected to give good results for vibrational and transitional nuclei, but not for rotational nuclei, where the small parameter used for the expansion of the various operators is not small any more.

In order to check the validity of the methods we have performed a calculation for a system of three bosons (six particles) occupying a single $j = \frac{23}{2}$ shell. We used a pairing plus quadrupole–quadrupole Hamiltonian of the form

$$H = -xA_0^+A_0 - 5(1-x)B^2 \cdot B^2, \tag{8}$$

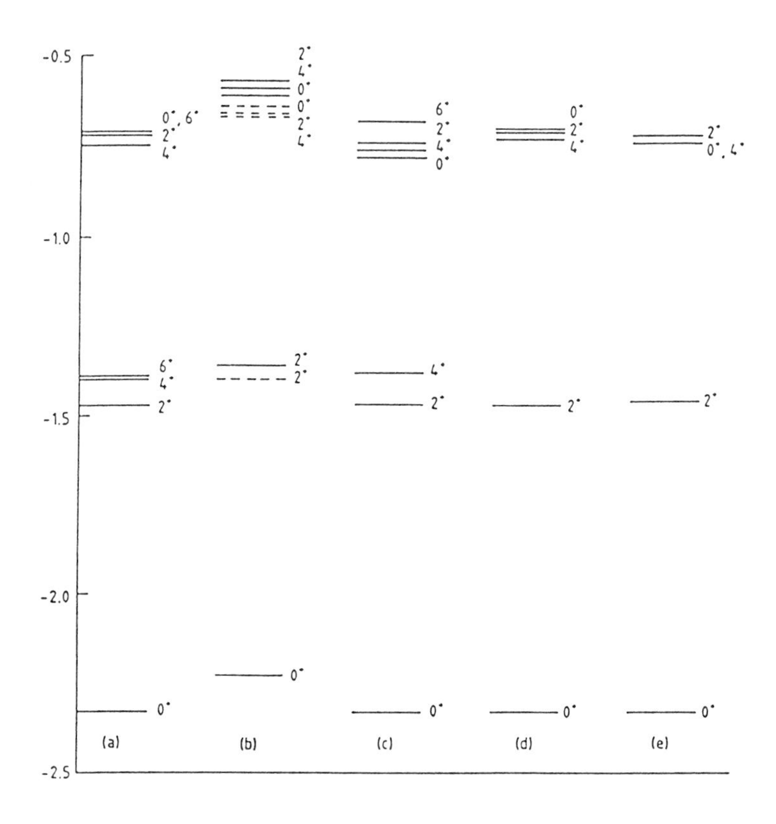

FIG. 8. Spectra for three bosons (six particles) occupying a single $j = \frac{23}{2}$ shell, obtained with the Hamiltonian of Eq. (8) for $x = 0.9$. Results are shown for (a) the shell model, (b) the lowest order BKL (s and d bosons only, solid line) and the SFM mapping (dashed line), (c) the next order BKL (s, d and g bosons), (d) the OAI mapping and (e) the Zirnbauer–Brink mapping. Binding energies have been pushed up by $15(1-x)/\Omega$, where $2\Omega = 2j + 1$. Unit is arbitrary.

where A_0 is the pairing operator, B^2 is the quadrupole operator and x is a free parameter, which can vary between 1 (spherical limit) and 0 (deformed limit). The results of each method are compared to the results of the exact shell model calculation for three different values of x in Figures 8, 9, 10, from which we draw the following conclusions:

i) All methods give very good results in the near-vibrational case of $x = 0.9$, still quite good results for $x = 0.7$, but they clearly fail in the near-rotational case of $x = 0.3$. In the case of $x = 0.3$ the gaps between the consequtive levels are largely oversized, indicating that the moment of inertia is too small.

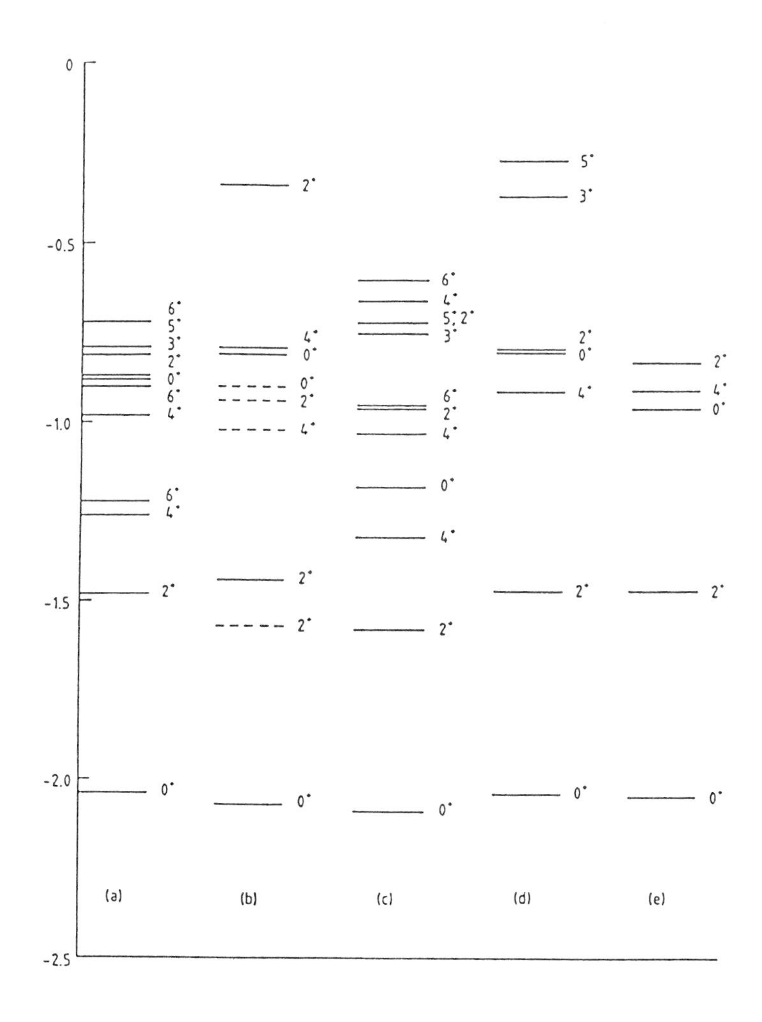

FIG. 9. Same as Figure 8, but for $x = 0.7$.

ii) All methods give results of comparable quality. This indicates that the violation of commutation relations occuring in the OAI method is not a serious disadvantage.

iii) In Figures 8 and 9 we see that the inclusion of g bosons does improve the BKL results. The failure of the BKL method including g bosons in the deformed region might be due to the omission (because of technical reasons) of the three-body terms arising in the quadrupole–quadrupole part of the Hamiltonian. It is essential in the BKL method to include all terms up to the order in use.

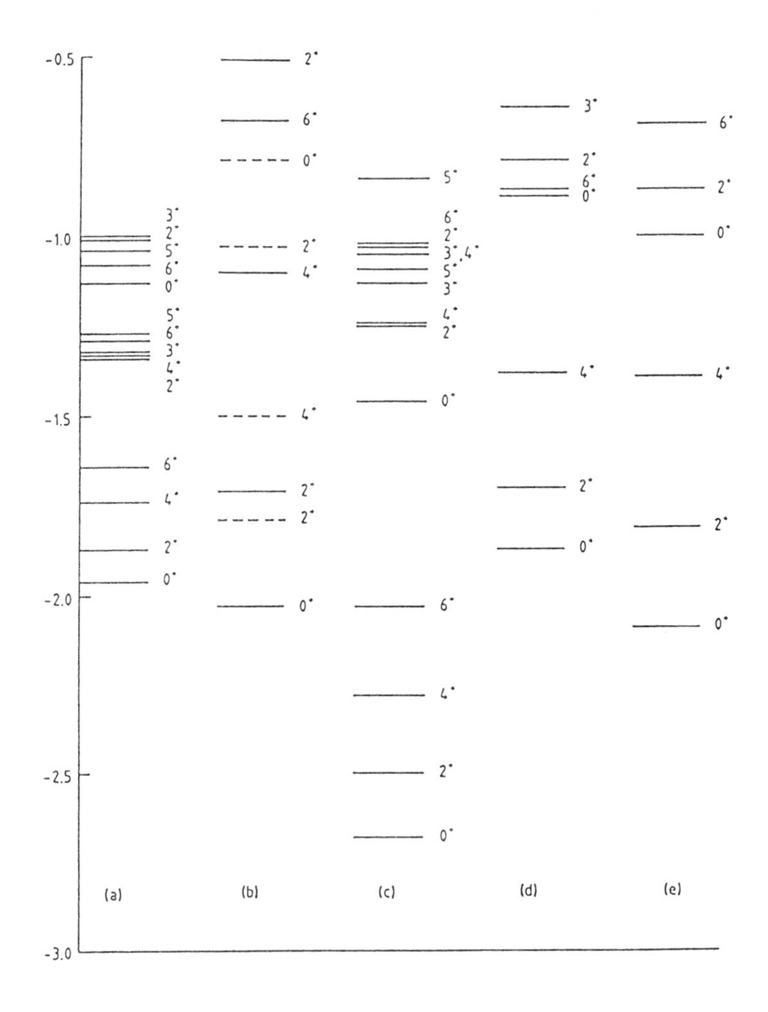

FIG. 10. Same as Figure 8, but for $x = 0.3$.

iv) It is known that one can get improved results if he calculates the boson image of the Hamiltonian directly and not through repeated use of the mapped operators. This can also be seen in Figures 8–10, where we have included the results of the mapping of Suzuki, Fuyuki and Matsuyanagi (SFM) (Suzuki *et al.* 1979). The mapped operators of SFM are the same as the mapped operators of lowest order BKL, but in SFM the image of the Hamiltonian is calculated directly, while in BKL it is calculated through repeated use of the mapped operators. The improvement provided by this modification is clear.

In addition to spectra, we have calculated B(E2) transition rates as well, which can be found in Menezes *et al.* (1989). It is clear that the BKL method including the g boson can reproduce some rates which are not present if only s and d bosons are included, $B(E2{:}4_1^+ \rightarrow 2_1^+)$ for example.

For applications to realistic nuclei it is essential to extend the above mentioned boson mappings to systems of several non-degenerate levels. This has already been done for the BKL method (Bonatsos and Klein 1987, Menezes *et al.* 1987). Of particular interest is a system of three shells with $j_1 = j_2 - 1 = j_3 - 3$, the last level having opposite parity than the other two. This is the pattern of levels occuring outside the closed shells of 50 or 82 protons, as well as outside the closed shells of 82 or 126 neutrons, i.e. the pattern which is of interest for several medium and heavy nuclei. It is clear that in this case one has bosons of negative parity in addition to the usual bosons of positive parity. The negative parity f boson is of particular interest (Bonatsos *et al.* 1988), since it has been used in the IBM framework for the description of low-lying collective octupole states. In the case of several non-degenerate j-levels the BKL mapping includes several s bosons, several d bosons, etc. Among these many bosons one needs to single out the collective s boson, the collective d boson, etc. This can be done, for example, through the use of the trace variational principle of Klein and Vallières (1981). Work in this direction is already in progress.

In conclusion, we have used a simple Hamiltonian in a single $j = \frac{23}{2}$ shell in order to check the accuracy of several boson mappings. In all cases good results were obtained in the vibrational and transitional regions, while in the rotational region problems arise. For applications to realistic nuclei one has to use several non-degenerate j-levels and take into account the parity of the levels. Using the BKL method one obtains many bosons, among which the collective ones can be selected through the use of the trace variational principle. The negative parity f bosons occuring in this case are related to low-lying collective octupole states.

Support from the Bundesministerium für Forschung und Technologie (DB), CNPq–Brazil (DPM), SERC (JR) and the Nishina Memorial Foundation (NY) is gratefully acknowledged.

REFERENCES

Bonatsos D 1988a *Interacting Boson Models of Nuclear Structure* (Oxford: Clarendon)
Bonatsos D 1988b *Phys. Lett.* 200B 1
Bonatsos D and Klein A 1987 *Nucl. Phys.* A469 253
Bonatsos D, Klein A and Li C T 1984 *Nucl. Phys.* A425 521
Bonatsos D, Menezes D P and Klein A 1988 *J. Phys.* G14 L45
Casten R F, Frank W and von Brentano P 1985 *Nucl. Phys.* A444 133
Cizewski J and Dieperink A E L 1985 *Phys. Lett.* 164B 236
Iachello F and Arima A 1987 *The Interacting Boson Model* (Cambridge: Cambridge University Press)
Jänecke J, Becchetti F D, Overway D, Cossairt J D and Spross R L 1981 *Phys. Rev.* C23 101

Klein A and Vallières M 1981 *Phys. Lett.* 98B 5
Maino G 1989 private communication
Maino G and Ventura A 1984 *Lett. Nuovo Cimento* 39 89
Maino G, Martinelli T, Menapace E and Ventura A 1981 *Lett. Nuovo Cimento* 32 235
Matsuyanagi K 1982 *Nuclear Physics* ed C H Dasso, R A Broglia and A Winther (Amsterdam: North Holland) p 29
Menezes D P, Bonatsos D and Klein A 1987 *Nucl. Phys.* A474 381
Menezes D P, Yoshinaga N and Bonatsos D 1989 *Phys. Lett.* 221B 103
Otsuka T, Arima A and Iachello F 1978 *Nucl. Phys.* A309 1
Raman S, Malarkey C H, Milner W T, Nestor Jr C W and Stelson P H 1987 *At. Data Nucl. Data Tables* 36 1
Rikovska J and Bonatsos D 1988 *Phys. Lett.* 211B 259
Sakai M 1984 *At. Data Nucl. Data Tables* 31 399
Suzuki T, Fuyuki M and Matsuyanagi K 1979 *Progr. Theor. Phys.* 61 1682
Van Isacker P, Heyde K, Waroquier M and Wenes G 1982 *Nucl. Phys.* A380 383
Warner D D and Casten R F 1983 *Phys. Rev.* C28 1798
Wu H C 1982 *Phys. Lett.* 110B 1
Zhang M, Vallières M, Gilmore R, Feng D H, Hoff R W and Sun H Z 1985 *Phys. Rev.* C32 1076
Zirnbauer M R 1984 *Nucl. Phys.* A419 241
Zirnbauer M R and Brink D M 1982 *Nucl. Phys.* A384 1

Complete spectroscopy and 0(6)-like structure in ^{126}Xe

W Lieberz, W Krips, A Dewald, W Frank, A Gelberg, R Wirowski, and P von Brentano

Institut für Kernphysik der Universität zu Köln, Zülpicher Str. 77, 5 Köln 41, W. Germany

ABSTRACT: Low spin states of ^{126}Xe have been investigated with the OSIRIS–12 spectrometer at the Cologne FN–Tandem accelerator. "Complete"–spectroscopy by means of the non–selective (α,n)–reaction turns out to be a powerful tool for nuclear structure studies as is demonstrated by comparing experimental data to the O(6)–limit of the IBM–1.

1. INTRODUCTION

Nuclei in the $A = 130$ region have been subject to intensive investigations in recent years. One reason for this is the proposed transition from a vibrator like structure in the vicinity of the closed neutron shell N = 82 through a γ–soft region (N $\approx$ 74) towards more rotational structure at N $\approx$ 66. In terms of the interacting boson model (IBM) this corresponds to a transition from the SU(5)– through the SO(6)– to the SU(3)–limit.

The nucleus ^{126}Xe with neutron number 72 is thus expected to exhibit an O(6)–like structure [1]. Since basically only the ground state band and the quasi–gamma band were known in this nucleus, more collective levels were needed to study ^{126}Xe in detail.

For this the non–selective (α,n)–reaction is an appropriate tool, since it is possible to populate highly excited side bands and to obtain a "complete" level scheme [2] in a given spin and energy window. This, together with the spin determination from the side–feeding excitation functions [3], allows a detailed comparison of the data and the IBM–1 O(6) predictions.

2. EXPERIMENTAL RESULTS

Our results are based on several experiments performed at the FN-Tandem accelerator at the Cologne University. In all measurements the ^{123}Te(α,nγ)–reaction was used with an enriched target (89.4%) of 1.6mg/cm^2 ^{123}Te on a Bi–backing [4]. The level scheme was constructed from the coincidence data of the OSIRIS–12 spectrometer [5], whereas spins and parities were determined from γ–angular distribution, γ–linear polarization and γ–excitation functions. In total 150 levels were observed, 25 being already known from earlier experiments [6–10].

The Yrast band could be followed up to the 10^+ state, in the quasi–γ–band we could establish the 2661.5 keV, 7^+ level and two new levels at 3061.8 resp. 3520.5 keV, which are candidates for the 8^+ and 9^+ state. Furthermore we found a "K"=4– and a "K"=0–band consisting of:

1903.1 keV, 4^+; 2363.0 keV, 5^+; 2622.9 keV, $5,6^{(+)}$ ("K"=4–band)

1313.9 keV, 0^+; 1678.5 keV, 2^+; 2042.1 keV, $4^{(+)}$; 2664.6 keV, $6^{(+)}$ ("K"=0–band).[1]

Figure 3 shows all relevant energy levels including a tentative 2^+, 2064 keV–, a 6^+, 2493 keV–state and the third 0^+–state, which has been taken from ref. [11] , since it was not sufficiently populated in our experiment. A corresponding selection of B(E2)–ratios is presented in table 1.

3. DISCUSSION

One of crucial points in comparing experimental data with model predictions is the test of a one to one correspondence between experimental and theoretically predicted levels. But this requires complete spectroscopy, at least in a certain energy and spin window. The classical example for such a reaction is average resonance neutron capture [12]. Recently it has been shown that also the $(\alpha,n\gamma)$–reaction has similar properties [2].

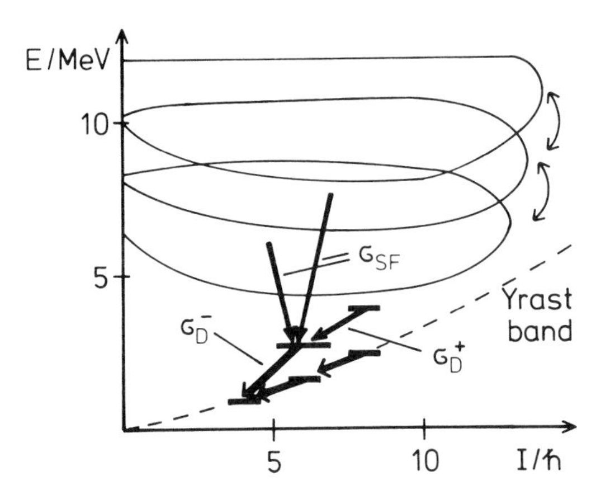

Fig. 1: Principal decay mechanism of the compound nucleus. After neutron evaporation the nucleus decays through few fast E1–transitions to either discrete levels or to continuum states forming a new population cloud. The side–feeding cross section ($\sigma_{S.F.}$) of a level is the difference of discrete depopulation (σ_D^-) and discrete population (σ_D^+) of this level. (From ref. [2])

The "completeness" can be deduced from the side–feeding excitation functions plotted in figure 2a and 2b. The side–feeding intensity (see figure 1) has been found to depend in the following way on spin I and excitation energy E_x:

[1]The 0^+– and 2^+–state have also been seen in β–decay measurements (e. g. ref. [10]).

- The slope of the excitation functions increases with increasing spin I and is equal for states with the same spin, thus allowing spin determination by comparing the side–feeding excitation functions of levels with unknown spin to those of levels where the spin is known.

- In general, the side–feeding intensities of levels with the same spin decrease monotonously with the excitation energy, i. e. if at a certain excitation energy and spin a level is observed all levels with the same spin and lower excitation energy should have a bigger amount of side–feeding population and thus should be visible. There may be exceptions due to e. g. E0–transitions or transitions with low energy.

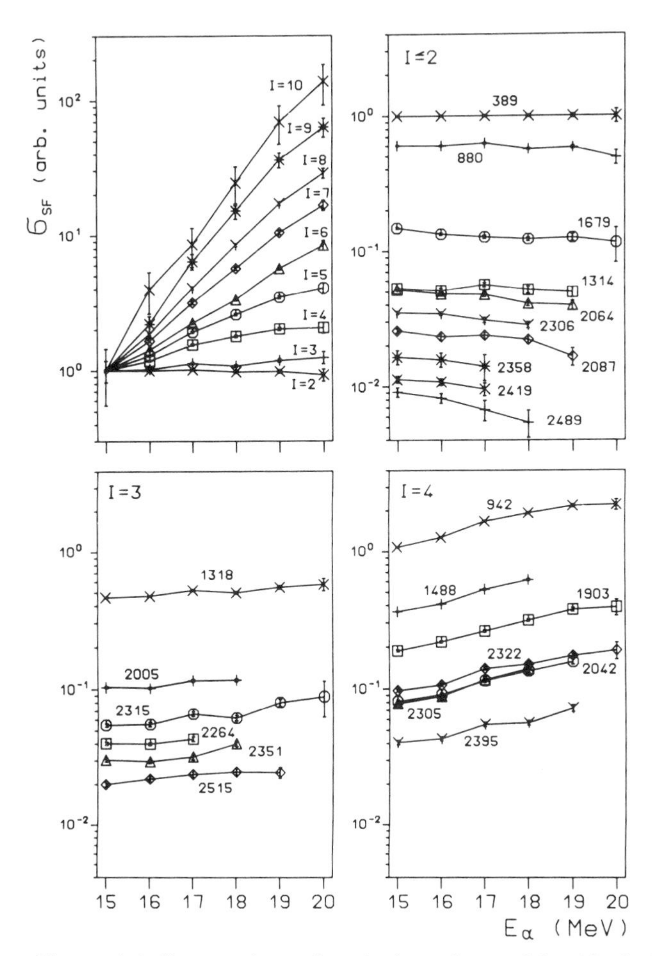

Fig. 2a: The top left diagram shows the spin dependence of the side–feeding cross section $\sigma_{S.F.}(I,E_x,E_\alpha)$ versus α–energy E_α in an overview. Its energy dependence for spins 2–4$\hbar$ is presented in the remaining drawings. Note, that the spin determination is not unique for all states. All data used in this and the following figure are compiled from ^{126}Xe. (Details see text)

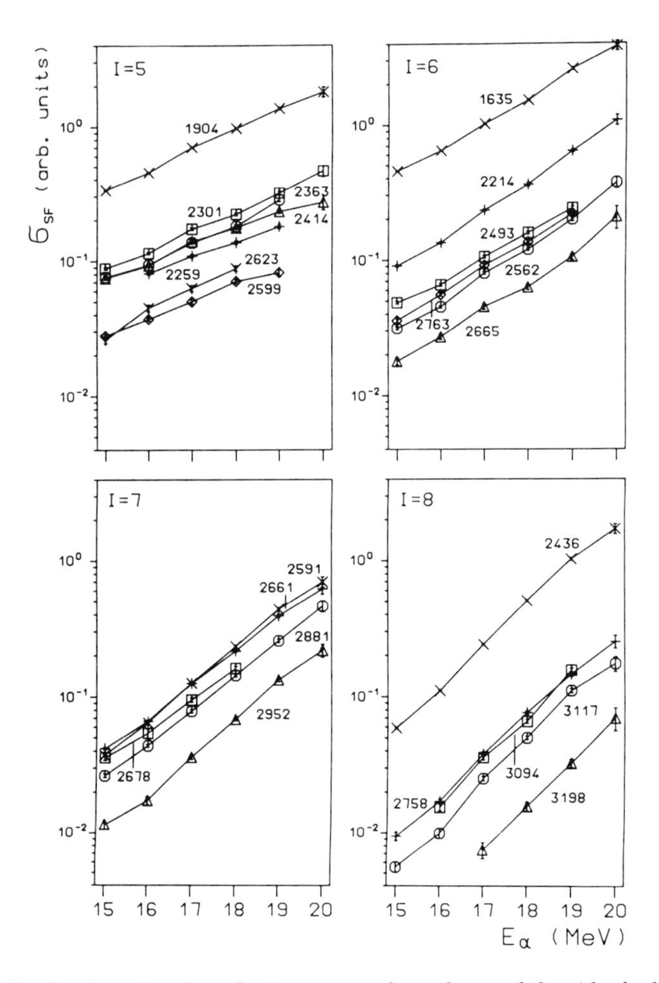

Fig. 2b: Continuation from fig. 2a: energy dependence of the side–feeding cross section for spins 5–8$\hbar$. See also fig. caption 2a.

With these remarks in mind a juxtaposition of experimental and IBM–1 O(6) predictions becomes highly interesting. Indeed, as can been seen from figure 3 all σ = N states up to $\tau = 5$ could be substantiated.[2]

Using the usual formula for the energy eigenvalues [1]

$$E(\sigma,\tau,I) = \frac{A}{4}(N-\sigma)(N+\sigma+4) + \frac{B}{6}\tau(\tau+3) + CI(I+1)$$

the best overall agreement with the experimental data is reached for A = 220 keV, B = 335 keV and C= 12.5 keV.

[2]Data of similar quality previously existed only for ^{128}Xe, where all states up to $\tau = 4$ in the σ = N representation were observed [13].

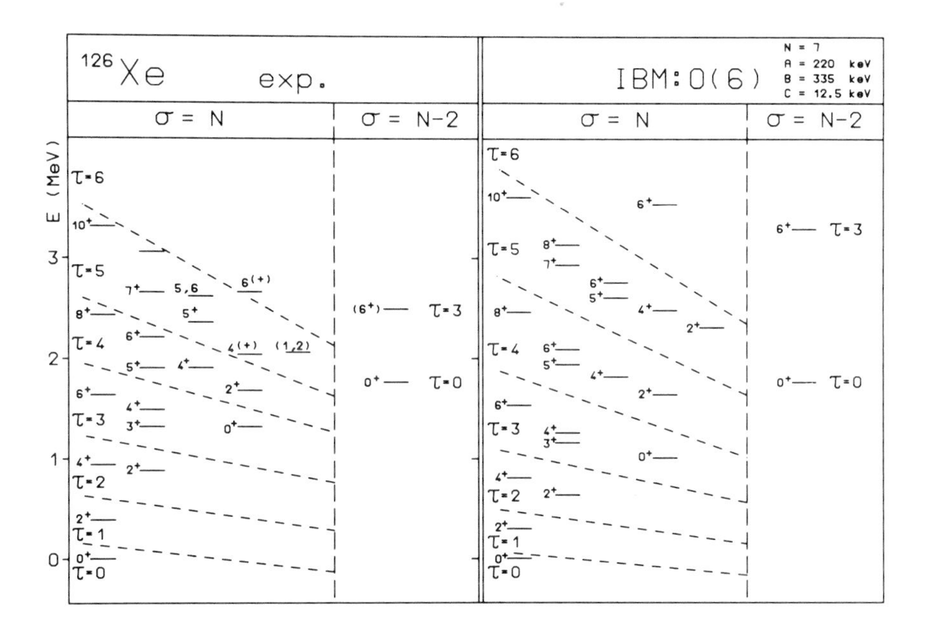

Fig. 3: Experimental and theoretically predicted energy levels of ^{126}Xe

Table 1: "B(E2)-ratios"[3] in ^{126}Xe. For the experimental values marked with a) the M1-portion is subtracted. All energies are given in keV.

level	spin	transition	"B(E2)-ratios"		
			exp.		O(6)
879.9	2_2^+	491.3→ 2_1^+	100	a)	100
		879.9→ 0_1^+	1.5±0.5		1.5
1313.9	0_2^+	434.0→ 2_2^+	100		100
		925.3→ 2_1^+	8.0±2.3		2.0
1317.7	3_1^+	375.7→ 4_1^+	33^{+11}_{-33}	a)	40
		437.8→ 2_2^+	100	a)	100
		929.0→ 2_1^+	$2.1^{+0.9}_{-1.7}$	a)	2.0
1488.4	4_2^+	546.4→ 4_1^+	76±25	a)	91
		608.5→ 2_2^+	100		100
		1099.8→ 2_1^+	0.4±0.1		2.0
1678.5	2_3^+	360.8→ 3_1^+	71±25		125
		364.2→ 0_2^+	100		100
		736.4→ 4_1^+	2.0±0.8		1.1
		798.8→ 2_2^+	2.2±1.2		3.5
		1290.0→ 2_1^+	0.20±0.07		0
		1678.2→ 0_1^+	0.12±0.04		0
1903.1	4_3^+	414.8→ 4_2^+	100		100
		585.3→ 3_1^+	35±11		115
		961.2→ 4_1^+	4.0±1.2		0.4
		1023.2→ 2_2^+	2.7±0.8		2.5

[3]Note, that the experimental ratios based on the intensities divided by E_γ^5, so that these ratios are B(E2)-ratios in the strict sense only if either no M1 part is involved or is taken into account.

Table 1 lists the experimental "B(E2)–ratios"[3] and the predicted values of the O(6)–limit. For the B(E2)–ratios of the O(6) the general quadrupole transition operator

$$T(E2) = e_b([s^\dagger \times \tilde{d} + s \times d^\dagger]^{(2)} + \chi[d^\dagger \times \tilde{d}]^{(2)})$$

with $\chi = -0.5$ was used.[4]

The agreement for the quasi–γ–band is very good, qualitatively and quantitatively. For the newly established "K=0"– and "K=4"–band at least a good qualitative agreement could be reached — an excellent result concerning the extremely simple approach we took.

We thank R. Reinhardt, A. v. d. Werth, S. Freund, A. Granderath and K. O. Zell for fruitful discussions. This work was supported by the German Federal Minister for Research and Technology (BMFT) under the contract number 06OK272.

REFERENCES

[1] Casten R F and Brentano P von 1985 Phys. Lett. **152** 22

[2] Brentano P von, Dewald A, Lieberz W, Reinhardt R, Zell K O, Zipper W, in: Nuclear Structure of the Zr Region, J Eberth, R A Meyer, K Sistemich (ed.), Springer Conf. Ser. (1988), p. 157

[3] Zipper W, Dewald A, Lieberz W, Reinhardt R, Dichter W, Seiffert F and Brentano P von, submitted to Nucl. Phys. A.

[4] Zell K O 1985 Nucl. Instr. Meth. A **236** 655

[5] Lieder R M, Jäger H, Neskakis A, Venkova T, Michel C 1984 Nucl. Instr. Meth. **220** 363

[6] Kusakari H, Kitao K, Sato K, Sugawara M, Katsuragawa H 1983 Nucl. Phys. A **401** 445

[7] Hattula J, Helppi H and Luukko A, 1982 Phys. Srcipta **26** 205

[8] Kusakari H, Yoshikawa N, Kawakami H, Ishihara M, Shida Y and Sakai M 1975 Nucl. Phys. A **242** 13

[9] Bergström I, Herlander C J and Kerek A 1969 Nucl. Phys. A **123** 99

[10] Pathak B P, Lessard L, Nikkinen L and Preiss I L 1976 Phys. Rev. C **14** 1573

[11] Alford W P, Anderson R E, Batay-Csorba P A, Emigh R A, Lind D A, Smith P A, Zafiratos C D 1979 Nucl. Phys. A **323** 339

[12] Casten R F, Warner D D, Stelts M L and Davidson W F 1980 Phys. Rev. Lett. **45** 1077

[13] Reinhardt R, Dewald A, Gelberg A, Lieberz W, Schiffer K, Schmittgen K P, Zell K O and Brentano P von 1988 Z. Phys. A **331** 113

[14] Isacker P van 1987 Nucl. Phys. A **465** 497

[4] An O(6) transition operator with $\chi \neq 0$ was first introduced by P. van Isacker [14] yielding static quadrupole moments and $\Delta\tau = 0, \pm 2$ transitions in the O(6)–limit.

Inst. Phys. Conf. Ser. No 105
Paper presented at Int. Conf. on Spectroscopy of Heavy Nuclei, Crete, Greece, 1989

Microscopic substantiation of the interacting Boson–Fermion model by using enforced symmetry conditions and SUSY concepts

G. Kyrchev*) and V. Paar**)

*) Institute of Nuclear Research and Nuclear Energy, Bulgarian Academy of Sciences, Sofia 1784, Bulgaria.
**) University of Zagreb, 41000 Zagreb, Yugoslavia.

ABSTRACT: The Interacting Boson-Fermion Model (IBFM) hamiltonian is rigorously deduced from the microscopic Quasiparticle Phonon Nuclear Model hamiltonian, describing an odd-mass nucleus. The derivation is based on boson-fermion extended RPA with SU(6) constraints. It is shown that IBFM hamiltonian is endowed with an SU(6/Ω) superalgebraic structure.

1. INTRODUCTION

In the framework of the IBFM (Iachello and Scholten 1979), the extention of the Interacting Boson Model (IBM) to encompass odd-mass nuclei, an experimental evidence for SU(6/Ω) dynamic supersymmetries (SUSY's) has been found (Vervier 1987) where the SU(6) symmetry is associated with the s- and d- bosons of IBM even-even core and the SU(Ω) symmetry - with the unpaired fermion in the valence shell orbits $[\Omega=\sum(2j_i+1)]$. There were a few attempts to investigate the microscopic origin of SU(6/Ω) SUSY and the microscopic aspect of IBFM (Yang et al 1984, Gambhir et al 1984, Bijker et al 1985). By all accounts, however, (cf. Vervier 1987 and references therein) there is by now nor a truly sound microscopic foundation of IBFM, neither an understanding (in the language of microscopic models) of the occurrence of traces of broken SUSY in certain nuclei.

The lack of a rigorous derivation of the IBFM hamiltonian and physical operators from the admittedly more fundamental shell model hamiltonian and physical operators seems to be having a deeper-going effect on the investigations of SUSY in nuclei, than hitherto suspected. It is the aim of this paper to advance an approach to the microscopic substantiation of IBFM and SU(6/Ω) superphenomenology and to demonstrate that the method is mathematically sound and computationally implementable. A sophisticated procedure has been elaborated to single out the fragment with the IBFM boson-fermion structure from the Quasiparticle Phonon Nuclear Model (QPNM) hamiltonian (Soloviev 1987), describing an odd mass nucleus. In this approach the key role is played by the constraint (due to the enforced closure of the SU(6) algebra) RPA phonon operators for boson system $\{Q_{2\mu 1}, Q^{+}_{2\mu 1}; \mu=0,\pm 1,\pm 2\}$ (Kyrchev et al 1988) and their respective modified (due to the presence of an odd fermOon) operators $\{\bar{Q}_{2\mu 1}, \bar{Q}^{+}_{2\mu 1}\}$ for boson-fermion system.

The present approach provides also microscopic justification of the SU(6/Ω) superphenomenology, since it has been established (Kyrchev et al 1987) that IBFM can be viewed as $SU(6/\Omega) \supset SU(6) \oplus SU(\Omega)$ - based model.

2. THE PHENOMENOLOGICAL IBFM HAMILTONIAN

The phenomenological model under consideration is defined by a hamiltonian which has the plausible tripartite form:

$$H_{IBFM} = H_{IBM} + H_{F} + H_{BF} \qquad (1)$$

where H_{IBM} is the well known SU(6) s-d boson hamiltonian (Arima et al 1976) while H_F, the term reflecting the degrees of freedom of the odd fermion, a^{+}_{jm}, can be written in the following coupled form (Iachello et al 1981):

$$H_F = \sum_{jm} E_j a^{+}_{jm} a_{jm} + \frac{1}{2} \sum_{j_1 j_2 j_3 j_4 L} U^{(L)}_{j_1 j_2 j_3 j_4} ((a^{+}_{j_1} \times a^{+}_{j_2})_{(L)} \times (\tilde{a}_{j_3} \times \tilde{a}_{j_4})_{(L)})_{(0)} \qquad (2)$$

with $\tilde{a}_{jm} = (-1)^{j-m} a_{j-m}$

The most important term, H_{BF}, is an interaction term, specifying the coupling of the unpaired fermion and the collective degree of freedom. It has been shown (Iachello et al 1979) that a boson-fermion interaction H_{BF}, which includes a direct monopole and quadrupole interactions and the so called exchange term, accounts quite well for the collective properties of odd-mass nuclei. The explicit form of H_{BF} reads (Iachello et al 1981):

$$H_{BF} = \sum_{j} A_j ((\alpha^{+}_{2} \times \tilde{\alpha}_{2})_{(0)} \times (a^{+}_{j} \times \tilde{a}_{j})_{(0)})_{(0)} + \sum_{jj'} \Gamma_{jj'} ((\tilde{d}^{+}_{2} \times s + s^{+} \times \tilde{d}_{2} + \chi d^{+}_{2} \times \tilde{d}_{2})_{(2)} \times (a^{+}_{j} \times \tilde{a}_{j'})_{(2)})_{(0)} + \sum_{jj'j''} \Lambda^{j''}_{jj'} :((a^{+}_{j} \times \tilde{d}_{2})_{(j'')} + (d^{+}_{2} \times \tilde{a}_{j'})_{(j'')})_{(0)}:, \qquad (3)$$

where $\{A_j, \Gamma_{jj'}, \Lambda^{j''}_{jj'}\}$ are free parameters. The goal is to infer

H_{IBFM} from an appropriate standard fermionic hamiltonian which will enable us to relate these parameters to the shell model hamiltonian parameters.

3. THE MICROSCOPIC QPNM HAMILTONIAN

We have chosen to utilize as a reference microscopic hamiltonian the hamiltonian of QPNM (Soloviev 1987). The Quasiparticle Phonon Nuclear Model is well established microscopic model and its hamiltonian is ideally suited for our purposes, since the collective degrees of freedom are explicitly introduced in the QPNM hamiltonian. To provide microscopic foundation of IBFM we employ the following QPNM hamiltonian:

$$H_{micr} = \sum_{jm} \varepsilon_j \overline{\alpha^+_{jm} \alpha_{jm}} - \frac{\varkappa_0^{(2)}}{2} \sum_{\mu} :\bar{M}^+_{2\mu} \bar{M}_{2\mu}: - G_\tau \Big(\sum_{jm} u_j v_j \overline{\alpha^+_{jm} \alpha_{jm}}\Big)^2 \qquad (4)$$

The first term representes the single quasiparticle hamiltonian for odd system. The second term is the part of the monopole pairing interaction which has not been accounted for in the single quasiparticle energies ε_j. The third term expresses the quadrupole-quadrupole interaction. It involves the normal product of modified quadrupole phonon operators

$$\bar{M}^+_{2\mu} = \frac{(-1)^\mu}{\sqrt{5}} \sum_{jj'} f^{(2)}_{jj'} [u^{(+)}_{jj'} \frac{1}{2}(\psi_{jj'} + \phi_{jj'}) (\bar{Q}^+_{2\mu} + (-1)^\mu \bar{Q}_{2-\mu}) + v^{(-)}_{jj'} \bar{B}(jj';2\mu)]. \qquad (5)$$

Here, $f^{(2)}_{jj}$ denote the reduced single-particle matrix elements of the quadrupole operator $r^2 Y_{2\mu}$; $u^{(+)}_{jj}$ and $v^{(-)}_{jj}$ are definite combinations of the Bogoliubov-Valatin u,v-factors,

$$\bar{B}(jj';2\mu) = \sum_{mm'} \langle jm\ j'm' | 2\mu \rangle \overline{\alpha^+_{jm} \tilde{\alpha}_{jm}}, \qquad (6)$$

where $\overline{\alpha^+_{jm} \tilde{\alpha}_{jm}}$ denote the modified density quasiparticle operators, referring to the odd-mass system. The operators $\bar{Q}^+_{2\mu}$ ($\tilde{\bar{Q}}_{2\mu} = (-1)^\mu \bar{Q}_{2-\mu}$) denote the modified quadrupole phonon creation (annihilation) operators corresponding to the first collective solution of the quasiparticle RPA secular equations for odd-mass nuclei (Vdovin et al 1985). These operators are defined in a straightforward way:

$$\bar{Q}^+_{2\mu} = \frac{1}{2} \sum_{jj'} \left[\psi_{jj'} \overline{(\alpha^+_{jm} \alpha^+_{j'm'})}_{(2\mu)} - \phi_{jj'} \overline{(\tilde{\alpha}_{j'm'} \tilde{\alpha}_{jm})}_{(2\mu)} \right]. \qquad (7)$$

4. ESSENTIAL STEPS TAKEN IN THE MICROSCOPIC DERIVATION OF IBFM HAMILTONIAN

As it is evident from eqs. (1)-(4) a special procedure should be worked out in order to deduce rigorously H_{IBFM}, given by eqs. (1)-(3) starting from H_{QPNM}, given by eq.(4). The basic idea is to construct boson-fermion equivalent of H_{QPNM}. There are several steps in picking out H_{IBFM} from the QPNM hamiltonian H_{micr}. Carrying out the microscopic derivation of H_{IBFM} we are guided by the symmetries underlying IBM and IBFM hamiltonians.

Our main objective is to construct explicitly the boson-fermion images of the set of operators $\{\bar{Q}^+_{2\mu}, \tilde{Q}_{2\mu}, \bar{B}(jj';2\mu), \bar{B}(jj;00)\}$. For the odd-fermion system, we have to add to the set of bifermion operators $\{\overline{\alpha^+\alpha^+}, \overline{\alpha^+\alpha}, \overline{\alpha\alpha}\}$ the set of single-fermion and unity operators $\{\bar{\alpha}^+, \bar{\alpha}, I\}$. The ensuing set of operators $\{I, \bar{\alpha}^+, \bar{\alpha}, (\overline{\alpha^+\alpha^+}), (\overline{\alpha^+\alpha}), (\overline{\alpha\alpha})\}$ forms the Lie algebra so(2n+1) of the special orthogonal group, where n is the number of relevant single particle levels. Taking advantage of the fact that an exact boson-fermion realization of so(2n+1) generators is known (Marshalek 1980, Okubo 1974), we construct explicitely, starting from definitions (6) and (7), the boson-fermion equivalents of the above mentioned set of operators:

$$\bar{Q}^+_{2\mu} = N^{-1/2} d^+_{2\mu} s + N^{1/2} \sum_{\lambda jm j'm'} \Gamma^{\mu\lambda jm}_{j'm'} d^+_{2\mu} s (s^+ s)^{-1} \beta^+_{jm} \tilde{\beta}_{j'm'} , \qquad (8)$$

$$\tilde{Q}_{2\mu} = N^{-1/2} \overset{+}{s} \tilde{d}_{2\mu} , \qquad (9)$$

$$\bar{B}(jj';LM) = (\beta^+_j \times \tilde{\beta}_{j'})_{(LM)} +$$

$$+ 5 \sum_{j''} (-1)^{j+j''} \begin{Bmatrix} 2 & 2 & L \\ j & j' & j'' \end{Bmatrix} (\psi_{jj''}\psi_{j'j''} - \phi_{jj''}\phi_{j'j''}) (d^+_2 \times \tilde{d}_2)_{(LM)} , \qquad (10)$$

$$\bar{B}(jj;00) = (\beta^+_j \times \tilde{\beta}_j)_{(0)} +$$

$$+ \sqrt{5}\, (2j+1)^{-1/2} \sum_{j''} (\psi^2_{jj''} + \phi^2_{jj''}) (d^+_2 \times \tilde{d}_2)_{(0)} , \qquad (11)$$

with: $N = \mathrm{Int}[C^{-1}]$. The quantity C, emerging as a result of the enforcement of the SU(6) symmetry, is defined in (Kyrchev et al 1988).

$$\Gamma^{\mu\nu jm}_{j'm'} = -\sum_{LMj''} \sqrt{5}\,(2L+1)^{1/2}\langle jmj''m''|LM\rangle\langle 2\nu LM|2\mu\rangle \times$$

$$\times (-1)^{j+j''}\begin{Bmatrix} 2 & 2 & L \\ j & j' & j'' \end{Bmatrix} \left[(-1)^{L}\psi_{jj''}\psi_{j'j''}-\phi_{jj''}\phi_{j'j''}\right] \quad (12)$$

One can check that substitution of $\{\bar{Q}^{+}_{2\mu},\tilde{Q}_{2\mu},\bar{B}(jj';LM)\}$ for their equivalents given by eqs. (8)-(11) into eq.(4) leads to the result that all terms, featuring in the IBFM hamiltonian, spring up. In addition, terms arise which do not have counterparts in H_{IBFM}. The term by term comparison of the terms which have counterparts in H_{IBFM}, has enabled us to relate the phenomenological parameters of H_{IBFM} to the microscopic quantities of H_{micr}, given by eq.(4). Since for the parameters of H_{IBM} in eq.(1) we obtain the same microscopic expressions, as in (Karadjov et al 1989),here we present the final results for the parameters of H_F and H_{BF} only:

$$E_j = \varepsilon_j \ , \quad (13)$$

$$U^{(L)}_{j_1j_2j_3j_4} = -\frac{\varkappa_o^{(2)}}{2}(-1)^{L}(2L+1) \times$$

$$\times f^{(2)}_{j_1j_2}V^{(-)}_{j_1j_2}f^{(2)}_{j_3j_4}V^{(-)}_{j_3j_4}(-1)^{j_2+j_3}\begin{Bmatrix} j_1 & j_2 & 2 \\ j_3 & j_4 & L \end{Bmatrix} \ , \quad (14)$$

$$A = -2G_{\tau}U_jV_j\sqrt{5(2j+1)}\sum_{j'j''}U_{j'}V_{j'}(\psi^2_{j'j''}+\phi^2_{j'j''}) \ , \quad (15)$$

$$\Gamma_{jj'} = N^{-1/2}Mf^{(2)}_{jj'}V^{(-)}_{jj'} \ , \quad \varkappa = N^{1/2}VM^{-1} \ , \quad (16)$$

$$\Lambda^{j''}_{jj'} = \varkappa_o^{(2)}\left[M^2(-1)^{j+j'}(2j''+1)^{-1/2}\psi_{jj''}\psi_{j''j'} + \right.$$

$$\left. + F^{(2)}_{jj''}U^{(+)}_{jj''}f^{(2)}_{j''j'}U^{(+)}_{j''j'}\right] \ , \quad (17)$$

where

$$M \equiv \frac{1}{2}\sum_{jj'} f^{(2)}_{jj'}U^{(+)}_{jj'}(\psi_{jj'}+\phi_{jj'}) \ , \quad (18)$$

$$V \equiv 5\sum_{jj'} f^{2)}_{jj'}V^{(-)}_{jj'}\sum_{j''}(1)^{j+j''}\begin{Bmatrix} 2 & 2 & 2 \\ j & j' & j'' \end{Bmatrix} \times (\psi_{jj''}\psi_{jj''}+\phi_{jj''}\phi_{jj''}) \quad (19)$$

From relations (15)-(19) it is apparent that the microscopically derived parameters depend, in particular, on $\{\psi_{jj'},\phi_{jj'}\}$ i.e. on quantities which are not determined as yet. The latter can be obtained by solving numerically nonlinear algebraic system of equations. This system has been derived from variational principle accounting for both the dynamics (governed by the microscopically derived IBFM hamiltonian) and the kinematical constraints dictated by the enforcement of the SU(6) symmetry (Kyrchev et al 1988) and antisymmetrization effects. The system of nonlinear equations in the present case is an extension for odd-mass nuclei of the system of equations derived firstly in (Karadjov et al 1989) for even-even nuclei. The main modifications in the former system of equations are due to the monopole-monopole and exchange boson-fermion interactions.
As pointed out in (Kyrchev et al 1987), the exchange term parametrized by $\{\Lambda^{j''}_{jj'}\}$ plays a distinguished role in models based on dynamical supersymmetries. Namely, it measures deviation from the usual dynamical symmetry SU(6)xSU(Ω), i.e. in the present framework the exchange term generates the supersymmetry. The presented approach opens up the possibility to investigate the microscopic aspect of dynamical symmetries and SUSY's.

Arima A and Iachello F 1976 Ann Phys 99 253
Bijker R and Scholten O 1985 Phys Rev C33 591
Gambhir Y K, Ring P and Schuck P Nucl Phys A423 35
Iachello F and Scholten O 1979 Phys Rev Lett 43 679
Iachello F and Kuyiycak S 1981 Ann Phys 136 19
Karadjov D, Kyrchev G and Voronov V V Proc. 2nd Int Spr Seminar on Nucl Phys ed A Covello (Singapore: World Scientific) pp 471-481
Kyrchev G and Paar V 1987 Phys Lett B195 107
Kyrchev G and Paar V 1988 Phys Rev C37 838
Marshalek E R 1980 Nucl Phys A347 253
Okubo S 1974 Phys Rev C10 2048
Soloviev V G 1987 Progr Part Nucl Phys 19 107
Vdovin A I, Voronov V V, Soloviev V G and Stoyanov Ch 1985 Sov J Part and Nucl 16 246
Verviev J 1987 La Rivista del Nuovo Cimento 10 N2
Yang L M, Lu D H and Zhou Z N 1984 A421 229C

Inst. Phys. Conf. Ser. No 105
Paper presented at Int. Conf. on Spectroscopy of Heavy Nuclei, Crete, Greece, 1989

Intruder states in $^{190, 192, 194}Pb$ populated in the α decay of $^{194, 196, 198}Po$

P. Van Duppen, P. Decrock, P. Dendooven, M. Huyse, G. Reusen and J. Wauters
L.I.S.O.L., K.U.Leuven, Celestijnenlaan 200 D, B-3030 Leuven, Belgium

Abstract.

The α decay of mass separated $^{194,196,198}Po$ is studied. The half life of the first excited 0^+ state in $^{190,192,194}Pb$ was measured using $\alpha.e^-.t$ coincidences. The deduced monopole strength increases by almost one order of magnitude going from ^{194}Pb to ^{190}Pb and the deduced α-hindrance factors show a dependence on neutron number.

1. Introduction.

In recent years we have collected an extensive systematics of shell-model intruder states in the Pb region (see fig. 1 ref. 1-5).

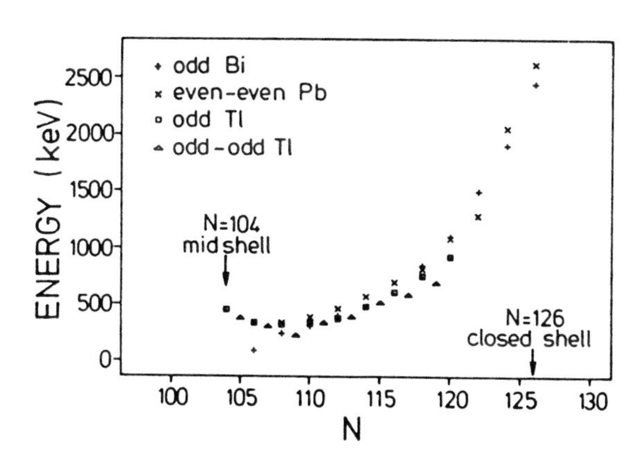

Fig.1. Systematics of the intruder-state excitation energies. The excitation energy of the 0^+ intruder states in the even-even Pb nuclei is divided by two. References to others work can be found in ref. 5.

A remarkable similar systematic behavior of the excitation energy of the intruder-based states as a function of the neutron number is evident for the odd Bi, odd Tl, odd-odd Tl and even-even Pb nuclei. This has led to several theoretical descriptions (ref. 6-9).

Around N=110, the intruder states come very low in excitation energy and can interact with the ground state. This will certainly be the case in the Pb isotopes where they have the same spin and parity: 0^+. Therefore, it is of interest to determine the monopole transition probability between these states.

We have measured in the α decay of $^{194,196,198}Po$ the half life of the 0^+ intruder states by collecting α-e-t coincidence events. Also the hindrance factor of the α decay to the intruder state, relative to the α decay to the ground state, was studied.

2. Experimental method.

The neutron-deficient isotopes ^{198}Po (1.76m), ^{196}Po (5.5s) and ^{194}Po (0.41s) were produced in the reaction of 240 MeV ^{20}Ne on ^{182}W (Ta foils were used to degrade the beam to optimal energy). A stack of three enriched ^{182}W foils (total thickness 2.1 mg/cm^2), was mounted inside our FEBIAD ion source. More details on the LISOL set-up can be found in ref. 10. The mass-separated Po beam was implanted on a 30$\mu g/cm^2$ C foil. For the study of ^{198}Po and ^{196}Po, this foil was mounted in a wheel that periodically moved the activity to a decay position in between an α detector (150 mm^2 PIPS, 15 keV resolution) and an e^- detector (5 mm thick NE104 plastic). An α-e coincidence efficiency of 4% was obtained. For the study of 0.41s ^{194}Po, the C foil was surrounded at the implantation station by the α and e detector. The resolution of the fast timing set-up was 840ps at FWHM.

3. Results.

3.1. E0 transition probability.

Figure 2 shows the α-singles spectrum at mass 196 together with three projections out of the α-e-t coincidence matrix of mass 196. The high selectivity of the α-e coincidences for the α branch to the intruder at 769 keV in ^{192}Pb is evident by comparing fig. 2a and 2b. The TAC spectrum of fig. 2d was gated by the α line of 5769 keV in the decay of ^{196}Po to the 0_2 intruder in ^{192}Pb and by the E0 (769 keV) electrons (see fig. 2c). A half-life value for the intruder state in ^{192}Pb of 750±100ps was obtained. Similar results were obtained for the decay of $^{198,194}Po$ to respectively ^{194}Pb and ^{190}Pb. Table I summarizes our results.

	$E(0_2^+)$ keV	$T_{1/2}(0_2^+)$ ps	$\rho^2_{exp}(E0)$ x10^3	$\rho^2_{SPU}(E0)$ x10^3	Hindrance factor	β	b^2	$\langle V \rangle$ keV
^{194}Pb	930.6(9)	1100±200	1.0±.2	14.6	14.6	0.17	0.003	51
^{192}Pb	768.5(17)	750±100	1.7±.2	14.7	8.6	0.175	0.005	52
^{190}Pb	658(4)	⩽220	⩾6	14.8	⩽3	0.18	⩾0.02	⩾80

Table I: Out of the measured half-life values ($T_{1/2}(0^+)$) the experimental strength $\rho^2_{exp}(E0)$ is determined. By comparing this value with the E0 single-particle unit $\rho^2_{SPU}(E0)$ the hindrance factor of the E0 transition can be obtained. Also the β-values used to calculate the mixing between intruder and ground state together with the obtained mixing amplitudes and -matrix elements are shown (eq. 2, 3, 4).

From the E0 half-lives the monopole strength was deduced. The absolute nuclear electric-monopole transition probability W(E0) is here:

$$W(E0) = \frac{\ln 2}{T_{1/2}(0^+)} = \rho^2(E0) \sum_j \Omega_j(Z,k) \qquad (1)$$

with $\rho^2(E0)$ the strength parameter, containing the nuclear matrix element and $\Omega_j(Z,k)$ the so-called electronic factors (j represents the electronic shells K, L_I, L_{II} ...). The Ω_K, Ω_{L_I} and $\Omega_{L_{II}}$ electronic factors are calculated with the method of Kantele (ref. 11).

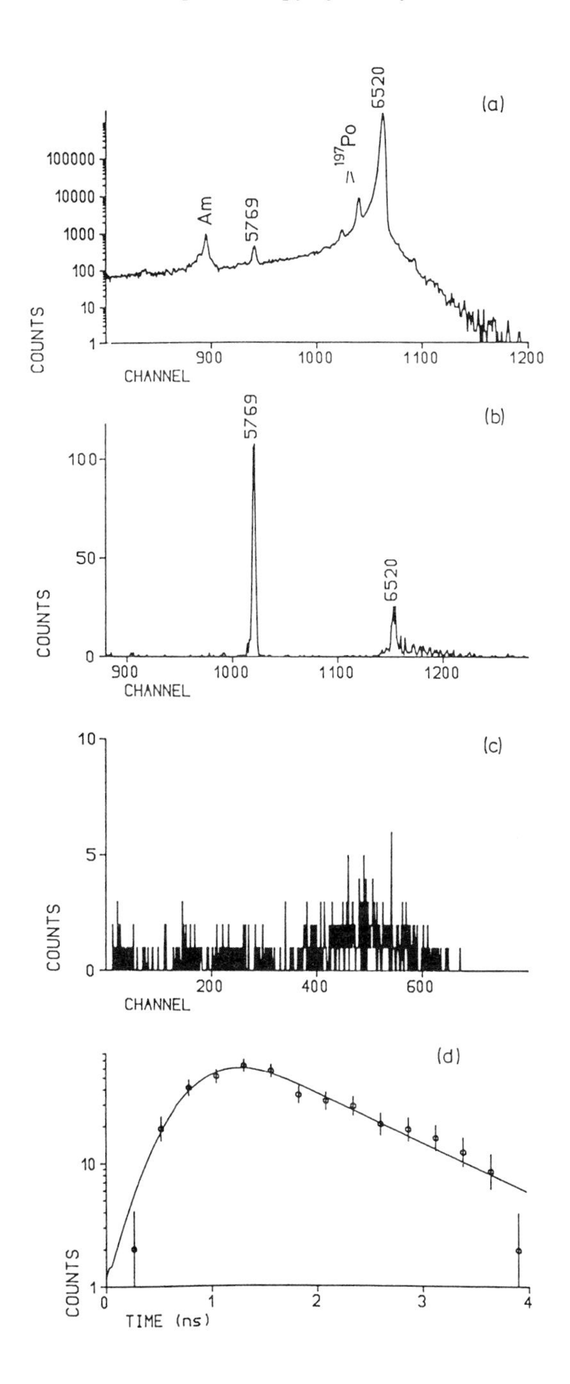

Fig. 2: Mass 196 spectra: (a) α-singles spectrum, (b) "prompt" projection without correction for random events of the α detector out of the α-e-t coincidences matrix, (c) electron spectrum gated by 5769 keV α line, (d) TAC spectrum gated by the 5769 keV α line and the 769 keV E0 electron line. The result of the half-life fit is given as a full line. The energy value of the ^{196}Po α lines are given in keV.

3.2. α-hindrance factors.

For ^{194}Po and ^{196}Po it is possible to obtain out of the α-singles spectrum the intensity of the α branch towards the intruder state, relative to the groundstate to groundstate α decay (see fig. 2a). As the α branch to the intruder is not visible anymore in the α-singles spectrum of ^{198}Po, the α-e coincidences have been used to deduce the relative intensity of the α branch to the intruder state.
With the method of Rasmussen [12]) it is now possible to deduce relative hindrance factors, taking unity for the groundstate to groundstate transition. The results are given in table II. A summary of our experimental results is also given in figure 3.

	$E_\alpha(0_1^+-0_2^+)$ (keV)	$I_\alpha(0_1^+-0_2^+)$ (%)	$H.F._\alpha(0_1^+-0_2^+)$
^{198}Po	5273±5	1.4 10^{-3}	2.8±0.5
^{196}Po	5769±5	2.2 10^{-2}	2.5±0.1
^{194}Po	6194±5	2.4 10^{-1}	1.1±0.1

Table II: Energy, relative intensity ($I_\alpha(0_1^+-0_1^+)$ = 100) and hindrance factor (relative to groundstate-groundstate α decay) of the α branch to the intruder state in ^{194,192,190}Pb.

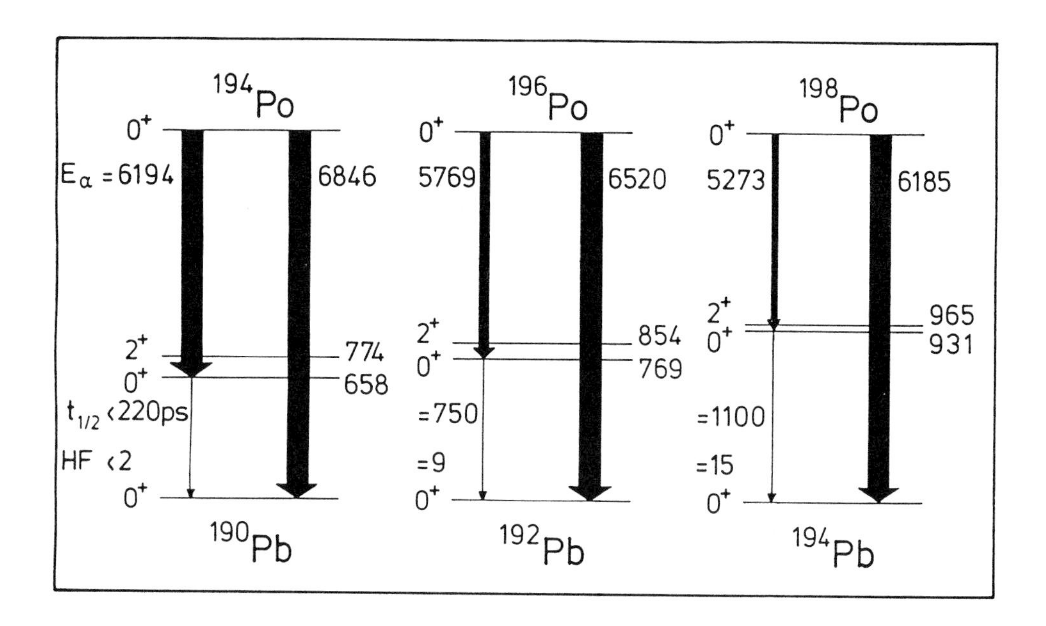

Fig.3. Summary of the experimental results on the decay of ^{198,196,194}Po. The energy of the α lines is given in keV, the half-lives in picoseconds and HF stands for Hindrance Factor.

4. Discussion

4.1. Monopole strengths

The strength parameter ρ^2 (EO) can be compared with the EO single-particle unit [13]:

$$\rho^2_{SPU} = 0.5\ A^{-2/3} \quad (1)$$

This single-particle unit is based on a 50% mixing of two shell-model states. The ratio between ρ^2_{SPU} AND ρ^2_{EXP} gives the hindrance for the EO transition (see table I). The hindrance goes down by almost one order of magnitude from ^{194}Pb to ^{190}Pb, indicating a structure change in the ground-/intruder state system.
One way of describing this phenomena is by considering the mixing between a spherical (g.s.) and a deformed (i.s.) state [14,15].

$$|0^+_{i.s.}\rangle = b\,|sph\rangle - a\,|def\rangle \quad (2)$$

$$|0^+_{g.s.}\rangle = a\,|sph\rangle + b\,|def\rangle \quad (3)$$

where $|sph\rangle$ and $|def\rangle$ are the unmixed spherical and deformed states respectively, $|0^+_{i.s.}\rangle$ and $|0^+_{g.s.}\rangle$ are the intruder and ground state respectively and a, b are the mixing amplitudes. The monopole strength can then be written as:

$$\rho^2_{i.s.\to g.s.} = a^2b^2 \left(\frac{3Z}{4\pi}\right)^2 \left(\langle\beta^2\rangle_{def}\right)^2 \quad (4)$$

where Z is the proton number and $\langle\beta\rangle_{def}$ is the difference between the $|sph\rangle$ and $|def\rangle$ states. In reference 4 and 6 we emphasized the strong resemblance between the bandstructure built upon the intruder state in Pb (a 2p-2h proton state) and the groundstate band in the Pt nuclei (a 4h proton state). The Pt isotopes with neutron number greater than 110 are known to be oblately deformed with $\beta \simeq 0.17$ (ref. 16). A similar deformation is expected for the intruder states in Pb, this is confirmed by calculations of R. Bengtsson and W. Nazarewicz (ref. 9 and 18).
The resulting mixing amplitudes and matrix elements together with the calculated $\langle\beta\rangle$ values from W. Nazarewicz (ref. 18) are given in table I.
Recently, the change in mean square radius $\delta\langle r^2\rangle$ has been measured for the neutron-deficient Pb nuclei down to mass 192 (ref. 17). The deviation from the droplet model, starting from mass 194 on, was thought to eventually arise from the mixing of the deformed intruder state into the spherical groundstate. We can use our results to estimate the influence of this mixing.
Starting from equations (2) and (3), we can calculate the deviation from the spherical value of the mean square radius,

$$\langle 0^+_{gs}|\,r^2\,|0^+_{gs}\rangle - \langle sph|\,r^2\,|sph\rangle = b^2(\langle r^2\rangle_{def} - \langle r^2\rangle_{sph}) \quad (5)$$

By using the 2-parameter formula, commonly used to evaluate laser spectroscopy data (18)

$$\langle r^2\rangle_{def} = \langle r^2\rangle_{sph}(1+ \frac{5}{4\pi}\,\beta^2) \quad (6)$$

expression (5) becomes:

$$\langle r^2\rangle_{gs} - \langle r^2\rangle_{sph} = b^2\,\frac{5}{4\pi}\,\beta^2\,\langle r^2\rangle_{sph} \quad (7)$$

The relative effect on the mean square radius is then:

$$\frac{\Delta\langle r^2\rangle}{\langle r^2\rangle_{sph}} = \frac{5}{4\pi}\,b^2\,\beta^2 \quad (8)$$

The effect of the mixing of the intruder and groundstate on the mean square radius of ^{194,192}Pb is, taking the results from the deformed approach in table I, 3 10^{-3} % and 6 10^{-3} % respectively. The observed deviation in the laser-spectroscopy data from the droplet model is roughly of the order of 0.3 % in the case of ^{192}Pb [17]) and thus can not be explained by the mixing of the deformed intruder state into the spherical groundstate.

4.2. α-decay hindrance factors

The α hindrance factors show a lowering trend when going from ^{198}Po (=2.8) to ^{194}Po (=1.1) (see table II). In the case of ^{194}Po, the α decay to the intruder state in ^{190}Pb is as fast as the α decay to the ground state. The mixing between intruder state and ground state in $^{192,194}Pb$ deduced from the monopole transition probability is less than 1%. Therefore the low α-hindrance factors might indicate a relatively strong mixing between the g.s. and i.s. in the corresponding Po isotopes. This can readely be understood by assuming that the α decay between a pure g.s. Po configuration and a pure i.s. Pb configuration resemblance to a Pb g.s. to Hg g.s. decay and will be strongly retarded compared to the Po g.s. to Pb g.s. decay.

5. Conclusion

We studied the fine structure of the α decay of $^{198,196,194}Po$ by measuring α-e-t coincidences on mass-separated sources. Only feeding to the first excited 0^+ intruder state has been observed. The half life of this intruder state in $^{194,192,190}Pb$ has been measured and by comparing the transition rate with theoretical estimates it has been possible to deduce the mixing between the intruder and groundstate. Although the mixing increases in function of decreasing neutron number (and thus decreasing excitation energy by the intruder state), the obtained mixing amplitudes can not explain the deviations of the mean square radii of the involved Pb groundstates from the droplet-model estimates. The hindrance factors of the α decay to the 0^+ intruder state, relative to the decay to the groundstate, indicate that, taking into account the small mixing in the Pb nuclei, the groundstate of the feeding Po nuclei are considerably mixed up.

Acknowledgements

We want to thank the technical staff of the LISOL separator and the CYCLONE cyclotron for the excellent beam quality. Many thanks to H. Folger and the crew of the GSI target laboratory for the development of the ^{182}W target. We wish to thank also K. Heyde and J. Wood for the many stimulating discussions. P.V.D. is a senior research assistent of the Nationaal Fonds voor Wetenschappelijk Onderzoek (Belgium).

References

1. P. Van Duppen et al. Phys. Rev. Lett. 52 (1984) 1974
2. P. Van Duppen et al. Phys. Lett. B154 (1985) 354
3. E. Coenen et al. Phys. Rev. Lett. 54 (1985) 1783
4. P. Van Duppen et al. Phys. Rev. C35 (1987) 1861
5. M. Huyse et al. Phys. Lett. B201 (1988) 293
6. K. Heyde et al. Nucl. Phys. A466 (1987) 189
7. K. Heyde et al. Nucl. Phys. A484 (1988) 275
8. G.E. Arenas Peris and P. Federman, Phys. Rev. C38 (1988) 493
9. R. Bergtsson and W. Nazarewicz Lund-MPh-87/08 preprint and W. Nazarewicz private communication
10. M. Huyse et al. Nucl. Instr. Meth. B26 (1987) 105
11. J. Kantele Nucl. Instr. Meth. A271 (1988) 625
12. J.O. Rasmussen Phys. Rev. 113 (1959) 1593

13. A. Bohr and B. Mottelson Nuclear Structure volume 2 (Benjamin Reading, Ma, 1975)
14. J. Kantele in "Heavy Ions and Nuclear Structure" Proceedings of teh XIV Summer School, Mikolajki 1984 edited by B. Sikora and Z. Wilhelmi (Harwood, Academic, New York 1984) p. 391
15. K. Heyde and R.A. Meyer, Phys. Rev. C37 (1988) 2170
16. J.K.P. Lee et al. Phys. Rev. C38 (1988) 2985
17. U. Dinger et al. Z. Phys. A328 (1987) 253
18. E. Otten, Nuclear Radii and Moments of Unstable Isotopes, to be published and A. Bohr and B. Mottelson, Nuclear Structure volume 1 (Benjamin Reading, MA, 1969).

Inst. Phys. Conf. Ser. No 105
Paper presented at Int. Conf. on Spectroscopy of Heavy Nuclei, Crete, Greece, 1989

Isomeric state at 945.2 keV in ^{237}Np

M. Steinmayer, K.E.G. Löbner, L. Corradi, U. Lenz, P. Pascholati, U. Quade, K. Rudolph, W.K. Schomburg, S.J. Skorka

Sektion Physik, Universität München, 8046 Garching, Fed. Rep. Germany

ABSTRACT: An isomeric state has been discovered in ^{237}Np at an excitation energy of 945.2 keV with a half-life of (711±40) ns. The comparison of the absolute transition probabilities of the 6 γ-rays depopulating the isomer to known rotational states in ^{237}Np with empirical values, suggest as most probable spin I and K-assignment I=K=13/2 for the isomer. In this case it should be a low-lying three-quasiparticle state. However, a $11/2^-$ [505] assignment for the isomer cannot be excluded.

1. EXPERIMENTAL EQUIPMENT

As a byproduct of the investigation of the γ-decay branch of the fissioning shape isomers ^{238}U and ^{237}Np an isomeric state has been discovered. This isomer has been produced with a pulsed 12.6 MeV proton beam of the Munich MP tandem accelerator and a highly enriched ^{238}U target of 15 $\mu g/cm^2$ UO_2 on a 10 $\mu g/cm^2$ carbon foil.

The first aim of the experimental set up has been to avoid the detection of prompt fission radiation. The second aim has been to have high detection efficiencies for the delayed decay of recoils. This is managed with the set-up sketched in Fig. 1. The detectors do not "see" the target directly. The highly charged recoils are transported by an electrostatic field near to 3 electron detectors (ion implanted solid-state detectors 10 mm x 20 mm) and a Ge(Li)-detector (180 cm^3). The detection efficiencies for γ-rays and electrons are extremely high for radiation emitted after the flight time of the recoils ($\approx$ 80 ns).

2. MEASUREMENT AND RESULTS

γ-rays in the Ge(Li) detector have been measured in prompt coincidence with events in one of the electron detectors and in delayed coincidence with the pulsed beam (pulse width = 1.5 ns and pulse distance 1600 ns). The 6 γ-ray transitions shown in Fig. 2 have been measured in this way. They all show a half-life of $T_{1/2}$ = (711±40) ns relative to the pulsed beam.

3. INTERPRETATION OF THE ISOMER

The question to which isotope the isomeric state belongs can be uniquely answered. The isomer must be in ^{237}Np since the 6 γ-ray lines fit very well into the known level scheme of ^{237}Np (see Ellis-Akovali (1986) and Fig. 3).

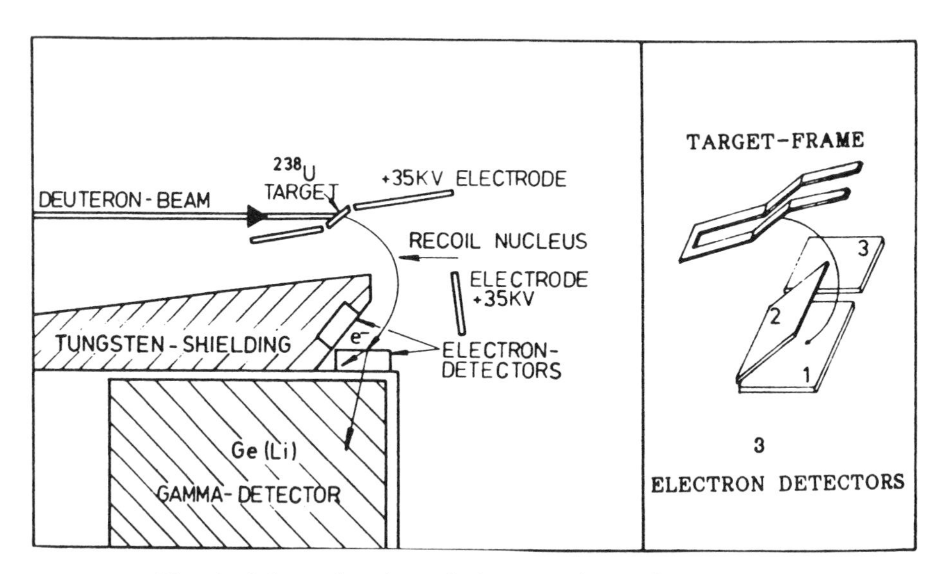

Fig. 1. Schematic view of the experimental set up.

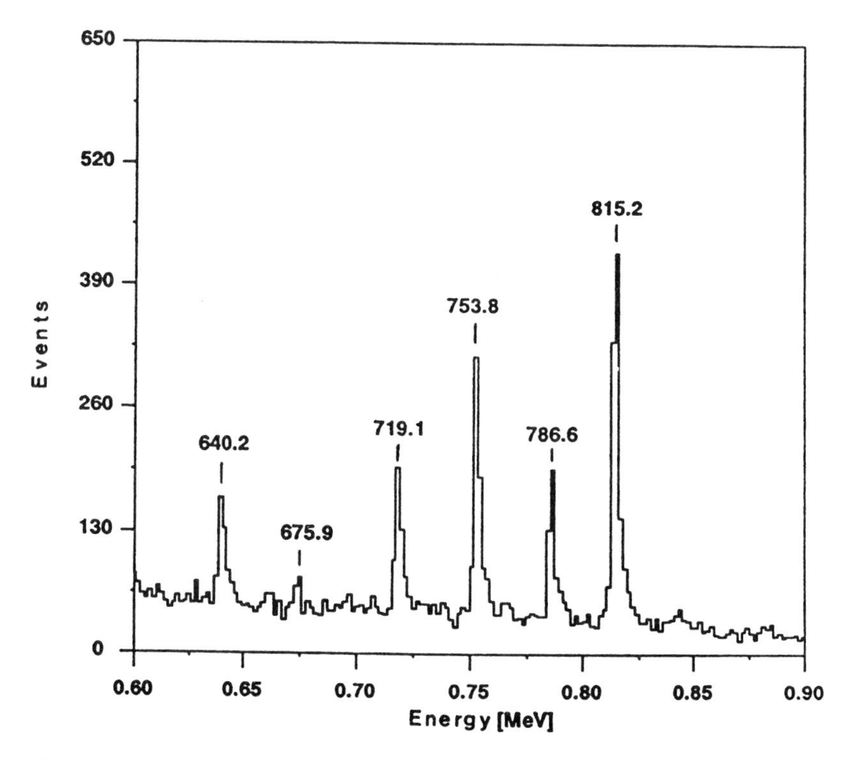

Fig. 2. Spectrum of γ-rays measured in prompt coincidence with electrons and delayed coincidence with the pulsed beam.

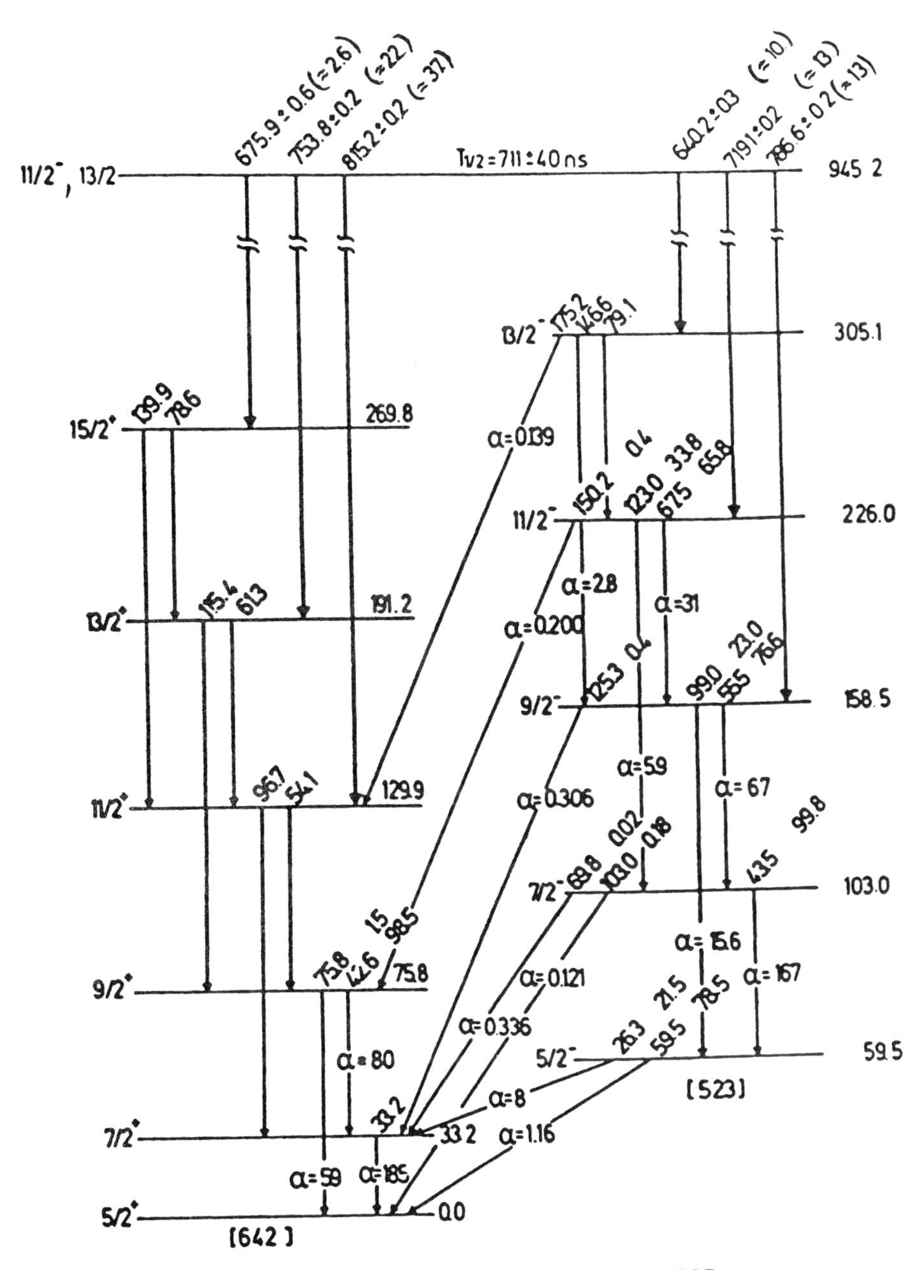

Fig. 3. Decay scheme of the (711 ± 40) ns isomer in ^{237}Np. The 6 γ-ray transitions depopulating the isomer fit very well into the known level scheme of ^{237}Np taken from Ellis-Akovali (1986). The relative intensities of the γ-rays are preliminary values.

The next question is: How can we obtain any information about the spin-parity assignment and the nuclear structure of the isomer ? Since the 6 gammay rays populate the 11/2, 13/2 and 15/2 rotational states of the $5/2^+[642]$ and the 9/2, 11/2 and 13/2 of the $5/2^-[533]$ Nilsson band, it seems reasonable to assume that the level at 945.2 keV is a K-isomer with I = K. Assuming different spin-parity assignments for the isomer, the Weisskopf hindrance factors

$$F_W(EL/ML) = T_{1/2\gamma}(exp)(EL/ML)/T_{1/2\gamma}(Weissk.)(EL/ML)$$

have been compared with empirical values for different ΔK-values compiled by Löbner (1968), Löbner (1975). I=K - assignments $I \leq 9/2$ and $I \geq 15/2$ seem to be completely out of the range of empirical values, that they can be excluded.

The following hindrance factors have been obtained for the
$I^\pi = K^\pi = 13/2^-$ assignment (ΔK = 4 transitions)
or $11/2^-$ assignment (ΔK = 3 transitions):

$$5 \times 10^9 \leq F_W\ (E1) \leq 5 \times 10^{10}$$
$$10^3 \leq F_W\ (M2)$$
$$10^8 \leq F_W\ (M1)$$
$$10^5 \leq F_W\ (E2) \leq 10^6$$
$$83 < F_W\ (E3)$$

and for the $13/2^+$ assignment (ΔK = 4 transitions)
or $11/2^+$ assignment (ΔK = 3 transitions):

$$10^{10} \leq F_W\ (E1) \leq 2 \times 10^{10}$$
$$10^3 \leq F_W\ (M2) \leq 3 \times 10^3$$
$$5 \times 10^7 \leq F_W\ (M1) \leq 10^9$$
$$10^5 \leq F_W\ (E2) \leq 10^6$$

Comparing those F_W-values with the empirical values suggest as most reasonable I-K-π assignment $I^\pi = K^\pi = 13/2^-$ or $13/2^+$.

Since no $13/2^-$ Nilsson orbit is available near to the Fermi surface for Z = 93 nuclei (cf. Gustafson et al (1967), the isomer in ^{237}Np should be a 3-quasiparticle state.

Several configurations would be possible

a) 3-proton level:
$13/2^+$: $5/2^+[642]$ + $5/2^-[523]$ + $3/2^-[521]$
$13/2^-$: $5/2^+[642]$ + $5/2^-[523]$ + $3/2^-[651]$
$13/2^-$: $5/2^+[642]$ + $7/2^+[633]$ + $1/2^-[530]$
$13/2^+$: $5/2^+[642]$ + $7/2^+[633]$ + $1/2^+[400]$

b) 1-proton-2-neutron-state:
$13/2^- \{\pi\ 5/2^+[642] + \nu\ 7/2^-[743] + \nu\ 1/2^+[631]\}$
$13/2^+ \{\pi\ 5/2^+[642] + \nu\ 7/2^+[624] + \nu\ 1/2^+[631]\}$

Since the isomer in ^{237}Np would be the 3-quasiparticle state at lowest excitation energy known in deformed nuclei, the assignment
$13/2^- \{\pi\ 5/2^+[642] + \nu\ 7/2^-[743] + \nu\ 1/2^+[631]$ is most probable.

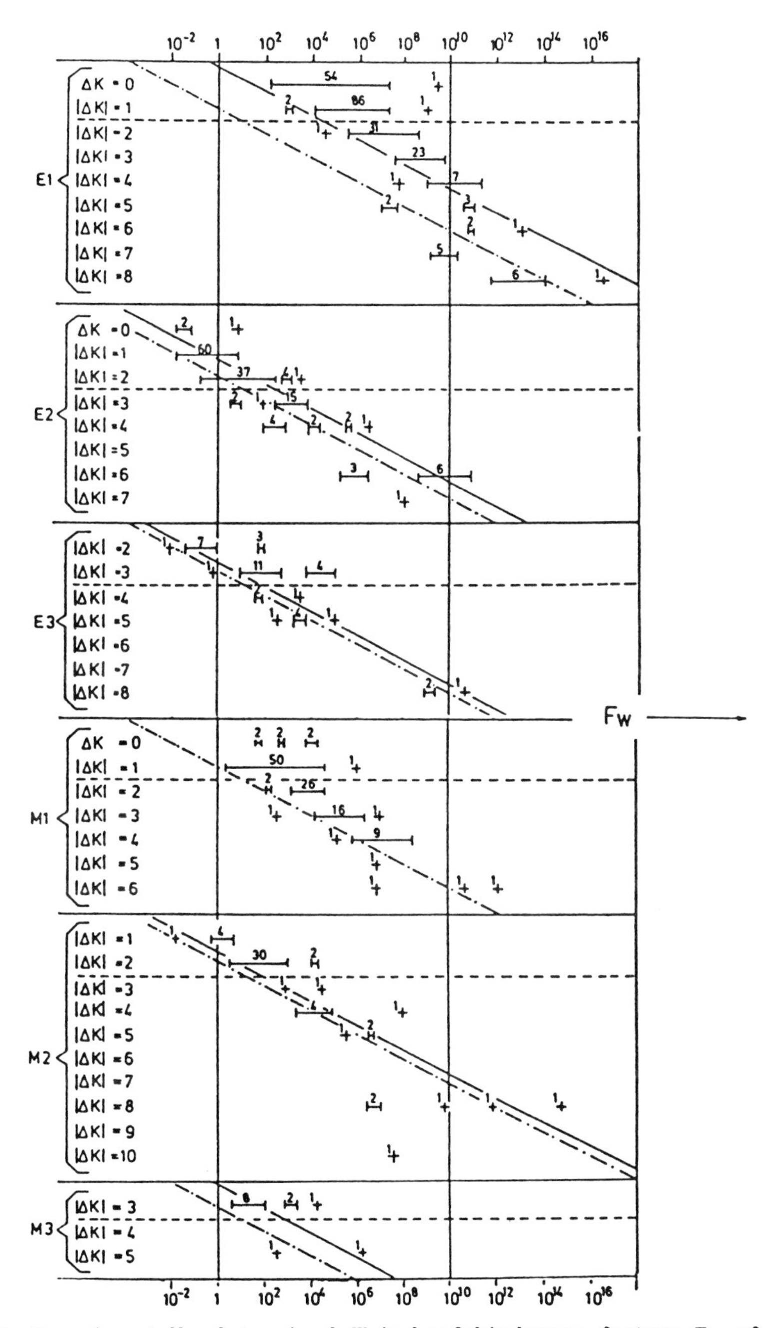

Fig. 4. Experimentally determined Weisskopf hindrance factors F_W of γ-ray transitions of different multipolarity and $|\Delta K|$-values taken from Löbner (1975). For details see Löbner (1975).

This interpretation is supported by two arguments:

a) In $^{235}_{92}U_{145}$ the $1/2^+[631]$ lies 0.08 keV above the $7/2^-[743]$ ground state.

b) Neutron pairing is generally smaller than proton pairing (Bohr and Mottelson (1969)).

However, since the excitation seems to be extremely low for a 3-quasiparticle state (the neutron pairing Δ_n has been calculated from the masses of the neighbouring nuclei to be 0.668 MeV with equation 2-92 of Bohr and Mottelson (1969)), we have also looked for other spin-parity assignments. From the Nilsson diagram a $11/2^-[505]$ excited level is expected for Z=93 nuclei (Ref. Gustafson et al (1967). Although the F_W-values for the $11/2^-$ spin-parity assignment are a little higher than the empirical data, this interpretation can certainly not be excluded.

4. SUMMARY

A new K-isomer has been found in ^{237}Np. The Weisskopf hindrance factors of the depopulating γ-rays suggest as spin-K-parity assignment $I^\pi = K^\pi = 13/2^+$ or $13/2^-$. Since no single particle state with K = 13/2 is available in this mass region for odd-proton nuclei, the isomer should be a 3-quasiparticle state. It would be the lowest lying 3-quasiparticle state found until now in deformed nuclei. However, a $11/2^-[505]$ configuration for the isomer cannot be excluded. Whereas the neutron state $11/2^-[505]$ is systematically known in many rare earth nuclei, it would be the first time that the proton state $11/2^-[505]$ has been observed in any actinide nucleus.

This work has been funded by the German Federal Minister for Research and Technology (BMFT) under the contract number 06 LM 171.

REFERENCES

Bohr A and Mottelson B R 1969 Nuclear Structure Volume 1 Benjamin Inc p 170
Ellis-Akovali Y A 1986 Nuclear Data Sheets 49 p 181
Gustafson C, Lamm L, Nilsson B and Nilsson S 1967 Ark. Fys. 36 p 617
Löbner K E G 1968 Physics Letters 26 B p 369
Löbner K E G 1975 "Gamma-Ray Transition Probabilities in Deformed Nuclei" Chapt. 5 of "The Electromagn. Interaction in Nucl. Spectroscopy" ed Hamilton W D (Amsterdam: North-Holland Publ. Comp.)

Inst. Phys. Conf. Ser. No 105
Paper presented at Int. Conf. on Spectroscopy of Heavy Nuclei, Crete, Greece, 1989

Twin rotational bands at the onset of deformation in neutron-rich Sr isotopes

G.Lhersonneau[1], H.Gabelmann[1,2], K.Heyde[3], N. Kaffrell[1], K.-L. Kratz[1], B.Pfeiffer[1]
and the ISOLDE[2] Collaboration
[1]Institut für Kernchemie, Universität Mainz, D-6500 Mainz, Federal Republic of Germany
[2]CERN, CH-1211 Geneva, Switzerland
[3]Laboratorium voor Kernfysica, Rijksuniversiteit Gent, B-9000 Gent, Belgium

ABSTRACT

The lifetimes of the first excited states in the very neutron rich isotopes $^{99}Sr_{61}$ and $^{100}Sr_{62}$ have been measured using the γ-γ delayed coincidence technique. The deduced corresponding quadrupole deformation parameters indicate a saturation of ground-state deformation at $\beta \sim 0.40$ already at the onset in $^{98}Sr_{60}$. Accordingly, the energies of the unperturbed $6^+ \rightarrow 4^+$ and $4^+ \rightarrow 2^+$ transitions of the rotational ground-state bands in ^{98}Sr and ^{100}Sr are equal within 1.6 keV. This is the first occurence of such a 'twin' character at the onset of deformation and well before the neutron midshell is reached.

1. INTRODUCTION

The neutron-rich isotopes of $_{38}Sr$, $_{39}Y$ and $_{40}Zr$ exhibit the fastest known shape transition when their neutron number reaches N=60. In particular, the change of shape in the Sr nuclei is remarkable; while $^{96}Sr_{58}$ has the character of an anharmonic vibrator with the first 2^+ state at 815 keV [Mül82], its neighbour, the odd-neutron nucleus $^{97}Sr_{59}$ shows shape coexistence, with a spherical ground state [Sil88] and a deformed band head at 585 keV [Lhe88]. The next even-even neighbour $^{98}Sr_{60}$ possesses a ground-state rotational band with the 2^+ state at 145 keV [Mül83]. The heaviest even-even Sr isotope for which experimental information on excited states is available, $^{100}Sr_{62}$, was measured at the ISOLDE facility [Azu79]. Its 2^+ energy is 129 keV. So, from the trend of the 2^+ energies, a larger deformation is expected for $^{100}Sr_{62}$ than for $^{98}Sr_{60}$. This agrees with the fact that ^{98}Sr is the first neutron-rich Sr isotope with a deformed ground state and that the N=66 neutron midshell would be reached at ^{104}Sr. The experimental data on the lifetimes of the 2^+ states in ^{98}Sr [Mül83] and ^{100}Sr [Azu79], resulting in quadrupole deformation parameters β of 0.32 and 0.35, respectively, indeed supported this assumption. However, after the recent lifetime measurement for ^{98}Sr at the separator OSTIS [Ohm87a], a deformation of β=0.39(1) was deduced which became the largest one in this region. The result of Azuma et al., as well as the older ^{98}Sr data, have been obtained using the delayed time distribution between β and γ-rays. This technique, which offers a good timing resolution suffers, however from the lack of energy selectivity. For instance, the measurement of [Azu79] might have been perturbed by the 5.4 ns halflife of the excited 0^+ state in ^{100}Zr [Lhe89], one of the decay products of ^{100}Rb. Thus, in view of the implications on the trend of deformation versus neutron number, a remeasurement of the ^{100}Sr halflife, using the more refined techniques available now, appeared of crucial importance.

2. EXPERIMENT

The isotope ^{100}Sr was produced at the ISOLDE facility as the β-decay daughter of 53 ms ^{100}Rb by bombarding a 12 g/cm^2 uranium carbide target with 600 MeV protons from the CERN synchrocyclotron. After fast diffusion from the target to a surface-ionisation source, the extracted Rb isotopes were mass separated. Apart from ^{100}Rb and its β-decay isobars, also activities of the A=99 chain were observed which originated from β-delayed neutron decay of ^{100}Rb. For the lifetime measurement the gamma rays were detected with an intrinsic Ge detector and a BaF_2 scintillator. Energies and time were recorded in listmode and sorted off-line. The time spectra, corrected for contributions from the Compton background in the gamma gates [Lhe78], were analysed by the centroid-shift and unfolding methods. Since these methods require the detailed knowledge of the position of the centroid of a prompt reference and of the electronic response function, calibration points had to be taken in order to allow the interpolation at the gamma energies in ^{100}Sr. These points included time distributions from off-line measurements with a ^{152}Eu standard source, from on-line measurements of the A=98 chain [Ohm87a] and in the A=99 and 100 chains [Lhe89]. In addition, γ-γ-t coincidences were recorded simultaneously using the above mentioned small Ge and a large volume Ge detector with the aim of extending the decay scheme of ^{100}Rb. Details about the time analysis and new information about the level scheme of ^{100}Sr will be presented in a forthcoming paper.

3. RESULTS

In a beam time period of 40 hours ~3.3x10^6 coincidence events belonging to ^{100}Rb decay or its daughter products have been accumulated with the Ge-BaF_2 detector system. The results on halflives are summarized in table 1. The agreement between the results of the centroid-shift and the unfolding method is evident. The weighted average for the halflife of the 2^+ state in ^{100}Sr from the 129-288 keV and 288-129 keV coincidences, (lines 1 and 3 of table 1), is $t_{1/2}$= 3.91(16)ns. This value is definitively shorter than that previously reported of $t_{1/2}$=5.15(20)ns [Azu79]. We further notice that coincidences with a broad gate on the BaF_2 detector suggest a delay for the 4^+ level. However, the apparent halflife of about 0.3 ns might be due to a newly identified high excited level with a halflife of ~110 ns, which is likely to interfere through Compton coincidences in the broad gates. We note that the rotational model, using the new value of 3.9 ns for the first 2^+ state, predicts $t_{1/2}(4^+)$=0.06 ns.
The analysis of the coincidences (~10^7 events were recorded) taken with the Ge-Ge detector system (last line) is also in agreement with the Ge-BaF_2 results. A rudimentary level scheme of ^{100}Sr, only showing the states relevant for the following discussion is shown in Fig.2. Up to now, 28 new gamma lines have been placed between 18 new levels in a preliminary scheme. A new level of particular interest is the one at 852 keV, which is very likely the 6^+ member of the ground-state band.
The lifetime for the first excited state in the nucleus $^{99}Sr_{61}$ could be measured to be $t_{1/2}$= 0.58(9) ns. This result, combined with the spectroscopic data from an experiment at the OSTIS separator [Pfe84], yields a deformation parameter of β=0.38(4) [Lhe89]. This value overlaps with the deformations for the even-even neighbours.

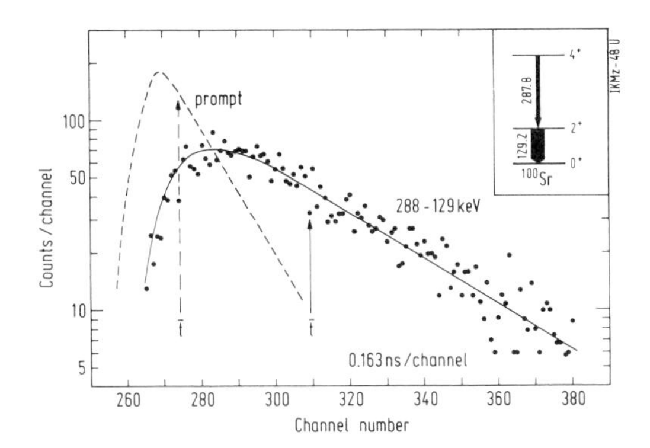

Figure 1: Time distribution of the 288 keV (Ge detector)-129 keV (BaF_2 detector) coincidences. The dashed line indicates the response function, obtained by interpolation between calibration points (see text). The tail at the right hand side is not due to a physical lifetime, but to an electronic effect. The resolution (FWHM) is 4.2 ns. The solid line is a fit of the time distribution, using all displayed experimental points. Centroids (t) of the prompt and delayed time spectra are indicated by vertical arrows.

Table 1: Results of the analysis of the time spectra gated by the $2^+ \rightarrow 0^+$ (129 keV) and $4^+ \rightarrow 2^+$ (288 keV) transitions in the small Ge detector [a], with various energy gates [b] in the BaF_2 counter (lines 1-4) and in the large Ge detector (line 5). Halflives in nanoseconds deduced from the centroid-shift method [c], from the unfolding method [d] and adopted values [e] are shown with their uncertainties (standard deviation plus systematical error).

Eγ [keV]		Halflives $t_{1/2}$ [ns]		
a	b	c	d	e
129	288	3.91(24)	3.86(19)	3.88(19)
129	500 - 1000	4.17(18)	4.18(16)	4.18(16)
288	129	3.93(30)	4.06(30)	4.00(30)
288	380 - 1000	0.30(18)		
129	288	3.7 (5)		

4. DISCUSSION

Our new value of β=0.40(1) implies that, in contrast to the former measurement of [Azu79] for ^{100}Sr, there is no decrease of deformation towards the neutron midshell, but - within the given uncertainties - rather a saturation reached already at ^{98}Sr with β=0.39. This trend has also been observed in laser spectroscopic studies [Sil88]. Moreover, $^{98}Sr_{60}$ and $^{100}Sr_{62}$ have very similar ground-state band properties (see Fig. 2). This is evident from the collectivity of the $2^+ \rightarrow 0^+$ transition, the deduced quadrupole moments and deformations (see table 2). The similarity of the inertial parameters is also striking.

The energies of the $4^+ \rightarrow 2^+$ transitions in ^{98}Sr and ^{100}Sr are equal within 1.6 keV while the energies of the $6^+ \rightarrow 4^+$ transitions differ by only 1 keV. Such a close similarity is encountered very seldom; there are examples in the rare-earth region for the $_{68}$Er and $_{70}$Yb isotopes [Led78] and the very recently studied neutron-rich $_{44}$Ru isotopes [Dur89]. But it must be stressed that in these cases the neutron number is at the midshell or very close to it and that these nuclei have undergone a smooth shape transition extending over a large number of neutrons. In the contrary, the Sr isotopes are the only example of such a situation at the onset of deformation and well before the midshell is reached.

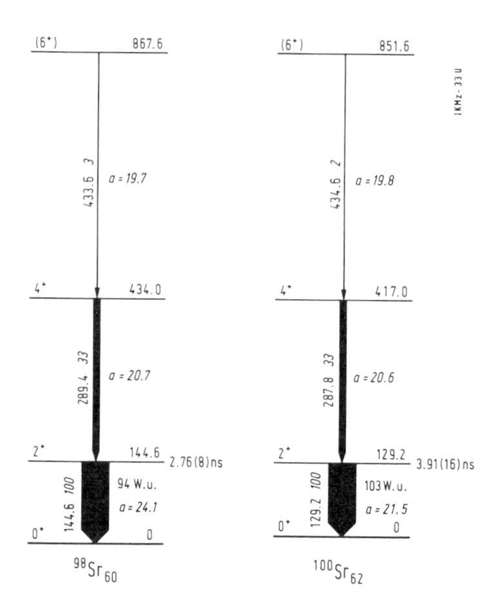

Figure 2: Ground-state bands of ^{98}Sr and ^{100}Sr. The level energies are adjusted to the 2^+ level. The values of the inertial parameter 'a' in keV show that both bands become more rigid with increasing angular momentum. The experimental B(E2) values, in Weisskopf units, are not corrected for band mixing.

The energies of the $2^+ \rightarrow 0^+$ transitions are 145 keV in ^{98}Sr and 129 keV in ^{100}Sr, so that at first glance one might conclude that ^{100}Sr is a more rigid rotor. The mixing calculation of [Mach] however, interprets the experimental energy of 145 keV as the result of mixing between the strongly deformed 0^+ ground state and the weakly deformed or spherical excited 0^+ state at 216 keV, which repel each other. On the other hand, the energy of 129 keV shows that the ground state band in ^{100}Sr is very little perturbed. Therefore, a coexisting 0^+ state, if present, must lie much higher in energy.
In the context of this study, it is worthwhile to compare these data to those for the isotopes of $_{40}$Zr with corresponding neutron numbers. The halflife for the first excited 2^+ state in ^{100}Zr$_{60}$ has been measured recently in the course of the present investigation [Lhe89] and at the separator TRISTAN [Mach]. Data for ^{101}Zr$_{61}$ are available from a study at the separator JOSEF [Ohm87b]. For ^{102}Zr, the isotone of ^{100}Sr, β=0.39 is deduced from the weighted average of two independent studies from the spontaneou

fission of ^{252}Cf [Jar74] and ^{254}Cf [Che80]. The E(4^+)/E(2^+) ratio of 3.15 in ^{102}Zr [Shi83] shows that this band, like the one in ^{100}Sr, also is close to the rotational limit. Thus, large deformations also occur in the Zr isotopes with N≥60. However, the trend of deformation in the Zr isotopes might be different from that in the Sr isotopes. The lifetime measurements rather indicate a small increase of β from $^{100}Zr_{60}$ to $^{102}Zr_{62}$. This assumption seems supported by the difference of the inertial parameters of the ground-state bands, namely the $4^+ \rightarrow 2^+$ transitions have energies of 352 and 325 keV respectively.

Table 2: Comparison of ground state band properties in ^{98}Sr and ^{100}Sr. Values in brackets are calculated by [Mach] for the unperturbed ground state band without mixing between the 0^+ states. For connecting Q_0 and β we use; $Q_0 = 3/\sqrt{5\pi}ZR^2\beta(1 + 0.158\beta)$. The moment of inertia is extracted from the energy of the $4^+ \rightarrow 2^+$ transition with $a_{rigid} = 36/(1 + 0.315\beta)A^{5/3}$ MeV. Sr isotopes are the most rigid even-even rotors known among the medium and heavy nuclei.

	$^{98}Sr_{60}$	$^{100}Sr_{62}$
B(E2,2→0) W.u.	94(3) [106(3)]	103(5)
Q_0 barn	3.56(6) [3.78(6)]	3.79(8)
β	0.39(1) [0.40(1)]	0.40(1)
E_γ ($2^+ \rightarrow 0^+$)	144.6 [121.3]	129.2
E_γ ($4^+ \rightarrow 2^+$)	289.4	287.8
E_γ ($6^+ \rightarrow 4^+$)	433.6	(434.6)
J/J_{rigid}	0.74	0.72

The saturation of deformation at N=60 in the Sr isotopes may be understood qualitatively in terms of the Quartet model [Dal86]. Counting the valence nucleons from the doubly closed shell nucleus $^{78}_{28}Ni_{50}$, the ground state deformation must reach a maximum in ^{98}Sr, which has the largest possible number of 10 quartetted bosons and no bosons left unquartetted. The saturation of the collectivity is also reproduced since, unlike in the conventional Interacting Boson Models, B(E2) values are proportional to the number of quartetted bosons rather than to the total number of bosons.
Alternatively, the saturation of deformation is interpreted in a microscopic way [Hey87]. In a deformed shell-model approach, due to the residual p-n quadrupole force, the binding energy associated with the occupation of certain Nilsson levels can become quite large. This is especially so when protons and neutrons occupy both up/down sloping Nilsson orbitals. [Cas88]. This particular situation exists for protons close to Z=40 (the down sloping $1/2^+$ orbital of $1g_{9/2}$ origin) and neutrons near N=58,60 (the down sloping $1/2^-$,$3/2^-$ orbitals arising from $1h_{11/2}$). Saturation of the p-n quadrupole force is reached when, at larger N, neutrons occupy orbitals with less slope so that they do not contribute significantly to the binding energy.
The saturation of deformation before the neutron midshell is also predicted, with the correct deformation parameter $\beta \sim 0.40$, by mass calculations [Möl88]. However, the onset of deformation is predicted at a too low neutron number.

5. CONCLUSION

The recent lifetime measurements for the neutron-rich Sr and Zr isotopes establish the existence of a narrow region of strong ($\beta \sim 0.4$) ground-state deformation at N$\geq$ 60. The new measurement of $t_{1/2}(2^+)$ in ^{100}Sr shows that in the Sr isotopes the deformation reaches its maximum immediately after its onset, in contrast to the usually expected gradual increase towards midshell. As a consequence the inertial parameters of the ground-state bands are so close that the transition energies of the unperturbed $6^+ \rightarrow 4^+$ and $4^+ \rightarrow 2^+$ transitions are equal within $\sim$ 1.6 keV. This situation is unique in the whole chart of nuclides. In addition, a drastic shape transition like the one in Sr and Zr isotopes is not observed in the neighbouring elements. Although quite strong ground-state deformations of $\beta \sim 0.32$ have been measured for $^{102}Mo_{60}$ [Jar74] and $^{104}Mo_{62}$ [Che80] nuclei, the shape transition is there smoother. Data for lighter elements e.g. $_{37}$Rb or $_{36}$Kr, at corresponding neutron numbers, are not yet available.

This work was supported by the German Federal Minister for Research and Technology (BMFT) under contract number 06 MZ 552.

REFERENCES

Azuma R E et al. 1979 *Phys.Lett.* **86B** 5
Casten R F, Heyde K and Wolf A 1988 *Phys.Lett.* **B208** 33
Cheifetz E et al. 1980 *Proc. Workshop on Nuclear Spectroscopy of Fission Products, Grenoble, France, May 1979 (von Egidy ed.) Inst. of Phys. Conf.Ser* **51** 193
Daley H J et al. 1986 *Phys.Rev.Lett.* **57** 198
Durell J L 1989 Contribution to these proceedings.
Heyde K 1987 *Nucl.Phys.* **A466** 189
Jared R C , Nifenecker H and Thompson S G 1974 *Proc. 3rd IAEA Symp. Phys. Chem. Fission Rochester (Vienna: IAEA)* 211
Lederer C M and Shirley V S 1978 *Table of Isotopes* (Wiley, New York)
Lhersonneau G 1978 *Nucl.Instr.Methods* **157** 349
Lhersonneau G et al. 1988 *Z.Phys.* **A330** 347
Lhersonneau G et al. 1989 *Z.Phys.* **A332** 243
Mach H preprint, submitted to *Phys.Lett.* **B**
Möller P and Nix R 1988 *Atomic Data and Nuclear Tables* **39** 213
Müller H W 1982 *Nucl.Data Sheets* **35** 281
Müller H W 1983 *Nucl.Data Sheets* **39** 467
Ohm H et al. 1987 *Z.Phys.* **A327** 483
Ohm H et al. 1987 *Z.Phys.* **A326** 233
Pfeiffer B et al. 1984 *Z.Phys.* **A317** 123
Shizuma K et al. 1983 *Phys.Rev.* **C27** 2869
Silverans R E et al. 1988 *Phys.Rev.Lett.* **60** 2607

Onset of deformation above and below N=82 in the even–even rare-earth nuclei

L.Goettig[1)] and W.Nazarewicz[2)]

[1)]Institute of Experimental Physics, Warsaw University, Warsaw
[2)]Institute of Physics, Warsaw Institute of Technology, Warsaw

ABSTRACT: A systematic disagreement has been found between the results of the Strutinsky- type calculations and the Grodzins' estimate for the nuclear deformation of the even-even rare- earth nuclei. As a rule, the Grodzins' formula overpredicts the deformation above N=82 and gives too small values below N=82. We have shown that the origin of these discrepancies lies in the neglection of the N and Z dependence of pairing. Assuming the calculated deformations and the microscopic values of the pairing gap, the experimentally observed behaviour of the nuclear moments of inertia has been reproduced.

The region of well deformed rare-earth nuclei below N=82 has been reached recently [1,2] in a series of experiments using the "Neutron Wall" and the recoil separator at the Daresbury Laboratory. Many exotic, very neutron-deficient Gd, Dy and Sm isotopes have been identified and the new spectroscopic data made it possible to study systematically the onset of nuclear deformation when moving away from the N=82 shell closure.

However, the experimental evidence for very neutron deficient nuclei only scarcely include the lifetime data. For the most exotic nuclei studied all what is usually known are the lowest excitation energies only. Because of that, to create the deformation systematics the so-called Grodzins' formula has been widely used. It relates the energy of the first 2^+ state in an even-even nucleus to the quadrupole deformation parameter β_2 by means of a simple formula:

$$E(2^+)=1225/(A^{7/3}\beta_2^2)[\mathrm{MeV}].$$

Based on the Grodzins' estimate it has been concluded[1,2] that the onset of the deformation for a given Z on each side of the N=82 shell is strikingly different (Fig.1, top). It increases with $n=|N-82|$ more rapidly above the N=82 gap than below. The spectacular jump seen at n=6, around N=88, is not present on the N<82 side.

Attempts have been made[3] to explain this behaviour in terms of the N_pN_n scheme. The rapid change of character from vibrational to rotational above N=82 was interpreted

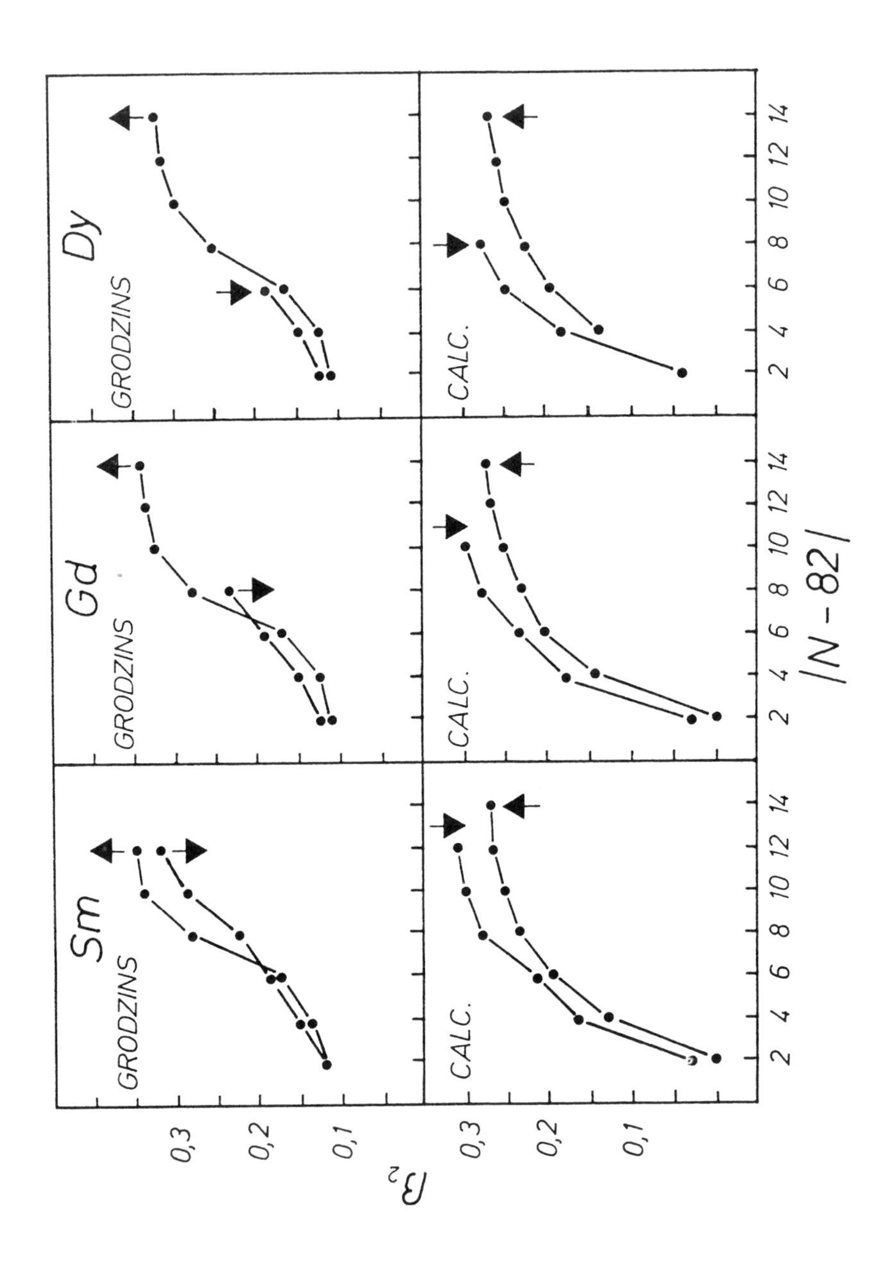

Fig.1. The deformation parameter β_2 extracted from the E(2^+_1) data using the Grodzins' estimate (top), and the results[6] of the shell-correction type calculations based on the Woods-Saxon average potential (bottom) above (🡅) and below (🡇) N=82.

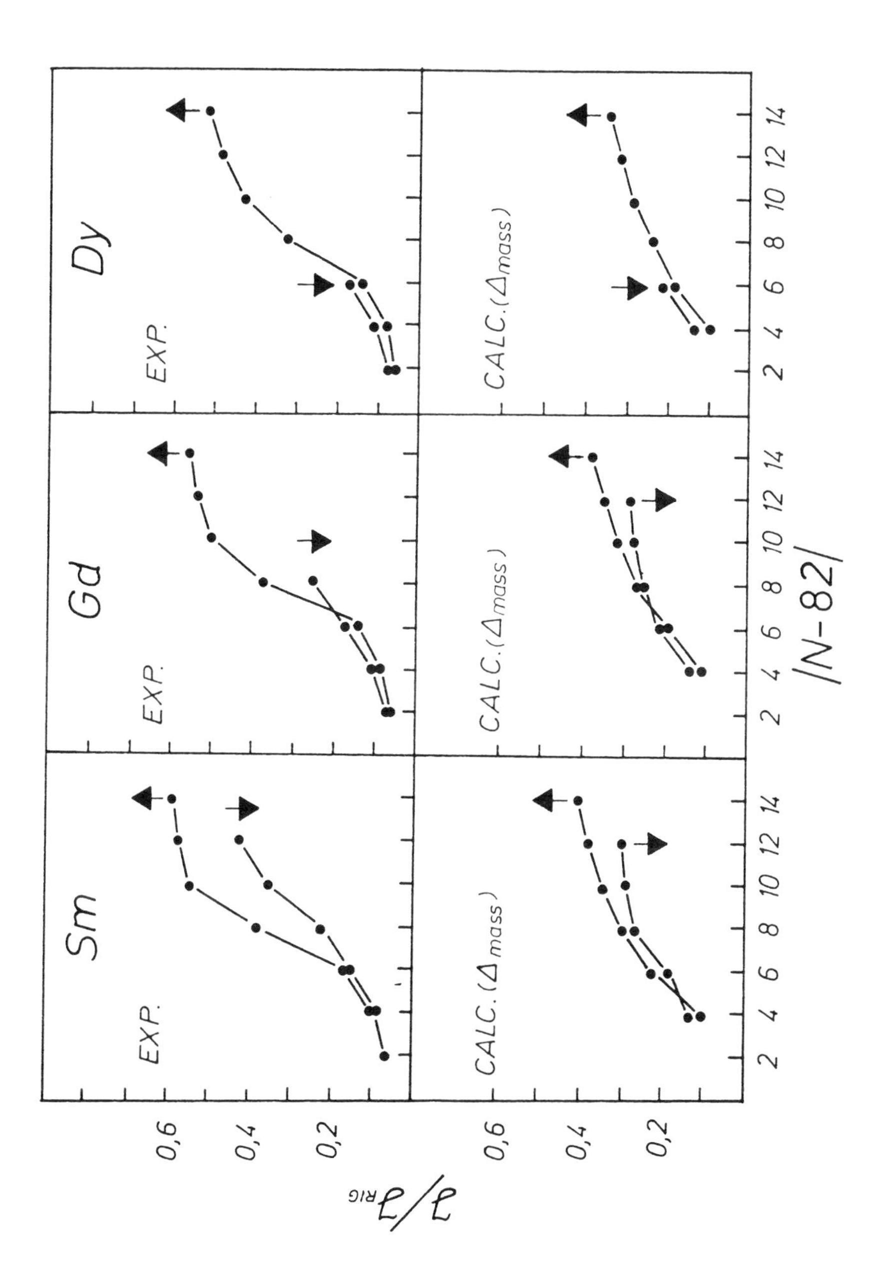

Fig.2. The ratio of the nuclear moment of inertia to its rigid body value extracted from the $E(2^{+}_{1})$ data (top), and the results of the present calculations (bottom) above (⬆) and below (⬇) N=82.

qualitatively as resulting from the T=0 n-p interaction which has been recognized as the driving force toward the nuclear deformation[4].

The effects of the n-p interaction are taken into account quantitatively in the mean-field theory calculations[5]. In the present work the results of such calculations based on the shell correction method[6] are compared with the Grodzins' estimates. As seen in Fig.1, we found a systematic disagreement between the "experimental" and calculated values. Firstly, the theoretical deformations tend to grow faster with n below N=82 as compared to the nuclei with N>82. Secondly, the characteristic jump in the deformation between N=88 and 90 given by the Grodzins' formula is not reproduced.

On the other hand, these theoretical results agree quite well with the existing data from the lifetime measurements[6], however very limited below N=82. Therefore we conclude that the Grodzins' estimate systematically overpredicts the deformation above N=82 and gives too small deformations below N=82.

In order to understand the origin of these systematic discrepancies we analyzed the influence of the N and Z dependence of pairing correlations on the deformation extracted from the $E(2^+_1)$ data. Recent study by Jensen et al.[7] revealed a very strong isovector dependence of the average pairing gap parameter Δ .For a given Z, Δ is a strongly decreasing function of N, e.g. for well deformed nuclei it is about 40% bigger below than above N=82. Such an effect is certainly neglected in the semi-empirical Grodzins' formula. However, it is well known that the position of the 2^+_1 state is strongly influenced by the interplay between the nuclear deformation and pairing (see e.g. ref.8)

The significance of the N and Z dependence of pairing becomes obvious when the $E(2^+_1)$ data, instead of being translated directly into deformation, are used to extract the moments of inertia. The ratio of the nuclear moment of inertia $\mathcal{J}$ to its rigid body value $\mathcal{J}_{RIG}$, extracted from the $E(2^+_1)$ data, is shown in Fig.2 (top). It is well known that this ratio depends both on the deformation and pairing, namely it increases with the deformation β_2 and decreases with the pairing gap.

Assuming the calculated deformations[6] and the following formulas for Δ given by Jensen et al.[7]:

$$\Delta_n = 7.36\ A^{-1/3}(1 - 8.15(N-Z)^2A^{-2})\ [\mathrm{MeV}]$$
$$\Delta_p = 7.55\ A^{-1/3}(1 - 6.07(N-Z)^2A^{-2})\ [\mathrm{MeV}]$$

we have calculated the ratio $\mathcal{J}/\mathcal{J}_{RIG}$ for even-even rare-earth nuclei above and below N=82 using the expression[9]:

$$\mathcal{J}/\mathcal{J}_{RIG} = 1 - g\left(\frac{\varepsilon\hbar\omega_0}{2\Delta}\right) \qquad \text{where} \quad g(x) = \frac{\ln(x+(1+x^2)^{1/2})}{x(1+x^2)^{1/2}},$$

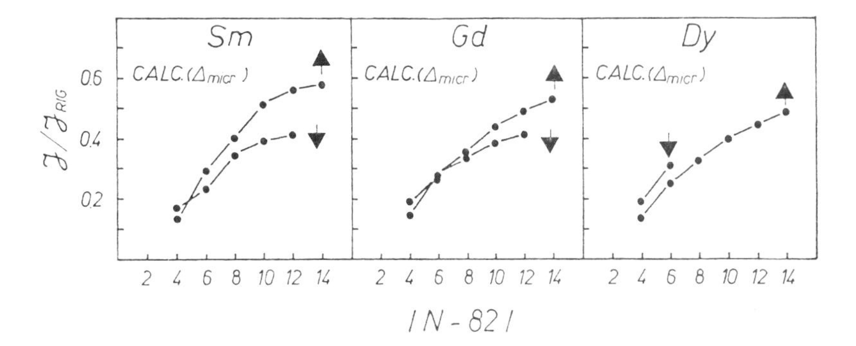

Fig.3. The ratio of the nuclear moment of inertia to its rigid body value calculated assuming the microscopic (BCS) pairing gap parameters.

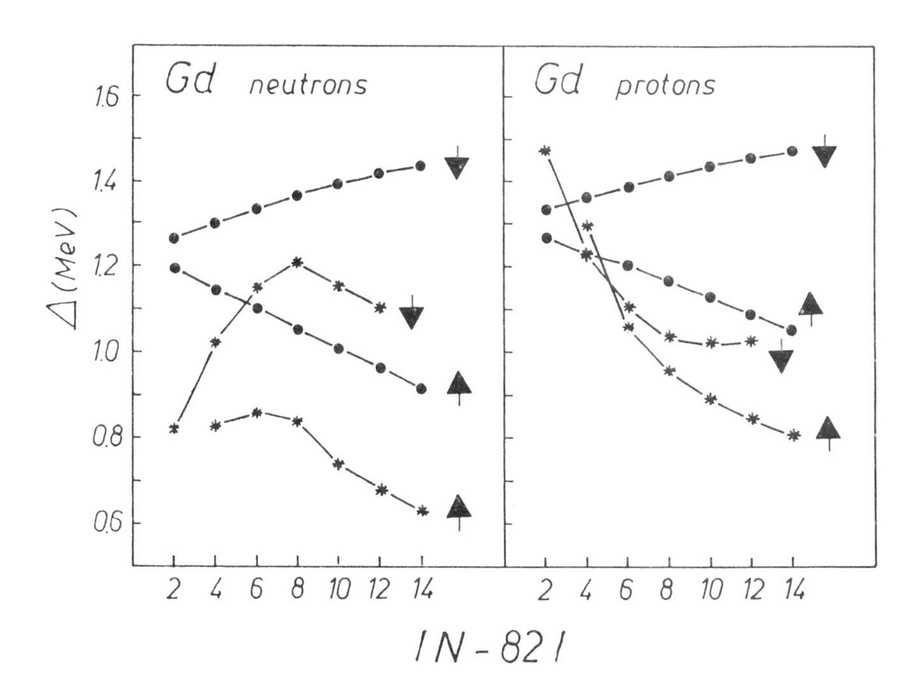

Fig.4. The Gd pairing gap parameters calculated within the BCS theory (stars), and extracted from the odd-even mass differences using the formulas by Jensen et al[7](dots).

$\hbar\omega_0 = 40/A^{1/3}$[MeV] and $\varepsilon \cong .944\beta_2 - 0.122\beta_2^2 + 0.154\beta_2\beta_4$. Because of different pairing the proton and neutron contributions were calculated separately and then added together to get the $\mathcal{J}/\mathcal{J}_{RIG}$ ratio as shown in Fig.2 (bottom). From the comparison in Fig.2 one can see that the results of our calculations display the same trends as the experimental data, though the absolute values are too big. It turns out that we can get a satisfactory agreement with the experimental values of the moments of inertia (Fig.3) using the pairing gap parameters calculated microscopically within the BCS theory. Such pairing gap parameters for Gd isotopes are shown in Fig.4, where they are compared with the ones extracted from the odd-even mass differences.

Thus, we have shown that the faster decrease of the $E(2^+_1)$ values above than below N=82, which previously was associated with the faster increase of the deformation, is in fact a combined result of a change in both the deformation and pairing. Above N=82, the decreasing Δ and increasing β_2, produce faster change of the $E(2^+_1)$ values, as n increases. On the contrary, below N=82 we have a compensating effect of pairing and deformation (both increase with n), which leads to more slowly changing $E(2^+_1)$ values.

In conclusion, the previous suggestions[1,2] about the difference in the onset of deformation below and above N=82 based on the Grodzins' estimate, should probably be revised. As a rule, the deformations resulting from the Grodzins' formula are systematically too big above, and too small below a closed shell because the formula does not take into account the N and Z dependence of pairing.

This research was partially sponsored by the Ministry of National Education under the contract CPBP 01.06.

REFERENCES:

1. C.J.Lister, B.J.Varley, R.Moscrop, W.Gelletly, P.J.Nolan, D.J.G.Love, P.J.Bishop, A.Kirwan, D.J.Thornley, L.Ying, R.Wadsworth, J.M.O'Donnell, H.G.Price and A.H.Nelson, Phys.Rev.Lett.55(1985)810
2. L.Goettig, W.Gelletly, C.J.Lister, R.Moscrop, B.J.Varley and R.Wadsworth, Nucl.Phys.A464(1987)159
3. R.F.Casten, Nucl.Phys.A443(1985)1
4. I.Talmi, Rev.Mod.Phys.34(1962)704
5. J.Dobaczewski, W.Nazarewicz, J.Skalski and T.Werner, Phys.Rev.Lett.60(1988)2254
6. W.Nazarewicz, J.Garrett, M.Riley, to be published; B.D.Kern, R.N.Mlekodaj, G.A.Leander, M.O.Kortelahti, E.F.Zganyar, R.A.Braga, R.W.Fink, C.P.Perez, W.Nazarewicz and R.B.Semmes, Phys.Rev.C36(1987)1514
7. A.S.Jensen,P.G.Hansen and B.Jonson, Nucl.Phys.A431(1984)393
8. J.A.Bounds, C.R.Bingham, P.Juncar, H.K.Carter, G.A.Leander, R.L.Mlekodaj, E.H.Spejewski and W.M.Fairbank,Jr., Phys.Rev.Lett.55(1985)2269
9. A.Bohr, B.R.Mottelson, Nuclear Structure vol.2, 4-128

Inst. Phys. Conf. Ser. No 105
Paper presented at Int. Conf. on Spectroscopy of Heavy Nuclei, Crete, Greece, 1989

Pseudo symplectic model and potential energy surfaces

J.P. Draayer

Department of Physics and Astronomy, Louisiana State University
Baton Rouge, Louisiana 70803-4001

O. Castaños and P. Hess

Instituto de Ciencias Nucleares, Universidad Nacional
Autónoma de México, Apartudo Postal 70-543, México, D.F. 01000, México

ABSTRACT: A shell-model scheme for heavy deformed nuclei is proposed. We begin by reviewing results that demonstrate the near equivalence of the quantum rotor and SU(3) pictures. Then we turn to a discussion of the pseudo SU(3) scheme which applies to rare earth and actinide nuclei. With this background established, we go on to show that the symplectic scheme allows one to extend the SU(3) picture, normal and pseudo, to include shell mixing effects generated by the quadrupole-quadrupole interaction. In the final section we show that the symplectic scheme can be used to provide a shell-model interpretation of the potential energy surface concept. Some preliminary results for ^{238}U are presented.

1. INTRODUCTION

There are many barriers to gaining a successful shell-model description of the structure of heavy deformed nuclei: huge shell-model spaces, an unknown nucleon-nucleon interaction, inadequate computer power, etc. Nevertheless, these nuclei remain one of nature's most challenging puzzles and therefore are deserving of the extraordinary effort an understanding of their structure requires. They are quantum systems in which single-particle, many-body and statistical features coexist with a minimum of interference.

In this paper we introduce the pseudo symplectic scheme which is a fully microscopic shell-model theory for heavy deformed nuclei. Our starting point is a review of results that demonstrate the near equivalence of the quantum rotor and SU(3) pictures. We then turn to a discussion of the pseudo SU(3) scheme which is an extension of the Elliott SU(3) Model that applies to rare earth and actinide nuclei. Next we show that the symplectic scheme can be used to extend the SU(3) picture, normal and pseudo, to include shell mixing effects generated by the quadrupole-quadrupole interaction. And this leads us, in the final section, to a shell-model interpretation of the potential energy surface concept.

2. ROTOR-SU(3) EQUIVALENCE

The pioneering work of Elliott (1958) established SU(3) as the underlying symmetry of observed rotational phenomena in light ds-shell nuclei. The success of the model follows from the fact that eigenvalues of the quadrupole-

quadrupole interaction ($Q^a \cdot Q^a$, "a" for algebraic) go as $4C_2 - 3L^2$ where C_2 is the second order invariant of SU(3) and L^2 is the square of the angular momentum operator. Not until recently, however, was the relationship between SU(3) and the rotor fully understood and appreciated. The thesis work of Leschber (1987) shows through numerous examples the close relationship between observables of the triaxial quantum rotor and a simple SU(3) → SO(3) integrity basis interaction. Subsequently, Leschber and Draayer (1987) derived analytic expressions that relate the parameters of these two theories.

A group theoretical reason for the correspondence can be given, namely, the symmetry group of the rotor is a contraction of the SU(3) algebra. However, there are important differences and these teach us a number of things about nuclei that display rotational behavior. For example, SU(3) representations are finite dimensional whereas those of the rotor are infinite. This yields significant differences in predicted E2 transition strengths between high-spin members of a rotational band. The SU(3) results fall off from rotor model values and, indeed, this is in agreement with experiment.

The correspondence we refer to goes well-beyond what can be deduced from the contraction process. Castaños, et al. (1988) have shown that a relationship between shape variables of the rotor and the SU(3) representation labels can be established. Since the SU(3) representation labels are dictated by the filling of shell-model orbitals, this implies a constraint on allowed nuclear shapes. Turning things around, one can use this to study the onset of triaxiality, the probability of finding oblate nuclear species, etc. In a recent paper Wu, et al. (1987) used this type of logic to provide a shell-model interpretation of the binding energy systematics of actinide nuclei.

3. PSEUDO SU(3) SCHEME

It is well-known that for nuclei beyond ^{40}Ca the spin-orbit interaction pushes the highest j member of a given shell down among orbitals of the next lower shell. This destroys the oscillator structure and with it the underlying SU(3) symmetry. Nonetheless, nuclei of the rare earth and actinide regions display rotational characteristics. How can this be? Though still somewhat of a mystery, a simple answer can be given. The normal parity levels that remain clustered together can be identified as members of an oscillator shell with one less quanta then that of the original set.

This mapping defines the pseudo SU(3) scheme (Raju, et al, 1973). It does not mortgage shell-model principles nor is it in any way whatsoever, as the name seems to suggest, a false or fake proposition. The scheme is a renaming and reorganization of the shell-model orbitals. The physics remains unchanged. Of course, the exercise would be meaningless if a description of the physics in terms of the new basis is as complicated as before. And herein lies the real value of the pseudo SU(3) scheme. When the hamiltonian is expressed in terms of the new labelling scheme, one finds that just as in the ds-shell, the quadrupole-quadrupole interaction dominates. The consequences are therefore known, namely, the hamiltonian favors representations of SU(3) that correspond to configurations of maximum intrinsic deformation as defined in terms of the new (pseudo) as opposed to the old (normal) oscillator.

To illustrate this point in more detail, in Table 1 we give the SU(3) tensor decomposition of Q^a for the $\eta = 4$ shell using both the normal and pseudo shell geometries. Since Q^a is a generator of the symmetry group of the normal oscillator it is a pure $(\lambda_0,\mu_0)\kappa_0,L_0,S_0 = (1,1)1,2,0$ tensor in that scheme. However, under a reorganization of the space this changes. In addition to a generator part, other SU(3) tensors appear. As the coefficients indicate, however, the generator part is not only the largest by far but it is also enhanced by the transformation. Indeed, we have found that the nongenerator forms make less than a two percent change in calculated E2 transition rates in applications to rare earth and actinide nuclei (Draayer and Weeks, 1984; Castaños, et al., 1987).

The pseudo SU(3) scheme has been used to study a variety of phenomena:

- Decoupling paramaters (Raju, et al., 1973)
- Alpha particle transfer strengths (Braunschweig, et al., 1978)
- Backbending in ^{126}Ba (Raju, et al., 1979)
- Band crossing in ^{126}Ba (Draayer, et al.,1981)
- Forking in ^{68}Ge (Weeks, et al., 1981)
- $Q_\pi \cdot Q_\nu$ and strong coupling (Draayer, et al., 1982)
- Unique parity spin sequences (Weeks, et al., 1983)
- Collective E2, M1 and M3 modes (Castaños, et al., 1987)

In each case the calculated results are in good agreement with the known experimental data.

4. SYMPLECTIC EXTENSION

The symplectic model extends the SU(3) scheme by including couplings to other shells generated by the quadrupole operator. In this case the symmetry group is Sp(3,R). It has 21 generators, the 8 of SU(3) plus an operator that counts the total number of oscillator quanta and six $2\hbar\omega$ raising and six $2\hbar\omega$ lowering operators (Rosensteel and Rowe, 1976, 1979 & 1980; Rowe, 1985):

L_μ	3	angular momentum	SU(3), U(3), Sp(3,R)
Q_μ	5	quadrupole ($0\hbar\omega$)	SU(3), U(3), Sp(3,R)
N	1	number operator	U(3), Sp(3,R)
B^+	6	$2\hbar\omega$ raising (L=0 & 2)	Sp(3,R)
B^-	6	$2\hbar\omega$ lowering (L=0 & 2)	Sp(3,R)
	21	total operators	

The application of these operators to any $0\hbar\omega$ SU(3) shell-model representation generates the corresponding Sp(3,R) shell-model representation. Because of the structure of the B^+ raising and B^- lowering operators, the symplectic representations have infinite dimensions. Actually, the lowering operators annihilate $0\hbar\omega$ states because quanta cannot be removed from the system without violating the Pauli Principle. This means that the $0\hbar\omega$ SU(3) representations serve as bandhead labels of symplectic shell-model representations.

The addition of the B^+ and B^- to the SU(3) algebra means that the full effect of the quadrupole-quadrupole interaction ($Q^c \cdot Q^c$, "c" for collective) between nucleons in a nucleus can be determined. The Q's are related to each other as follows: $Q^c = Q^a + \sqrt{3/2}\,(B^+ + B^-)$. In particular, an effective charge is not required to get correct E2 strengths, even in cases like ^{238}U which require enhancement factors on the order of 100 single-particle units. It also means deformations can be correctly determined and, as we demonstrate in Figure 1, the theory can be used to probe the microscopic structure of the giant quadrupole resonance.

For applications to heavy deformed nuclei it is important to know that the expansion of the B^+ and B^- operators in terms of their pseudo operator counterparts goes like Q^a, namely, the main contribution is the corresponding pseudo $B^+$and B^- operators. This is shown in Table 2. The pseudo symplectic scheme is the pseudo SU(3) scheme enhanced by the addition of the pseudo symplectic raising and lowering operators. As in the pseudo SU(3) case, it is anticipated that differences due to the use of pseudo rather than normal operators will be at most a five percent effect. However, this has not been verified by example.

TABLE 1. Tensor expansion (η=4) of the algebraic part of the quadrupole operator $Q_\mu = \sqrt{16\pi/5}\,\Sigma_i r_i^2\, Y_{2\mu}(r_i)$.

(λ_0,μ_0)	κ_0	L_0	S_0	Coefficient
Normal				
(1,1)	1	2	0	20.49390
Pseudo				
(1,1)	1	1	1	-0.76536
(1,1)	1	2	0	24.44570
(1,1)	1	2	1	0.0
(2,2)	1	2	0	-0.23130
(2,2)	1	2	1	1.73270
(2,2)	2	2	0	0.47402
(2,2)	2	2	1	0.84548
(2,2)	1	3	1	0.78448
(3,3)	1	1	1	-0.02848
(3,3)	1	2	0	0.65376
(3,3)	1	2	1	0.0
(3,3)	1	3	1	-0.06952
(3,3)	2	3	1	0.10176
(4,4)	1	2	0	-0.01392
(4,4)	1	2	1	0.16024
(4,4)	2	2	0	0.02666
(4,4)	2	2	1	0.08368
(4,4)	1	3	1	0.08522

5. POTENTIAL ENERGY SURFACES

The potential energy surface concept comes from the generalized collective model which describes nuclei in terms of rotational and vibrational degrees of freedom (Gneuss and Greiner, 1971; Hess, et al., 1981). The hamiltonian consists of kinetic and potential parts built out of scalars in the collective coordinates. This hamiltonian is diagonalized in a five-dimensional harmonic oscillator basis (angular momentum two phonons) with a fixed maximum number of quanta. The parameters of the potential are determined by a least-squares procedure that fits calculated excitation energies and E2 transition strengths to the corresponding experimental numbers. The potential energy surface is a plot of this potential as a function of the (β,γ) shape variables.

To get stable minima and insure against divergent behavior for large values of the deformation it is necessary to use at least a sixth-order polynomial form for the collective model potential: $V(\beta,\gamma) = \Sigma_{\rho\sigma}\, V_{\rho\sigma}(\beta^2)^\rho(\beta^3\cos 3\gamma)^\sigma$ with $2\rho + 3\sigma \leq 6$. In contrast, we find that within the framework of the symplectic model a simple quadrupole-quadrupole interaction suffices to account for the dynamics. The reason for this difference is a simple but nonetheless profound matter. Specifically, the collective model places no restriction on allowed configurations. This means an attractive quadrupole-quadrupole interaction ($V_{\rho\sigma} = 0$ except for $V_{10} < 0$) favors configurations of maximum deformation, that is, those with the maximum possible number of quanta and all of these quanta aligned along an intrinsic symmetry axis. Higher order terms in the potential, like β^4, cancel this divergence ($V_{20} > 0$).

Since the symplectic model is a fermion based theory, not all configurations are allowed. In particular, the Pauli Principle forces an increase in the system's energy for large values of the deformation. (Draayer, et al., 1989). This can be seen by noting that large deformations come about by exciting valence particles out of the $0\hbar\omega$ space into higher shells. But this adds multiples of $2\hbar\omega$ in energy to the system. Here $\hbar\omega$ is the shell separation energy. So binding energy gains are offset by kinetic energy gains. This is shown in Figure 2. A rise in the total energy (expectation value of the hamiltonian) is due to the Pauli Principle. Higher order terms that enter into the collective model description are required to correct for the fact that there is no direct way in that theory to take account of the Exclusion Principle.

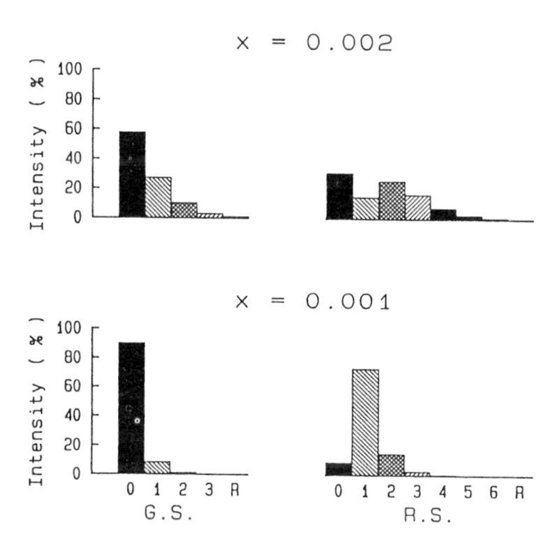

FIGURE 1. Percent analysis of the ground state (0_1) and resonant mode (0_2) in ^{238}U for two values of χ, the $Q^c \cdot Q^c$ coupling strength, $H = H_0 - \frac{1}{2}\chi Q^c \cdot Q^c$. The numbers along the abscissa are in units of $2\hbar\omega$. So for $\chi = 0.002$, the ground state is about 60% in the $0\hbar\omega$ space, $(\lambda_0,\mu_0) = (54,0)$, while the resonant state has less than 15% weight in the $2\hbar\omega$ space and reaches out with non-negligible contributions beyond $10\hbar\omega$. The deformation of the resonant mode is about 1.1 times the ground-state deformation.

6. CONCLUSION

We have proposed a scheme called the pseudo symplectic model for studying the structure of heavy deformed nuclei. The underlying symmetry is pseudo SU(3), a scheme that has been shown to give a good account of a number of interesting phenomena in strongly deformed nuclei. The symplectic enhancement allows one to incorporate collective correlations so calculated eigenstates give correct values for E2 transition rates without requiring an effective charge. We showed preliminary results for ^{238}U that indicate very clearly the inadequacy of a $0\hbar\omega$ shell-model theory. These results also suggest that the pseudo symplectic model will prove useful for describing giant resonance phenomena.

We used the symplectic scheme in its simplest form to give a shell-model interpretation of the collective-model potential energy surface concept. To conclude, we want to point out what we see as further applications of the theory. First of all, we believe the symplectic scheme holds promise for gaining a deeper understanding of the true nature of rotations: the fraction of the total number of nucleons that participate in the motion, questions concerning transverse versus longitudinal current flows, whether vorticity degrees of freedom play an important role, etc. These are questions that can be probed by experimentalists at facilities like CEBAF. Another topic of interest is superdeformation. We find it difficut to imagine a shell-model scheme contributing much to our understanding of this phenomena if it does not include vertical mixing of the symplectic type. Indeed, the percentage analysis of the calculated resonant state in ^{238}U suggests the symplectic model may well play an important role in our gaining a deeper understanding of this important physical phenomena. And lastly, we believe the symplectic model, extended to include many bandhead configurations mixed through pairing and other coupling modes offers real hope for gaining a good understanding of the very excellent and detailed data being collected at the various x-sphere facilities, current and future, that we have heard about at this conference.

Our program is supported by the U.S. National Science Foundation and Conacyt in México. Travel expenses were supplied under the U.S.-México Cooperative Science Program. We also acknowledge discussions with G. Rosensteel, Tulane University, and C. Bahri and S. Park, graduate students at Louisiana State University, for their help in generating the ^{238}U results. And last, but by no means least, we want to thank our Greek hosts for the wonderful hospitality extended to us and all the hard work that went into making the conference a success.

TABLE 2. Tensor expansion (η=4) of the $2\hbar\omega$ raising, B^+ $[B^- = (B^+)^+]$, generators of the symplectic algebra.

(λ_0,μ_0)	κ_0	L_0	S_0	Coefficient
Normal				
(2,0)	1	0	0	-16.73320
Pseudo				
(2,0)	0	1	0	-19.89008
(3,1)	1	1	1	- 1.95141
(4,2)	1	0	0	- 0.69593
(5,3)	1	1	1	- 0.29740
(6,4)	1	0	0	- 0.06295
Normal				
(2,0)	1	2	0	16.73320
Pseudo				
(2,0)	1	2	0	18.54388
(2,0)	1	2	1	2.41666
(3,1)	1	1	1	0.65306
(3,1)	1	2	0	0.84002
(3,1)	1	2	1	0.82559
(3,1)	1	3	1	0.35250
(4,2)	1	2	0	0.37102
(4,2)	1	2	1	0.35373
(4,2)	2	2	0	-0.08262
(4,2)	2	2	1	0.03246
(4,2)	1	3	1	0.16523
(5,3)	1	1	1	0.09171
(5,3)	1	2	0	0.12822
(5,3)	1	3	1	0.06454
(5,3)	2	3	1	-0.01430
(6,4)	1	2	0	0.03749
(6,4)	1	2	1	0.03438
(6,4)	2	2	0	-0.00518
(6,4)	2	2	1	0.00607
(6,4)	1	3	1	0.01584

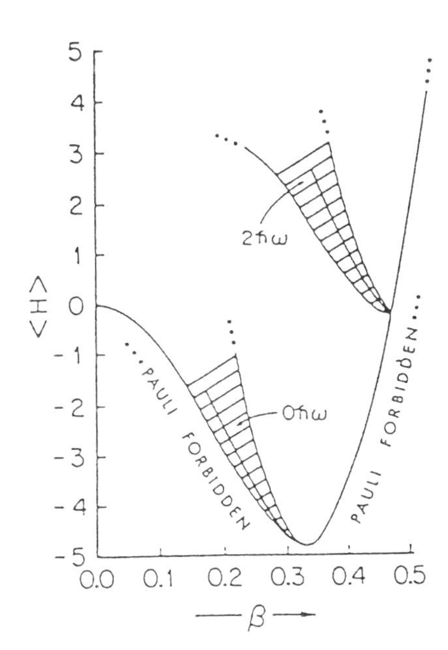

FIGURE 2. A schematic diagram that illustrates a shell-model realization of the potential energy surface. All allowed configurations of each major shell fall in a conical strip or a V-shaped band with terminus defined by the representation of SU(3) that gives the maximum deformation in that shell. The potential energy surface is set for $\beta < \beta_0$ by the representations of the 0ħω shell that lie lowest in energy and for $\beta > \beta_0$ by the leading SU(3) representations of the higher shells. The rise for $\beta > \beta_0$ is simply a result of competition between additional binding energy from $Q^c \cdot Q^c$ and shell effects.

REFERENCES:

D. Braunschweig, K. T. Hecht and J. P. Draayer, Phys. Lett. 76B (1978) 538.
O. Castaños, J. P. Draayer and Y. Leschber, Ann. Phys. 180 (1987) 290.
O. Castaños, J. P. Draayer and Y. Leschber, Z. Phys. A329 (1988) 33.
O. Castaños and J. P. Draayer Nucl Phys. A491 (1989) 349.
J. P. Draayer, C. S. Han and K. J. Weeks, Nucl. Phys. A365 (1981) 127.
J. P. Draayer, K. J. Weeks and K. T. Hecht, Nucl. Phys. A381 (1982) 1.
J. P. Draayer and K. J. Weeks, Ann. Phys. 156 (1984) 41.
J. P. Draayer, S. C. Park and O. Castaños, Phys. Rev. Lett. 62 (1989) 20.
J. P. Elliott, Proc. Roy. Soc. A245 (1958) 128, 562.
G. Gneuss and W. Greiner, Nucl. Phys. A171 (1971) 449.
P. O. Hess, J. Maruhn and W. Greines, J. Phys. G7 (1981) 737.
Y. Leschber, Hadronic J. Supp. 3 (1987) 95.
Y. Leschber and J. P. Draayer, Phys. Lett. 190B (1987) 1.
G. Rosensteel and D. J. Rowe, Ann. Phys. 96 (1976) 1; 123 (1979) 36; 126 (1980) 198, 343.
D. J. Rowe, Rep. Prog. Phys. 48 (1985) 1419.
R. D. Ratna Raju, J. P. Draayer and K. T. Hecht, Nucl. Phys. A202 (1973) 433.
R. D. Ratna Raju, K. T. Hecht, B. D. Chang and J. P. Draayer, Phys. Rev. 20C (1979) 2397.
K. J. Weeks, C. S. Han and J. P. Draayer, Nucl. Phys. A371 (1981) 19.
K. J. Weeks and J. P. Draayer, Nucl. Phys. A393 (1983) 69.
Cheng-Li Wu, Xiao-Ling Han, Zhen-Ping Li, Mike W. Guidry and Da Hsuan Feng, Phys. Lett. 194B (1987) 447.

Paper presented at Int. Conf. on Spectroscopy of Heavy Nuclei, Crete, Greece, 1989

Nuclear structure at the drip lines

P.G. Hansen

The Institute of Physics, University of Aarhus
DK-8000 Aarhus C

1. INTRODUCTION

This paper deals with nuclei that have a composition (N,Z) in the nuclear chart (Fig.1) that places them near to or at the limits of prompt instability with respect to proton or neutron emission, the so-called "drip lines". We refer to several recent conference proceedings and reviews[1-4] for discussions of the properties and structure of drip-line nuclei. It is characteristic of beta decays at both drip lines that the excited levels generally are unbound with respect to particle emission, and have particle widths that exceed the gamma widths, so that the beta decay pattern becomes detectable via the energy spectra of the emitted particles. This decay mode is traditionally referred to as "beta-delayed particle emission".

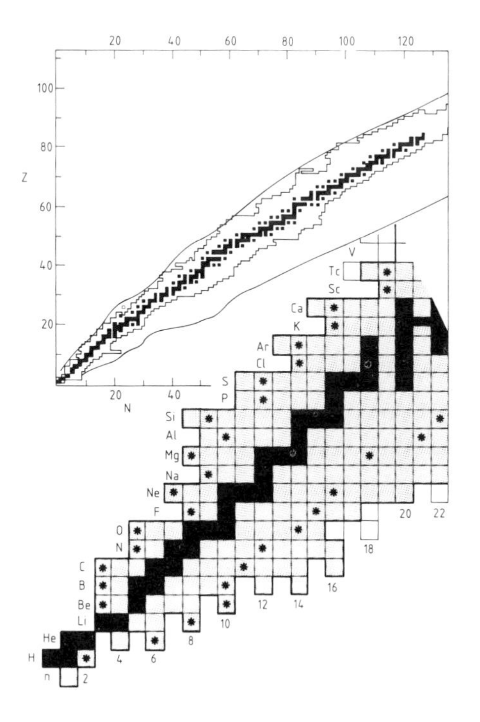

Fig. 1
The nuclear chart with stable isotopes shown as black squares. In the upper part of the figure the zig-zag lines are the experimental limits of known isotopes and the smooth curves indicate the estimated position of the drip lines. In the lower part of the figure known isotopes are shown shaded, while predicted but as yet unobserved isotopes are shown as white squares surrounded by thin lines. Stars indicate the heaviest and lightest isotopes, for which the half-life is known experimentally.

Already from Fig. 1 it can be seen that the experimental situation is not symmetrical at the two drip lines. The proton drip line has been reached and even overstepped in experiments, even for quite heavy nuclei. The extremely neutron-rich isotopes, on the other hand, represent a major remaining challenge in the study of far-unstable nuclei. It seems literally certain that of the order of 2000 undiscovered neutron-rich and beta-unstable isotopes must exist, but we seem to lack the tools for producing them. In all but the very lightest elements, the neutron drip line appears to be out of the reach of all existing techniques.

The situation is also very asymmetrical with respect to our understanding of nuclear structure at the two drip lines. In the following Section 2 we shall give examples to illustrate that the shell-model provides a solid grasp of the structure of proton-rich nuclei. For nuclei at the neutron drip-line, however, there has been surprises. From studies that are necessarily limited to isotopes of the lightest elements has come evidence for the existence of an extended spatial distribution, the neutron "halo" and for surprisingly fast electric dipole and Gamow-Teller (GT) transitions. This is the subject of Section 3, which also reports a new decay mode: the emission of beta-delayed deuterons.

2. THE LIGHT PROTON-RICH NUCLEI

We begin by recalling some elementary features of nuclear beta decay. The Fermi transition strength is exhausted by the transition connecting a nuclear state to its isobaric analogue state. Owing to the Coulomb energy shifts in the isospin multiplet, the Fermi decays are to a good approximation possible only in nuclei with Z>N, so that they are extremely weak over most of the nuclear chart. The Coulomb force gives rise to small admixtures into other states, admixtures which can be traced by measuring the Fermi decays to these states. The Gamow-Teller (GT) operator depends on spin and isospin coordinates and usually connects to a number of final states with appreciable transition matrix elements. Still, according to a bold conjecture by Fujita and Ikeda and later confirmed by experiment, the major part of the GT strength is found in a broad resonance called the Gamow-Teller resonance (GTR). This is situated somewhat above the isobaric analogue state, so that one has a situation in which the major part of the allowed strength, the so-called superallowed transitions, can be observed only in decays of nuclei with Z>N. The neutron-rich, light nuclei offer some interesting exceptions to this rule, a problem to which we shall return in sub-section 3.3.

The most proton-rich nuclei are observed in the mass range A=20-50. A favourable case for a detailed study of their beta decay is provided by the light argon isotopes[5,6], which can be produced in large yields at ISOLDE. The spectra of beta-delayed protons from the short-lived nuclei ^{33}Ar and for ^{32}Ar with isospin projections $T_z = (N-Z)/2$ of -3/2 and -2, respectively, provide as illustrated in Fig. 2 a rich amount of detail on energies and transition probabilities for excited states up to 10 MeV Excitation energy. The experimental strength functions were compared with a large shell-model calculations[7,8] of the Brown-Wildenthal type. The agreement is striking and shows the remarkable predictive power of modern shell-model calculations. The results for the overall quenching of the strength (relative to a calculation based on free-nucleon coupling constants) are 0.49 and 0.58, respectively, in good agreement with other types of experiments.

In a recently completed, precise study[9)] of an isotope at the boundary between the p and sd shells, the T_z=-3/2 nucleus ^{17}Ne, proton-emitting excited levels could be observed up to 12 MeV excitation energy in ^{17}F. These levels decay by proton and alpha emission. A shell model calculation with two particles in the sd shell and one in the p shell was used to calculate level energies and GT transition probabilities. Fig 3 shows that the agreement between the measured and calculated strengths is excellent up to the highest energies. On the basis of free-nucleon coupling constants, an experimental GT quenching of 0.82 was obtained, in good agreement with the general trend that the reduction in strength is smaller in the p shell than in the sd shell. A surprising new feature in this kind of a study was the appearance of strong interference effects between the broad particle-emitting resonances (Fig. 4).

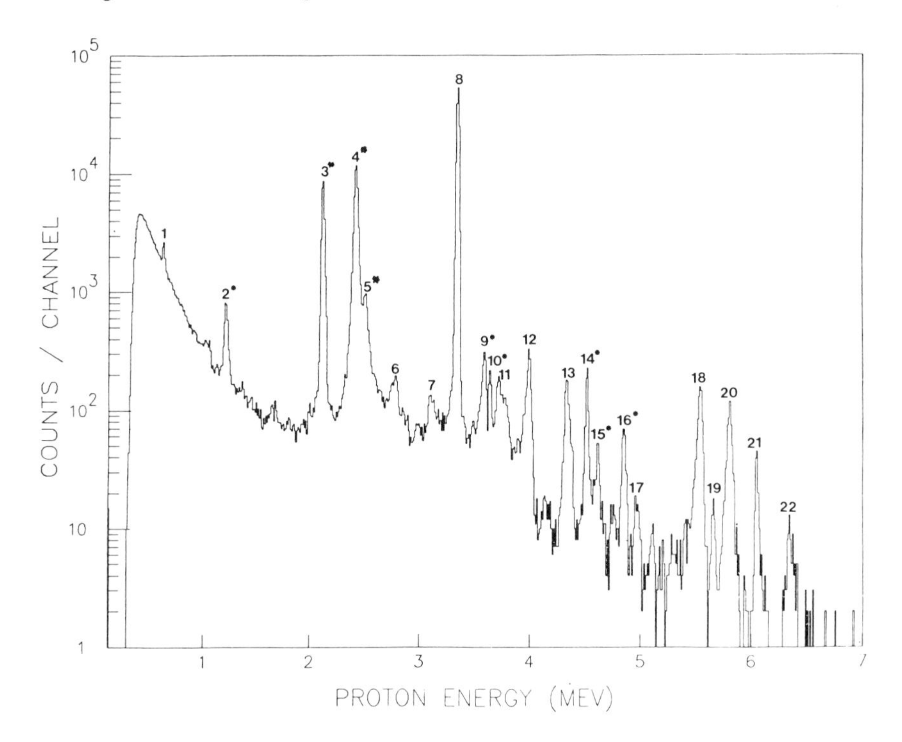

Fig. 2
Beta-delayed protons from the decay of ^{32}Ar measured with a Si detector in coincidence with beta particles to reduce summing. The strongest peak (#8) represents the Fermi transition to the isobaric analogue state in ^{32}Cl. Peaks marked with an asterisk have been identified as feeding the 1248 keV first excited state in ^{31}S.

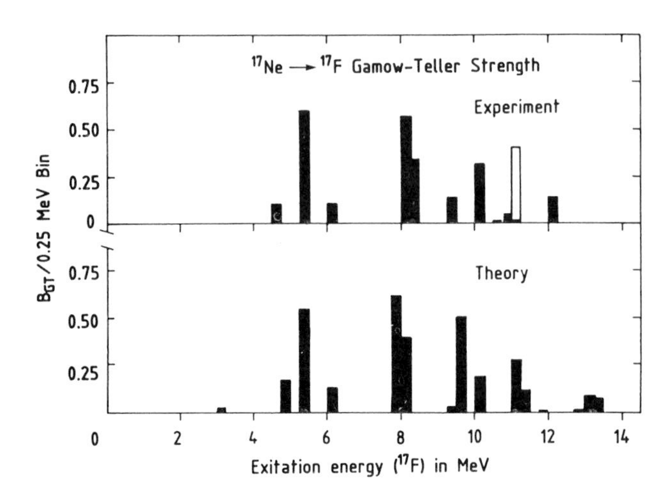

Fig. 3
Comparison of the measured Gamow-Teller strength function in the ^{17}Ne decay with the result of a shell-model calculation. The open bar near 11 MeV represents the error limits in the estimate of the Gamow-Teller strength to the isobaric analogue state.

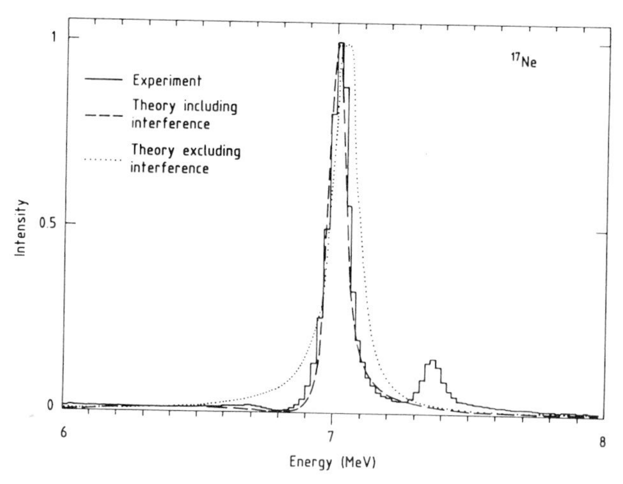

Fig. 4
The histogram shows for the decay of ^{17}Ne the beta-delayed proton spectrum near 7 MeV recorded in coincidence with beta particles to suppress summing. The peak near 7.4 MeV corresponds to a resonance in ^{17}F with spin 1/2, while the remaining structure is due to two spin 3/2 resonances, which interfere destructively near 6.8 MeV.

3. THE LIGHT NEUTRON-RICH NUCLEI

The only drip-line nuclei that have been investigated in some detail are ^{6}He, ^{8}He, ^{9}Li and ^{11}Li. We discuss in this section some recent progress in this area, and refer to a recent review[10] for a more detailed account of our knowledge of very neutron-rich systems.

3.1 The Neutron Halo

In a remarkable experimental development that has stimulated much new thinking about the structure of nuclei at the neutron dripline, Tanihata et al.[11] initiated experiments to study nuclear reactions caused by fast-moving short-lived radioactive fragments. Fig. 5 illustrates schematically their experimental technique and some of the key results. In the first experiments they found that the nuclear interaction cross-sections and hence the matter radii of ^{6,8}He and of ^{11}Li were surprisingly large. They interpreted this as evidence either for a large quadrupole deformation or for a long tail in the matter distribution, a neutron "halo". Evidence in support of the second hypothesis came from a sophisticated optical experiment[12]

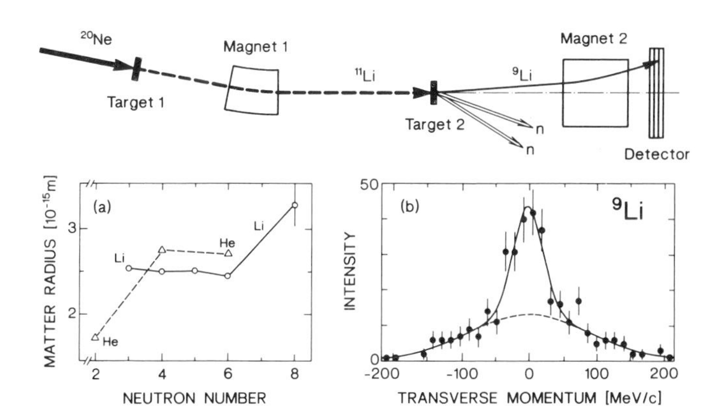

Fig. 5
The upper part of the figure illustrates the experimental principle of the experiment of Tanihata et al.[11]*. Ions of* ^{20}*Ne at 790 MeV/nucleon interact in a first target and produces a shower of fragments, from which a secondary beam of* ^{11}*Li is separated magnetically. After a second target one can either determines the transmission of the primary beam or one can study the 2n removal reaction by observing* ^{9}*Li. The experiment provides two important clues to the structure of* ^{11}*Li, (a) an increase in the matter radii determined from the total cross-section, and (b) a narrow component*[16] *in the* ^{9}*Li transverse momentum distribution.*

carried out at ISOLDE and which showed that the proton system of ^{11}Li, at least, has the properties expected from the spherical shell model.

It was suggested[13] that the increase in the matter radius can be understood in terms of the extremely low binding energy (250±80 keV, only) of the last two neutrons in ^{11}Li, which gives rise to a long tail in their wave function, so that it has a radius of 8 fm as compared with 2.5 fm for the core nucleons. It is possible to perform simple estimates of the properties of the halo by noting that it is characteristic of nuclei such as ^{9}Li that they will not bind one neutron but that they will bind two. This suggests that the neutron-neutron force in the singlet state is essential to the stability of ^{11}Li and, presumably, to that of other drip line nuclei, and revives an idea of Migdal[14] that the dineutron being very close to having a bound state can be stabilized through the attractive interaction with a distant nucleus. This picture can be used[13] to obtain a semi-quantitative estimate of the matter radius from the known binding energy. For a dineutron in an s state, the external wave function is of a Yukawa type characterized by the decay length

$$\varrho = \hbar/(2\mu B)^{1/2} \tag{1}$$

expressed in terms of the reduced mass and the binding energy. Although the increased stability of nuclei containing an even number of neutrons may be viewed as an effect of pairing, the wave function in the neutron-drip-line nuclei is probably more complex as the n-n and n-nucleus binding have roughly equal importance. This suggests that correlations could play an important role. The problem is closely related to that of the general quantum-mechanical three-body problem with (weak) attractive forces, discussed by Efimov[15], who showed that it is possible to have one or several bound states even if neither of the three two-body sub-systems have bound states.

A second direct manifestation of the halo was observed in an experiment by Kobayashi et al.[16], who found (Fig.4) that the transverse momentum distribution of ^{9}Li recoils from fragmentation of ^{11}Li on a carbon target has a narrow component with a full width at half maximum of 53 MeV/c, only. From Heisenberg's uncertainty principle it is clear that this is evidence for a wide spatial component, the neutron halo. With the external wave function given above, the momentum distribution corresponding to the square of the Fourier transform to be

$$|A_p|^2 = \frac{\varrho^3}{\pi^2(1+(p\varrho/\hbar)^2)^2} \tag{2}$$

which corresponds to a calculated FWHM for the ^{9}Li recoils of about 30 MeV/c in qualitative agreement with the experiment. Similar results have been obtained by Uchiyama and Masuda[17]. The interaction cross-section and momentum distribution have been analyzed by Harvey[18], who confirms that the neutron distribution in ^{11}Li extends to very large radii.

An interesting case, in which the halo is formed by a single neutron loosely bound to a nucleus, is provided by the nucleus ^{11}Be with a neutron separation energy of 504 keV. Here, the ground state is known to be $1s_{1/2}$ while the $0p_{1/2}$ state lies at 320 keV. It was pointed out a long time ago[19] that this is due to a systematic shift arising from the proton-neutron interaction, and that one can follow the 0p-1s spacing from the value of 5 MeV in ^{15}O to the crossing of the two levels in ^{11}Be, where 8 presumably has ceased to be a magic neutron number. Sato and Ogawa[20] have emphasized the importance of this reversal of the s and p states for the development of the ^{11}Li halo.

A series of experiments at GANIL deals with cross-sections for reactions of neutron-rich nuclei and the connection to the question of strong-interaction radii. Since the results of this work are being discussed by Yves Schutz in a parallel contribution to this conference, I shall say no more about this.

3.2 Soft Coulomb Dissociation

It has been pointed out[13] that the low binding and large size of the neutron halo imply a large electric polarizability and hence large cross-sections for Coulomb dissociation via a "soft" E1 mode. The simple wave function discussed above leads to an approximate ^{2}n removal cross-section of

$$\sigma_c \cong \frac{\pi Z_1^2 Z_2^2 \mu e^4 (1+x)}{3v^2 M^2 B} \ln \frac{\mu v^2}{B} \tag{3}$$

where Z_1 and v are charge and velocity of the projectile, and M is the mass of the projectile residue.The small parameter x is the ratio of core radius to halo radius. The softness of the E1 mode can be illustrated by using a sum-rule to estimate the average excitation energy imparted to the system after break-up. This turns out to be 6B/(1+x), or 1 MeV only, in striking contrast to the usual typical energy of the E1 giant resonance, around 25 MeV in a light nucleus. More precise estimates of Coulomb dissociation cross sections have been discussed by Baur[21]. The existence of these large cross-sections have been reported in a new experiment[22], in which ^{11}Li ions at 790 Mev/nucleon were brought to collide with a lead target. After subtraction of the estimated contribution from the nuclear interaction, the total and 2n-removal Coulomb cross-sections were found to be 1.8 and 0.9 barns, respectively, qualitatively in good agreement with the value 1.9 barns calculated from the expression above.

The approximate $1/v^2$ dependence of the Coulomb dissociation cross-section suggests that more selective studies of the soft Coulomb dissociation mode can be made by going to lower energies. We have undertaken an experiment[23] together with colleagues from GANIL to use the LISE spectrometer for an investigation of Coulomb dissociation at 30 MeV/nucleon with simultaneous detection of a neutron in coincidence with the nuclear residue. In another experiment[24], planned for the GSI, we intend to study momentum distributions and n-n correlations using the fragment separator FRS and the large neutron detector LAND, now under construction.

Finally we draw attention to an experimental result of considerable relevance to the structure of drip-line nuclei namely the lifetime of the $1/2^-$ state in ^{11}Be, measured by Millener et al.[25], who found that this 320 keV gamma ray is the fastest E1 known in nuclear physics with a reduced transition probability of 0.36 Weisskopf units. They point out that this can only be understood by "taking into account the actual binding energies of the single-particle orbits", providing thus a complete parallel to the halo and to the large cross-section for electrodissociation of ^{11}Li.

3.3 Beta-delayed particle emission

We point here to some evidence that super-allowed beta decays, that is decays with a reduced transition probability comparable to or greater than that of the free neutron, occur systematically in the light neutron drip-line nuclei. Here, the beta decays are dominated by neutron emission from mostly very broad states, often with complex decays. As furthermore neutron detection is more difficult than charged particle detection, it is often hard to obtain a precise picture of the beta strength function. An example is provided by ^{11}Li, for which time correlation techniques have led to the identification[26] of single, double and triple neutron emission from ^{11}Li in the intensities 84.9±0.8%, 4.1±0.4% and 1.9±0.2%, respectively. In the same decay Langevin et al.[27] found triton emission in an intensity of 0.010±0.004% and break-up into $n+\alpha+^{6}$He in an intensity[28] of 0.9±0.3%. The triton branch alone provides a lower limit on the reduced Gamow-Teller transition probability B_{GT} of 0.4 for a transition to a resonance at 18.5 MeV in ^{11}Be, but as we expect neutron emission to be more abundant than triton emission, the true value could be much higher. Better evidence comes from lighter drip-line nuclei.

In the ^{8}He decay one also observes[29] beta-delayed tritons in the remarkable intensity of 0.9±0.1%. Barker and Warburton[30] used this and other experimental information to fit a many-level, many-channel R-matrix calculation and found feeding to four 1^{+} levels. The highest of these, at about 9 MeV, has a reduced Gamow-Teller transition probability B_{GT} of about 3, clearly a superallowed transition. A similar analysis[31] of new delayed-neutron and -alpha data for ^{9}Li shows a super-allowed branch with a B_{GT} of about 7 to resonances in ^{9}Be at 11.28 and 11.81 MeV.

The spatially extended neutron wave function of drip-line nuclei discussed in sub-section 3.2 suggests an especially simple model of a super-allowed beta decay, namely the decay of the quasi-free neutrons. For the case of a neutron pair in a ^{1}S state this could then lead to an unbound deuteron. We have undertaken a search for this decay mode in ^{11}Li, but the low production rate and complex decay of this rare radioactivity renders the experiment difficult. We have a short time ago succeded in observing beta-delayed deuteron emission in a technically much more favourable case, that of 0.807 s ^{6}He, which, however, hardly can be viewed as having a neutron halo.

In the experiment[32] 10^{7} atoms/s of ^{6}He were produced at ISOLDE and collected on-line on a carbon foil. Coincidences were observed between two charged-particle detctors placed at 180°, each subtending a solid angle of $7*10^{-4}$ sr, only. The count rate vanished when one spectrometer arm was turned by 6°, which shows that the observed events must arise from two-body break-up and hence correspond to the only beta-delayed branch that is energetically possible for ^{6}He, namely the disintegration of ^{6}Li into a deuteron and an alpha particle. The energy spectrum is shown in Fig. 6 and corresponds to a branching ratio of $(2.8\pm0.5)*10^{-6}$. Both the spectral shape and the intensity can be accounted for in an R-matrix calculation, which includes contributions from two 1^{+} states in ^{6}Li, the ground state and the unbound 5.7 MeV broad resonance. As can be seen from Fig. 7, both states are outside the energy window for the process.

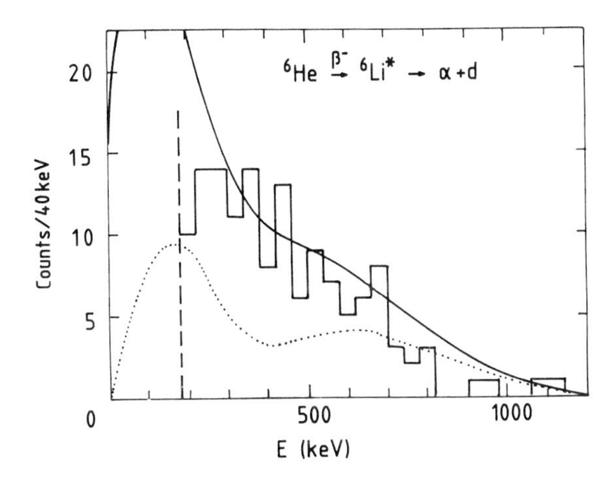

Fig. 6
The histogram shows the energy spectrum of coincident events from ^6He observed in a Si detector. As the experiment had no particle identification, the spectrum represents the sum of the alpha and deuteron contributions. The experimental cut-off in energy is indicated by the dashed line at 180 keV. The dotted line is a calculation with no free parameters, which assumes that the spectrum is due to the 5.7 MeV resonance (Fig. 7) alone. The full drawn line includes a contribution from the ground state.

4. CONCLUDING REMARK

The present paper suggests that while the structure and properties of nuclei at the proton drip line seem to be well in hand, the nuclei at the neutron drip line present novel features. These are linked to the low binding of the neutron system and to the absence of a Coulomb barrier, and have as a consequence that certain processes, normally dominated by a giant-resonance behaviour, regain some of their single-particle freedom. The experimental evidence is still preliminary, both for the existence of soft El mode, holding a sizeable fraction of the Giant-Dipole sum rule, and for the fast beta decays, which maybe by analogy could be called a "soft Gamow-Teller mode".

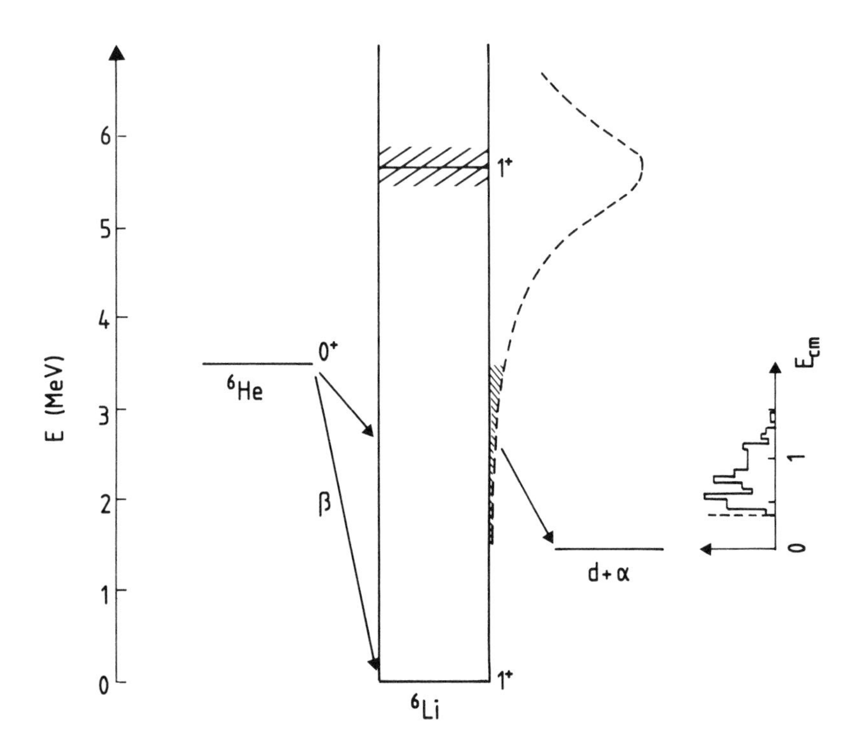

Fig. 7
Schematic level scheme for the beta decay of ^{6}He to the d + α channel. The histogram to the right shows the beta-decay intensity deduced in the analysis.

5. REFERENCES

1. Proc. of the 4th Int. Symposium on Heavy Ion Physics and Nuclear Astrophysical Problems, Tokyo 1988 (S. Kubono, M. Ishihara and T. Nomura eds.), (World Scientific 1989).
2. Proc. of the 5th Int. Conf. on Nuclei Far From Stability, Lake Rosseau 1987, AIP Conf. Proc. **164** (1988), eds. E. Hagberg, J. Hardy and I.S. Towner.
3. Particle Emission From Nuclei, Vol. I-III, (M.S. Ivascu and D.N. Poenaru eds.), (CRC Press, Boca Raton, Florida, 1989).
4. J.H. Hamilton, P.G. Hansen and E.F. Zganjar, Rep. Progress in Physics **48** (1985) 708.
5. T. Bjørnstad, M.J.G Borge, P. Dessagne, R.D. von Dincklage, G.T. Ewan, P.G. Hansen, A. Huck, B. Jonson, G. Klotz, A. Knipper, P.O. Larsson, G. Nyman, H.L. Ravn, C. Richard-Serre, K. Riisager, D. Schardt and G. Walter, Nucl. Phys. **A443** (1985) 283; M.J.G. Borge, P. Dessagne, G.T. Ewan, P.G. Hansen, A. Huck, B. Jonson, G. Klotz, A. Knipper, S. Mattsson, G. Nyman, C. Richard-Serre, K. Riisager and G. Walter, Phys. Scripta **36** (1987) 218.
6. M.J.G. Borge, P.G. Hansen, B. Jonson, S. Matsson, G. Nyman, A. Richter and K. Riisager, Z. Phys. A **332** (1989) 413.
7. W. Müller, B.C. Metsch, W. Knüpfer and A. Richter, Nucl. Phys. **A430** (1984) 61.
8. B.A. Brown and B.H. Wildenthal, Ann. Rev. Nucl. Part. Sci. **38** (1988) 29.
9. M.J.G. Borge, H. Cronberg, M. Cronqvist, H. Gabelmann, P.G. Hansen, L. Johannsen, B. Jonson, S. Mattsson, G. Nyman, A. Richter, K. Riisager, O. Tengblad and M. Tomaselli, Nucl. Phys. **A490** (1988) 287.
10. P.G. Hansen and B. Jonson in ref. 3, pp 157-201.
11. I. Tanihata,H. Hamagaki, O. Hashimoto, Y. Shida, N. Yoshikawa, K. Sugimoto, O. Yamakawa, T. Kobayashi and N. Takahashi, Phys. Rev. Lett. **55** (1985) 2676.
12. E. Arnold, J. Bonn, R. Gegenwart, W. Neu, R. Neugart, E.-W. Otten, G. Ulm and K. Wendt, Phys. Lett. **197B** (1987) 311.
13. P.G. Hansen and B. Jonson, Europhys. Lett. **4** (1987) 409.
14. A.B. Migdal, Yad. Fiz. **16** (1972) 427; English tr. Sov. J. Nucl. Phys. **16** (1973) 238.
15. V.M. Efimov, Sovi. J. Nucl. Phys. **10** (1970) 62 and **12** (1970) 1080.
16. T. Kobayashi, O. Yamakawa, K. Omata, K. Sugimoto, T. Shimoda, N. Takahashi and I. Tanihata, Phys. Rev. Lett. **60** (1988) 2599.
17. F. Uchiyama and N. Masuda, Phys. Rev. C **38** (1988) 2670.
18. B.G. Harvey, to be published.
19. I. Talmi and I. Unna, Phys. Rev. Lett. **4** (1960) 469.
20. H. Sato and K. Ogawa, Study of the anomalous behaviour of the interaction radii of extremely neutron-rich nuclei, INS-Report-679 (1988).
21. G. Baur in ref. 1, p. 225.
22. T. Kobayashi in ref. 1, p. 217; I. Tanihata, Nucl. Phys. **A488** (1988) 113c.

23. R. Anne, R. Bimbot, P. Bricault, M. Cronqvist, C. Détraz, H. Emling, D. Guillemaud-Mueller, P.G. Hansen, L. Johannsen, B. Jonson, M. Lewitowicz, S. Mattsson, A.C. Mueller, R. Neugart, G. Nyman, F. Pougheon, A. Richter, K. Riisager, M.G. Saint-Laurent, G. Schrieder and Ö. Skeppstedt, "A Measurement of the Coulomb dissociation cross-section of extremely neutron-rich nuclei at 30 MeV/u", Proposal to GANIL 1989, (unpublished).

24. M. Cronqvist, P.G. Hansen, L. Johannsen, B. Jonson, G. Nyman, A. Richter, K. Riisager, G. Schrieder, O. Tengblad and Members of the Aladin-, FRS- and LAND-Collaborations, "Structure of extremely neutron-rich nuclei: A proposal to study soft Coulomb dissociation", Proposal to the G.S.I. 1988 with addendum 1989, (unpublished).

25. D.J. Millener, J.W. Olness, E.K. Warburton and S.S. Hanna, Phys. Rev. C **28** (1983) 497.

26. R.E. Azuma, T. Bjørnstad, H.Å. Gustafsson, P.G. Hansen, B. Jonson, S. Mattsson, G. Nyman, A.M. Poskanzer and H.L. Ravn, Phys. Lett. **96B** (1980) 31.

27. M. Langevin, C. Détraz, M. Epherre, D. Guillemaud-Mueller, B. Jonson, and C. Thibault, Phys. Lett. **146B** (1984) 176.

28. M. Langevin, C. Détraz, D. Guillemaud, F. Naulin, M. Epherre, R. Klapisch, S.K.T. Mark, and M. de Saint-Simon, C. Thibault and F. Touchard, Nucl. Phys. **A366** (1981) 449.

29. M.J.G. Borge, M. Epherre-Rey-Campagnolle, D. Guillemaud-Mueller, B.Jonson, M. Langevin, G. Nyman and C. Thibault, Nucl. Phys. **A460** (1986) 373.

30. F.C. Barker and E.K. Warburton, Nucl. Phys. **A487** (1988) 269.

31. G. Nyman, private communication and to be published.

32. K. Riisager, M.J.G. Borge, H. Gabelmann, P.G. Hansen, L. Johannsen, B. Jonson, W. Kurcewicz, G. Nyman, A. Richter, O. Tengblad, K. Wilhelmsen, and the ISOLDE Collaboration, CERN Preprint, May 1989 and to be published.

Inst. Phys. Conf. Ser. No 105
Paper presented at Int. Conf. on Spectroscopy of Heavy Nuclei, Crete, Greece, 1989

Gamma yields of nuclei produced in reactions of ^{74}Ge and ^{93}Nb with thin ^{197}Au and thick ^{181}Ta targets

P.R.Adžić, I.V.Aničin, A.H.Kukoč, D.S.Toprek, R.B.Vukanović, M.T.Župančić
The B.Kidrič Institute, P.O.Box 522, 11001 Belgrade, Yugoslavia

Abstract

Low-background off-line gamma-ray spectroscopic measurements of the thin Au and thick Ta targets have been performed to determine or estimate yields of long-lived isotopes created in the following reactions: ^{74}Ge + ^{197}Au and ^{93}Nb + ^{181}Ta. The main goal of these experiments was to investigate and compare mechanisms dominating the reactions in the two similar compound systems for a thin and thick target at approximately the same incident energy of 18 MeV/u. The cumulative cross-section distribution for the formation of proton-rich, neutron-rich chains and independent isotopes in 495 μg/cm^2 thick Au target after bombardment with ^{74}Ge ions, has been determined for about 90 ($T_{1/2}$>1.5 d) nuclei. The similar cumulative cross-section distribution of about 50 ($T_{1/2}$>20 d) reaction products in 0.5 mm thick natural Ta target after bombardment with ^{93}Nb, has been determined and compared with the former one.

EXPERIMENTS AND RESULTS

The high resolution, low background HP-Ge spectrometer[1)] was used in off-line gamma-spectroscopic measurements of long-lived reaction products which remained in the thin Au and thick Ta targets after the following reactions: 1) ^{74}Ge + ^{197}Au and 2) ^{93}Nb + ^{181}Ta. Both experiments have been performed at GSI UNILAC in Darmstadt.

In the first experiment[2)], the thin Au (495 μg/cm^2) target was irradiated during 24 hours with ^{74}Ge ions of 18.8 MeV/u and 0.03 pnA. In the beam direction the target is transparent for the projectile-like nuclei with TKE>1 MeV and for the target-like nuclei with TKE>3 MeV. The reaction products, about 90 isotopes, with half-lives longer than 1.5 days and in the region 7<A<206 which we observe, are either those emitted in the transversal direction or those which originate from very strongly damped processes. This means that if reaction products were scattered at angles close to 90° they would remain even in a very thin target which is normally, at smaller angles, transparent for the same ions causing negligible energy loss at these energies. The other interaction that might have occured is the deep inelastic scattering. Nuclei suffering these strongly damped processes are incapable to come out of a thin target due to great kinetic energy loss. The cumulative cross-section distribution for the formation of proton-rich (□), neutron-rich chains (◇) and independent (shielded) isotopes (●) with respective estimated upper limits (∇) in the reaction of ^{74}Ge and ^{197}Au, is shown in Fig.1. The dashed line shows a typical mass distribution for a such system, indicating projectile-like and target-like nuclei.

In the second experiment [3], the thick natural Ta target (0.5 mm) was irradiated with ^{93}Nb ions during different time periods and with maximal incident energy of 18.0 MeV/u. Sandwiched behind was another identical Ta target which was used later for the estimation of the neutron flux coming out of the first target[4]. The time weighted average energy was 16.2 MeV/u while the beam current of 0.2 pnA was kept throughout 101 hours of the total time irradiation. The cumulative cross-section distribution of about 50 isotopes in the first target, with half-lives longer than 20 days, is shown in Fig.2. As in the previous case, the dashed line indicates the projectile and target-like products.

It should be noticed that to make comparison easier both figures are drawn in the same scale. The visual inspection shows that the general character of the two distributions is basically the same. Bearing in mind that only deep inelastic processes are supposed to contribute significantly to the first distribution and that all possible processes must have contributed to the second one, this is perhaps somewhat surprising. The absolute values of the cross-sections, however, seem strikingly disproportionate. That the cross-section for the given reaction product to be formed and to remain in the target is roughly an order of magnitude bigger in a thin than in an infinitely thick target we find quite outstanding. Although in the Ta case the cross-sections are averaged over the entire energy range, one feels that this should be at least partly compensated by the small thickness of the Au target (which is only about 1/150 of the thickness of the Ta interaction layer). We are now trying to find the reasons why the two effects remain so uncompensated.

REFERENCES

1) Adžić et al.,Environment International,14,(1988)295.

2) Kukoč et al.,GSI Sci.Rep.,87-1,60,1987.

3) Kukoč et al.,GSI Sci.Rep.,88-1,58,1988.

4) Toprek et al.,GSI SCI.Rep.,89-1,54,1989.

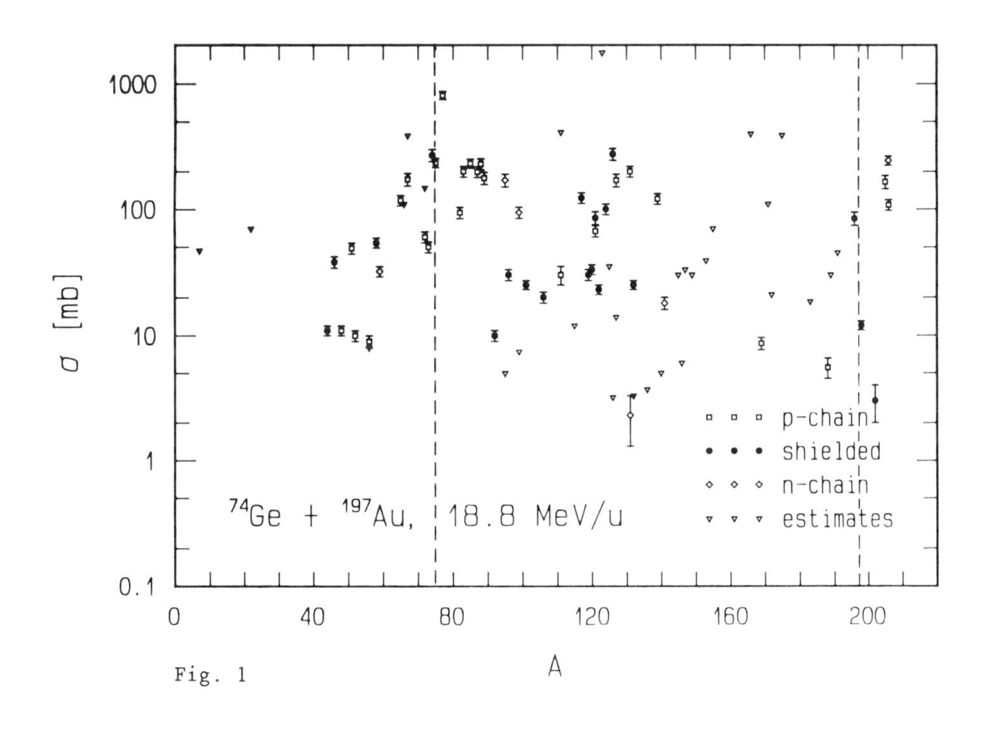
1000
100
10
1
0.1
σ [mb]
0
40
80
120
160
200
A
p-chain
shielded
n-chain
estimates
^{74}Ge + ^{197}Au, 18.8 MeV/u

Fig. 1

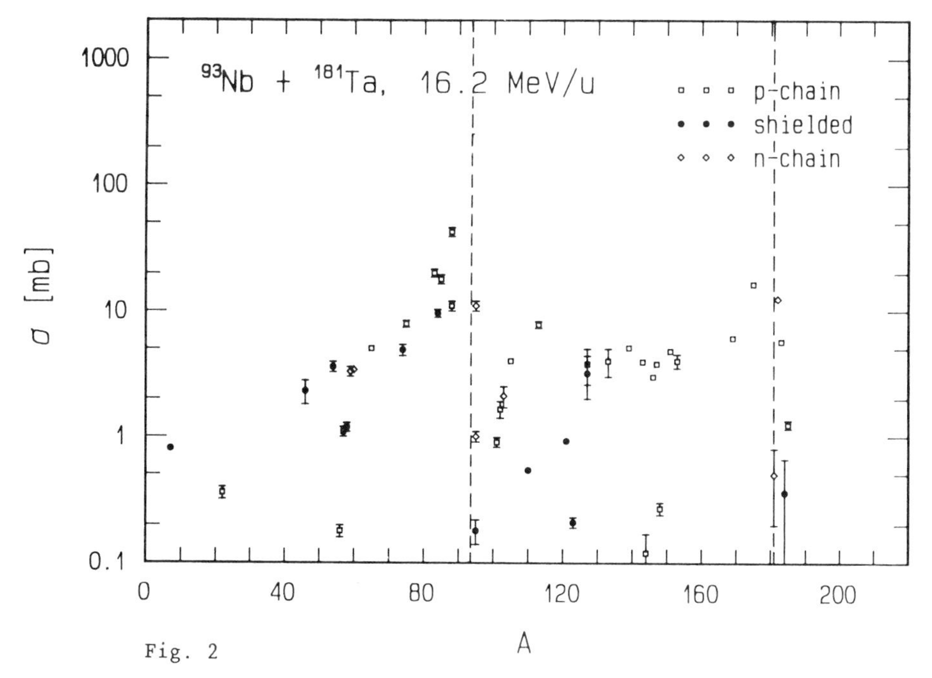
1000
100
10
1
0.1
σ [mb]
0
40
80
120
160
200
A
^{93}Nb + ^{181}Ta, 16.2 MeV/u
p-chain
shielded
n-chain

Fig. 2

Inst. Phys. Conf. Ser. No 105
Paper presented at Int. Conf. on Spectroscopy of Heavy Nuclei, Crete, Greece, 1989

Summary talk

J M Irvine

Department of Physics, The University, Manchester M13 9PL, U.K.

As I arrived at the registration desk I met a delegate who was looking at the programme with a puzzled expression. When he saw me he asked "didn't you talk about astrophysics at the conference in Peniscola last month?" When I replied "yes" he said "I see that you are giving the summary talk here. What do you know about this subject?" I can only suppose that the organisers were looking for a fresh and unbiased view.

Let me begin by echoing the words of **John Simpson** in the opening talk of the meeting when he said he "could not imagine a more idyllic setting". My pleasure at being here is only slightly marred by the task now before me.

I asked the organisers what form they wanted this talk to take and received the suggestion "just pick out a few high-lights of the week". So I am happy to report that the tennis and the swimming were excellent. A Montreal-Munich collaboration had great success on the courts while a UK team with access to much greater momentum transfer were handicapped by a lack of theoretical support. We British are infamous for our relatively poor language skills. At this meeting my feelings of inferiority reached new depths when I heard a French-Canadian colleague say "gosh! I missed that shot" in more languages than I knew existed.

There are four standard approaches to giving summary talks:

(i) Uncritically mention every contribution to the conference. This pleases all the authors and is likely to lead to invitations to give summary talks at other conferences. A second advantage is that with over 90 papers submitted one can spend less than 30 seconds per paper, thus minimising the opportunity for misrepresentation. The disadvantages are that it is extremely unlikely that the rapporteur will say anything that has not already been said to greater effect by the primary authors and it is as boring as hell.

(ii) Present a truly critical review mentioning only those papers of essential interest. This means that 75% of the contributions are ignored leaving their authors disgruntled and convinced that you failed to grasp the gems of wisdom they sought to impart, while 20% of the papers are seriously criticised and their authors become your bitter enemies. This leaves the 5% of the authors you praised and who think you did a wonderful job, but since they inevitably come from your own laboratory the chances of being invited to future conferences is remote.

(iii) Totally ignore the topic of the conference and give the talk of your choice. This was a very tempting option and had I decided to follow it I would have told you why the standard cosmology, the standard model of particle physics and the recommended neutron half-life are incompatible. This would have undoubtedly been the star contribution to the conference, but, since the organisers did not ask me to talk about it and I would like to be invited back to Crete, I decided to follow the fourth approach.

iv) Ramble on in a self-indulgent manner hoping to make contact with the subject matter of the conference from time-to-time.

Let me begin by quoting **Tord Bengtsson** who, in the second contribution to the conference, pointed out that there is no theory of nuclear physics only extremely clever models. In sessions 10 and 11 we were exposed to many of these clever models and were particularly fortunate to hear the excellent review of models based upon dynamical symmetries by **Da Hsuan Feng** (Da Hsuan said he would sulk all through dinner if I didn't say something nice about him). In session 6 we heard from **Costas Papanicolas** (I have now mentioned four contributors, I hope someone is keeping count) about "The Limits of Mean Field Theory Explored by Electron Scattering". How can a theory be explored if no theory exists? I thought I might begin by exploring this apparent paradox.

The basis of most nuclear models is the idea that the nucleus can be considered as an assembly of interacting neutrons and protons. I cannot resist showing the beautiful picture of the density difference between $^{205}T\ell$ and ^{206}Pb revealed by electron scattering. This clearly shows the contribution of the 3S proton wave-function to the charge density. In the discussion session **John Durell** was quick to remind us of the evidence for the validity of the neutron-proton model coming from particle exchange reactions. Listening to the discussion I was reminded of Paul Mathew's observation that when you kick a dog out come barks. It is, however, very difficult to construct a dog from barks.

Let us consider the model hamiltonian

$$H = \sum_{i=1}^{A} p_i^2/2m_i + \sum_{i>j} V_{ij} \qquad (1)$$

where the one-body data determines the mass of the neutron and proton and the two-body data (phase shifts and deuteron properties) determine a non relativistic internucleon potential. There is no unique fit to the two-body data and hence we have a set of 'phase equivalent' potentials from which to choose.

We are interested in approximate eigenfunctions of the hamiltonian (1). The wave-functions are required because they contain the information which relates one kind of measurement to another. This is an essential feature of spectroscopy where energy levels are given spin and parity assignments and are connected by decay schemes.

Unable to solve the many-body problem exactly we turn to the variational principle.

$$\Phi = \Phi([\phi_i]) \qquad (2)$$

where the ϕ_i are a set of single particle orbital functions. Minimising

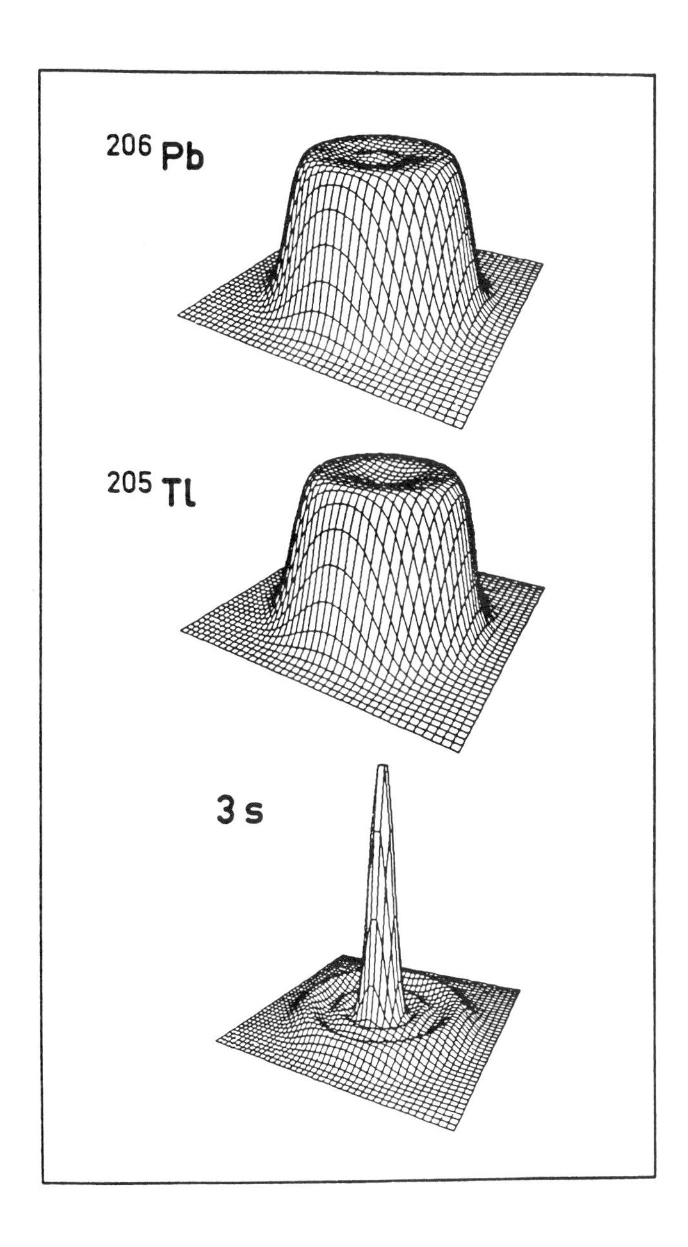

Figure 1 The charge distributions of ^{206}Pb and ^{205}Tℓ together with the difference, characteristic of a 3S proton contribution (**Costas Papanicolas**)

$\langle\Phi|H|\Phi\rangle$ with respect to the ϕ_i subject to natural constraints, eg wave-function normalisation, angular momentum conservation etc is called the Hartree-Fock approximation. This does not work for nuclei.

(ii) $$\Psi = \sum_{n=1}^{N} C_n \Phi_n([\phi_i]) \tag{3}$$

and minimise with respect to the coefficients C_n. This is called configuration mixing. Configuration mixing Hartree-Fock ((i) + (ii)) is the bread and butter of atomic and molecular electronic structure calculations. In practice the configuration space can never be large enough to obtain a realistic solution to the nuclear problem

(iii) $$\Psi = F\,\Phi([\phi_i]) \tag{4}$$

where F contains all the correlations induced by the internucleon potential. To give substance to this approach we expand F

$$F = \sum_{n=2}^{A} F_n \tag{5}$$

where F_n represents an n-body correlation operator. There are many variations on such expansion schemes, fermi hypernetted chain, exp(s), coupled clusters etc. Terminating the expansion at F_2 requires the introduction of further constraints to ensure the rapid convergence of the cluster expansion. Such approximations are known as lowest order constrained variational calculations. The simplest scheme is one based upon

$$F = \prod_{i>j} f_{ij} \tag{6}$$

where

$$f_{ij} = f(r_{ij}) \tag{7}$$

is a simple central function of the interparticle separation distance. This is known as the Jastrow approximation and its application to nuclear physics was pioneered by the Bethe and Pandharipande. It clearly has the ability to remove the short range repulsive correlations induced by the hard-core of realistic N-N interactions.

Applied to nuclear matter the LOCV-Jastrow approximation gives results very similar to those of the Brueckner-Bethe-Goldstone perturbative hole-line expansion. For various phase equivalent potentials the results lie along the line AB in figure 3, known as the Coester line. Not only does this line not go through the region of nuclear saturation at a density $\rho_o \simeq 0.16\mathrm{Nf}_m^{-3}$ with a

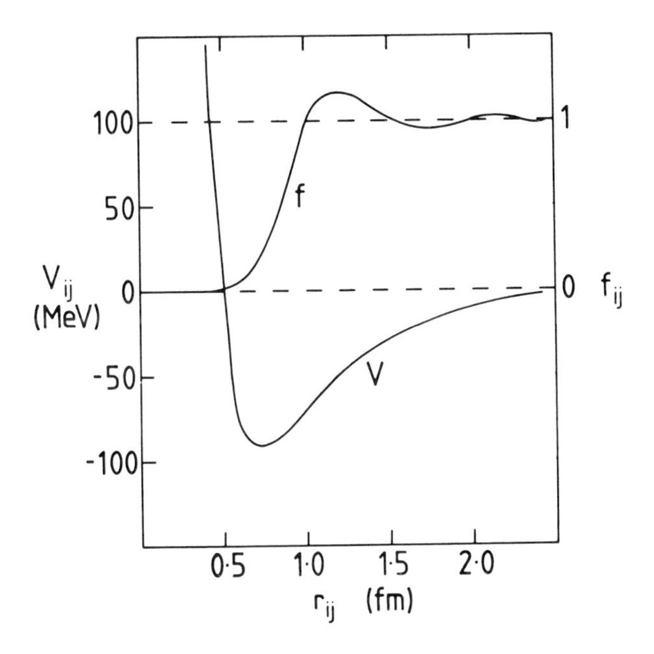

Figure 2 Typical realistic N-N potential form factor and Jastrow correlation function

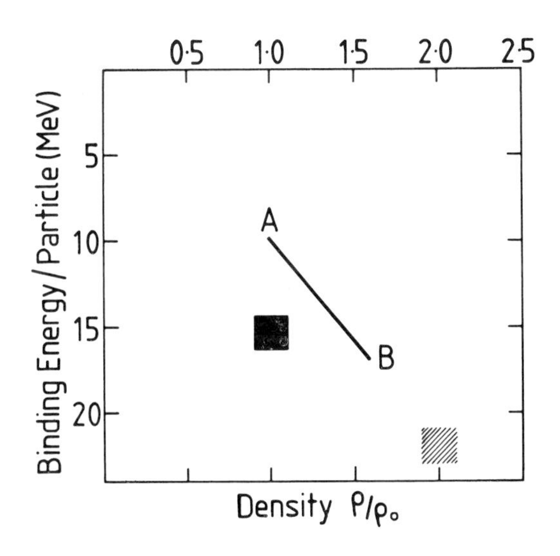

Figure 3 Results of nuclear matter saturation calculations. Results from the Brueckner-Bethe-Goldstone hole-line expansion and Jastrow approximation calculations for different phase equivalent potentials lie on Coester line AB. Full solutions of the neutron-proton model lie in the hatched area while the semiempirical mass formula predicts saturation in the region of the solid black square.

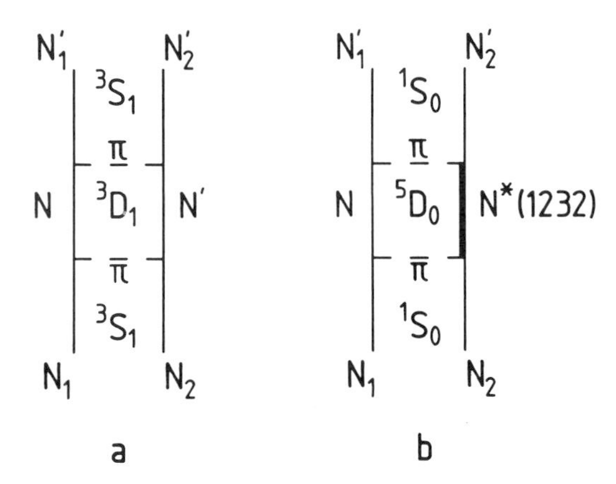

Figure 4 a) A typical two-pion exchange amplitude contributing to the tensor correlations in the $^3S_1$ channel

b) The analogue of (a) contributing in the ^1S_o channel to the isobar correlations

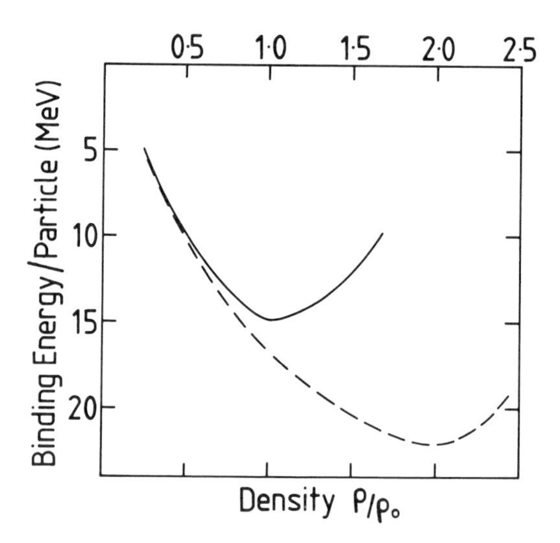

Figure 5 The dashed curve represents a typical neutron-proton model saturation calculation. This becomes the solid curve when isobar corrections are allowed for

binding energy per particle $E_o \simeq 16$ MeV. It is not even a good approximation to a solution of the model.

Owen, Bishop and Irvine [1] generalised the two-body correlation operator

$$f_{ij} = f_c(r_{ij}) + f_T(r_{ij})\, S_{ij} + f_{LS}(r_{ij})\, \underline{L}.\, \underline{S} \quad (8)$$

to have the same degrees of freedom as the realistic two-body potential V_{ij}. The results for all the phase equivalent potentials now lie in the shaded region in figure 3 which can be shown to be a good approximation to the true solution of the model by demonstrating that higher cluster terms only become important for densities $\rho \gtrsim 3\rho_o$ or simply by a brute force supercomputer Monte Carlo calculation. To understand why results which were strung out along the Coester line now all cluster together we have to examine the effect of the generalisation (8) of the correlation operator. The important ingredient is the tensor correlation. Phase equivalent potentials at the end A of the Coester line have strong tensor forces and induce strong tensor correlations. Potentials at the end B have weaker tensor forces and hence induce weaker correlations.

The tragedy is that the shaded region in figure 3 does not overlap the empirical saturation region, ie the neutron-proton model has failed its first and simplest test.

A guide to the missing ingredient is provided by an understanding of the role of the tensor correlations. A typical contribution to the $^3S_1$ amplitude is given in figure 4a. The analogous effect in the next most important channel, the ^1S_o, requires the introduction of the intermediate N*(1232), figure 4b. The phenomenological phase equivalent potentials have these amplitudes subsumed in their parameters. However, when the interactions take place in nuclear matter the intermediate nucleon states must be excluded from the occupied fermi sea. Thus, we must take away the effect of these amplitudes for intermediate momenta $k < k_F$, effectively introducing a density dependent repulsion. The result is illustrated in figure 5 (ref [1]).

The introduction of the N* correlations together with configuration mixing also helps to resolve the quenching problem mentioned by **Otto Hauser** and **Caroline De Coster**.

Why then does the neutron-proton model appear to work so well? The correlation effects we have been discussing appear principally in relative S-channels where we would expect short range interactions to be effective. The effect is to enhance the attraction in the $^3S_1$ channel (remember all the fudge factors to increase the attraction in this channel in Brueckner g-matrix calculations) and to introduce a density dependent repulsion in the ^1S_o channel. We note that the Skyrme interactions are of just such a form to reproduce these effects and in particular a zero range three-body force is equivalent to a two-body density dependent interaction. Thus, we can look upon mean-field calculations based upon Skyrme interactions as an intermediate phenomenology (ref [2]).

Why does the shell-model work so well? This is still somewhat of a mystery but we have demonstrated that in the p- and sd- shells the effective two-body matrix elements of the phase equivalent potentials are remarkably similar to those fitted to spectroscopic data, especially to those of Wildenthal and his collaborators (ref [3]).

Two words of warning: first, distrust any theorist who tells you their calculation is based upon a Brueckner-Hartree-Fock approach. It is almost certainly not the solution of the problem at hand. Second, the important elements for resolving the Coester line problem all involved two-pion exchange. Thus, be sceptical of all one-boson exchange calculations. The introduction of a mythical scalar σ-meson cannot do justice to the rich physics that in reality must be there.

Now we can begin to make contact with topics discussed in the conference. **Costas Papanicolas** and **Louk Lapikas** told us about the exciting opportunities in electron-nucleus physics and of the power of these studies to explore nuclear correlations. Here I might point out that the Soviet group in the EMC have published preliminary data on the ^{12}C structure function for Bjorken variable x>1 and we have shown [4] that this seems to require the tensor correlations discussed above.

Having a basis for discussing nuclear models, let me make a comment about intruder states and shape coexistence. I remind you that this arose first, not in a discussion of heavy nuclei but in the search for an explanation of the low lying 6.06 MeV 0^+ state in ^{16}O (ref [5]). This is a story that **Leonidas Skouras** knows very well. The first excited 0^+ state in ^{16}O is heavily deformed and in the spherical basis of the closed-shell configuration is predominately a 4 particle - 4 hole state. Configuration mixing then results in a ground state which is only 70% closed-shell. When we move up to ^{40}Ca we find the ground state is only 50% closed-shell. For a long time ^{40}Ca was the heaviest N=Z 'closed-shell' nucleus known to experimentalists. This did not stop theoreticians carrying out calculations on the closed-shell ^{80}Zr. Now in heavy ion studies ^{80}Zr has been identified and its properties are being investigated. It appears to be heavily deformed suggesting a ground state which is much less than 50% closed shell, thus continuing the trend seen in the lighter N=Z 'closed shell' nuclei. The message is, do not take closed-shells too literally. Here I would echo the comments of **Da Hsuan Feng** that dynamical symmetries are a better basis for choosing configuration spaces than energy level cut-offs. This is what lies behind the success of BCS theory and the IBM's.

The identification of ^{80}Zr brings me to a strong theme in the conference, the study of exotic nuclei. First let me mention the beautiful talk by **Yves Schutz** (he didn't say he would sulk if I didn't say something nice about him). In figure 6 I reproduce his transparency showing the richness of exotic fragments that can be produced in intermediate energy, inverse, heavy ion reactions. This whets my appetite for a proposal currently circulating in the U.K. to use the spallatron neutron source ISIS at RAL as a source for exotic radioactive beams. While the idea sounds simple, I am informed that the problems of dealing with the radioactivity may require costly management needing European scale investment.

The story of exotic nuclei was continued in **Gregor Hansen's** account of drip-line studies. He convincingly argued that the coulomb barrier contained nucleons along the proton drip-line in structures of a familiar shell-model nature while the absence of a barrier on the neutron-drip-line leads to the possibility of many new soft modes giving rise to exotic shapes and transition amplitudes.

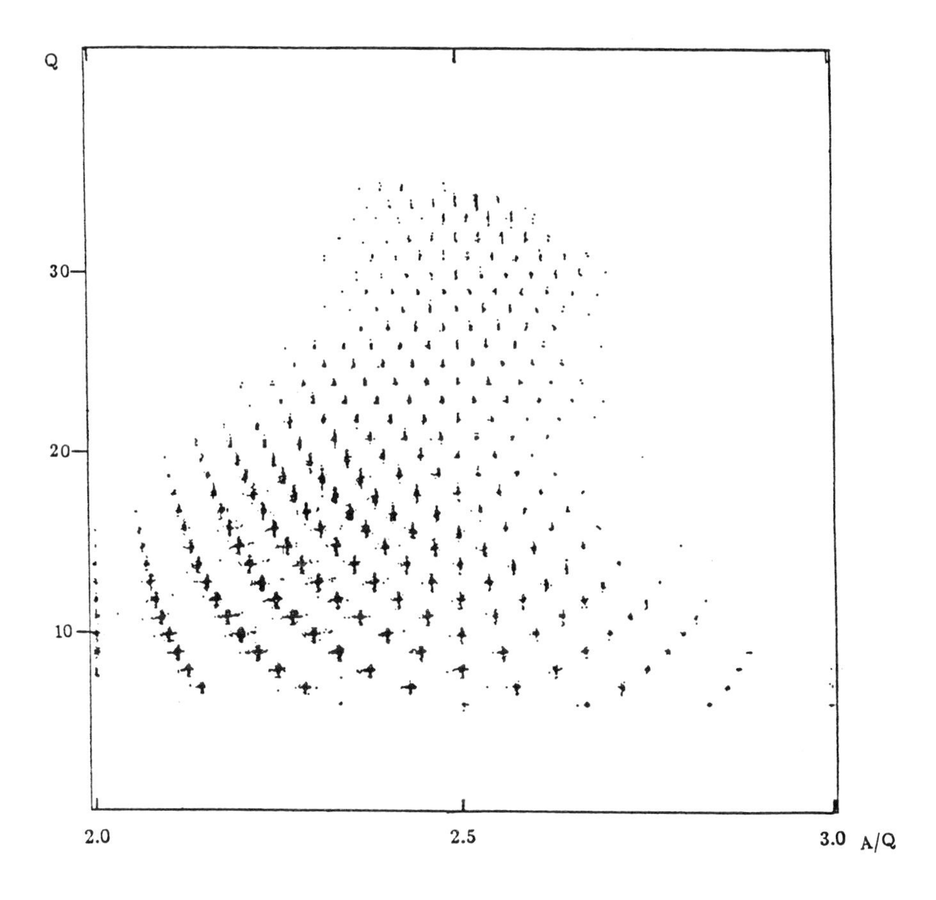

Figure 6 The richness of exotic fragments produced in an intermediate energy, inverse, heavy-ion reaction (**Yves Schutz**)

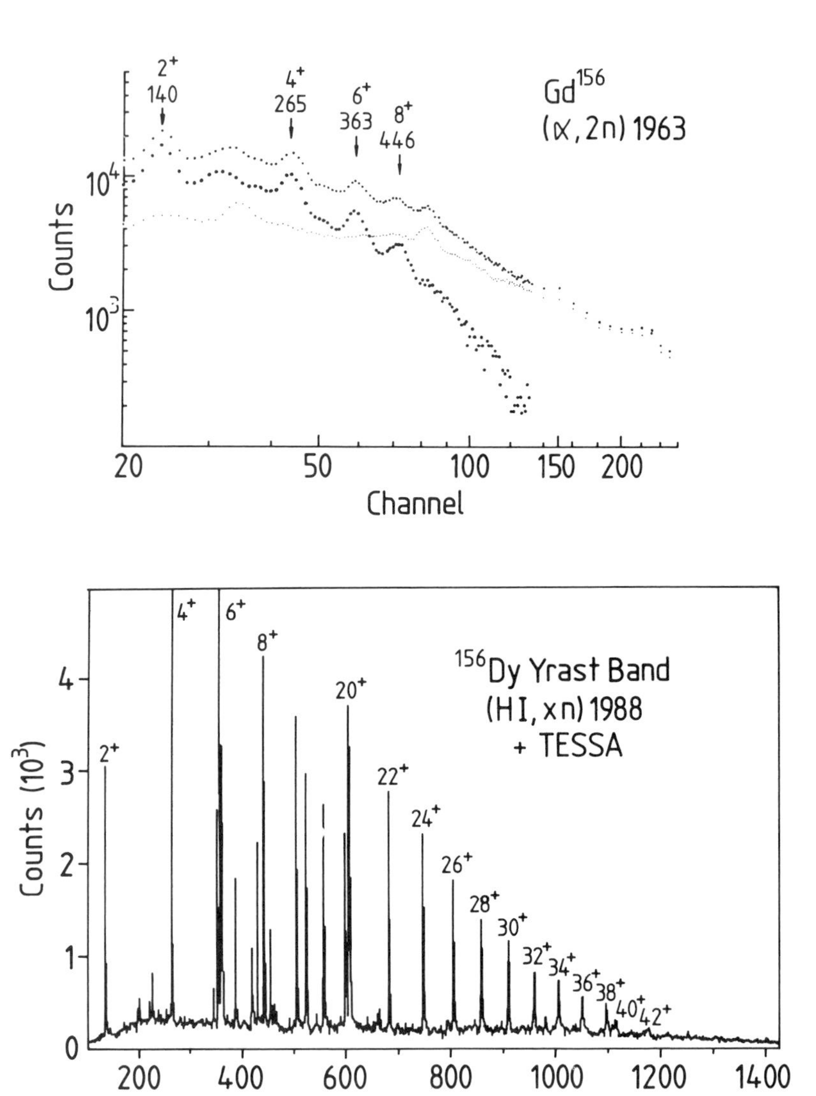

Figure 7 A comparison of the state-of-the-art A=156 spectrum in 1963 and 1988 (**John Simpson**)

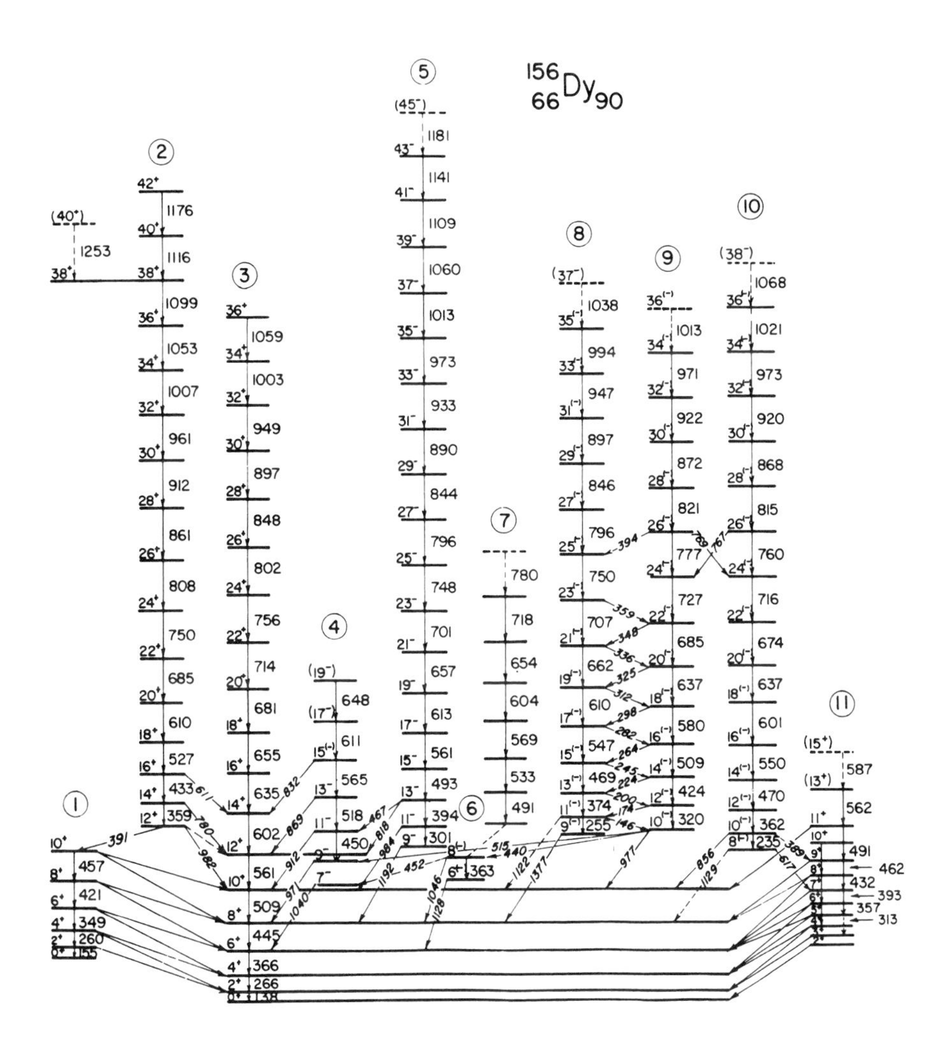

Figure 8 Part of the gamma-ray spectrum of ^{156}Dy (**John Simpson**)

These exotic nuclei studies are essential for the advancement of astrophysics. We had another example of this in **Hans Borners** delightful account of very high resolution crystal gamma-ray spectroscopy. The resulting improvement in our knowledge of the ^{176}Lu decay scheme allowed the abundance of this isotope to be used as an accurate s-process cosmic clock.

Moving to a central theme of the meeting, we had a review of the development of the TESSA family of Compton suppressed γ-ray spectrometer arrays from **John Simpson** culminating in ESSA30 and of the first physics with the NORDBALL by **Geirr Sletten.** We heard of various applications of the CHATEAU DE CRISTAL from **Asaiez, Gizon, Nyakō, Roberts, Sharpey-Schaffer** and **Schuck** and a view of the future with the GAMMASHERE and EUROBALL I and II. I have neither the time knor the skill to review these developments in detail. For me the progress of the last twenty-five years was encapsulated in a single transparency from **John Simpson.** I was left with three impressions:

(i) These large arrays only achieve their full potential when used in conjunction with a suitable channel selector, eg recoil separator, on-line isotope separator, charged particle detector, etc.

(ii) There is a growing problem with data handling and reduction with increasing multiplicity of coincidence. Here we had a poster from the Stockholm group (**Lars Hildingsson**) dealing with triple γ-coincidences and I know of the invaluable work done by Dave Radford initially at Chalk River and now at NBI. A similar situation arises in the field of relativistic heavy-ion physics which was graphically illustrated by **Rainer Santo.** Here it seems to me there are opportunities for European collaborations in the development of data manipulation codes.

(iii) Whatever the value of ESSA30 (**Hildingsson, Gizon, Hübel, Kalfas, Kossionides, Lisle** and **Papadopoulos**) and the NORDBALL in terms of immediate physics, which I am sure is excellent, the true worth of these projects will be seen in the new levels of European collaboration in nuclear physics.

To illustrate the power of these spectrometer arrays I once more reproduce the familiar level scheme of ^{156}Dy and ask "is it reasonable to expect simple model builders to describe such detailed spectroscopy?" Here indeed is a challenge to the skills of **Dennis Bonatsos, Jerry Drayer, Da Hsuan Feng, Pete Van Isacker** and **Neil Rowley.** Not content with simple energy levels, one Liverpool, Weizmann, Rochester, Argonne, NBI, DL collaboration (**Alderson, Roberts, Sharpey-Schafer** and **Sletten**) reported a measurement of the g-factor of the K=25 isomer in ^{182}Os lamely concluding that these results appeared to be incompatible with cranked shell-model predictions. What did they expect? The simple fact is that the experimental progress in recent years has far outstripped the predictive powers of simple models.

Theoreticians view the results from the new gamma-detector arrays like the British working classes view Margaret Thatcher, they might not like it but it keeps the unemployment figures down.

We also heard of the power of more modest arrays with a judicious choice of channel selector. **John Durell** told us how $\gamma.\gamma$ coincidences in the prompt decay of fission fragments allowed measurements of nanosecond lifetimes. He was followed by **Henner Ohm's** account of how $\beta.\gamma.\gamma$ coincidences in delayed decays took us into the sub-nanosecond region. **Peter Butler** reported on how

$\gamma.\gamma$ coincidences with the similtaneous detection of back scattered daughter nuclei from heavy ion reactions could be used to probe exotic shapes and **Graham Jones** how ingenuity in designing targets meant that γ-α coincidences could be used to study the spectroscopy of the actinides. This demonstration that arrays were not everything was reinforced by **Hans Borner's** account of (n,γ) high-resolution crystal spectroscopy.

The meeting was indeed far ranging. We were introduced to the whole new field of polarised heavy ion physics by **Otto Karban** and warned that progress was hampered by a lack of theoretical input. **Otto Hauser** told us about spin physics with polarised nucleons, even if it did sound like a commercial for a kaon factory. He also detoured into the exciting new muon collaboration results on nucleon spin structure pointing out how these and the high-energy ($\vec{p}$,$\vec{p}$) studies caused some difficulties for the simplest quark-gluon models of the nucleon. In the session on electron physics **Costas Papanicolas** and **Louk Lapikas** discussed the need for a European CEBAF style facility. May I remind you of the exciting new 840 MeV Mainz racetrack microtron which is just becoming available and of the physics that can still be done with the muon beams at CERN. I have a concern that results from CEBAF may be drowned in flood of vector meson production. I do recognise that the dynamical freedom introduced in virtual photon (real electron) scattering may allow an unravelling of these results.

The difficulties in interpreting the quark-gluon structure of the nucleon high-lights the problems we may face in extracting an unambiguous signal of quark-gluon plasma production in relativistic heavy-ion collisions (**Rainer Santo** and **Nicos Antoniou**). The properties of such a plasma are crucial to the behaviour of the early universe and at this point I should tell you why the standard big-bang cosmology, the standard model of particle physics and the recommended value of the neutron half-life are inconsistent. However, as I stated at the beginning of my talk the organisers did not invite me to speak on that subject and I should like to be invited back to Crete, thus, I will finish with an apology to those contributors I failed to mention.

References

[1] J M Irvine, Prog. Part. and Nucl. Phys. 5, (1980), 1.

[2] T H R Skyrme, Phil. Mag. 1 (1956) 1043; Nucl. Phys. 9 (1959) 615
D Vautherin and D M Brink Phys. Rev. C5 (1972) 626

[3] J M Irvine et al, Annals of Phys 102, (1976), 129
J M Irvine and F Yazici, J. Phys. G13, (1987), 615
J M Irvine et al, J. Phys. G14, (1988), 27.

[4] J M Irvine and Hu Guoju, J. Phys. G15 (1989) 147

[5] J M Irvine, V Pucknell and C Latorre, Adv. Phys. 20 (1971) 661.

Author Index